Topics in Physical Mathematics

Kishore Marathe

Topics in Physical Mathematics

 Springer

Prof. Kishore Marathe
City University of New York
Brooklyn College
Bedford Avenue 2900
11210-2889 Brooklyn
New York
USA
KMarathe@brooklyn.cuny.edu

Whilst we have made considerable efforts to contact all holders of copyright material contained in this book, we have failed to locate some of them. Should holders wish to contact the Publisher, we will make every effort to come to some arrangement with them.

ISBN 978-1-4471-6121-9   ISBN 978-1-84882-939-8 (eBook)
DOI 10.1007/978-1-84882-939-8
Springer London Dordrecht Heidelberg New York

British Library Cataloguing in Publication Data
A catalogue record for this book is available from the British Library

Mathematics Subject Classification (2010): 53C25, 57M25, 57N10, 57R57, 58J60, 81T13, 81T30

Cover design: deblik

Printed on acid-free paper

Springer is part of Springer Science+Business Media (www.springer.com)

Dedicated to the memory of my mother,
Indumati (1920 – 2005),
who passed on to me her love of learning.
==

Memories

Your voice is silent now.
But the sound of your soft
Music will always resonate
In my heart.
The wisp of morning incense
Floats in the air no more.
It now resides only in my
Childhood memories.

Contents

Preface

<table>
<tr><td>Physikalische Mathematik</td><td>Physical Mathematics</td></tr>
</table>

Die meisten		Most
Mathematiker		Mathematicians
glauben.		believe.
Aber alle		But all
Physiker		Physicists
wissen.		know.

The marriage between gauge theory and the geometry of fiber bundles from the sometime warring tribes of physics and mathematics is now over thirty years old. The marriage brokers were none other than Chern and Simons. The 1978 paper by Wu and Yang can be regarded as the announcement of this union. It has led to many wonderful offspring. The theories of Donaldson, Chern–Simons, Floer–Fukaya, Seiberg–Witten, and TQFT are just some of the more famous members of their extended family. Quantum groups, CFT, supersymmetry, string theory and gravity also have close ties with this family. In this book we will discuss some topics related to the areas mentioned above where the interaction of physical and mathematical theories has led to new points of view and new results in mathematics. The area where this is most evident is that of geometric topology of low-dimensional manifolds. I coined the term "physical mathematics" to describe this new and fast growing area of research and used it in the title of my paper [265]. A very nice discussion of this term is given in Zeidler's book on quantum field theory [417], which is the first volume of a six-volume work that he has undertaken (see also [418]).

Historically, mathematics and physics were part of what was generally called "natural philosophy." The intersection of ideas from different areas of natural philosophy was quite common. Perhaps the earliest example of this is to be found in the work of Kepler. Kepler's laws of planetary motion caused a major sensation when they were announced. Newton's theory of gravitation and his development of the calculus were the direct result of his successful attempt to provide a mathematical explanation of Kepler's laws. We may consider this the beginning of modern mathematical physics or, in the spirit of this book, **physical mathematics**.

Kepler was an extraordinary observer of nature. His observations of snowflakes, honeycombs, and the packing of seeds in various fruits led him to his lesser known study of the sphere packing problem. For dimensions 1, 2, and 3 he found the answers to be 2, 6, and 12 respectively. The lattice structures on these spaces played a crucial role in Kepler's "proof." The three-dimensional problem came to be known as Kepler's conjecture. The slow progress in the solution of this problem led John Milnor to remark that here was a problem that nobody could solve but its answer was known to every schoolboy. It was only solved in 1998 by Tom Hales and the problem in higher dimensions is still wide open. It was the study of the symmetries of a special lattice (the 24-dimensional Leech lattice) that led John Conway to the discovery of his sporadic simple groups. Conway's groups and other sporadic simple groups are closely related to the automorphisms of lattices and algebras. The study of representations of the largest of these sporadic groups (called the Friendly Giant or Fischer–Griess Monster) has led to the creation of a new field of mathematics called vertex algebras. They turn out to be closely related to the chiral algebras in conformal field theory.

It is well known that physical theories use the language of mathematics for their formulation. However, the original formulation of a physical law often does not reveal its appropriate mathematical context. Indeed, the relevant mathematical context may not even exist when the physical law is first formulated. The most well known example of this is Maxwell's equations, which were formulated as a system of partial differential equations for the electric and magnetic fields. Their formulation in terms of the electromagnetic field tensor came later, when Minkowski space and the theory of special relativity were introduced. The classical theory of gravitation as developed by Newton offers another example of a theory that found later mathematical expression as a first approximation in Einstein's work on gravitation. Classical Riemannian geometry played a fundamental role in Einstein's general theory of relativity, and the search for a unified theory of electromagnetism and gravitation led to continued interest in geometrical methods for some time.

However, communication between physicists and mathematicians has been rather sporadic. Indeed, one group has sometimes developed essentially the same ideas as the other without being aware of the other's work. A recent example of this missed opportunity (see [115] for other examples) for communication is the development of Yang–Mills theory in physics and the theory of connections in a fiber bundle in mathematics. Attempts to understand the precise relationship between these theories has led to a great deal of research by mathematicians and physicists. The problems posed and the methods of solution used by each have led to significant contributions towards better mutual understanding of the problems and the methods of the other. For example, the solution of the positive mass conjecture in gravitation was obtained as a result of the mathematical work by Schoen and Yau [339]. Yau's solution of the Calabi conjecture in differential geometry led to the definition of Calabi–Yau manifolds. Manifolds are useful as models in superstring compactification in string theory.

A complete solution for a class of Yang–Mills instantons (the Euclidean BPST instantons) was obtained by using methods from differential geometry by Atiyah, Drinfeld, Hitchin, and Manin (see [19]). This result is an example of a result in mathematical physics. Donaldson turned this result around and studied the topology of the moduli space of BPST instantons. He found a surprising application of this to the study of the topology of four-dimensional manifolds. The first announcement of his results [106] stunned the mathematical community. When combined with the work of Freedman [136,137] one of its implications, the existence of exotic $\mathbf{R}^4$ spaces, was a surprising enough piece of mathematics to get into the New York Times. Since then Donaldson and other mathematicians have found many surprising applications of Freedman's work and have developed a whole area of mathematics, which may be called **gauge-theoretic mathematics**. In a series of papers, Witten has proposed new geometrical and topological interpretations of physical quantities arising in such diverse areas as supersymmetry, conformal and quantum field theories, and string theories. Several of these ideas have led to new insights into old mathematical structures and some have led to new structures. We can regard the work of Donaldson and Witten as belonging to physical mathematics.

Scientists often wonder about the "unreasonable effectiveness of mathematics in the natural sciences." In his famous article [402] Wigner writes:

> The first point is that the enormous usefulness of mathematics in the natural sciences is something bordering on the mysterious and that there is no rational explanation for it. Second, it is just this uncanny usefulness of mathematical concepts that raises the question of the uniqueness of our physical theories.

It now seems that mathematicians have received an unreasonably effective (and even mysterious) gift of classical and quantum field theories from physics and that other gifts continue to arrive with exciting mathematical applications.

Associated to the Yang–Mills equations by coupling to the Higgs field are the Yang–Mills–Higgs equations. If the gauge group is non-abelian then the Yang–Mills–Higgs equations admit smooth, static solutions with finite action. These equations with the gauge group $G_{ew} = U(1) \times SU(2)$ play a fundamental role in the unified theory of electromagnetic and weak interactions (also called the **electroweak theory**), developed in major part by Glashow [155], Salam [333], and Weinberg [397]. The subgroup of G_{ew} corresponding to $U(1)$ gives rise to the electromagnetic field, while the force of weak interaction corresponds to the $SU(2)$ subgroup of G_{ew}. The electroweak theory predicted the existence of massive vector particles (the intermediate bosons W^+, W^-, and Z^0) corresponding to the various components of the gauge potential, which mediate the weak interactions at short distances. The experimental verification of these predictions was an important factor in the renewed interest in gauge theories as providing a suitable model for the unification of fundamental forces of nature. Soon thereafter a theory was proposed to unify the electromagnetic, weak and strong interactions by adjoining the group $SU(3)$ of quantum chromodynamics to the gauge group of the electroweak theory. The resulting theory is called the standard model. It has had great success in describing the known fundamental particles and their interactions. An essential feature of the standard model is symmetry breaking. It requires the introduction of the Higgs field. The corresponding Higgs particle is as yet unobserved. Unified theory including the standard model and the fourth fundamental force, gravity, is still a distant dream. It seems that further progress may depend on a better understanding of the mathematical foundations of these theories.

The gulf between mathematics and physics widened during the first half of the twentieth century. The languages used by the two groups also diverged to the extent that experts in one group had difficulty understanding the work of those in the other. Perhaps the classic example of this is the following excerpt from an interview of Dirac by an American reporter during Dirac's visit to Chicago.

Reporter: I have been told that few people understand your work. Is there anyone that you do not understand?

Dirac: Yes.

Reporter: Could you please tell me the name of that person?
Dirac: Weyl.

Dirac's opinion was shared by most physicists. The following remark by Yang made at the Stoney Brook Festschrift honoring him illustrates this: Most physicists had a copy of Hermann Weyl's "Gruppentheorie und Quantenmechanik" in their study, but few had read it.

On the mathematical side the great emphasis on generality and abstraction driven largely by the work of the Bourbaki group and its followers further widened the gulf between mathematics and science. Most of them viewed the separation of mathematics and science as a sign of maturity for mathematics: It was becoming an independent field of knowledge. In fact, Dieudonné (one of the founders of the Bourbaki group) expressed the following thoughts in [99]:

The nay-sayers who predicted that mathematics will be doomed by its separation from science have been proven wrong. In the sixty years or so after early 1900s, mathematics has made great progress, most of which has little to do with physical applications. The one exception is the theory of distributions by Laurent Schwartz, which was motivated by Dirac's work in quantum theory.

These statements are often quoted to show that mathematicians had little interest in talking to scientists. However, in the same article Dieudonné writes:

> I do not intend to say that close contact with other fields, such as theoretical physics, is not beneficial to all parties concerned.

He did not live to see such close contact and dialogue between physicists and mathematicians and to observe that it has been far more beneficial to the mathematicians than to the physicists in the last quarter century.

Gauss called mathematics the queen of sciences. It is well known that mathematics is indispensable in the study of the sciences. Mathematicians often gloat over this. For example, Atiyah has said that he and other mathematicians were very happy to help physicists solve the pseudoparticle (now called the **Euclidean instanton**) problem. His student, Donaldson, was not happy. He wanted to study the geometry and topology of the moduli space of instantons on a 4-manifold M and to find out what information it might provide on the topology of M. Donaldson's work led to totally unexpected results about the topology of M and made gauge theory an important tool for studying low-dimensional topology. At about the same time, the famous physicist Ed Witten was using ideas and techniques from theoretical physics to provide new results and new ways of understanding old ones in mathematics. It is this work that ushered in the study of what we have called "physical mathematics."

Nature is the ultimate arbiter in science. Predictions of any theory have to be tested against experimental observations before it can be called a physical theory. A theory that makes wrong predictions or no predictions at all must be regarded as just a toy model or a proposal for a possible theory. An appealing (or beautiful) formulation is a desirable feature of the theory, but it cannot sustain the theory without experimental verification. The equations of Yang–Mills gauge theory provide a natural generalization of Maxwell's equations. They have a simple and elegant formulation. However, the theory predicted massless bosons, which have never been observed. Yang has said that this was the reason he did not work on the problem for over two decades. Such a constraint does not exist in "Physical Mathematics." So the nonphysical pure Yang–Mills theory has been heartily welcomed, forming the basis for Donaldson's theory of 4-manifolds and Floer's instanton homology of 3-manifolds. However, it was Witten who brought forth a broad spectrum of physical theories to obtain new results and new points of view on old results in mathematics. His work created a whole new area of research that led me to coin the term "physical mathematics" to describe it. Perhaps we can now reverse Dirac's famous statement and say instead "Mathematics is now. Physics can wait." The mathematicians can now say to physicists, "give us your rejects, toy models and nonpredictive theories and we will see if they can give us new mathematics and let us hope that some day they may be useful in physics."

The starting point of the present monograph was *The Mathematical Foundations of Gauge Theories* [274], which the author cowrote with Prof. Martucci (Firenze). That book was based in part on a course in differential geometric methods in physics" that the author gave at CUNY and then repeated at the Dipartimento di Fisica, Università di Firenze in 1986. The course was attended by advanced graduate students in physics and research workers in theoretical physics and mathematics. This monograph is aimed at a similar general audience. The author has given a number of lectures updating the material of that book (which has been out of print for some time) and presenting new developments in physics and their interaction with results in mathematics, in particular in geometric topology. This material now forms the basis for the present work. The classical and quantum theory of fields remains a very active area of research in theoretical physics as well as mathematics. However, the differential geometric foundations of classical gauge theories are now firmly established.

The latest period of strong interaction between theoretical physics and mathematics began in the early 1980s with Donaldson's fundamental work on the topology of 4-manifolds. A look at Fields Medals since then shows several going for work closely linked to physics. The Fields Medal is the highest honor bestowed by the mathematics community on a young (under 40 years of age) mathematician. The Noble Prize is the highest honor in physics but is often given to scientists many years after a work was done and there is no age bar. Appendix B contains more information on the Fields Medals.

Our aim in this work is to give a self-contained treatment of a mathematical formulation of some physical theories and to show how they have led to new results and new viewpoints in mathematical theories. This includes a differential geometric formulation of gauge theories and, in particular, of the theory of Yang–Mills fields. We assume that the readers have had a first course in topology, analysis and abstract algebra and an acquaintance with elements of the theory of differential manifolds, including the structures associated with manifolds such as tensor bundles and differential forms. We give a review of this mathematical background material in the first three chapters and also include material that is generally not covered in a first course.

We discuss in detail principal and associated bundles and develop the theory of connections in Chapter 4.

In Chapter 5 we introduce the characteristic classes associated to principal bundles and discuss their role in the classification of principal and associated bundles. A brief account of K-theory and index theory is also included in this chapter. The first five chapters lay the groundwork for applications to gauge theories, but the material contained in them is also useful for understanding many other physical theories.

Chapter 6 begins with an introduction and a review of the physical background necessary for understanding the role of gauge theories in high-energy physics. We give a geometrical formulation of gauge potentials and fields on a principal bundle P over an arbitrary pseudo-Riemannian base manifold M with the gauge group G. Various formulations of the group of gauge transformations are also given here. Pure gauge theories cannot describe interactions that have massive carrier particles. A resolution of this problem requires the introduction of matter fields. These matter fields arise as sections of bundles associated to the principal bundle P. The base manifold M may also support other fields such as the gravitational field. We refer to all these fields as **associated fields**. A Lagrangian approach to associated fields and coupled equations is also discussed in this chapter. We also discuss the generalized gravitational field equations, which include Einstein's equations with or without the cosmological constant, as well as the gravitational instanton equations as special cases.

Quantum and topological field theories are introduced in Chapter 7. Quantization of classical fields is an area of fundamental importance in modern mathematical physics. Although there is no satisfactory mathematical theory of quantization of classical dynamical systems or fields, physicists have developed several methods of quantization that can be applied to specific problems. We discuss the Feynman path integral method and some regularization techniques briefly.

In Chapter 8 we begin with some historical observations and then discuss Maxwell's electromagnetic theory, which is the prototype of gauge theories. Here, a novel feature is the discussion of the geometrical implications of Maxwell's equations and the use of universal connections in obtaining their solutions. This last method also yields solutions of pure (or source-free) Yang–Mills fields. We then discuss the most extensively studied coupled system, namely, the system of Yang–Mills–Higgs fields. After a brief discussion of various couplings we introduce the idea of spontaneous symmetry breaking and discuss the standard model of electroweak theory. The idea of spontaneous symmetry breaking was introduced by Nambu (who received the Nobel prize in Physics in 2008) and has been extensively studied by many physicists. Its most spectacular application is the Higgs mechanism in the standard model. A brief indication of some of its extensions is also given there.

Chapter 9 is devoted to a discussion of invariants of 4-manifolds. The special solutions of Yang–Mills equations, namely the instantons, are discussed separately. We give an explicit construction of the moduli space $\mathcal{M}_1$ of the BPST-instantons of instanton number 1 and indicate the construction of the moduli space $\mathcal{M}_k$ of the complete $(8k - 3)$-parameter family of instanton solutions over S^4 with gauge group $SU(2)$ and instanton number k. The moduli spaces of instantons on an arbitrary Riemannian 4-manifold with a semisimple Lie group as gauge group are then introduced. A brief account of Donaldson's theorem on the topology of moduli spaces of instantons and its implications for smoothability of 4-manifolds and Donaldson's polynomial invariants is then given. We then discuss Seiberg–Witten monopole equations. The study of $N = 2$ supersymmetric Yang–Mills theory led Seiberg and Witten to the now well-known monopole, or SW equations. The Seiberg–Witten theory provides new tools for the study of 4-manifolds. It contains all the information provided by Donaldson's theory and is much simpler to use. We discuss some applications of the SW invariants and their relation to Donaldson's polynomial invariants.

Chern–Simons theory and its application to Floer type homologies of 3-manifolds and other 3-manifold invariants form the subject of Chapter 10. Witten has argued that invariants obtained via Chern–Simons theory be related to invariants of a string theory. We discuss one particular example of such a correspondence between Chern–Simons theory and string theory in the last section. (String theory is expected to provide unification of all four fundamental forces. This expectation is not yet a reality and the theory (or its different versions) cannot be regarded as a physical theory. However, it has led to many interesting developments in mathematics.)

Classical and quantum invariants of 3-manifolds and knots and links in 3-manifolds are considered in Chapter 11. The relation of some of these invariants with conformal field theory and TQFT are also indicated there. The chapter concludes with a section on Khovanov's categorification of the Jones' polynomial and its extensions to categorification of other link invariants. The treatment of some aspects of these theories is facilitated by the use of techniques from analytic (complex) and algebraic geometry. A full treatment of these would have greatly increased the size of this work. Moreover, excellent monographs covering these areas are available (see, for example, Atiyah [15], Manin [257], Wells [399]). Therefore, topics requiring extensive use of techniques from analytic and algebraic geometry are not considered in this monograph.

There are too many other topics omitted to be listed individually. The most important is string theory. There are several books that deal with this still very active topic. For a mathematical treatment see for example, [95,96] and [9,212].

The epilogue points out some highlights of the topics considered. We note that the last three chapters touch upon some areas of active current research where a final definitive mathematical formulation is not yet available. They are intended as an introduction to the ever growing list of topics that can be thought of as belonging to physical mathematics. There are four appendices. Appendix A is a dictionary of terminology and notation between that used in physics and in mathematics. Background notes including historical and biographical notes are contained in Appendix B. The notions of categories and chain complexes are fundamental in modern mathematics. They are the subject of Appendix C. The cobordism category originally introduced and used in Thom's work is now the basis of axioms for TQFT. The general theory of chain complexes is basic in the study of any homology theory. Appendix D contains a brief discussion of operator theory and a more detailed discussion of the Dirac type operators.

Remark on References and Notation

Even though the foundations of electromagnetic theory (the prototype of gauge theories) were firmly in place by the beginning of nineteenth century, the discovery of its relation to the theory of connections and subsequent mathematical developments occurred only during the last three decades. As of this writing field theories remain a very active area of research in mathematical physics. However, the mathematical foundations of classical field theories are now well understood, and these have already led to interesting new mathematics. But we also use theories such as QFT, supersymmetry, and string theory for which the precise mathematical structure or experimental verification is not yet available. We have tried to bring the references up to date as of June 2009. In addition to the standard texts and monographs we have also included some books that give an elementary introductory treatment of some topics. We have included an extensive list of original research papers and review articles that have contributed to our understanding of the mathematical aspects of physical theories. However, many of the important results in papers published before 1980 and in the early 1980s are now available in texts or monographs and hence, in general, are not cited individually. The references to e-prints and private communication are cited in the text itself and are not included in the references at the end of the book.

As we remarked earlier, gauge theories and the theory of connections were developed independently by physicists and mathematicians, and as such have no standard notation. This is also true of other theories. We have used notation and terminology that is primarily used in the mathematical literature, but we have also taken into account the terminology that is most frequently used in physics. To help the reader we have included in Appendix A a correlation of terminology between physics and mathematics prepared along the lines of Trautman [378] and [409].

Acknowledgements

A large part of my work over the last 40 years was carried out in Italy, mostly at Florence University. A number of institutions, including Consiglio Nazionale delle Ricerche (Italy), the Dipartimento di Fisica, Università di Firenze, and the INFN (sezione di Firenze), have supported my work and I am thankful to them. I have had a number of collaborators during this period. I would like to mention in particular Giovanni Martucci (Firenze) for his friendship and long term collaboration. I would also like to thank Madre Superiora and Suore Stabilite nella Carità for gracious hospitality over the last three decades at their monumental Villa Agape (a villa that once belonged to Galileo's friend and disciple, Senator Arrighetti). Their beautiful gardens and quiet study hall provided a perfect setting for contemplation and writing.

The Einstein chair seminars on Topology and Quantum Objects organized by friend and colleague Dennis Sullivan at the CUNY Graduate Center is a continuing source of new ideas and information. Dennis' incisive questions and comments at these all-day seminars make them more accessible to nonexperts. I have enjoyed listening to and occasionally giving a substitute lecture in these seminars. Over the last 12 years I have held the MPG (Max Planck Gesselschaft) research fellowship at the MPI-MIS (Max Planck Institute for Mathematics in the Sciences) in Leipzig. I would like to thank my friend Eberhard Zeidler (founding director of MPI-MIS) for his interest in this work. I have attended many seminars by Prof. Jürgen Jost (director of MPI-MIS) on a wide range of topics in geometry and physics. I thank him for numerous discussions and for giving me a copy of his forthcoming book on geometry and physics before publication. The institute staff made this place a real ivory tower for my work. This work has required much more time and effort than I had anticipated. This would not have been possible without the continued support of MPI-MIS. I am deeply indebted to my colleague and friend Attila Mate who took over my duties on several occasions and who helped me with correcting the copy edited manuscript. His expertise with the typesetting system was also very useful in the production of this document. I would like to thank Stefan Wagner and Julia Plehnert, doctoral students at TU Darmstadt for their careful reading of some chapters of the first draft of the manuscript and for catching many errors of commission and omission. I would like to thank Dr. J. Heinze for his continued interest in my work. I met him about ten years ago when he asked me to write up my Shlosmann lecture for inclusion in Springer's Mathematics Unlimited. After reading my IMPRS lecture notes he suggested that I should write a new book updating my out of print book (with Prof. Martucci) on gauge theories. The present book is the outcome of this suggestion. After the project shifted to Springer UK, I dealt at first with Ms. Karen Borthwick and then with Ms. Lauren Stoney. My special thanks go to Karen and Lauren for providing several reviews of the manuscript at various stages and for making a number of useful suggestions that helped me navigate through a rather long period from initial to the final draft of the book. Thanks are also due to the reviewers for catching errors and obscure statements.

As a child I always learned something new during visits with my grandparents. My grandfather Moreshwar Marathe was a highly respected lawyer with a wide range of interests. He always had some article or poem for me to read. I was also strongly influenced by the great interest in learning that my mother, Indumati, and my grandmother, Parvatibai Agashe, exhibited. Neither of them had university education but they exemplified for me what a truly educated person should be. My grandmother passed away a long time ago. My mother passed away on January 26, 2005. This book is dedicated to her memory.

Kishore B. Marathe
Brooklyn, New York, May 2010

Chapter 1
Algebra

1.1 Introduction

We suppose that the reader is familiar with basic structures of algebra such as groups, rings, fields, and vector spaces and their morphisms, as well as the elements of representation theory of groups. Theory of groups was discovered by Cauchy. He called it "theory of substitutions." He found it so exotic that he is said to have remarked: "It is a beautiful toy, but it will not have any use in the mathematical sciences." In fact, quite the opposite was revealed to be true. The concept of group has proved to be fundamental in all mathematical sciences. In particular, the theory of Lie groups enjoys wide applicability in theoretical physics. We will discuss Lie groups in Chapter 3. Springer has started to reissue the volumes originally published under the general title "Éléments de mathématique" by Nicholas Bourbaki (see the note in Appendix B). The volumes dealing with Lie groups and Lie algebras are [57,56]. They can be consulted as standard reference works, even though they were written more than 20 years ago.

In the rest of this chapter we discuss some algebraic structure that may not be included in a first year course in algebra. These are some of the structures that appear in many physical theories. Section 1.2 considers the general structure of algebras, including graded algebras. Kac–Moody algebras are discussed in Section 1.3. Clifford algebras are introduced in Section 1.4. The gamma matrices in Dirac's equation for the electron wave function generate one such special Clifford algebra. Section 1.5 is devoted to the classification of finite simple groups and in particular to some strange coincidences dubbed "monstrous moonshine" related to the largest sporadic group called the monster. The quantum dimension of representations of the monster are encoded in various classical Hauptmoduls. Surprising relations between the monster and vertex algebras, conformal field theory, and string theory have emerged, and these remain a very active area of research.

K. Marathe, *Topics in Physical Mathematics*, DOI 10.1007/978-1-84882-939-8_1,

1.2 Algebras

Let K denote a field of characteristic zero. All the structures considered
in this section are over K, and hence we will often omit explicit reference
to K. In most applications K will be either the field **R** of real numbers
or the field **C** of complex numbers. Recall that an **algebra** A over K (or
simply an algebra) is a vector space with a bilinear function from $A \times A$
to A (multiplication) and denoted by juxtaposition of elements of A. Note
that in general the multiplication in A need not be associative. A is called
an **associative algebra** (resp., a **commutative algebra**) if A has a two-
sided multiplicative identity (usually denoted by 1) and the multiplication
is associative (resp., commutative). A vector subspace B of an algebra A is
called a **subalgebra** if it is an algebra under the product induced on it by the
product on A. A subalgebra I of A is called a **left ideal** if $xI \subset I$, $\forall x \in A$.
Right ideal and **two-sided ideal** are defined similarly.

If A, B are algebras, then a map $f : A \to B$ which preserves the algebra
structure is called an (algebra) **homomorphism**; i.e., f is a linear map of the
underlying vector spaces and $f(xy) = f(x)f(y)$, $\forall x, y \in A$. For associative
algebras we also require $f(1) = 1$. If f has an inverse then the inverse is
also a homomorphism and f is called an **isomorphism**. A homomorphism
(resp., an isomorphism) $f : A \to A$ is called an **endomorphism**. (resp. an
automorphism). A **derivation** $d : A \to A$ is a linear map that satisfies the
Leibniz product rule, i.e.,

$$d(xy) = d(x)y + xd(y), \qquad \forall x, y \in A.$$

The set of all derivations of A has a natural vector space structure. However,
the product of two derivations is not a derivation.

Example 1.1 *The set of all endomorphisms of a vector space V, denoted by*
End(V) has the natural structure of an associative algebra with multiplication
defined by composition of endomorphisms. A choice of a basis for V allows
one to identify the algebra End(V) with the **algebra of matrices** *(with the*
usual matrix multiplication). Recall that the set $M_n(K)$ of $(n \times n)$ matrices
with coefficients from the field K form an associative algebra with the usual
operations of addition and multiplication of matrices. Any subalgebra of this
algebra is called a **matrix algebra** *over the field K.*

The set of all automorphisms of a vector space V, denoted by Aut(V) or
$GL(V)$, has the natural structure of a group with multiplication defined by
*composition of automorphisms. If $K = $ **R** (resp., $K = $ **C**) and $\dim(V) = n$*
then $GL(V)$ can be identified (by choosing a basis for V) with the group of
invertible real (resp., complex) matrices of order n. These groups contain all
the **classical groups** *(i.e., orthogonal, symplectic, and unitary groups) as*
subgroups. The group operations are continuous in the topology on the groups
induced by the standard topology on V (identified with $\mathbf{R}^n$ or $\mathbf{C}^n$). This makes
them topological groups. In fact, the classical groups are Lie groups (i.e., they

are differentiable manifolds and the group operations are differentiable maps). These groups play a fundamental role in the study of global and local symmetry properties of physical systems in classical as well as quantum theories. The tangent space to a Lie group G can be given a natural structure of a Lie algebra LG. This Lie algebra LG carries most of the local information about G. All groups with the same Lie algebra are locally isomorphic and can be obtained as quotients of a unique simply connected group modulo discrete central subgroups. Thinking of LG as a linearization of G allows one to study analytic and global properties of G by algebraic properties of LG. (This is discussed in greater detail in Chapter 3.)

In physical applications, the most extensively used algebra structure is that of a **Lie algebra**. It is customary to denote the product of two elements x, y by the **bracket** $[x, y]$. Recall that an algebra $\mathfrak{g}$ is called a Lie algebra if its product is **skew-symmetric** and satisfies the well known **Jacobi identity**, i.e.

$$[x, y] = -[y, x], \qquad \forall x, y \in \mathfrak{g}, \tag{1.1}$$

and

$$[x, [y, z]] + [y, [z, x]] + [z, [x, y]] = 0, \qquad \forall x, y, z \in \mathfrak{g}, \tag{1.2}$$

The skew-symmetry property is equivalent to the following **alternating property** of multiplication.

$$[x, x] = 0, \qquad \forall x \in \mathfrak{g}, \tag{1.3}$$

This property is an immediate consequence of our assumption that the field K has characteristic zero. We say that the Lie algebra $\mathfrak{g}$ is m-dimensional if the underlying vector space is m-dimensional. If E_i , $1 \leq i \leq m$, is a basis for the Lie algebra $\mathfrak{g}$ then we have

$$[E_j, E_k] = c^i_{jk} E_i,$$

where we have used the Einstein summation convention of summing over repeated indices. The constants c^i_{jk} are called the **structure constants** of $\mathfrak{g}$ with respect to the basis $\{E_i\}$. They characterize the Lie algebra $\mathfrak{g}$ and satisfy the following relations:

1. $c^i_{jk} = -c^i_{kj}$,
2. $c^i_{jk} c^l_{im} + c^i_{km} c^l_{ij} + c^i_{mj} c^l_{ik} = 0$ (Jacobi identity).

The basis $\{E_i\}$ is called an **integral basis** if all the structure constants are integers. A vector subspace $\mathfrak{h}$ of a Lie algebra $\mathfrak{g}$ is called a **subalgebra** if it is a Lie algebra under the product (i.e., bracket) induced on it by the product on $\mathfrak{g}$. A subalgebra $\mathfrak{i}$ of $\mathfrak{g}$ is called an **ideal** if $x \in \mathfrak{g}, y \in \mathfrak{i}$ implies that $[x, y] \in \mathfrak{i}$. A Lie algebra ideal is always two-sided.

Given an associative algebra A, we can define a new product on A that gives it a Lie algebra structure. The new product, denoted by $[\, . \, , \, . \,]$, is defined by

$$[x, y] := xy - yx , \qquad \forall x, y \in A. \tag{1.4}$$

The expression on the right hand side of (1.4) is called the **commutator** of x and y in A. It is easy to verify that the new product defined by (1.4) is skew-symmetric and satisfies the Jacobi identity. We denote this Lie algebra by $\text{Lie}(A)$.

Example 1.2 *Let V be a vector space. The Lie algebra obtained by the above construction from $\text{End}(V)$ is denoted by $\mathfrak{gl}(V)$ with multiplication defined by the commutator of endomorphisms. If $K = \mathbf{R}$ (resp., $K = \mathbf{C}$) and $\dim(V) = n$ then $\mathfrak{gl}(V)$ can be identified (by choosing a basis for V) with the Lie algebra of all real (resp., complex) matrices of order n. These Lie algebras contain all the classical (orthogonal, symplectic, and unitary) Lie algebras as Lie subalgebras.*

Given a Lie algebra $\mathfrak{g}$, there exists a unique (up to isomorphism) associative algebra $U(\mathfrak{g})$ called the **universal enveloping algebra** of $\mathfrak{g}$ such that $\text{Lie}(U(\mathfrak{g})) = \mathfrak{g}$.

Example 1.3 *Let A be an associative algebra. A **derivation** of A is a linear map $d : A \to A$ satisfying the **Leibniz rule***

$$d(xy) = x dy + (dx)y, \qquad \forall x, y \in A.$$

Let $\mathfrak{d}(A)$ denote the vector space of all derivations of A. It can be given a Lie algebra structure by defining the product of two derivations to be their commutator; i.e.,

$$[d_1, d_2] := d_1 d_2 - d_2 d_1, \qquad \forall d_1, d_2 \in \mathfrak{d}(A).$$

The commutator $[\mathfrak{g}, \mathfrak{g}]$ of $\mathfrak{g}$ with itself is called the **derived algebra** of $\mathfrak{g}$. The commutator $[\mathfrak{g}, \mathfrak{g}]$ is an ideal of $\mathfrak{g}$ which is zero if and only if $\mathfrak{g}$ is abelian. By induction one defines the **derived series** $\mathfrak{g}^{(k)}, k \in \mathbf{N}$, by

$$\mathfrak{g}^{(1)} := [\mathfrak{g}, \mathfrak{g}] \quad \text{and} \quad \mathfrak{g}^{(k)} := [\mathfrak{g}^{(k-1)}, \mathfrak{g}^{(k-1)}], \quad k > 1.$$

A Lie algebra $\mathfrak{g}$ is called **solvable** if $\mathfrak{g}^{(k)} = 0$, for some $k \in \mathbf{N}$. The **lower central series** $\mathfrak{g}^k, k \in \mathbf{N}$, is defined by

$$\mathfrak{g}^1 := [\mathfrak{g}, \mathfrak{g}] \quad \text{and} \quad \mathfrak{g}^k := [\mathfrak{g}^{(k-1)}, \mathfrak{g}], \quad k > 1.$$

A Lie algebra $\mathfrak{g}$ is called **nilpotent** if $\mathfrak{g}^k = 0$ for some $k \in \mathbf{N}$. Definitions given earlier for morphisms of algebras have their natural counterparts for Lie algebras. A **representation** of a Lie algebra $\mathfrak{g}$ on a vector space V is a homomorphism $\rho : \mathfrak{g} \to \mathfrak{gl}(V)$. The vector space V becomes a left $\mathfrak{g}$-module under the action of $\mathfrak{g}$ on V induced by ρ. Conversely, given a **Lie algebra module** V we can obtain the representation ρ of the Lie algebra $\mathfrak{g}$ on V. In view of this observation we can use the language of representations and modules interchangeably. The dimension of V is called the **degree** of the representation. We now recall some basic facts about representations. A representation is called **faithful** if ρ is injective (i.e., a monomorphism). A **submodule** W of V is a subspace of V that is left invariant under the action of $\mathfrak{g}$ on V. It is called an **invariant subspace** of V. Clearly the zero subspace and V are invariant subspaces. A representation is called **irreducible** if zero and V are its only invariant subspaces. Otherwise, it is called **reducible**. A representation is called **fully reducible** if V is a direct sum of irreducible $\mathfrak{g}$-modules.

Given an element x in a Lie algebra L we define the map

$$\operatorname{ad} x : L \to L \text{ by } (\operatorname{ad} x)(y) := [x, y], \qquad \forall y \in L.$$

It is easy to check that the map ad x is a linear transformation of the vector space L. The bilinear form on L defined by

$$\langle x, y \rangle := \operatorname{Tr}(\operatorname{ad} x \ \operatorname{ad} y) \tag{1.5}$$

is called the **Killing form** of L. We define the **adjoint map**

$$\operatorname{ad} : L \to \mathfrak{gl}(L) \text{ by } \operatorname{ad}(x) := \operatorname{ad} x, \qquad \forall x \in L.$$

A simple calculation shows that the adjoint map is a homomorphism of Lie algebras. It is called the **adjoint representation** of L. The kernel of the adjoint representation is the center $Z(L)$ (i.e., $\operatorname{Ker} \operatorname{ad} = Z(L)$). The **center** $Z(L) := \{x \in L \mid [x, y] = 0, \forall y \in L\}$ is an ideal of L.

A non-Abelian Lie algebra $\mathfrak{g}$ is called **simple** if its only ideals are zero and itself. A Lie algebra $\mathfrak{g}$ is called **semi-simple** if it can be written as a direct sum of simple Lie algebras. **Elie Cartan** (1869-1951) obtained a characterization of semi-simple Lie algebras in terms of their Killing form called the **Cartan criterion**. The Cartan criterion states:

A Lie algebra $\mathfrak{g}$ is semi-simple if and only if its Killing form is non-degenerate. This is equivalent to saying that the Killing form is an inner product on $\mathfrak{g}$.

The simple summands of a semi-simple Lie algebra $\mathfrak{g}$ are orthogonal with respect to the inner product defined by the Killing form. A Lie group G is called semi-simple (resp., simple) if LG is semi-simple (resp., simple).

The classification of semi-simple Lie groups was initiated by **Wilhelm Killing** (1847–1923) at the end of the nineteenth century. It was completed by E. Cartan at the beginning of the twentieth century. The main tool in this classification is the classification of finite-dimensional complex, simple Lie algebras. We give a brief discussion of the basic structures used in obtaining this classification. They are also useful in the general theory of representations. Let $\mathfrak{g}$ be a finite dimensional complex, simple Lie algebra. A nilpotent subalgebra $\mathfrak{h}$ of $\mathfrak{g}$ that is **self-centralizing** is called a **Cartan subalgebra**. It can

be shown that a non-zero Cartan subalgebra exists and is abelian. Any two
Cartan subalgebras are isomorphic. The dimension of a Cartan subalgebra is
an invariant of $\mathfrak{g}$. It is called the **rank** of $\mathfrak{g}$. Let $\mathfrak{h}$ be a Cartan subalgebra and
$x \in \mathfrak{h}$. Then $\mathrm{ad}(x)$ is a **diagonalizable** linear transformation of $\mathfrak{g}$. Moreover,
all these linear transformation are simultaneously diagonalizable. Let $\mathfrak{h}^*$ be
the dual vector space of $\mathfrak{h}$. For $\lambda \in \mathfrak{h}^*$ define the space $\mathfrak{g}_\lambda$ by

$$\mathfrak{g}_\lambda := \{x \in \mathfrak{g} \mid [a, x] = \lambda(a)x, \qquad \forall a \in \mathfrak{h}. \tag{1.6}$$

We say that λ is a **root** of $\mathfrak{g}$ relative to the Cartan subalgebra $\mathfrak{h}$ if the space
$\mathfrak{g}_\lambda$ is non-zero. There exist a set of non-zero roots $A := \{\alpha_i, 1 \le i \le s\}$ such
that

$$\mathfrak{g} = \mathfrak{h} \oplus \left(\bigoplus_1^s \mathfrak{g}_{\alpha_i} \right), \tag{1.7}$$

where each space $\mathfrak{g}_{\alpha_i}$ is one-dimensional. The decomposition of $\mathfrak{g}$ given in (1.7)
is called a **root space decomposition**. Let r denote the rank of $\mathfrak{g}$ (dimension
of $\mathfrak{h}$). Then we can find a set $B := \{\beta_j, 1 \le j \le r\} \subset A$ satisfying the following
properties:

1. B is a **basis** for the space $\mathfrak{h}^*$.
2. Every root in A can be written as an integral linear combination of the
   elements of B, i.e.,
   $$\alpha_i = k_i^j \beta_j, \qquad 1 \le i \le s.$$
3. For a given i all the coefficients k_i^j are either in $\mathbf{Z}+$ (non-negative) or are
   in $\mathbf{Z}-$ (non-positive). In the first case we say that α_i is a **positive root**
   (resp. **negative root**).

If B satisfies the above properties then we say that B is a set of **simple
roots** of $\mathfrak{g}$ with respect to the Cartan subalgebra $\mathfrak{h}$. The positive and negative
roots are in one-to-one correspondence. Let $\mathfrak{g}_+$ denote the direct sum of
positive root spaces. The algebra $\mathfrak{g}_+ \oplus \mathfrak{h}$ is called the **Borel subalgebra**
relative to the basis of simple roots B. The classification is carried out by
studying root systems that correspond to distinct (non-isomorphic) simple
Lie algebras.

The finite dimensional complex, simple Lie algebras were classified by
Killing and Cartan into four families of **classical algebras** and five **exceptional algebras**. The classical algebras are isomorphic to subalgebras of
the matrix algebras $\mathfrak{gl}(n, \mathbf{C})$. Each exceptional Lie algebra is the Lie algebra
of a unique simple Lie group. These Lie groups are called the **exceptional
groups**. We list the classical Lie algebras, their dimensions, and a matrix
representative for each in Table 1.1.

Table 1.1 Classical Lie algebras

Type	Dimension	Matrix algebra
$A_n, n \geq 1$	$n(n+2)$	$\mathfrak{sl}(n+1, \mathbf{C})$
$B_n, n \geq 2$	$n(2n+1)$	$\mathfrak{so}(2n+1, \mathbf{C})$
$C_n, n \geq 3$	$n(2n+1)$	$\mathfrak{sp}(2n, \mathbf{C})$
$D_n, n \geq 4$	$n(2n-1)$	$\mathfrak{so}(2n, \mathbf{C})$

The exceptional Lie groups are listed in Table 1.2 in increasing order of dimension.

Table 1.2 Exceptional Lie groups

Type	G_2	F_4	E_6	E_7	E_8
Dimension	14	52	78	133	248

We conclude this section with a discussion of weights for a finite dimensional $\mathfrak{g}$-module V with corresponding representation ρ. Our starting point is an important theorem due to Hermann Weyl.

Theorem 1.1 *If $\mathfrak{g}$ is a complex semi-simple Lie algebra, then every finite dimensional representation of $\mathfrak{g}$ is fully reducible.*

It follows from Weyl's theorem that $\rho(x), x \in \mathfrak{h}$ ($\mathfrak{h}$ a Cartan subalgebra) is a diagonalizable linear transformation of V. Moreover, all these linear transformations are simultaneously diagonalizable. We say that $\lambda \in \mathfrak{h}^*$ is a **weight** of the $\mathfrak{g}$-module V if the space

$$V_\lambda := \{v \in V \mid \rho(a)v = \lambda(a)v\}, \qquad \forall a \in \mathfrak{h} \tag{1.8}$$

is non-zero. It can be shown that the space V is the direct sum of all the **weight spaces** V_λ and that $\mathfrak{g}_\alpha V_\lambda \subset V_{\alpha+\lambda}$ whenever α is a root. A non-zero vector $v_0 \in V_\lambda$ is called a **highest weight vector** or a **vacuum vector** if $\mathfrak{g}_\alpha v_0 = 0$ for all positive roots α of $\mathfrak{g}$. The weight λ is then called a **highest weight**. The highest weight λ is maximal with respect to the partial order on $\mathfrak{h}^*$ defined by $\mu > \nu$ if $\mu - \nu$ is a sum of positive roots. It is easy to check that the highest weight vector is a simultaneous eigenvector of the Borel algebra of $\mathfrak{g}$. Given a vacuum vector v_0 we can generate an irreducible submodule V_0 of V as follows. Let $\{\alpha_1, \ldots, \alpha_k\}$ be a finite collection of negative roots (not necessarily distinct). Let V_o be the vector space generated by the vectors $(\mathfrak{g}_{\alpha_1} \cdots \mathfrak{g}_{\alpha_k})v_0$ obtained by the successive application of the negative root spaces to v_0. It can be shown that V_0 is an irreducible submodule of V. In particular, if V itself is irreducible, then we have $V_0 = V$. The module V_0 is called the **highest weight module** and the corresponding representation is called the **highest weight representation**. A highest weight vector always exists for a finite-dimensional representation of a semi-simple Lie algebra. It plays a fundamental role in the theory of such representations. In the following example we describe all the irreducible representations of the simple complex Lie algebra $\mathfrak{sl}(2, \mathbf{C})$ as highest weight representations.

Example 1.4 *The Lie algebra* $\mathfrak{sl}(2, \mathbf{C})$ *consists of 2-by-2 complex matrices with trace zero. A standard basis for it is given by the elements* h, e, f *defined by*

$$h := \begin{pmatrix} 1 & 0 \\ 0 & -1 \end{pmatrix}, \quad e := \begin{pmatrix} 0 & 1 \\ 0 & 0 \end{pmatrix}, \quad f := \begin{pmatrix} 0 & 0 \\ 1 & 0 \end{pmatrix}.$$

The commutators of the basis elements are given by

$$[h, e] = 2e, \quad [h, f] = -2f, \quad [e, f] = h.$$

We note that the basis and commutators are valid for $sl(2, K)$ *for any field* K. *The Cartan subalgebra is one dimensional and is generated by* h. *The following theorem gives complete information about the finite dimensional irreducible representations of the Lie algebra* $\mathfrak{sl}(2, \mathbf{C})$.

Theorem 1.2 *For each* $n \in \mathbf{N}$ *there exists a unique (up to isomorphism) irreducible representation* ρ_n *of* $\mathfrak{sl}(2, \mathbf{C})$ *on a complex vector space* V_n *of dimension* n. *There exists a basis* $\{v_0, \ldots, v_{n-1}\}$ *of* V_n *consisting of eigenvectors of* $\rho_n(h)$ *with the vacuum vector (highest weight vector)* v_0 *satisfying the following properties:*

1. $h.v_i = (n - 1 - 2i)v_i$,
2. $c.v_i = (n - i)v_{i-1}$,
3. $f.v_i = v_{i+1}$,

where we have put $\rho_n(x)v = x.v, x \in \mathfrak{sl}(2, \mathbf{C}), v \in V, v_{-1} = 0 = v_n$ *and where* $0 \le i \le n - 1$.

1.2.1 Graded Algebras

Graded algebraic structures appear naturally in many mathematical and physical theories. We shall restrict our considerations only to $\mathbf{Z}$- and $\mathbf{Z}_2$-gradings. The most basic such structure is that of a graded vector space which we now describe. Let V be a vector space. We say that V is $\mathbf{Z}$-**graded** (resp., $\mathbf{Z}_2$-**graded**) if V is the direct sum of vector subspaces V_i, indexed by the integers (resp., integers mod 2), i.e.

$$V = \bigoplus_{i \in \mathbf{Z}} V_i \quad (\text{resp., } V = V_0 \oplus V_1).$$

The elements of V_i are said to be **homogeneous of degree** i. In the case of $\mathbf{Z}_2$-grading it is customary to call the elements of V_0 (resp. V_1) **even** (resp. **odd**). If V and W are two $\mathbf{Z}$-graded vector spaces, a linear transformation $f : V \to W$ is said to be **graded of degree** k if $f(V_i) \subset W_{i+k}$, $\forall i \in \mathbf{Z}$. If V and W are $\mathbf{Z}_2$-graded, then a linear map $f : V \to W$ is said to be **even** if $f(V_i) \subset W_i$, $i \in \mathbf{Z}_2$, and **odd** if $f(V_i) \subset W_{i+1}$, $i \in \mathbf{Z}_2$. An algebra A is said to be **Z-graded** if A is $\mathbf{Z}$-graded as a vector space; i.e.

$$A = \bigoplus_{i \in \mathbf{Z}} A_i$$

and $A_i A_j \subset A_{i+j}$, $\forall i, j \in \mathbf{Z}$. An ideal $I \subset A$ is called a **homogeneous ideal** if

$$I = \bigoplus_{i \in \mathbf{Z}} (I \cap A_i).$$

Other algebraic structures (such as Lie, commutative, etc.) have their graded counterparts. It was **Hermann Grassmann** (1809–1877) who first defined the structure of an **exterior algebra** associated to a finite-dimensional vector space. Grassmann's work was well ahead of his time and did not receive recognition for a long time. In the preface to his 1862 book he wrote: [T]here will come a time when these ideas, perhaps in a new form, will enter into contemporary developments. Indeed, Grassmann's expectation has come to fruition and his work has found many applications in mathematics and physics. An example of a **Z**-graded algebra is given by the exterior algebra of differential forms $\Lambda(M)$ of a manifold M if we define $\Lambda^i(M) = 0$ for $i < 0$. It is called the **Grassmann algebra** of the manifold M. (We discuss manifolds and their associated structures in Chapter 3.) The exterior differential d is a graded linear transformation of degree 1 of $\Lambda(M)$. The transformation d is **nilpotent** of order 2, i.e., it satisfies the condition $d^2 = 0$. A graded algebra together with a **differential** d (i.e. a graded linear transformation d of degree 1 satisfying the Leibniz rule and $d^2 = 0$) is called a **differential graded algebra** or **DGA**. Thus the Grassmann algebra is an example of a DGA. Similarly, we can define a **differential graded Lie algebra**, or **DGLA**. The DGA is a special case of the A^∞ algebra, also known as strong homotopy associative algebra. Similarly, the DGLA is a special case of an infinite-dimensional algebra called the L^∞-algebra. These algebras appear in open-closed string field theory.

The **graded** or **quantum dimension** of a $\mathbf{Z}$-graded vector space V is defined by

$$\dim_q V = \sum_{i \in \mathbf{Z}} q^i (\dim(V_i)) \ ,$$

where q is a formal variable.

If A and B are graded algebras, their **graded tensor product**, denoted $A \hat{\otimes} B$, is the tensor product $A \otimes B$ of A and B as vector spaces with the product defined on the homogeneous elements by

$$(u \otimes v)(x \otimes y) = (-1)^{ij} ux \otimes vy, \ v \in B_i \ , \ x \in A_j.$$

A similar definition can be given for $\mathbf{Z}_2$-graded algebras. In the physics literature a $\mathbf{Z}_2$-**graded algebra** is referred to as a **superalgebra**. For example, a (complex) **Lie superalgebra** is a $\mathbf{Z}_2$-graded vector space $g = g_0 \oplus g_1$ with a product satisfying the superanticommutativity and super Jacobi identity.

An important example of a Lie superalgebra is the space of all linear endomorphisms of a $\mathbf{Z}_2$-graded vector space; an important example of graded algebras is provided by **Vertex operator algebras**, or **VOA**, introduced in [139]. This definition of vertex operator algebras was motivated by the Conway–Norton conjectures about properties of the monster sporadic group. There are now a number of variants of these algebras, including the **vertex algebras** and the closely related Lie algebras now called the **Borcherds algebras** introduced by **Richard Ewen Borcherds** in his proof of the Conway-Norton conjectures. (Borcherds received a Fields Medal at the ICM 1998 in Berlin for his work in algebra, the theory of automorphic forms and mathematical physics.) These vertex algebras are closely related to algebras arising in conformal field theory and string theory. Vertex operators arise in 2-dimensional Euclidean conformal field theory from field insertions at marked points on a Riemann surface. A definition of VOA is obtained by isolating the mathematical properties related to interactions represented by operator product expansion in chiral CFT. An algebraic description of such interactions and their products codify these new graded algebras. These infinite-dimensional symmetry algebras were first introduced in [37]. Borcherds definition of vertex algebras is closely related to these. These algebras are also expected to play a role in the study of the geometric Langlands conjecture. We discuss the VOA in more detail in the last section.

1.3 Kac–Moody Algebras

Kac–Moody algebras were discovered independently in 1968 by Victor Kac and Robert Moody. These algebras generalize the notion of finite-dimensional semi-simple Lie algebras. Many properties related to the structure of Lie algebras and their representations have counterparts in the theory of Kac–Moody algebras. In fact, they are usually defined via a generalization of the Cartan matrix associated to a Lie algebra. We restrict ourselves to a special class of Kac–Moody algebras that arise in many physical and mathematical situations, such as conformal field theory, exactly solvable statistical models, and combinatorial identities (e.g., Rogers–Ramanujan and Macdonald identities). They are called **affine algebras**.

Let $\mathfrak{g}$ be a Lie algebra with a symmetric bilinear form $\langle \ \cdot \ , \ \cdot \ \rangle$. This form is said to be $\mathfrak{g}$-invariant (or simply **invariant**) if

$$\langle [x,y],z \rangle = \langle x,[y,z] \rangle, \qquad \forall x,y,z \in \mathfrak{g}. \tag{1.9}$$

In the rest of this section we assume that a symmetric, $\mathfrak{g}$-invariant, bilinear form on $\mathfrak{g}$ is given. Let t be a formal variable and let $K[t,t^{-1}]$ be the algebra of Laurent polynomials in t. Then the vector space $\mathfrak{g}_t$ defined by

$$\mathfrak{g}_t := \mathfrak{g} \oplus K[t,t^{-1}] \tag{1.10}$$

can be given a Lie algebra structure by defining the product

$$[x \oplus t^m, y \oplus t^n] := [x,y] \oplus t^{m+n}, \text{ for } x,y \in \mathfrak{g}, \ m,n \in \mathbf{Z}. \tag{1.11}$$

When the base field $K = \mathbf{C}$, the algebra $\mathfrak{g}_t$ can be regarded as the Lie algebra of polynomial maps of the unit circle into $\mathfrak{g}$. For this reason the algebra $\mathfrak{g}_t$ is called the **loop algebra** of $\mathfrak{g}$. The **affine Kac–Moody algebra**, or simply the **affine algebra**, is a central extension of $\mathfrak{g}_t$. Let c be another formal variable. It is called the **central charge** in physics literature. Using the one-dimensional space Kc we obtain a central extension $\mathfrak{g}_{t,c}$ of $\mathfrak{g}_t$ defined by

$$\mathfrak{g}_{t,c} := \mathfrak{g} \oplus K[t,t^{-1}] \oplus Kc, \tag{1.12}$$

where c is taken as a **central element**. The product on $\mathfrak{g}_{t,c}$ is defined by extending the product on $\mathfrak{g}_t$ defined in (1.11) as follows:

$$[x \oplus t^m, y \oplus t^n] := [x,y] \oplus t^{m+n} + \langle x,y \rangle m \delta_{m,-n} c, \text{ for } x,y \in \mathfrak{g}. \tag{1.13}$$

The choice of c as a central element means that c commutes with every element of the algebra $\mathfrak{g}_{t,c}$. The algebra $\mathfrak{g}_{t,c}$, with product defined by extending (1.13) by linearity, is called the **affine Lie algebra**, or simply the affine algebra associated to the Lie algebra $\mathfrak{g}$ with central charge c and the given invariant bilinear form. Similarly, we can define the affine algebra $\mathfrak{g}_{\sqrt{t},c}$ by letting m,n take half-integer values in the definition (1.13).

Example 1.5 *(Virasoro algebra) Applying the construction of example (1.3) to the associative algebra $K[t,t^{-1}]$, we obtain the Lie algebra of its derivations $\mathfrak{d}(K[t,t^{-1}])$. It can be shown that this Lie algebra has a basis d_n, $n \in \mathbf{Z}$, defined by*

$$d_n(f(t)) := -t^{n+1} f'(t), \qquad \forall f \in K[t,t^{-1}], \tag{1.14}$$

where $f'(t)$ is the derivative of the Laurent polynomial $f(t)$. The product of the basis elements is given by the formula

$$[d_m, d_n] = (m-n)d_{m+n}, \qquad \forall m,n \in \mathbf{Z}. \tag{1.15}$$

*The algebra $K[t,t^{-1}]$ has a unique nontrivial one-dimensional central extension (up to isomorphism). It is called the **Virasoro algebra** and is denoted by $\mathfrak{v}$. It is generated by the **central element** (also called the **central charge** C) and the elements L_n corresponding to the elements d_n defined* *above. The product in $\mathfrak{v}$ is given by*

$$[L_m, L_n] = (m - n)L_{m+n} + \frac{1}{12}(m^3 - m)\delta_{m,-n}C, \ \forall m, n \in \mathbf{Z}. \quad (1.16)$$

The constant $1/12$ is chosen for physical reasons. It can be replaced by an arbitrary complex constant giving an isomorphic algebra. We note that the central term is zero when $m = -1, 0, 1$. Thus the central extension is trivial when restricted to the subalgebra of $\mathfrak{v}$ generated by the elements L_{-1}, L_0, and L_1. This subalgebra is denoted by $\mathfrak{p}$ and it can be shown to be isomorphic to the Lie algebra $\mathfrak{sl}(2, K)$ of matrices of order 2 and trace zero. The Virasoro algebra is an example of an infinite-dimensional Lie algebra. It was defined by physicists in their study of conformal field theory for the case $K = \mathbf{C}$. Gelfand and Fuchs have shown that the Virasoro algebra can be realized as a central extension of the algebra of polynomial vector fields on the unit circle S^1 by identifying the derivation d_n used in our definition with a vector field on S^1.

An infinite-dimensional algebra does not, in general, have a highest weight representation. However, the Virasoro algebra does admit such modules. We now describe its construction. Let U denote the universal enveloping algebra of the Virasoro algebra $\mathfrak{v}$ defined by the relations (1.16) and let $(h, c) \in \mathbf{C}^2$ be a given pair of complex numbers. Let I be the left ideal in U generated by $L_0 - h\iota, C - c\iota, L_i, i \in \mathbf{N}$, where ι denotes the identity. Let $V(h, c)$ be the quotient of U by the ideal I and let $v \in V(h, c)$ be the class of the identity (i.e., $v = \iota + I$). Then it can be shown that $V(h, c)$ is a highest weight module for $\mathfrak{v}$ with highest weight vector v and highest weight (h, c) satisfying the following conditions:

1. $L_0 v = hv, Cv = cv, L_i v = 0, \ i \in \mathbf{N}$;
2. the set of vectors $L_{i_1} \ldots L_{i_k} v$, where $i_1, \ldots, i_k$ is a decreasing sequence of negative integers, generates $V(h, c)$.

Any $\mathfrak{v}$-module V satisfying the above two conditions with a nonzero vector v defines a highest weight module with highest weight (h, c). A highest weight module is called a **Verma module** if the vectors in condition 2 form a basis of V. It can be shown that this is the case for the module $V(h, c)$ constructed above. Thus $V(h, c)$ is a Verma module for the Virasoro algebra $\mathfrak{v}$. Moreover, every highest weight module is a quotient of $V(h, c)$ by a submodule with the same highest weight (h, c). Verma modules have many other interesting properties. For example, their homomorphisms are closely related to the invariant differential operators on homogeneous manifolds obtained as quotients of Lie groups by their subgroups. Verma modules were defined by D. N. Verma[1] in his study of representations of semi-simple Lie algebras over forty years ago.

Example 1.6 (Heisenberg algebras) *In classical mechanics the state of a particle at time t is given by its position and momentum vectors (q, p)*

---

[1] My long-time friend whose enthusiasm and interest in mathematics is still strong.

in Euclidean space. The evolution of the system is governed by Hamilton's equations (see Chapter 3). Heisenberg's fundamental idea for quantization of such a system was to take the components of q and p to be operators on a Hilbert space satisfying the canonical commutation relations

$$[q_j, q_m] = 0, \ [p_j, p_m] = 0, \ [p_j, q_m] = -i\hbar\delta_{j,m}, \ 1 \le j, m \le n,$$

where n is the dimension of the Euclidean space. Heisenberg's uncertainty principle is closely related to the noncommuting of the position and its conjugate momentum. If $-i\hbar.1$ is replaced by a basis vector c taken as a central element, then we obtain a $(2n + 1)$-dimensional real Lie algebra with basis q_j, p_j, c. This algebra is denoted by $\mathfrak{h}_n$ and is called a **Heisenberg algebra** *with central element c. This Heisenberg algebra is isomorphic to an algebra of upper triangular matrices. If we define an index set I to be the set of the first n natural numbers, then the algebra $\mathfrak{h}_n$ can be called the (I, c)-Heisenberg algebra. This definition can be generalized to arbitrary index set I. If I is infinite we get an infinite-dimensional Heisenberg algebra. Another example of an infinite-dimensional Heisenberg algebra is given by a Lie algebra with a basis a_n, $n \in \mathbf{Z}$, together with the central element b. The product of the basis elements is given by*

$$[a_m, a_n] = m\delta_{m,-n}b, \quad [a_m, b] = 0, \quad \forall m, n \in \mathbf{Z}. \tag{1.17}$$

This Heisenberg algebra is also referred to as the **oscillator algebra**. *It arises as an algebra of operators on the* **bosonic Fock space** $\mathcal{F}$. *The space $\mathcal{F}$ is defined as the space of complex polynomials in infinitely many variables z_i, $i \in \mathbf{N}$. The operators on $\mathcal{F}$ defined by*

$$a_n := \frac{\partial}{\partial z_n}, \ a_{-n} := nz_n, \ \forall n \in \mathbf{N}, \ and \ a_0 = id,$$

together with the central element b generate an algebra isomorphic to the Heisenberg algebra. It is also possible to generate a copy of the Virasoro algebra by a suitable combination of the operators defined above.

The original Heisenberg commutation relations were applied successfully in producing the spectrum of the quantum harmonic oscillator (see Appendix B) and the hydrogen atom, even though the Hilbert space on which the operators act was not specified. The following theorem explains why this did not cause any problems.

Theorem 1.3 (Stone–von Neumann theorem) *Fix a nonzero scalar with which the central element c acts. Then there is a unique irreducible representation of the Heisenberg algebra $\mathfrak{h}_n$.*

The experimental agreement of the spectra resulted in the early acceptance of quantum mechanics based on canonical quantization. It is important to note that canonical quantization does not apply to all classical systems and there is no method of quantizing a general classical mechanical system at this time.

1.4 Clifford Algebras

Important examples of graded and superalgebras are furnished by the Clifford algebras of pseudo-inner product spaces. In defining these algebras, William Kingdon Clifford (1845–1879) generalized Grassmann's work on exterior algebras. Clifford algebras arise in many physical applications. In particular, their relation to the Pauli spin matrices and the Dirac operator on spinorial spaces and to the Dirac γ matrices is well known. Let V be a real vector space and $g : V \times V \to \mathbf{R}$ a bilinear symmetric map. A **Clifford map** for (V, g) is a pair (A, ϕ), where A is an associative algebra with unit 1_A (often denoted simply by 1) and $\phi : V \to A$ a linear map satisfying the condition

$$\phi(u)\phi(v) + \phi(v)\phi(u) = 2g(u,v)1_A, \qquad \forall u, v \in V.$$

Such pairs are the objects of a category whose morphisms are, for two objects $(A, \phi), (B, \psi)$, the algebra homomorphisms $h : A \to B$ such that $\psi = h \circ \phi$; i.e., the following diagram commutes:

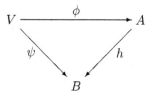

This category has a universal initial object $(C(V, g), \gamma)$. The algebra $C(V, g)$, or simply $C(V)$ is called the **Clifford algebra** of (V, g). In other words given any Clifford map (A, ϕ), there is a unique algebra homomorphism $\Phi : C(V) \to A$ such that $\phi = \Phi \circ \gamma$, i.e. the following diagram commutes.

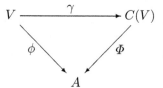

There are several ways of constructing a model for $C(V)$. For example, let

$$T(V) = \bigoplus_{r \geq 0} T^r(V)$$

and let J be the two-sided ideal generated by elements of the type

$$u \otimes v + v \otimes u - 2g(u,v)1_{T(V)}, \qquad \forall u,v \in V.$$

We note that every element of J can be written as a finite sum

$$\sum_i \lambda_i \otimes (v_i \otimes v_i - g(v_i,v_i)1) \otimes \mu_i,$$

where $\lambda_i, \mu_i \in T(V)$ and $v_i \in V$. Then we define $C(V)$ to be the quotient $T(V)/J$ and $\gamma = \pi \circ \iota$, where $\iota : V \to T(V)$ is the canonical injection and $\pi : T(V) \to T(V)/J$ is the canonical projection. $C(V)$ is generated by the elements of $\gamma(V)$. The tensor multiplication on $T(V)$ induces a multiplication on the quotient algebra $C(V)$ that is uniquely determined by the products

$$\gamma(u)\gamma(v) = u \otimes v + J, \qquad \forall u,v \in V.$$

We often write $\gamma(v) \in C(V)$ as simply v. Then an element of $C(V)$ can be written as a finite sum of products of elements $v \in V$. In particular, $v^2 = v \otimes v + J = g(v,v)1_{C(V)}$ and, by polarization,

$$uv + vu = 2g(u,v)1_{C(V)}, \tag{1.18}$$

which shows that γ is a Clifford map. We observe that the tensor algebra $T(V)$ is $\mathbf{Z}$-graded but the ideal J is not homogeneous. Therefore, the $\mathbf{Z}$-grading does not pass to the Clifford algebra. However, there exists a natural $\mathbf{Z}_2$-grading on $C(V)$, which makes it into a superalgebra. This $\mathbf{Z}_2$-grading is defined as follows. Let $C_0(V)$ (resp., $C_1(V)$) be the vector subspace of $C(V)$ generated by the products of an even (resp., odd) number of elements of V. Then

$$C(V) = C_0(V) \oplus C_1(V)$$

and the elements of $C_0(V)$ (resp. $C_1(V)$) are the even (resp. odd) elements of $C(V)$. We observe that, when $g = 0$, $C(V)$ reduces to the exterior algebra $\Lambda(V)$ of V. An important result is given by the following theorem.

Theorem 1.4 *Let $(V,g), (V',g')$ be two pseudo-inner product spaces. Then there exists a natural Clifford algebra isomorphism*

$$C(V \oplus V', g \oplus g') \cong C(V,g) \hat{\otimes} C(V',g'),$$

where $\hat{\otimes}$ is the graded tensor product of $\mathbf{Z}_2$-graded algebras.
Proof: Let $\gamma : V \to C(V)$, $\gamma' : V' \to C(V')$ be the canonical inclusion map and define

$$\tilde{\gamma} : V \oplus V' \to C(V) \hat{\otimes} C(V')$$

by

$$\tilde{\gamma}(v,v') = \gamma(v) \otimes 1 + 1 \otimes \gamma'(v').$$

It is easy to check that $\tilde{\gamma}(v,v')^2 = [g(v,v) + g'(v',v')]1$ and hence by the universal property of Clifford algebras $\tilde{\gamma}$ induces an algebra homomorphism

$$\psi : C(V \oplus V') \to C(V) \hat{\otimes} C(V').$$

It can be shown that ψ is a Clifford algebra isomorphism (see M. Karoubi [216] for further details).   □

Infinite-dimensional Clifford algebras have been studied and applied to the problem of describing the quantization of gauge fields and BRST cohomology in [236]. The structure of finite-dimensional Clifford algebras is well known and has numerous applications in mathematics as well as in physics. We now describe these algebras in some detail.

Recall that a subspace $U \subset V$ is called g-**isotropic** or simply **isotropic** if $g_{|U} = 0$. Factoring out the maximal isotropic subspace of V, we can restrict ourselves to the case when g is **non-degenerate** (i.e., $g(x,y) = 0, \forall x \in V$ implies $y = 0$). In the finite-dimensional case a pseudo-inner product space (V,g), with **pseudo-metric** or **pseudo-inner product** g of **signature** $(k, n-k)$, is isometric to $(\mathbf{R}^n, g_k)$, where g_k is the canonical pseudo-metric on $\mathbf{R}^n$ of **index** $n-k$. We denote by $C(k, n-k)$ the Clifford algebra $C(\mathbf{R}^n, g_k)$. Let $\{e_i \mid i = 1, \ldots, n\}$ be a g_k-orthonormal basis of $\mathbf{R}^n$. To emphasize the Clifford algebra aspect we denote the images $\gamma(e_i)$ by γ_i; this is also in accord with the notation used in the physical literature. Then the Clifford algebra $C(k, n-k)$ has a basis consisting of 1 and the products

$$\gamma_{i_1} \gamma_{i_2} \cdots \gamma_{i_r}, \quad 1 \le i_1 < i_2 < \cdots < i_r \le n,$$

where $r = 1, \ldots, n$ and hence has dimension 2^n. The element $\gamma_1 \gamma_2 \cdots \gamma_n$ is usually denoted by γ_{n+1}. From equation (1.18) it follows that the product is subject to the relations

$$\gamma_i^2 = 1, \quad i = 1, \ldots, k \ , \ \gamma_j^2 = -1, \quad j = k+1, \ldots, n \qquad (1.19)$$

and

$$\gamma_i \gamma_j = -\gamma_j \gamma_i, \qquad \forall \, i \ne j. \qquad (1.20)$$

Conversely, the product is completely determined by these relations. In physical applications one often starts with matrices satisfying the product relations (1.19) to construct various Clifford algebras. In particular, starting with the **Pauli spin matrices** one can construct various Clifford algebras as is shown below. Recall that the Pauli spin matrices are the complex 2×2 matrices

$$\sigma_1 = \begin{pmatrix} 0 & 1 \\ 1 & 0 \end{pmatrix}, \quad \sigma_2 = \begin{pmatrix} 0 & -i \\ i & 0 \end{pmatrix}, \quad \sigma_3 = \begin{pmatrix} 1 & 0 \\ 0 & -1 \end{pmatrix}.$$

Taking $\gamma_1 = \sigma_1$, $\gamma_2 = i\sigma_2$, we observe that I, γ_1, γ_2, $\gamma_1\gamma_2 = -\sigma_3$ form a basis of $M_2(\mathbf{R})$. The resulting algebra is isomorphic to the Clifford algebra $C(1,1)$. If on the other hand we choose $\gamma_1 = \sigma_1$, $\gamma_2 = \sigma_3$, then I, γ_1, γ_2, $\gamma_1\gamma_2 = -i\sigma_2$ is a basis of $M_2(\mathbf{R})$. The resulting algebra is isomorphic to the Clifford algebra $C(2,0)$. Thus, we have two non-isomorphic Clifford algebras with the same underlying vector space. Taking $\gamma_1 = i\sigma_1$, $\gamma_2 = i\sigma_2$, we obtain the

Clifford algebra $C(0,2)$ with basis I, γ_1, γ_2, $\gamma_1\gamma_2 = -i\sigma_3$. It is easy to verify that $C(0,2) \cong \mathbf{H}$. Consider now the case $C(m,n)$ where $m + n = 2p$. Then $(\gamma_{2p+1})^2 = (-1)^{n+p}1$. The following theorem is useful for the construction of Clifford algebras.

Theorem 1.5 *Let $m + n = 2p$; then we have the isomorphisms*

$$C(m,n) \otimes C(m',n') \cong C(m+m',n+n') \quad if \ (-1)^{n+p} = 1, \tag{1.21}$$

$$C(m,n) \otimes C(m',n') \cong C(m+n',n+m') \quad if \ (-1)^{n+p} = -1. \tag{1.22}$$

Proof: Consider the set

$$B = \{\gamma_i \otimes 1', \ \gamma_{2p+1} \otimes \gamma_j\}, \quad i = 1,\ldots,m+n, \ j = 1,\ldots,m'+n'.$$

Observe that

$$(\gamma_i \otimes 1')^2 = (1 \otimes 1')g(e_i,e_i),$$

$$(\gamma_{2p+1} \otimes \gamma_j)^2 = (-1)^{n+p}(1 \otimes 1')g(e_j,e_j)$$

and that different elements of B anticommute. The number of elements of B whose square is $-1 \otimes 1'$ is $n + n'$ (resp., $n + m'$) if $(-1)^{n+p} = 1$ (resp., $(-1)^{n+p} = -1$). If $(-1)^{n+p} = 1$, consider the map

$$\phi : \mathbf{R}^{n+m} \oplus \mathbf{R}^{n'+m'} \to C(m,n) \otimes C(m',n')$$

such that

$$\phi(v,v') = v \otimes 1 + \gamma_{2p+1} \otimes v'.$$

By reasoning similar to that in the proof of Theorem (1.4) it can be shown that ϕ induces an isomorphism of $C(m+m',n+n')$ with $C(m,n) \otimes C(m',n')$. This proves (1.21). Similar reasoning, in the case that $(-1)^{n+p} = -1$, establishes the isomorphism indicated in (1.22). $\square$

The theorem (1.5) and the structure of matrix algebras lead to the isomorphism

$$C(p+n,q+n) \cong C(p,q) \otimes M_{2^n}(\mathbf{R}) \cong M_{2^n}(C(p,q)), \tag{1.23}$$

where $M_k(A)$ is the algebra of $k \times k$-matrices with coefficients in the algebra A. In particular we have

$$C(n,n) \cong M_{2^n}(\mathbf{R}), \tag{1.24}$$

$$C(p,q) \cong C(p-q,0) \otimes M_{2^q}(\mathbf{R}), \quad \text{if } p > q \tag{1.25}$$

and

$$C(p,q) \cong C(0,q-p) \otimes M_{2^p}(\mathbf{R}), \quad \text{if } p < q. \tag{1.26}$$

Example 1.7 *In this example we consider the important special case of the Clifford algebra $C(p,q)$ where $p = 3$ and $q = 1$. Our discussion is based on* *the isomorphism*

$$C(3,1) \cong C(2,0) \otimes C(1,1).$$

The generating elements of $C(3,1)$ correspond to a representation of the classical Dirac matrices. From the construction outlined above we obtain the following well-known expressions for the **Dirac matrices** *in terms of the Pauli spin matrices.*

$$\gamma_1 = \sigma_2 \otimes \sigma_2, \ \ \gamma_2 = \sigma_3 \otimes 1, \ \ \gamma_3 = -\sigma_1 \otimes 1, \ \ \gamma_4 = i\sigma_2 \otimes \sigma_1.$$

They are referred to as the **real representation** *or the* **Majorana representation** *of the Dirac γ matrices. Let M denote the Minkowski space of signature (3, 1) with coordinates (x_1, x_2, x_3, x_4) and let ∂_i denote the partial derivative with respect to x_i. Then the* **Dirac operator** *D is defined by*

$$D := \gamma_1\partial_1 + \gamma_2\partial_2 + \gamma_3\partial_3 + \gamma_4\partial_4 \qquad (1.27)$$

It acts on wave functions $\psi : M \to \mathbf{C}^4$. The **Dirac equation** *for a* **free electron** *is then given by $D\psi = m\psi$, where m is the mass of the electron (we have set the velocity of light in vacuum and the Planck's constant each to 1). This equation is invariant under Lorentz transformation if ψ transforms as a 4-component spinor. Thus the Dirac equation is a relativistic equation. Dirac operator is a first order differential operator. We will discuss it and other differential operators in Appendix D. Using the properties of the γ matrices one can show that the square of the Dirac operator is given by*

$$D^2 = \partial_1^2 + \partial_2^2 + \partial_3^2 - \partial_4^2 \ . \qquad (1.28)$$

This operator is the well known **D'Alembertian** *or the wave operator. It was in attempting to find a square root of this operator that Dirac discovered the γ matrices and his operator. For an electron moving in an electromagnetic field the partial derivatives are replaced by covariant derivatives by minimal coupling to the electromagnetic gauge potential. This equation also admits a solution where the particle has the same mass as the electron but positive charge of the same magnitude as the electron charge. This particle, called the* **positron** *was the first experimentally discovered* **anti-particle***. It is known that every charged fundamental particle has its anti-particle partner.*

Another representation which is not real and is frequently used is the following:

$$\gamma_1 = \sigma_2 \otimes \sigma_1, \ \ \gamma_2 = \sigma_2 \otimes \sigma_2, \ \ \gamma_3 = \sigma_2 \otimes \sigma_3, \ \ \gamma_4 = -i\sigma_3 \otimes 1.$$

A systematic development of the theory of spinors was given by E. Cartan in [71]. He obtained the expression $\gamma_i\partial_i\xi$ for the Dirac operator acting on the spinor ξ in Minkowski space. We give below these γ matrices in block matrix form:

$$\gamma_1 = \begin{pmatrix} \theta & \sigma_3 \\ \sigma_3 & \theta \end{pmatrix}, \quad \gamma_2 = \begin{pmatrix} \theta & iI_2 \\ -iI_2 & \theta \end{pmatrix}, \quad \gamma_3 = \begin{pmatrix} \theta & \sigma_1 \\ \sigma_1 & \theta \end{pmatrix}, \quad \gamma_4 = \begin{pmatrix} \theta & -i\sigma_2 \\ -i\sigma_2 & \theta \end{pmatrix},$$

where θ is the zero matrix and I_2 is the identity matrix. The components of the Cartan spinor ξ are linear combinations of the components of the Dirac spinor ψ. Substituting these in Cartan's expression of the Dirac operator we obtain the expression of the operator used in Dirac's equation for an electron in an electromagnetic field. For further details see the classical work of E. Cartan [71] and the contemporary book by Friedrich [144].

Using the algebras $C(0,2)$, $C(2,0)$ and the isomorphisms of (1.21), (1.22), and their special cases discussed after that theorem, we obtain the following periodicity relations

$$C(n + 8, 0) \cong C(n, 0) \otimes M_{16}(\mathbf{R}), \qquad (1.29)$$

$$C(0, n + 8) \cong C(0, n) \otimes M_{16}(\mathbf{R}). \qquad (1.30)$$

These periodicity relations of Clifford algebras can be used to prove the Bott periodicity theorem for stable real homotopy via K-theory. Similar results also hold for the complex case. Thus, we can consider these periodicity relations as a version of Bott periodicity. From the periodicity relations of Clifford algebras, it follows that the construction of all real Clifford algebras can be obtained by using the relations in the following table.

Table 1.3 Real Clifford algebras

n	$C(n, 0)$	$C(0, n)$
0	$\mathbf{R}$	$\mathbf{R}$
1	$\mathbf{R} \oplus \mathbf{R}$	$\mathbf{C}$
2	$M_2(\mathbf{R})$	$\mathbf{H}$
3	$M_2(\mathbf{C})$	$\mathbf{H} \oplus \mathbf{H}$
4	$M_2(\mathbf{H})$	$M_2(\mathbf{H})$
5	$M_2(\mathbf{H}) \oplus M_2(\mathbf{H})$	$M_4(\mathbf{C})$
6	$M_4(\mathbf{H})$	$M_8(\mathbf{R})$
7	$M_8(\mathbf{C})$	$M_8(\mathbf{R}) \oplus M_8(\mathbf{R})$
8	$M_{16}(\mathbf{R})$	$M_{16}(\mathbf{R})$

1.5 Hopf Algebras

Let A be an algebra with unit 1_A over a commutative ring K with unit 1. Denote by $m : A \times A \to A$, the multiplication in A and by $I : A \to A$, the identity homomorphism. Let (Δ, ϵ, s) denote the K-linear homomorphisms

$$\Delta : A \to A \otimes A, \quad \epsilon : A \to K, \quad s : A \to A.$$

A quadruple (A, Δ, ϵ, s) or simply A when $(\Delta, \epsilon, \sigma)$ are understood, is called a **Hopf algebra** if the following conditions are satisfied:
HA1. $(I \otimes \Delta)\Delta = (\Delta \otimes I)\Delta$,
HA2. $m(I \otimes \sigma)\Delta = m(\sigma \otimes I)\Delta$,
HA3. $(I \otimes \epsilon)\Delta = (\epsilon \otimes I)\Delta = I$.
Note that in the above conditions we have used the identification

$$(A \otimes A) \otimes A = A \otimes (A \otimes A),$$

via the natural isomorphism

$$(a \otimes b) \otimes c \mapsto a \otimes (b \otimes c), \qquad \forall a, b, c \in A$$

and

$$A \otimes K = K \otimes A = A,$$

via the isomorphisms

$$a \otimes 1 \mapsto 1 \otimes a \mapsto a , \qquad \forall a \in A.$$

The map Δ is called the **co-multiplication**. The map ϵ is called the **co-unit** and the map σ is called the **antipode**. We define the map $P : A \otimes A \to A \otimes A$ by

$$P(a \otimes b) = b \otimes a, \qquad \forall a, b \in A.$$

The map P is an isomorphism called the **flip**. The definition of A implies that the antipode σ is an anti-automorphism of both the algebra and the coalgebra structures on A.

Define **opposite co-multiplication** Δ' in A by $\Delta' := P \circ \Delta$.

Definition 1.1 *Let R be an invertible element of $A \otimes A$. Define elements* $R_{12}, R_{23}, R_{13} \in A \otimes A \otimes A$ *by*

$$R_{12} = R \otimes I, R_{23} = I \otimes R, R_{13} = (I \otimes P)R_{12} = (P \otimes I)R_{23}.$$

The pair (A, R) is called a **quasitriangular Hopf algebra** *if the following conditions are satisfied:*
QHA1. $\Delta'(a) = R\Delta(a)R^{-1}, \quad \forall a \in A$,
QHA2. $(I \otimes \Delta)(R) = R_{13}R_{12}$,
QHA3. $(\Delta \otimes I)(R) = R_{13}R_{23}$.

The matrix R of the above definition is called the **universal R-matrix** of A. The R-matrices satisfy the **Yang–Baxter equation**

$$R_{12}R_{13}R_{23} = R_{23}R_{13}R_{12}.$$

There are a number of other special Hopf algebras such as ribbon Hopf algebras and modular Hopf algebras. Associated with each Hopf algebra there is the category of its representations. Such categories are at the heart of Atiyah–Segal axioms for TQFT. These and other related structures are playing an increasingly important role in our understanding of various physical theories such as conformal field theories and statistical mechanics and their relation to invariants of low-dimensional manifolds. An excellent reference for this material is Turaev's book [381].

1.5.1 Quantum Groups

Quantum groups were introduced independently by Drinfeld and Jimbo. They were to be used as tools to produce solutions of Yang–Baxter equations, which arise in the theory of integrable models in statistical mechanics. Since then quantum groups have played a fundamental role in the precise mathematical formulation of invariants of links, and 3-manifolds obtained by physical methods applied to Chern–Simons gauge theory. The results obtained are topological. In view of this, topologists refer to this area as **quantum topology**. There are now several different approaches and corresponding definitions of quantum groups. We give below a definition of the quantum group associated to a simple complex Lie algebra in terms of a deformation of its universal enveloping algebra.

Definition 1.2 *Let $\mathfrak{g}$ denote a simple complex Lie algebra. The* **quantum group** $U_q(\mathfrak{g})$ *is a Hopf algebra obtained by deformation of the universal enveloping algebra $U(\mathfrak{g})$ of the Lie algebra $\mathfrak{g}$ by a formal parameter q. In the applications we have in mind, we take q to be an rth root of unity for $r > 2$. We call these* **quantum groups at roots of unity.** *They are closely related to modular categories and three-dimensional TQFT.*

The general definition of a quantum group is rather long. We will consider in detail the case when $\mathfrak{g} = \mathfrak{sl}(2, \mathbf{C})$ in our study of quantum invariants of links and 3-manifolds in Chapter 11.

1.6 Monstrous Moonshine

As evidence for the existence of the largest sporadic simple group F_1 predicted in 1973 by Fischer and Griess mounted, several scientists put forth conjectures that this exceptional group should have relations with other areas of mathematics and should even appear in some natural phenomena. The results that have poured in since then seem to justify this early assessment. Some strange coincidences noticed first by McKay and Thompson were investigated by Conway and Norton. The latter researchers called these unbelievable set of conjectures "**monstrous moonshine**" and the Fischer–Griess group F_1 the **monster** and denoted it by $\mathbb{M}$. Their paper [82] appeared in the Bulletin of the London Mathematical Society in 1979. The same issue of the Bulletin contained three papers by Thompson [372, 370, 371] discussing his observations of some numerology between the Fischer–Griess monster and the elliptic modular functions and formulating his conjecture about the relation of the characters of the monster and Hauptmoduls for various modular groups. He also showed that there is at most one group that satisfies the properties expected of F_1 and has a complex, irreducible representation of degree $196{,}883 = 47 \cdot 59 \cdot 71$ (47, 59 and 71 are the three largest prime divisors of the order of the monster group). Conway and Norton had conjectured earlier that the monster should have a complex, irreducible representation of degree 196,883. Based on this conjecture, Fischer, Livingstone, and Thorne (Birmingham notes 1978, unpublished) computed the entire character table of the monster.

The existence and uniqueness of the monster was the last piece in the classification of finite simple groups. This classification is arguably the greatest achievement of twentieth century mathematics. In fact, it is unique in the history of mathematics, since its completion was the result of hundreds of mathematicians working in many countries around the world for over a quarter century. This global initiative was launched by Daniel Gorenstein and we will use his book [159] as a general reference for this section. Further details and references to works mentioned here may be found in [159]. Another important resource for this section is Mark Ronan's book [327] *Symmetry and the monster*. The book is written in a nontechnical language and yet conveys the excitement of a great mathematical discovery usually accessible only to professional mathematicians (see also [262]). We describe the highlights of this fascinating story below. Finite simple groups are discussed in the subsections 1.6.1. Subsection 1.6.2 introduces modular groups and modular functions. Numerology between the monster and Hauptmoduls is given in subsection 1.6.3. The moonshine conjectures are also stated here. Indication of the proof of the moonshine conjectures using infinite-dimensional algebras such as the vertex operator algebra is given in subsection 1.6.4. This proof is inspired in part by structures arising in theoretical physics, for example, in conformal field theory and string theory.

1.6.1 Finite Simple Groups

Recall that a group is called simple if it has no proper nontrivial normal subgroup. Thus, an abelian group is simple if and only if it is isomorphic to one of the groups $\mathbf{Z}_p$, for p a prime number. This is the simplest example of an infinite family of finite simple groups. Another infinite family of finite simple groups is the family of alternating groups A_n, $n > 4$, which are studied in a first course in algebra. These two families were known in the nineteenth century. The last of the families of finite groups, called groups of Lie type, were defined by Chevalley in the mid twentieth century. We now give a brief indication of the main ideas in Chevalley's work.

By the early twentieth century, the Killing–Cartan classification of simple Lie groups, defined over the field $\mathbf{C}$ of complex numbers, had produced four infinite families and five exceptional groups. Mathematicians began by classifying simple Lie algebras over $\mathbf{C}$ and then constructing corresponding simple Lie groups. In 1955, using this structure but replacing the complex numbers by a finite field, Chevalley's fundamental paper showed how to construct finite groups of Lie type. Every finite field is uniquely determined up to isomorphism by a prime p and a natural number n. This field of p^n elements is called the **Galois field** and is denoted by $GF(p^n)$. Évariste Galois introduced and used these fields in studying number theory. Galois is, of course, best known for his fundamental work on the solvability of polynomial equations by radicals. This work, now called **Galois theory**, was the first to use the theory of groups to completely answer the long open question of the solvability of polynomial equations by radicals.

Chevalley first showed that every complex semi-simple Lie algebra has an **integral basis**. Recall that an integral basis is a basis such that the Lie bracket of any two basis elements is an integral multiple of a basis element. Such an integral basis is now called a **Chevalley basis** of L. Using his basis Chevalley constructed a Lie algebra $L(K)$ over a finite field K. He then showed how to obtain a finite group from this algebra. This group, denoted by $G(L, K)$, is called the **Chevalley group** of the pair (L, K). Chevalley proved that the groups $G(L, K)$ (with a few well defined exceptions) are simple, thereby obtaining several new families of finite simple groups. This work led to the classification of all infinite families of finite simple groups.

However, it was known that there were finite simple groups that did not belong to any of these families. Such groups are called **sporadic groups**. The first sporadic group was constructed by Mathieu in 1861. In fact, he constructed five sporadic groups, now called Mathieu groups. They are just the tip of an enormous iceberg of sporadic groups discovered over the next 120 years. There was an interval of more than 100 years before the sixth sporadic group was discovered by Janko in 1965. Two theoretical developments played a crucial role in the search for new simple groups. The first of these was Brauer's address at the 1954 ICM in Amsterdam, which gave the definitive indication of the surprising fact that general classification theorems would have to include sporadic groups as exceptional cases. For example, Fischer discovered and constructed his first three sporadic groups in the process of proving such a classification theorem. Brauer's work made essential use of elements of order 2. Then, in 1961 Feit and Thompson proved that every non-Abelian simple finite group contains an element of order 2. The proof of this one-line result occupies an entire 255-page issue of the *Pacific Journal of Mathematics* (volume 13, 1963). Before the Feit–Thompson theorem, the classification of finite simple groups seemed to be a rather distant goal. This theorem and Janko's new sporadic group greatly stimulated the mathematics community to look for new sporadic groups. Thompson was awarded a Fields Medal at ICM 1970 in Nice for his work towards the classification of finite simple groups. John Conway's entry into this search party was quite accidental. He was lured into it by Leech, who had discovered his 24-dimensional lattice (now well known as the **Leech lattice**) while studying the problem of sphere-packing. The origin of this problem can be traced back to Kepler.

Kepler was an extraordinary observer of nature. His observations of snowflakes, honeycombs, and the packing of seeds in various fruits led him to his lesser known study of the **sphere-packing problem**. The sphere-packing problem asks for the densest packing of standard unit spheres in a given Euclidean space. The answer is arrived at by determination of the number of spheres that touch a fixed sphere. For dimensions 1, 2, and 3 Kepler found the answers to be 2, 6, and 12 respectively. The lattice structures on these spaces played a crucial role in Kepler's "proof." The three dimensional problem came to be known as **Kepler's (sphere-packing) conjecture**. The slow progress in the solution of this problem led John Milnor to remark that here was a problem nobody could solve, but its answer was known to every schoolboy. It was only solved recently (1998, Tom Hales). The Leech lattice provides the tightest sphere-packing in a lattice in 24 dimensions (its proof was announced by Cohn and Kumar in 2004), but the sphere-packing problem in most other dimensions is still wide open. In the Leech lattice each 24-dimensional sphere touches 196,560 others. Symmetries of the Leech lattice contained Mathieu's largest sporadic group and it had a large number of symmetries of order 2. Leech believed that the symmetries of his lattice contained other sporadic groups. Leech was not a group theorist and he could not get other group theorists interested in his lattice. But, he did find a young mathematician, who was not a group theorist to study his work. In 1968, John Conway was a junior faculty member at Cambridge. He quickly became a believer in Leech's ideas. He tried to get Thompson (the great guru of group theorists) interested. Thompson told him to find the size of the group of symmetries and then call him. Conway later remarked that he did not know that Thompson was referring to a folk theorem, which says:

The two main steps in finding a new sporadic group are

1. find the size of the group of symmetries, and
2. call Thompson.

Conway worked very hard on this problem and soon came up with a number. This work turned out to be his big break. It changed the course of his life and made him into a world class mathematician. He called Thompson with his number. Thompson called back in 20 minutes and told him that half his number could be a possible size of a new sporadic group and that there were two other new sporadic groups associated with it. These three groups are now denoted by $Co1, Co2, Co3$ in Conway's honor. Further study by Conway and Thompson showed that the symmetries of the Leech lattice give 12 sporadic groups in all, including all five Mathieu groups, the first set of sporadic groups discovered over a hundred years ago. In the early 1970s Conway started the **ATLAS project**. The aim of this project was to collect all essential information (mainly the character tables) about the sporadic and some other groups. The work continued into the early 1980s when all the sporadic groups were finally known. It was published in 1985 [84]. This important reference work incorporates a great deal of information that was only available before through unpublished work.

After Conway's work the next major advance in the discovery of new sporadic groups came through the work of Berndt Fischer. As with so many mathematicians, Fischer had decided to study physics. He had great interest and ability in mathematics from childhood, but he was influenced by his high school teacher, who showed him how mathematics can be used to solve physical problems. Sl Fischer's physics plans changed once he started to study mathematics under Baer. Prof. Baer had returned to Germany after a long stay in America. The main reason he went back to his home country was his love of the German higher education system (universities and research institutes).[2] Fischer became interested in groups generated by transpositions. Recall that in a permutation group a transposition interchanges two elements, so that the product of two transpositions is of order 2 or 3. Fischer first proved that a group G generated by such transpositions falls into one of six types. The first type is a permutation group and the next four lead to known families of simple groups. It was the sixth case that led to three new sporadic groups each related to one of the three largest Mathieu groups. The geometry underlying the construction of G is that of a graph associated to generators of G. Permutation groups and the classical groups all have natural representations as automorphism groups of such graphs. Fischer's graphs give some known groups but also his three new sporadic groups. Fischer published this work in 1971 as the first of a series of papers. No further papers in the series after the initial one ever appeared. In fact most of his work is not published. Fischer continued studying other transposition groups. This led him first to a new sporadic group B, now called the **baby monster** and to conjecture the existence of an even larger group: the monster.

---

[2] I have had a very pleasant firsthand experience with this system since 1998, when I bacame a Max Planck Gesselschaft Fellow at the Max Planck Institute for Mathematics in the Sciences in Leipzig, Germany.

By 1981, twenty new sporadic groups were discovered bringing the total to twenty five. The existence of the 26th and the largest of these groups was conjectured, independently, by Fischer and Griess in 1973. The construction of this "friendly giant" (now the group finally known as the monster) was announced by Griess in 1981 and the complete details were given in [168]. Griess first constructs a commutative, nonassociative algebra A of dimension $196,884$ (now called the **Griess algebra**). He then shows that the monster group is the automorphism group of the Griess algebra. In the same year, the final step in the classification of finite simple groups was completed by Norton by establishing that the monster has an irreducible complex representation of degree 196,883 (the proof appeared in print later in [300]). Combined with the earlier result of Thompson, this work proves the uniqueness of the Fischer–Griess sporadic simple group. So the classification of finite simple groups was complete. It ranks as the greatest achievement of twentieth century mathematics. Hundreds of mathematicians contributed to it. The various parts of the classification proof together fill thousands of pages. The project to organize all this material and to prepare a flow chart of the proof is expected to continue for years to come.

1.6.2 Modular Groups and Modular Functions

The search for sporadic groups entered its last phase with John McKay's observation about the closeness of the coefficients of Jacobi's Hauptmodul and the character degrees of representations of the monster. This observation is now known as McKay correspondence. To understand this as well as the full moonshine conjectures we need the classical theory of modular forms and functions. We now discuss parts of this theory.

The modular group $\Gamma = SL(2, \mathbf{Z})$ acts on the upper half plane, called $\mathbb{H}$, by fractional linear transformations as follows:

$$Az = \frac{az+b}{cz+d}, \quad z \in \mathbb{H}, \quad A = \begin{pmatrix} a & b \\ c & d \end{pmatrix} \in \Gamma = SL(2, \mathbf{Z}) . \tag{1.31}$$

For positive integers k, n we define the **congruence subgroups** $\Gamma(k)$ and $\Gamma_0(n)$ of Γ as follows:

$$\Gamma(k) := \left\{ \begin{pmatrix} a & b \\ c & d \end{pmatrix} \in \Gamma \mid \begin{pmatrix} a & b \\ c & d \end{pmatrix} = \begin{pmatrix} 1 & 0 \\ 0 & 1 \end{pmatrix} \bmod k \right\};$$

i.e., $\Gamma(k)$ is the kernel of the canonical homomorphism of $SL(2, \mathbf{Z})$ onto $SL(2, \mathbf{Z}_k)$ and

$$\Gamma_0(n) := \left\{ \begin{pmatrix} a & b \\ c & d \end{pmatrix} \in \Gamma \mid c = 0 \bmod n \right\} .$$

For a fixed prime p, we define the **extended congruence subgroup** $\Gamma_0(p)+$ by

$$\Gamma_0(p)+ := \left\langle \Gamma_0(p), \begin{pmatrix} 0 & 1 \\ -p & 0 \end{pmatrix} \right\rangle .$$

To describe the classical construction of the **Jacobi modular function** j, we begin by defining **modular forms**.

Definition 1.3 *A function f that is analytic (i.e., holomorphic) on $\mathbb{H}$ and at ∞ is called a* **modular form of weight** k *if it satisfies the following conditions:*

$$f(Az) = (cz + d)^k f(z), \qquad \forall A \in \Gamma.$$

The **Eisenstein series** provide many interesting examples of modular forms. The Eisenstein series E_k, where $k \geq 4$ is an even integer is defined by

$$E_k(z) := \frac{1}{2\zeta(k)} \sum_{m,n} (nz + m)^{-k} ,$$

where the sum taken is over all integer pairs $(m, n) \neq (0, 0)$ and where $\zeta(k) = \sum_{r=1}^{\infty} r^{-k}$ is the **Euler zeta function**. The series E_k can also be expressed as a function of $q = e^{2\pi i z}$ by

$$E_k(z) := 1 - \left(\frac{2k}{B_k}\right) \sum_{r=1}^{\infty} \sigma_{k-1}(r)q^r,$$

where B_k is the kth **Bernoulli number** (see Appendix B) and $\sigma_t(r) = \sum d^t$, where t is a natural number and the sum runs over all positive divisors d of r. Thus, E_4 and E_6 are modular forms of weights 4 and 6, respectively. Therefore, the function Δ defined by

$$\Delta = \frac{E_4^3 - E_6^2}{1728} = q \prod_{n=1}^{\infty} (1 - q^n)^{24}$$

is a modular form of weight 12. It is related to the **Dirichlet function** η by

$$\eta = \Delta^{(1/24)} = q^{(1/24)} \prod_{n=1}^{\infty} (1 - q^n) .$$

Note that the η function is a modular form of fractional weight $1/2$. Jacobi's modular function also called a Hauptmodul for the modular group Γ, is defined by

$$j := E_4^3/\Delta = q^{-1} + 744 + \sum_{n=1}^{\infty} c(n)q^n . \tag{1.32}$$

Since Δ is never zero, j is a modular form of weight zero, i.e., a modular function In this work we use the modular function J defined by $J(z) :=$

$j(z) - 744$. The first few terms in the Fourier expansion of the function J are given by

$$J = q^{-1} + 196{,}884q + 21{,}493{,}760q^2 + 864{,}299{,}970q^3 + \cdots . \qquad (1.33)$$

We note that the quotient space $\mathbb{H}/\Gamma$ under the action of Γ on $\mathbb{H}$ is a surface of genus zero. In general, we define the **genus** of a congruence subgroup as the genus of the surface obtained as indicated above for Γ. If G is a congruence subgroup such that the quotient space $\mathbb{H}/G$ has genus zero then it can be shown that every holomorphic function on $\mathbb{H}/G$ can be expressed as a rational function of a single function j_G. This function j_G is called a Hauptmodul for G. The fact that the coefficients of the J-function are integers has had several arithmetic applications, for example in the theory of complex multiplication and class field theory. The p-adic properties and congruences of these coefficients have been studied extensively. But their positivity brings to mind the following classical folk principle of representation theory: *If you meet an interesting natural number, ask if it is the character degree of some representation.* This is exactly the question that John McKay asked about the number $196{,}884$. He sent his thoughts about it to Thompson in November 1978, and this led to the McKay–Thompson series and the monstrous moonshine. Correspondences of this type are also known for some other finite and infinite groups. For example, McKay had found a relation between the character degrees of irreducible representations of the complexified exceptional Lie group E_8 and the q-coefficients of the cube root of the J-function.

1.6.3 The monster and the Moonshine Conjectures

Before discussing the relation of the monster with the modular functions we give some numerology related to this group. This largest of the sporadic groups is denoted by $\mathbb{M}$, and its order $o(\mathbb{M})$ equals

$$808{,}017{,}424{,}794{,}512{,}875{,}886{,}459{,}904{,}961{,}710{,}757{,}005{,}754{,}368 \times 10^9 .$$

The prime factorization of $o(\mathbb{M})$ is given by

$$o(\mathbb{M}) = 2^{46} \cdot 3^{20} \cdot 5^9 \cdot 7^6 \cdot 11^2 \cdot 13^3 \cdot 17 \cdot 19 \cdot 23 \cdot 29 \cdot 31 \cdot 41 \cdot 47 \cdot 59 \cdot 71.$$

It has more elements (about 10^{54}) than number of atoms in the earth. We call the prime factors of the monster the **monster primes**. We denote by $\pi_{\mathbb{M}}$, the set of all monster primes. Thus,

$$\pi_{\mathbb{M}} = \{2, 3, 5, 7, 11, 13, 17, 19, 23, 29, 31, 41, 47, 59, 71\}.$$

The set $\pi_{\mathbb{M}}$ contains 15 of the first 19 primes. The primes 37, 43, and 67 also occur as divisors of the order of some sporadic groups. The only prime that does not divide the order of any of the sporadic groups among the first 19 primes is 61. Thus, 61 can be characterized as the smallest prime that does not divide the order of any of the sporadic groups. There are a number of characterizations of the set $\pi_{\mathbb{M}}$ of all monster primes. The theorem below gives an indication of some deep connection between the monster and the modular groups.

Theorem 1.6 *The group $\Gamma_0(p)+$ has genus zero if and only if p is a monster prime, i.e., p divides $o(\mathbb{M})$. Moreover, if the element $g \in \mathbb{M}$ has order p then its McKay–Thompson series T_g is a Hauptmodul for the group $\Gamma_0(p)+$.*

Remark 1.1 The 15 monster primes are singled out in other situations. For example, the primes p dividing the $o(\mathbb{M})$ are exactly the primes for which every supersingular elliptic curve in characteristic p is defined over the field $\mathbf{Z}_p$. They also appear in the proof of the **Hecke conjecture** by Pizer [318]. In 1940 Hecke made a conjecture concerning the representation of modular forms of weight 2 on the congruence group $\Gamma_0(p)$ (p a prime) by certain theta series determined by the quaternionic division algebra over the field of rational numbers, ramified at p and ∞. Pizer proved that the conjecture is true if and only if p is a monster prime. The 15 monster primes also arise in Ogg's work on modular groups. Andrew Ogg was in the audience at a seminar on the monster given by Jacques Tits in 1975. When he wrote down the prime factorization of $o(\mathbb{M})$, Ogg was astounded. He knew this list of 15 primes very well through his work on modular groups. He offered a small prize for anyone who could explain this remarkable coincidence. In spite of a great deal of work on the properties of the monster and its relations to other parts of mathematics and physics, this prize remains unclaimed.

Recall that the quantum dimension of a graded vector space was defined as a power series in the formal variable q. If we write $q = \exp(2\pi i z), z \in \mathbf{C}$ then $\dim_q V$ can be regarded as the Fourier expansion of a complex function. A spectacular application of this occurs in the study of finite groups. The study of representations of the largest of these groups has led to the creation of a new field of mathematics called vertex algebras; these turn out to be closely related to the chiral algebras in conformal field theory. These and other ideas inspired by string theory have led to a proof of Conway and Norton's moonshine conjectures (see, for example, Borcherds [48], and the book [139] by Frenkel, Lepowski, and Meurman). The monster Lie algebra is the simplest example of a Lie algebra of physical states of a chiral string on a 26-dimensional orbifold. This algebra can be defined by using the infinite-dimensional graded representation V of the monster simple group. Its quantum dimension is related to Jacobi's $SL(2, \mathbf{Z})$ Hauptmodul (elliptic modular function of genus zero) $j(q)$, where $q = e^{2\pi i z}, z \in \mathbb{H}$, by

$$\dim_q V = J(q) := j(q) - 744 = q^{-1} + 196{,}884q + 21{,}493{,}760q^2 + 864{,}299{,}970q^3$$
$$+ \dots .$$

The coefficient $196,884$ in the above formula attracted John McKay's attention, it being very close to $196,883$, the character degree of the smallest nontrivial irreducible representation of the monster. McKay communicated his observation to Thompson. We summarize below Thomson's observations on the numerology between the monster and the Jacobi modular function J. Let $c(n)$ denote the coefficient of the nth term in $J(q)$ and let χ_n be the nth irreducible character of the monster group. Then the character degree $\chi_n(1)$ is the dimension of the nth irreducible representation of M. We list the first few values of $c(n)$ and $\chi_n(1)$ in Table 1.4.

Table 1.4 A strange correspondence

n	$c(n)$	$\chi_n(1)$
1	1	1
2	196,884	196,883
3	21,493,760	21,296,876
4	864,299,970	842,609,326
5	20,245,856,256	18,538,750,076

A short calculation using the values in Table 1.4 gives the following formulas for the first few coefficients of the J-function in terms of the character degrees of the monster:

$$c(1) = \chi_1(1)$$
$$c(2) = \chi_1(1) + \chi_2(1)$$
$$c(3) = \chi_1(1) + \chi_2(1) + \chi_3(1)$$
$$c(4) = 2\chi_1(1) + 2\chi_2(1) + \chi_3(1) + \chi_4(1)$$
$$c(5) = 3\chi_1(1) + 3\chi_2(1) + \chi_3(1) + 2\chi_4(1) + \chi_5(1).$$

These relations led Thompson to ask the following questions:

1. Is there a function theoretic description where the coefficients $c(n)$ equal the dimension of some vector space V_n?
2. Is there a group G containing the monster as a subgroup that admits V_n as a representation space such that the restriction of this representation to the group M provides the explanation of the entries in Table 1.4.

These two questions are a special case of Conway and Norton's monstrous moonshine, the moonshine conjectures, which we now state.

1. For each $g \in M$ there exists a function $T_g(z)$ with normalized Fourier series expansion given by

$$T_g(z) = q^{-1} + \sum_1^\infty c_g(n)q^n. \qquad (1.34)$$

There exists a sequence H_n of representation of $\mathbb{M}$ called the **head representations** such that

$$c_g(n) = \chi_n(g), \qquad (1.35)$$

where χ_n is the character of H_n.

2. For each $g \in \mathbb{M}$, there exists a Hauptmodul J_g for some modular group of genus zero such that $T_g = J_g$. In particular,

   a. $T_1 = J_1 = J$, the Jacobi Hauptmodul for the modular group Γ.
   b. If g is an element of prime order p, then T_g is a Hauptmodul for the modular group $\Gamma_0(p)+$.

3. Let $[g]$ denote the set of all elements in $\mathbb{M}$ that are conjugate to g^i, $i \in \mathbf{Z}$. Then T_g depends only on the class $[g]$. Note that from equation (1.35), it follows that T_g is a class function in the usual sense. However, $[g]$ is not the usual conjugacy class. There are 194 conjugacy classes of $\mathbb{M}$ but only 171 distinct McKay–Thompson series.

Conway and Norton calculated all the functions T_g and compared their first few coefficients with the coefficients of known genus zero Hauptmoduls. Such a check turns out to be part of Borcherds's proof, which he outlined in his lecture at the 1998 ICM in Berlin [48]. The first step was the construction of the moonshine module. The entire book [139] by Frenkel, Lepowsky and Meurman is devoted to the construction of this module, denoted by $V^\natural$. It has the structure of an algebra called the **Moonshine vertex operator algebra** (also denoted by $V^\natural$). Frenket et al. proved that the automorphism group of the infinite-dimensional graded algebra $V^\natural$ is the largest of the finite, sporadic, simple groups, namely, the monster. We now describe some parts in the construction of the graded algebra $V^\natural$ which can be written as

$$V^\natural = \bigoplus_{i=-1}^\infty V_i .$$

All the V_i are finite dimensional complex vector spaces with V_{-1} one-dimensional and $V_0 = 0$ (corresponding to zero constant term in the Hauptmodul J). There are two distinguished homogeneous elements:

i) the **vacuum vector**, denoted by $\mathbf{1}$ in V_{-1}, is the identity element of the algebra;

ii) the **conformal vector** (also called the Virasoro vector), denoted by $\omega \in V_2$.

For each $v \in V^\natural$ there is given a linear map $Y(v,z)$ with formal parameter z defined by

$$Y(v,z) = \sum_{n\in\mathbf{Z}} v_n z^{-n-1}, \quad v_n \in End(V^\natural).$$

The operator $Y(v, z)$ is called the vertex operator associated with v. The map $v \mapsto Y(v, z)$ is called the **state-field correspondence**. These operators are subject to a number of axioms. For example, the endomorphisms ω_n appearing in the operator $Y(\omega, z)$ corresponding to the Virasoro vector ω generate a copy of the Virasoro algebra with central charge 24, acting on $V^\natural$. The magical number 24 appearing here is the same as the dimension of the Leech lattice. It also plays an important role in number theory and in the theory of modular forms.

The second step was the construction by Borcherds of the **monster Lie algebra** using the moonshine vertex operator algebra $V^\natural$. He used this algebra to obtain combinatorial recursion relations between the coefficients $c_g(n)$ of the McKay–Thompson series. It was known that the Hauptmoduls satisfied these relations and that any function satisfying these relations is uniquely determined by a finite number of coefficients. In fact, checking the first five coefficients is sufficient for each of the 171 distinct series. The j-function belongs to a class of functions called **replicable functions**. This class also includes the three classical functions, the exponential (q), the sine ($(q - 1/q)/(2i)$) and the cosine ($(q + 1/q)/2$). McKay has called these three functions the **modular fictions**. Modular fictions are known to be the only replicable functions with finite Laurent series. All modular functions given by the 171 distinct series are replicable functions. Thus, all the monstrous moonshine conjectures are now parts of what we can call the "**Moonshine Theorem**" (for more details, see [91], [151], [152]). Its relation to vertex operator algebras, which arise as chiral algebras in conformal field theory and string theory has been established. In spite of the great success of these new mathematical ideas, many mysteries about the monster are still unexplained.

We end this section with a comment which is a modification of the remarks made by Ogg in [305], when the existence of the monster group and its relation to modular functions were still conjectures (strongly supported by computational evidence). Its deep significance for theoretical physics is still emerging; so mathematicians and physicists young and old may find exciting emergence of a new subject, guaranteed to be rich and varied and deep, with many new questions to be asked and many of the conjectured results yet to be proved. It is indeed quite extraordinary that new light should be shed on the theory of modular functions, one of the most beautiful and extensively studied areas of classical mathematics, by the largest and the most exotic sporadic group, the monster. That its interaction goes beyond mathematics, into areas of theoretical physics such as conformal field theory, chiral algebras,[3] and string theory, may indicate that the field of physical mathematics is rich and worthy of deeper study.

---

[3] A discussion of chiral algebras from a mathematical point of view may be found in the book by Beilinson and Drinfeld [35]. For vertex operator algebras and their relation to CFT see, for example, [138, 197, 250].

Chapter 2
Topology

2.1 Introduction

Several areas of research in modern mathematics have developed as a result
of interaction between two or more specialized areas. For example, the subject of algebraic topology associates with topological spaces various algebraic
structures and uses their properties to answer topological questions. An elegant proof of the theorem that $\mathbf{R}^m$ and $\mathbf{R}^n$ with their respective standard
topologies, are not homeomorphic for $m \neq n$ is provided by computing the
homology of the one point compactification of these spaces. Indeed, the problem of classifying topological spaces up to homeomorphism was fundamental
in the creation of algebraic topology. In general, however, the knowledge of
these algebraic structures is not enough to decide whether two topological
spaces are homeomorphic. The equivalence of algebraic structures follows
from a weaker relation among topological spaces, namely, that of homotopy
equivalence. In fact, homotopy equivalent spaces have isomorphic homotopy
and homology structures. Equivalence of algebraic structures associated to
two topological spaces is a necessary but not sufficient condition for their
homeomorphism. Thus, one may think of homotopy and homology as providing obstructions to the existence of homeomorphisms. As we impose further
structure on a topological space such as piecewise linear, differentiable, or
analytic structures other obstructions may arise.

For example, it is well known that $\mathbf{R}^n, n \neq 4$, with the standard topology
admits a unique compatible differential structure. On the other hand, as a
result of the study of the moduli spaces of instantons by Donaldson and the
classification of four-dimensional topological manifolds by Freedman, it follows that $\mathbf{R}^4$ admits an uncountable number of non-diffeomorphic structures.
In the case of the standard sphere $S^n \subset \mathbf{R}^{n+1}$, the generalized Poincaré conjecture states that a compact n-dimensional manifold homotopically equivalent to S^n is homeomorphic to S^n. This conjecture is now known to be
true for all n and is one of the most interesting recent results in algebraic

K. Marathe, *Topics in Physical Mathematics*, DOI 10.1007/978-1-84882-939-8_2, 33
© Springer-Verlag London Limited 2010

topology. The case $n = 2$ is classical. For $n > 4$ it is due to Stephen Smale. Smale (b. 1930) received a Fields Medal at the ICM 1966 held in Moscow for his contributions to various aspects of differential topology and, in particular, to his novel use of Morse theory, which led him to his solution of the generalized Poincaré conjecture for $n > 4$. Smale has extensive work in the application of dynamical systems to physical processes and to economic equilibria. His discovery of strange attractors led naturally to chaotic dynamical systems. The result for $n = 4$ is due to Michael Hartley Freedman (b. 1951), who received a Fields Medal at ICM 1986[1] held in Berkeley for his complete classification of all compact simply connected topological 4-manifolds, which leads to his proof of the Poincaré conjecture. The original Poincaré conjecture was recently proved by Grigory Yakovlevich Perelman (b. 1966). Perelman received a Fields Medal at ICM 2006 held in Madrid for his fundamental contributions to geometry and for his revolutionary insights into the analytical and geometric structure of the Ricci flow. He studied the geometric topology of 3-manifolds by extending Hamilton's Ricci flow ideas. While he did not publish his work in a final form, it contains all the essential ingredients of a proof of the Thruston geometrization conjectures and in particular of the original Poincaré conjecture (the case $n = 3$). This problem is one of the seven, million dollar Clay Prize problems. As of this writing, it is not known if and when he will get this prize. We will discuss the topology of 3- and 4-manifolds later in this chapter.

In the category of differentiable manifolds, it was shown by **John Willard Milnor** that S^7 admits an exotic differential structure, i.e., a structure not diffeomorphic to the standard one. This work ushered in the new field of differential topology. Milnor was awarded a Fields Medal at the ICM 1962 held in Stockholm for his fundamental work in differential geometry and topology. Using homotopy theory, Kervaire and Milnor proved the striking result that the number of distinct differentiable structures on S^n is finite for any $n \neq 4$. For $n = 1, 2, 3, 5, 6$, there is a unique differential structure on the standard n-sphere. As of this writing (May 2010) there is no information on the number of distinct differentiable structures on S^4. The following table gives a partial list of the number of diffeomorphism classes $[S^n]$ of n-spheres.

Table 2.1 Number of diffeomorphism classes of n-**spheres**

n	7	8	9	10	11	12	13	14
$\#[S^n]$	28	2	8	6	992	1	3	2

---

[1] The year was the 50th anniversary of the inception of Fields medals. However, several mathematicians including invited speakers were denied U.S. visas. Their papers were read by other mathematicians in a show of solidarity.

We note that the set of these equivalence classes can be given a structure of a group denoted by θ_n. Milnor showed that θ_7 is cyclic group of order 28. Brieskorn has constructed geometric representatives of the elements of θ_7 as 7-dimensional **Brieskorn spheres** $\Sigma(6m-1,3,2,2,2)$, $1 \le m \le 28$, by generalizing the Poincaré homology spheres in three dimensions. Thus the mth sphere is the intersection of $S^9 \subset \mathbf{C}^5$ with the space of solutions of the equation

$$z_1^{6m-1} + z_2^3 + z_3^2 + z_4^2 + z_5^2 = 0, \quad 1 \le m \le 28, \quad z_i \in \mathbf{C}, \ 1 \le i \le 5.$$

Until recently, such considerations would have seemed too exotic to be of utility in physical applications. However, topological methods have become increasingly important in classical and quantum field theories. In particular, several invariants associated to homotopy and homology of a manifold have appeared in physical theories as topological quantum numbers. In the remaining sections of this chapter and in the next chapter is a detailed account of some of the most important topics in this area.

2.2 Point Set Topology

Point set topology is one of the core areas in modern mathematics. However, unlike algebraic structures, topological structures are not familiar to physicists. In this appendix we collect some basic definitions and results concerning topological spaces. Topological concepts are playing an increasingly important role in physical applications, some of which are mentioned here.

Let X be a set and $\mathcal{P}(X)$ the **power set** of X, i.e., the class of all subsets of X. $\mathcal{T} \subset \mathcal{P}(X)$ is called a **topology** on X if the following conditions are satisfied:

1. $\{\emptyset, X\} \subset \mathcal{T}$;
2. if $A, B \in \mathcal{T}$, then $A \cap B \in \mathcal{T}$;
3. if $\{U_i \mid i \in I\} \subset \mathcal{T}$, then $\bigcup_{i \in I} U_i \in \mathcal{T}$, where I is an arbitrary indexing set.

The pair $(X, \mathcal{T})$ is called a **topological space**. It is customary to refer to X as a topological space when the topology $\mathcal{T}$ is understood. An element of $\mathcal{T}$ is called an open set of $(X, \mathcal{T})$. If $W \subset X$, then $\mathcal{T}_W := \{W \cap A \mid A \in \mathcal{T}\}$ is a topology on W called the **relative topology** on W induced by the topology $\mathcal{T}$ on X.

Example 2.1 *Let $\mathcal{T} = \{\emptyset, X\}$. Then $\mathcal{T}$ is called the* **indiscrete** *topology on X. If $\mathcal{T} = \mathcal{P}(X)$, then $\mathcal{T}$ is called the* **discrete** *topology on X. If $\{\mathcal{T}_i \mid i \in I\}$ is a family of topologies on X, then $\bigcap\{\mathcal{T}_i \mid i \in I\}$ is also a topology on X.*

The indiscrete and the discrete topologies are frequently called trivial topologies. An important example of a nontrivial topology is given by the metric topology.

Example 2.2 *Let* $\mathbf{R}_+$ *denote the set of nonnegative real numbers. A metric or a distance function on* X*, is a function* $d : X \times X \to \mathbf{R}_+$ *satisfying,* $\forall x, y, z \in X$*, the following properties:*

1. $d(x, y) = d(y, x)$, **symmetry;**
2. $d(x, y) = 0$ *if and only if* $x = y$, **non-degeneracy;**
3. $d(x, y) \leq d(x, z) + d(z, y)$, **triangle inequality**.

The pair (X, d) *is called a metric space. If* (X, d) *is a metric space, we can make it into a topological space with topology* $\mathcal{T}_d$ *defined as follows.* $\mathcal{T}_d$ *is the class of all subsets* $U \subset X$ *such that*

$$\forall x \in U, \exists \epsilon > 0, \quad \text{such that} \quad B(\epsilon, x) := \{y \in X \mid d(x, y) < \epsilon\} \subset U.$$

The set $B(\epsilon, x)$ *is called an* ϵ-**ball** *around* x *and is itself in* $\mathcal{T}_d$*.* $\mathbf{R}^n$ *with the usual Euclidean distance function is a metric space. The corresponding topology is called the standard topology on* $\mathbf{R}^n$*. The relative topology on* $S^{n-1} \subset \mathbf{R}^n$ *is called the standard topology on* S^{n-1}*.*

A topological space $(X, \mathcal{T})$ is said to be metrizable if there exists a distance function d on X such that $\mathcal{T} = \mathcal{T}_d$. It is well known that Riemannian manifolds are metrizable. It is shown in [256] that pseudo-Riemannian manifolds, and in particular, space-time manifolds, are also metrizable.

Let $(X, \mathcal{T}_X)$ and $(Y, \mathcal{T}_Y)$ be topological spaces. A function $f : X \to Y$ is said to be continuous if $\forall V \in \mathcal{T}_Y, f^{-1}(V) \in \mathcal{T}_X$. If f is a continuous bijection and f^{-1} is also continuous, then f is called a homeomorphism between X and Y. Homeomorphism is an equivalence relation on the class of topological spaces. A property of topological spaces preserved under homeomorphisms is called a topological property. For example, metrizability is a topological property.

Let $(X_i, \mathcal{T}_i), i \in I$, be a family of topological spaces and let $X = \prod_{i \in I} X_i$ be the Cartesian product of the family of sets $\{X_i \mid i \in I\}$. Let $\pi_i : X \to X_i$ be the canonical projection. Let $\{\mathcal{S}_j \mid j \in J\}$ be the family of all topologies on X such that π_i is continuous for all $i \in I$. If $\mathcal{S} = \bigcap\{\mathcal{S}_j \mid j \in J\}$, then $\mathcal{S}$ is called the product topology on X. We observe that it is the smallest topology on X such that all the π_i are continuous.

Let $(X, \mathcal{T})$ be a topological space, Y a set, and $f : X \to Y$ a surjection. The class $\mathcal{T}_f$ defined by

$$\mathcal{T}_f := \{V \subset Y \mid f^{-1}(V) \in \mathcal{T}\}$$

is a topology on Y called the quotient topology on Y defined by f. $\mathcal{T}_f$ is the largest topology on Y with respect to which f is continuous. We observe that if ρ is an equivalence relation on X, $Y = X/\rho$ is the set of equivalence classes and $\pi : X \to Y$ is the canonical projection, then Y with the quotient topology $\mathcal{T}_\pi$ is called quotient topological space of X by ρ.

Let $(X, \mathcal{T})$ be a topological space and let A, B, C denote subsets of X. C is said to be closed if $X \setminus C$ is open. The closure $\bar{A}$ or $cl(A)$ of A is defined by

$$\bar{A} := \bigcap \{F \subset X \mid F \text{ is closed and } A \subset F\}.$$

Thus, $\bar{A}$ is the smallest closed set containing A. It follows that C is closed if and only if $C = \bar{C}$. Let $f : X \to Y$ be a function. We define $\operatorname{supp} f$, the support of f to be the set $cl\{x \in X \mid f(x) \neq 0\}$. A subset $A \subset X$ is said to be dense in X if $\bar{A} = X$. X is said to be separable if it contains a countable dense subset. A is said to be a neighborhood of $x \in X$ if there exists $U \in \mathcal{T}$ such that $x \in U \subset A$. We denote by $\mathcal{N}_x$ the class of neighborhoods of x. A subclass $\mathcal{B} \subset \mathcal{T} \cap \mathcal{N}_x$ is called a local base at $x \in X$ if for each neighborhood A of x there exists $U \in \mathcal{B}$ such that $U \subset A$. X is said to be first countable if each point in X admits a countable local base. A subclass $\mathcal{B} \subset \mathcal{T}$ is called a base for $\mathcal{T}$ if $\forall A \in \mathcal{T}, \forall x \in A$, there exists $U \in \mathcal{B}$ such that $x \in U \subset A$. X is said to be second countable if its topology has a countable base. A subclass $\mathcal{S} \subset \mathcal{T}$ is called a subbase for $\mathcal{T}$ if the class of finite intersections of elements of $\mathcal{S}$ is a base for $\mathcal{T}$. Any metric space is first countable but not necessarily second countable. First and second countability are topological properties. We now give some further important topological properties.

X is said to be a **Hausdorff space** if $\forall x, y \in X$, there exist $A, B \in \mathcal{T}$ such that $x \in A, y \in B$ and $A \cap B = \emptyset$. Such a topology is said to separate points and the Hausdorff property is one of a family of separation axioms for topological spaces. The Hausdorff property implies that finite subsets of X are closed. A metric space is a Hausdorff space.

A family $\mathcal{U} = \{U_i \mid i \in I\}$ of subsets of X is said to be a **cover** or a **covering** of $A \subset X$ if $A \subset \bigcup \mathcal{U}$. A cover $\{V_j \mid j \in J\}$ of $A \subset X$ is called a **refinement** of $\mathcal{U}$ if, for all $j \in J$, $V_j \subset U_i$ for some $i \in I$. A covering by open sets is called an **open covering**. $A \subset X$ is said to be **compact** if every open covering of A has a finite refinement or, equivalently, if it has a finite subcovering. The continuous image of a compact set is compact. It follows that compactness is a topological property. The **Heine–Borel theorem** asserts that a subset of $\mathbf{R}^n$ is compact if and only if it is closed and bounded. A consequence of this is the **extreme value theorem**, which asserts that every continuous real-valued function on a compact space attains its maximum and minimum values. A Hausdorff space X is said to be **paracompact** if every open covering of X has a **locally finite** open refinement, i.e., each point has a neighborhood that intersects only finitely many sets of the refinement. A family $\mathcal{F} = \{f_i : X \to \mathbf{R} \mid i \in I\}$ of functions is said to be **locally finite** if each $x \in X$ has a neighborhood U such that $f_i(U) = 0$, for all but a finite subset of I. A family $\mathcal{F}$ of continuous functions is said to be a **partition of unity** if it is a locally finite family of nonnegative functions and

$$\sum_{i \in I} f_i(x) = 1, \quad \forall x \in X.$$

If X is a paracompact space and $\mathcal{U} = \{U_i \mid i \in I\}$ is an open covering of X, then there exists a partition of unity $\mathcal{F} = \{f_i : X \to \mathbf{R} \mid i \in I\}$ such that $\mathrm{supp}\, f_i \subset U_i$. $\mathcal{F}$ is called a partition of unity **subordinate** to the cover $\mathcal{U}$. The existence of such a partition of unity plays a crucial role in showing the existence of a Riemannian metric on a paracompact manifold. The concepts of paracompactness and partition of unity were introduced into topology by Dieudonné. It was shown by the author in [265] that pseudo-Riemannian manifolds are paracompact. In particular, this implies that space-time (a Lorentz manifold) is topologically a metric space. X is said to be **locally compact** if each point has a compact neighborhood.

Let $(X, \mathcal{T})$ be a topological space and $A \subset X$. $U, V \in \mathcal{T}$ are said to form a **partition** or a **disconnection** of A if the following conditions are satisfied:

1. $A \subset U \cup V$,
2. $A \cap U \neq \emptyset, A \cap V \neq \emptyset$,
3. $A \cap U \cap V = \emptyset$.

The set A is said to be **connected** if there does not exist any disconnection of A. This is equivalent to saying that A is connected as a topological space with the relative topology. It follows that X is connected if and only if the only subsets of X that are both open and closed are $\emptyset$ and X. If X is not connected then it can be partitioned into maximal connected subsets called the **connected components** of X. Each connected component is a closed subset of X. The set of all connected components is denoted by $\pi_0(X)$. The cardinality of $\pi_0(X)$ is a topological invariant. The continuous image of a connected set is connected. Since the connected subsets of $\mathbf{R}$ are intervals, it follows that every real-valued continuous function f on a connected subset of X satisfies the **intermediate value property**, i.e., f takes every value between any two values. X is said to be **locally connected** if its topology has a base consisting of connected sets. The commonly used concepts of path connected and simply connected are discussed in the next section.

2.3 Homotopy Groups

In homotopy theory the algebraic structures (homotopy groups) associated with a topological space X are defined through the concept of homotopy between maps from standard sets (intervals and spheres) to X. The fundamental group or the first homotopy group of a topological space was introduced by H. Poincaré (1895), while the idea of the higher homotopy groups is principally due to W. Hurewicz (1935). All the homotopy groups arise naturally in the mathematical formulation of classical and quantum field theories. To make our treatment essentially self-contained, we have given more details than are strictly necessary for the physical applications.

Let X and Y be topological spaces and h a map from $\mathcal{D}(h) \subset Y$ to X that can be extended to a continuous map from Y to X. Let $C(Y, X; h)$ be the set

$$C(Y, X; h) = \{f \in C(Y, X) \mid f_{|\mathcal{D}(h)} = h\}$$

where $C(Y, X)$ is the set of continuous maps from Y to X. We say that $f \in C(Y, X)$ is **homotopic** to $g \in C(Y, X)$ **relative** to h and write $f \sim_h g$ if there exists a continuous map $H : Y \times I \to X$, where $I := [0, 1]$, such that the following conditions hold:

$$H(y, 0) = f(y), \quad H(y, 1) = g(y), \quad \forall y \in Y, \tag{2.1}$$

$$H(y, t) = h(y), \quad \forall y \in \mathcal{D}(h), \quad \forall t \in I. \tag{2.2}$$

H is called a **homotopy relative** to h from f to g. Observe that condition (2.2) implies that $f, g \in C(Y, X; h)$. We may think of H as a family $\{H_t := H(\cdot, t) \mid t \in I\} \subset C(Y, X; h)$ of continuous maps from Y to X parametrized by t, which deforms the map f continuously into the map g, keeping fixed their values on $\mathcal{D}(h)$, i.e., $H_t \in C(Y, X; h), \forall t \in I$. It can be shown that the relation $\sim_h$ is an equivalence relation in $C(Y, X; h)$. We denote the equivalence class of f by $[f]$. If h is the empty map, i.e., $\mathcal{D}(h) = \emptyset$ so that $C(Y, X; h) = C(Y, X)$, then we will simply write $f \sim g$ and say that f and g are **homotopic**. We observe that in this case there is no condition (2.2) but only the condition (2.1). A topological space X is **contractible** if $id_X \sim c_a$, where id_X is the identity map on X and $c_a : X \to X$ is the constant map defined by $c_a(x) = a$, $\forall x \in X$ and for some fixed $a \in X$.

Let X be a topological space. A **path** in X from $a \in X$ to $b \in X$ is a map $\alpha \in C(I, X)$ such that $\alpha(0) = a$, $\alpha(1) = b$. We say that X is **path connected** if there exists a path from a to b, $\forall a, b \in X$. X is **locally path connected** if its topology is generated by path connected open sets. A path connected topological space is connected, but the converse is not true. However, a connected and locally path connected topological space *is* path connected, and hence connected manifolds are path connected. In what follows, we take all topological spaces to be connected manifolds unless otherwise indicated.

Let α be a path in X from a to b; the **opposite path** of α is the path $\overleftarrow{\alpha}$ in X from b to a such that $\overleftarrow{\alpha}(t) = \alpha(1 - t)$, $\forall t \in I$. A **loop** in X at $a \in X$ is a path in X from a to a. The set of loops in X at a is

$$P(X, a) := C(I, X; h_a),$$

where $\mathcal{D}(h_a) = \partial I = \{0, 1\}$ and $h_a(0) = h_a(1) = a$. Let $[\alpha]$ be the equivalence class of the loops at a that are homotopic to α relative to h_a and let $E_1(X, a)$ be the set of equivalence classes of homotopic loops at a, i.e.,

$$E_1(X, a) := \{[\alpha] \mid \alpha \in P(X, a)\}.$$

If $\alpha, \beta \in P(X, a)$, then we denote by $\alpha * \beta \in P(X, a)$ the loop defined by

$$(\alpha * \beta)(t) = \begin{cases} \alpha(2t), & 0 \le t \le 1/2 \\ \beta(2t - 1), & 1/2 \le t \le 1. \end{cases} \tag{2.3}$$

The operation $*$ induces an operation on $E_1(X, a)$, which we denote by juxtaposition. This operation makes $E_1(X, a)$ into a group with identity the class $[c_a]$ of the constant loop at a, the class $[\overleftarrow{\alpha}]$ being the inverse of $[\alpha]$. This group is called the **fundamental group** or the **first homotopy group** of X at a and is denoted by $\pi_1(X, a)$. A topological space X with a distinguished point a is called a **pointed topological space** and is denoted by (X, a). Thus, we have associated with every pointed topological space (X, a) a group $\pi_1(X, a)$. Let (X, a), (Y, b) be two pointed topological spaces and $f : X \to Y$ a morphism of pointed topological spaces; i.e., f is continuous and $f(a) = b$. Then the map

$$\pi_1(f) : \pi_1(X, a) \to \pi_1(Y, b)$$

defined by $[\alpha] \mapsto [f \circ \alpha]$ is a homomorphism. π_1 turns out to be a covariant functor from the category of pointed topological spaces to the category of groups (see Appendix C). If X is path connected then $\pi_1(X, a) \cong \pi_1(X, b), \forall a, b \in X$ (the isomorphism is induced by a path from a to b and hence is not canonical). In view of this result we sometimes write $\pi_1(X)$ to indicate the fundamental group of a path connected topological space X. A topological space X is said to be **simply connected** if it is path connected and $\pi_1(X)$ is the trivial group consisting of only the identity element. A contractible space is simply connected.

We now introduce the notion of n-connected, which allows us to give an alternative definition of simply connected. Let $X_n := C(S^n, X)$ be the space of continuous maps from the n-sphere S^n to X. We say that X is n-**connected** if the space X_n with its standard (compact open) topology is path connected. Thus 0-**connected** is just path connected and 1-**connected** is simply connected as defined above. The fundamental group is an important invariant of a topological space, i.e.,

$$X \cong Y \Rightarrow \pi_1(X) \cong \pi_1(Y).$$

A surprising application of the non-triviality of the fundamental group is found in the Bohm–Aharonov effect in Abelian gauge theories. We discuss this application in Chapter 8.

The topological spaces X, Y are said to be **homotopically equivalent** or of the same **homotopy type** if there exist continuous maps $f : X \to Y$ and $g : Y \to X$ such that

$$f \circ g \sim id_Y, \qquad g \circ f \sim id_X.$$

The relation of homotopy equivalence is, in general, weaker than homeomorphism. The following discussion of the Poincaré conjecture and its generalizations illustrate this.

Poincaré Conjecture: *Every closed (i.e., compact and without boundary) simply connected 3-manifold is homeomorphic to S^3.*

For $n > 3$ the conjecture is not true, as shown by our discussion after Example 2.6. However, we have the following.

Generalized Poincaré Conjecture: *Every closed n-manifold homotopically equivalent to the n-sphere S^n is homeomorphic to S^n.*

As we remarked in the introduction, this generalized conjecture was proved to be true for $n > 4$ by Smale in 1960. The case $n = 4$ was settled in the affirmative by Freedman [136] in 1981 and the case $n = 3$ by Perelman (see section 6.8 for Perelman's work).

Let $p : E \to B$ be a continuous surjection. We say that the pair (E, p) is a **covering** of B if each $x \in B$ has a path connected neighborhood U such that each pathwise connected component of $p^{-1}(U)$ is homeomorphic to U. In particular, p is a local homeomorphism. E is called the **covering space**, B the **base space**, and p the **covering projection**. It can be shown that if B is path connected, then the cardinality of the fibers $p^{-1}(x)$, $x \in B$, is the same for all x. If this cardinality is a natural number n, then we say that (E, p) is an n-fold covering of B.

Example 2.3 *Let $U(1) := \{z \in \mathbf{C} \mid |z| = 1\}$.*

1. Let $q_n : U(1) \to U(1)$ be the map defined by

$$q_n(z) = z^n,$$

where $z \in U(1) = \{z \in \mathbf{C} \mid |z| = 1\}$. Then $(U(1), q_n)$ is an n-fold covering of $U(1)$.

2. Let $p : \mathbf{R} \to U(1)$ be the map defined by

$$p(t) = \exp(2\pi i t).$$

Then $(\mathbf{R}, p)$ is a simply connected covering of $U(1)$. In this case the fiber $p^{-1}(1)$ is $\mathbf{Z}$.

3. Let $n > 1$ be a positive integer and $\pi : S^n \to \mathbf{RP}^n$ be the natural projection

$$x \mapsto [x].$$

This is a 2-fold covering. In the special case $n = 3$, $S^3 \cong SU(2) \cong Spin(3)$ and $\mathbf{RP}^3 \cong SO(3)$. This covering $\pi : Spin(3) \to SO(3)$ is well known in physics as associating two spin matrices in $Spin(3)$ to the same rotation matrix in $SO(3)$. The distinction between spin and angular momentum is related to this covering map.

A covering space (U, q) with U simply connected is called a **universal covering space** of the base space B. A necessary and sufficient condition for the existence of a universal covering space of a path connected and locally path connected topological space X is that X be **semi locally simply connected**, i.e., $\forall x \in X$ there should exist an open neighborhood A of x such that any loop in A at x is homotopic in X to the constant loop at x. All connected manifolds are semi locally simply connected. If (E, p) is a covering of B and (U, q) is a universal covering of B and $u \in U, x \in E$ are such that $q(u) = p(x)$, then there exists a unique covering (U, f) of E such that $f(u) = x$ and $p \circ f = q$, i.e., the following diagram commutes.

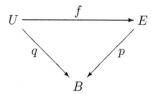

From this it follows that, if $(U_1, q_1), (U_2, q_2)$ are two universal covering spaces of B and $u_1 \in U_1, u_2 \in U_2$ are such that $q_1(u_1) = q_2(u_2)$, then there exists a unique homeomorphism $f : U_1 \to U_2$, such that $f(u_1) = u_2$ and $q_2 \circ f = q_1$, i.e., the following diagram commutes.

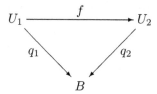

Thus a universal covering space is essentially unique, i.e., is unique up to homeomorphism. Let (U, q) be a universal covering of B. A **covering** or **deck transformation** f is an automorphism of U such that $q \circ f = q$ or the following diagram commutes:

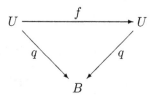

It can be shown that the set $C(U, q)$ of all covering transformations is a subgroup of $\mathrm{Aut}(U)$ isomorphic to $\pi_1(B)$. This observation is useful in computing fundamental groups of some spaces as indicated in the following example.

Example 2.4 *The covering $(U(1), q_n)$ of Example (2.3) above is not a universal covering while the coverings $(\mathbf{R}, p)$ and (S^n, π) discussed there are universal coverings. Every deck transformation of $(\mathbf{R}, p)$ has the form $f_n(t) = t + n, \ n \in \mathbf{Z}$. From this it is easy to deduce the following:*

$$\pi_1(\mathbf{RP}^1) \cong \pi_1(S^1) \cong \pi_1(U(1)) \cong \mathbf{Z}.$$

The only deck transformation of (S^n, π) different from the identity is the antipodal map α defined by $\alpha(x) = -x$, $\forall x \in S^n$. It follows that

$$\pi_1(\mathbf{RP}^n) \cong \mathbf{Z}_2, \quad n > 1.$$

The fundamental group can be used to define invariants of geometric structures such as knots and links in 3-manifolds.

Example 2.5 *An embedding $k : S^1 \to \mathbf{R}^3$ is called a **knot** in $\mathbf{R}^3$. Two knots k_1, k_2 are said to be **equivalent** if there exists a homeomorphism $h : \mathbf{R}^3 \to \mathbf{R}^3$ which is the identity on the complement of some disk $D_n = \{x \in \mathbf{R}^3 \mid \|x\| \leq n\}$, $n \in \mathbf{N}$ and such that $h \circ k_1 = k_2$, i.e., the following diagram commutes.*

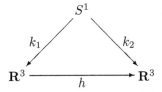

*We define the **knot group** $\nu(k)$ by*

$$\nu(k) := \pi_1(\mathbf{R}^3 \setminus k(S^1)).$$

*It is easy to verify that equivalent knots have isomorphic knot groups. An algebraic structure preserved under knot equivalence is called a **knot invariant**. Thus, the fundamental group provides an important example of a knot invariant.*

If X, Y are topological spaces, then $\pi_1(X \times Y) \cong \pi_1(X) \times \pi_1(Y)$. In particular, if X and Y are simply connected, then $X \times Y$ is simply connected. From this result it follows, for example, that $\pi_1(\mathbf{R}^n) = id$ and

$$\pi_1(T^n) \cong \mathbf{Z}^n \quad \text{where} \quad T^n = \underbrace{S^1 \times \cdots \times S^1}_{n \text{ times}} \quad \text{is the real } n\text{-torus.}$$

If B is a connected manifold then there exists a universal covering (U, q) of B such that U is also a manifold and q is smooth. If G is a connected Lie group then there exists a universal covering (U, p) of G such that U is a simply connected Lie group and p is a local isomorphism of Lie groups. The pair (U, p) is called the **universal covering group** of the group G. In particular G and all its covering spaces are locally isomorphic Lie groups and hence have the same Lie algebra. This fact has the following application in representation theory. Given a representation r of a Lie algebra L on V, there exists a unique simply connected Lie group U with Lie algebra $u \cong L$ and a representation ρ of U on V such that its induced representation $\hat{\rho}$ of u on V is equivalent to r. If G is a Lie group with Lie algebra $\mathbf{g} \cong L$, then we get a representation of G on V only if the representation ρ of U is equivariant under the action of $\pi_1(G)$. Thus, from a representation r of the angular momentum algebra $so(3)$ we get a unique representation of the group $Spin(3) \cong SU(2)$ (spin representation). However, r gives a representation of $SO(3)$ (an angular momentum representation) only for even parity, r in this case, being invariant under the action of $\pi_1(SO(3)) \cong \mathbf{Z}_2$. A similar situation arises for the case of the connected component of the Lorentz group $SO(3,1)$ and its universal covering group $SL(2, \mathbf{C})$. In general, the universal covering group of $SO(r, s)_0$ (the connected component of the identity of the group $SO(r, s)$) is denoted by $Spin(r, s)$ and is called the spinor group (see Chapter 3).

There are several possible ways to generalize the definition of π_1 to obtain the higher homotopy groups. We list three important approaches.

(1) Let

$$I^n = \{t = (t_1, \ldots, t_n) \in \mathbf{R}^n \mid 0 \le t_i \le 1, \quad 1 \le i \le n\}.$$

Define the boundary of I^n by

$$\partial I^n = \{t \in I^n \mid t_i = 0 \text{ or } t_i = 1 \text{ for some } i, \ 1 \le i \le n\}.$$

Consider the homotopy relation in

$$P_n(X, a) := C(I^n, X; h)$$

where $\mathcal{D}(h) = \partial I^n$ and $h(\partial I^n) = \{a\} \subset X$. Let $E_n(X, a)$ be the set of equivalence classes in $P_n(X, a)$. Observe that $P_n(X, a)$ is the generalization of $P(X, a) = P_1(X, a)$ for $n > 1$. We generalize the product $*$ in $P(X, a)$ with the following definition. Let

$$R_1 = \{(t_1, \ldots, t_n) \in I^n \mid 0 \le t_1 \le 1/2\},$$

$$R_2 = \{(t_1, \ldots, t_n) \in I^n \mid 1/2 \le t_1 \le 1\}$$

and $j_i : R_i \to I^n$, $i = 1, 2$, be the maps such that

$$j_1(t) = (2t_1, t_2, \ldots, t_n), \quad j_2(t) = (2t_1 - 1, t_2, \ldots, t_n).$$

For $\alpha, \beta \in P_n(X, a)$ we define $\alpha * \beta$ by

$$(\alpha * \beta)(t) = \begin{cases} \alpha(j_1(t)), \ t \in R_1 \\ \beta(j_2(t)), \ t \in R_2. \end{cases}$$

Then $\alpha * \beta \in P_n(X, a)$ and $\alpha \sim_h \alpha_1$, $\beta \sim_h \beta_1$ implies $\alpha * \beta \sim_h \alpha_1 * \beta_1$. This allows us to define a product in $E_n(X, a)$, denoted by juxtaposition, by

$$[\alpha][\beta] = [\alpha * \beta].$$

With this product $E_n(X, a)$ is a group. It is called the nth **homotopy group** of X at a and is denoted by $\pi_n(X, a)$.

(2) The second definition is obtained with S^n in the place of I^n and $e_1 = (1, 0, \ldots, 0)$ in the place of ∂I^n. Let us consider the space

$$P'_n(X, a) := C(S^n, X; h_0),$$

where $\mathcal{D}(h_0) = \{e_1\}$ and $h_0(e_1) = a$. Let $q : I^n \to S^n$ be a continuous map that identifies ∂I^n to e_1. Then $F_1 = q(R_1), F_2 = q(R_2)$ are hemispheres whose intersection $A = q(R_1 \cap R_2)$ is homeomorphic to S^{n-1} and contains e_1. The quotient spaces of F_1 and F_2 obtained by identifying A to e_1 are homeomorphic to S^n. Let r_1 (resp., r_2) be a continuous map of F_1 (resp., F_2) to S^n that identifies A to e_1. One can take q, r_1, r_2 so that

$$q \circ j_i = r_i \circ q_{|R_i}, \quad i = 1, 2.$$

Let us define a product $*'$ in $P'_n(X, a)$ by

$$(\alpha *' \beta)(u) = \begin{cases} \alpha(r_1(u)), \ u \in F_1 \\ \beta(r_2(u)), \ u \in F_2. \end{cases}$$

Let $E'_n(X, a)$ be the set of equivalence classes of homotopic maps in $P'_n(X, a)$. The operation $*'$ induces a product on $E'_n(X, a)$ which makes $E'_n(X, a)$ into a group, which we denote by $\pi'_n(X, a)$. Let $\phi : P'_n(X, a) \to P_n(X, a)$ be the map defined by

$$\alpha \mapsto \phi(\alpha) = \alpha \circ q.$$

One can verify that $\alpha \sim \beta \implies \phi(\alpha) \sim \phi(\beta)$ and $\phi(\alpha *' \beta) = \phi(\alpha) * \phi(\beta)$. Then ϕ induces a map $\tilde{\phi} : \pi'_n(X, a) \to \pi_n(X, a)$ which is an isomorphism. Thus, we can identify $\pi'_n(X, a)$ and $\pi_n(X, a)$.

(3) The third definition considers loops on the space of loops. We give only a brief indication of the construction of $\pi_n(X, a)$ using loop spaces. In order to consider loops in the space $P(X, a)$, we have to define a topology on this set. $P(X, a)$ is a function space and a standard topology on $P(X, a)$ is the **compact-open topology** defined as follows. Let $W(K, U) \subset P(X, a)$ be the set

$$W(K, U) := \{\alpha \in P(X, a) \mid \alpha(K) \subset U, \ K \subset I, \ U \subset X\} \, .$$

The compact-open topology is the topology that has a subbase given by the family of subsets $W(K, U)$, where K varies over the compact subsets of I and U over the open subsets of X. Then we define

$$\pi''_2(X, a) := \pi_1(P(X, a), c_a),$$

where c_a is the constant loop at a. We inductively define

$$\pi''_n(X, a) := \pi''_{n-1}(P(X, a), c_a). \tag{2.4}$$

It can be shown that $\pi_n''(X,a)$ is isomorphic to $\pi_n(X,a)$ and thus can be identified with $\pi_n(X,a)$.

The space $P(X,a)$ with the compact-open topology is called the **first loop space** of the pointed space (X,a) and is denoted by $\Omega(X,a)$, or simply by $\Omega(X)$ when the base point is understood. With the constant loop c_a at a as the base point, the loop space $\Omega(X)$ becomes the pointed space $(\Omega(X),c_a)$. We continue to denote by $\Omega(X)$ this pointed loop space. The nth **loop space** $\Omega^n(X)$ of X is defined inductively by

$$\Omega^n(X) = \Omega(\Omega^{n-1}(X)).$$

From this definition and equation (2.4) it follows that

$$\pi_n(X) = \pi_1(\Omega^{n-1}(X)).$$

Thus one can calculate all the homotopy groups $\pi_n(X)$ of any space if one can calculate just the fundamental group of all spaces. However, very little is known about the topology and geometry of general loop spaces. A loop space carries a natural structure of a Hopf space in the sense of the following definition.

Definition 2.1 *A pointed topological space (X,e) is said to be a* **Hopf space** *(or simply an H-space) if there exists a continuous map*

$$\mu : X \times X \to X$$

of pointed spaces called **multiplication** *such that the maps defined by $x \mapsto \mu(x,e)$ and $x \mapsto \mu(e,x)$ are homotopic to the identity map of X.*

One can verify that the map μ induced by the operation $*$ defined by equation (2.3) makes the pointed loop space $\Omega(X)$ into an H-space. Iterating this construction leads to the following theorem:

Theorem 2.1 *The loop space $\Omega^n(X)$, $n \geq 1$, is an H-space.*

We note that loop spaces of Lie groups have recently arisen in many mathematical and physical calculations (see Segal and Presley [321]). A general treatment of loop spaces can be found in Adams [4]. For a detailed discussion of the three definitions of homotopy groups and their applications see, for example, Croom [90]. In the following theorem we collect some properties of the groups π_n.

Theorem 2.2 *Let X denote a path connected topological space; then*

1. *$\pi_n(X,a) \cong \pi_n(X,b)$, $\forall a,b \in X$. In view of this we will write $\pi_n(X)$ instead of $\pi_n(X,a)$.*
2. *If X is contractible by a homotopy leaving x_0 fixed, then $\pi_n(X) = id$.*
3. *$\pi_n(X)$ is Abelian for $n > 1$.*
4. *If (E,p) is a covering space of X, then p induces an injective homomorphism $p_* : \pi_n(E) \to \pi_n(X)$ for $n > 1$.*

Furthermore, if Y is another path connected topological space, then

$$\pi_n(X \times Y) \cong \pi_n(X) \times \pi_n(Y).$$

The computation of homotopy groups is, in general, a difficult problem. Even in the case of spheres not all the higher homotopy groups are known. The computation of the known groups is facilitated by the following theorem:

Theorem 2.3 (Freudenthal) *There exists a homomorphism*

$$F : \pi_k(S^n) \to \pi_{k+1}(S^{n+1}),$$

called the **Freudenthal suspension homomorphism**, *with the following properties:*

1. *F is surjective for $k = 2n - 1$;*
2. *F is an isomorphism for $k < 2n - 1$.*

The results stated in the above theorems are useful in computing the homotopy groups of some spaces that are commonly encountered in applications.

Example 2.6 *In this example we give the homotopy groups of some important spaces that are useful in physical applications.*

1. *$\pi_n(\mathbf{R}^m) = id$.*
2. *$\pi_k(S^n) = id, \ k < n$.*
3. *$\pi_n(S^n) \cong \mathbf{Z}$.*

If G is a Lie group then $\pi_2(G) = 0$. In many physical applications one needs to compute the homotopy of semi-simple Lie groups such as the groups $SO(n), SU(n), U(n)$. If G is a semi-simple Lie group then $\pi_3(G) \cong \mathbf{Z}$. An element $\alpha \in \pi_3(G)$ often arises in field theories as a **topological quantum number**. *It arises in the problem of extending a G-gauge field from $\mathbf{R}^4$ to its compactification S^4 (see Chapter 8 for details).*

From Theorems 2.1 and 2.2 and Example 2.6 it follows that $\pi_2(S^4) = id$ and

$$\pi_2(S^2 \times S^2) \cong \pi_2(S^2) \times \pi_2(S^2) \cong \mathbf{Z} \times \mathbf{Z}.$$

Also $\pi_1(S^4) = id$ and $\pi_1(S^2 \times S^2) \cong \pi_1(S^2) \times \pi_1(S^2) = id$. Thus S^4 and $S^2 \times S^2$ are both closed simply connected manifolds that are not homeomorphic. This illustrates the role that higher homotopy groups play in the generalized Poincaré conjecture.

All the homotopy groups of the circle S^1 except the first one are trivial. A path connected topological space X is said to be an **Eilenberg–MacLane space** for the group π if there exists $n \in \mathbf{N}$ such that

$$\pi_n(X) = \pi \quad \text{and} \quad \pi_k(X) = id, \qquad \forall k \neq n.$$

Note that π must be Abelian if $n > 1$. It is customary to denote such a space by $K(\pi, n)$. Thus, S^1 is a $K(\mathbf{Z}, 1)$ space. The construction of Eilenberg–MacLane spaces in the late 1940s is considered a milestone in algebraic topology. In 1955, Postnikov showed how to construct a topological space starting with an Eilenberg–MacLane space as a base and building a succession of fiber spaces with other Eilenberg–MacLane spaces as fibers. This construction is known as the **Postnikov tower** construction and allows us to construct a model topological space having the homotopy type of a given space.

Example 2.7 *(Hopf Fibration) An important example of computation of higher homotopy groups was given by H. Hopf in 1931 in his computation of $\pi_3(S^2)$. Consider the following action $h : U(1) \times \mathbf{C}^2 \to \mathbf{C}^2$ of $U(1)$ on $\mathbf{C}^2$ defined by*

$$(z, (z_1, z_2)) \mapsto (zz_1, zz_2).$$

This action leaves the unit sphere $S^3 \subset \mathbf{C}^2$ invariant and hence induces an action on S^3 with fibers isomorphic to S^1 and quotient $\mathbf{CP}^1 \cong S^2$, making S^3 a principal fiber bundle over S^2. We also denote by $h : S^3 \to S^2$ the natural projection. The above construction is called the Hopf fibration of S^3. Hopf showed that $[h] \in \pi_3(S^2)$ is non-trivial, i.e., $[h] \neq id$, and generates $\pi_3(S^2)$ as an infinite cyclic group, i.e. $\pi_3(S^2) \cong \mathbf{Z}$. This class $[h]$ is essentially the invariant that appears in the Dirac monopole quantization condition (see Chapter 8). The Hopf fibration of S^3 can be extended to the unit sphere $S^{2n-1} \subset \mathbf{C}^n$. The quotient space in this case is the complex projective space $\mathbf{CP}^{n-1}$ and the fibration is called the complex Hopf fibration. This fibration arises in the geometric quantization of the isotropic harmonic oscillator.

One can similarly consider the real, quaternionic and octonionic Hopf fibrations. For example, to study the quaternionic Hopf fibration we begin by observing that

$$SU(2) \cong \{x = x_0 + x_1 i + x_2 j + x_3 k \in \mathbf{H} \mid |x| = 1\}$$

acts as the group of unit quaternions on $\mathbf{H}^n$ on the right by quaternionic multiplication. This action leaves the unit sphere $S^{4n-1} \subset \mathbf{H}^n$ invariant and induces a fibration of S^{4n-1} over the quaternionic projective space $\mathbf{HP}^{n-1}$. For the case $n = 2$, $\mathbf{HP}^1 \cong S^4$ and the Hopf fibration gives S^7 as a nontrivial principal $SU(2)$ bundle over S^4. This bundle plays a fundamental role in our discussion of the BPST instanton in Chapter 9.

We conclude this section with a brief discussion of a fundamental result in homotopy theory, namely, the Bott periodicity theorem.

2.3.1 Bott Periodicity

The higher homotopy groups of the classical groups were calculated by Bott
[49] in the course of proving his well known periodicity theorem. An excellent
account of this proof as well as other applications of Morse theory may be
found in Milnor [284]. We comment briefly on the original proof of the Bott
periodicity theorem for the special unitary group. Bott considered the space
S of parametrized smooth curves $c : [0, 1] \to SU(2m)$, joining $-I$ and $+I$ in
$SU(2m)$, and applied Morse theory to the total kinetic energy function $\mathcal{K}$ of
c defined by

$$\mathcal{K}(c) = \frac{1}{2} \int_0^1 v^2 dt,$$

where $v = \dot{c}$ is the velocity of c. The Euler–Lagrange equations for the functional $\mathcal{K} : S \to \mathbf{R}$ are the well known equations of geodesics, which are the
auto-parallel curves with respect to the Levi-Civita connection on $SU(2m)$.
Now $SU(2m)/SU(m) \times SU(m)$ can be identified with the complex Grassmannian $G_m(2m)$ of m-planes in $2m$ space. The gradient flow of $\mathcal{K}$ is a homotopy
equivalence between the loop space on $SU(2m)$ and the Grassmannian up to
dimension $2m$, i.e.,

$$\pi_{i+1} SU(2m) = \pi_i(\Omega SU(2m)) = \pi_i G_m(2m), \qquad 0 \le i \le 2m.$$

This result together with the standard results from algebraic topology on the
homotopy groups of fibrations imply the periodicity relation

$$\pi_{i-1} SU(k) = \pi_{i+1} SU(k), \qquad i \le 2m \le k.$$

We give below a table of the higher homotopy groups of $U(n)$, $SO(n)$, and
$SP(n)$ and indicate the **stable range** of values of n in which the periodicity
appears.

Table 2.2 Stable homotopy of the classical groups

π_k	$U(n), 2n > k$	$SO(n), n > k+1$	$SP(n), 4n > k - 2$
π_1	$\mathbf{Z}$	$\mathbf{Z}_2$	0
π_2	0	0	0
π_3	$\mathbf{Z}$	$\mathbf{Z}$	$\mathbf{Z}$
π_4	0	0	$\mathbf{Z}_2$
π_5	$\mathbf{Z}$	0	$\mathbf{Z}_2$
π_6	0	0	0
π_7	$\mathbf{Z}$	$\mathbf{Z}$	$\mathbf{Z}$
π_8	0	$\mathbf{Z}_2$	0
period	2	8	8

Table 2.2 of **stable homotopy groups** of the classical groups we have
given is another way of stating the Bott periodicity theorem. More generally, Bott showed that for sufficiently large n the homotopy groups of the
n-dimensional unitary group $U(n)$, the rotation group $SO(n)$ and the symplectic group $Sp(n)$ do not depend on n and that they exhibit a certain
periodicity relation. To state the precise result we need to define the infinite-
dimensional groups $U(\infty), SO(\infty)$, and $Sp(\infty)$. Recall that the natural embedding of $\mathbf{C}^n$ into $\mathbf{C}^{n+1}$ induces the natural embedding of $U(n)$ into $U(n+1)$
and defines the inductive system (see Appendix C)

$$U(1) \subset U(2) \subset \cdots \subset U(n) \subset U(n+1) \subset \cdots$$

of unitary groups. We define the infinite-dimensional unitary group $U(\infty)$ to
be the inductive limit of the above system. The groups $SO(\infty)$ and $Sp(\infty)$
are defined similarly. Using these groups we can state the following version
of the **Bott periodicity theorem.**

Theorem 2.4 *The homotopy groups of the infinite-dimensional unitary, rotation, and symplectic groups satisfy the following relations:*

1. $\pi_{k+2}(U(\infty)) = \pi_k(U(\infty))$
2. $\pi_{k+8}(SO(\infty)) = \pi_k(SO(\infty))$
3. $\pi_{k+8}(Sp(\infty)) = \pi_k(Sp(\infty))$

We already indicated how the statements of this theorem are related to
the periodicity relations of Clifford algebras in Chapter 1. We will give the K-
theory version of Bott periodicity in Chapter 5. The Bott periodicity theorem
is one of the most important results in mathematics and has surprising connections with several other fundamental results, such as the Atiyah–Singer
index theorem. Several of the groups appearing in this theorem have been
used in physical theories. Some homotopy groups outside the stable range also
arise in gauge theories. For example, $\pi_3(SO(4)) = \mathbf{Z} \oplus \mathbf{Z}$ is closely related to
the self-dual and the anti-self-dual solutions of the Yang–Mills equations on
S^4. $\pi_7(SO(8)) = \mathbf{Z} \oplus \mathbf{Z}$ arises in the solution of the Yang–Mills equations on
S^8 (see [175, 243] for details). It can be shown that this solution and simi-
lar solutions on higher-dimensional spheres satisfy certain generalized duality
conditions.

2.4 Singular Homology and Cohomology

In homology theory the algebraic structures (homology modules) associated
with a topological space X are defined through the construction of chain
complexes (see Appendix D) related to X. If one uses simplexes related to X,
one has simplicial homology, which was introduced by Poincaré (see [90] for
a very accessible introduction). There are, however, other homology theories

that give rise to isomorphic homology modules under fairly general conditions on X. We will discuss only the singular homology theory, whose introduction is usually attributed to Lefschetz. For other approaches see, for example, Eilenberg, Steenrod [118], and Spanier [357].

Let q be a nonnegative integer and $\Delta^q \subset \mathbf{R}^{q+1}$ be the set

$$\Delta^q := \{(x_0, \ldots, x_q) \in \mathbf{R}^{q+1} \mid \sum_{i=0}^q x_i = 1, \ x_i \geq 0, \ i = 0, 1, \ldots, q\}.$$

The set Δ^q with the relative topology is called the **standard q-simplex**. Let X be a topological space. A **singular q-simplex** in X is a continuous map $s : \Delta^q \to X$. We denote by $\Sigma_q(X)$ the set of all singular q-simplexes in X. If $\mathbf{P}$ is a principal ideal domain, we denote by $\mathcal{S}_q(X; \mathbf{P})$ the free $\mathbf{P}$-module generated by $\Sigma_q(X)$ and we will simply write $\mathcal{S}_q(X)$ when the reference to $\mathbf{P}$ is understood. By definition of a free module it follows that every element of $\mathcal{S}_q(X)$ can be regarded as a function $c : \Sigma_q(X) \to \mathbf{P}$ such that $c(s) = 0$ for all but finitely many singular q-simplexes s in X. An element of $\mathcal{S}_q(X)$ is called a **singular q-chain** and $\mathcal{S}_q(X)$ is called the qth **singular chain module** of X. If $s \in \Sigma_q(X)$, let χ_s denote the singular q-chain defined by

$$\chi_s(s') = \delta_{ss'}, \ \forall s' \in \Sigma_q(X),$$

where $\delta_{ss'} = 0$ for $s \neq s'$ and $\delta_{ss} = 1$ (1 is the unit element of $\mathbf{P}$). χ_s is called an **elementary** singular chain. It is customary to write s instead of χ_s. Thus, any element $c \in \mathcal{S}_q(X)$ can be expressed uniquely as

$$c = \sum_{s \in \Sigma_q(X)} g_s s, \qquad g_s \in \mathbf{P},$$

where $g_s = 0$ for all but finitely many s.

Let q be a positive integer, $s \in \Sigma_q(X)$ be a singular q-simplex and $i \leq q$ a nonnegative integer. The map

$$s^{(i)} : \Delta^{q-1} \to X$$

defined by

$$s^{(i)}(x_0, \ldots, x_{q-1}) = s(x_0, \ldots, x_{i-1}, 0, x_i, \ldots, x_{q-1})$$

is a singular $(q-1)$-simplex, called the ith **face** of s. Let us denote by δ_q the unique linear map

$$\delta_q : \mathcal{S}_q(X) \to \mathcal{S}_{q-1}(X)$$

such that

$$\delta_q(s) = \sum_{i=0}^q (-1)^i s^{(i)}.$$

One can show that

$$\delta_{q-1} \circ \delta_q = 0.$$

Let $q \in \mathbf{Z}$. For $q < 0$ we define $\mathcal{S}_q(X) = 0$ and for $q \leq 0$ we define $\delta_q = 0$. With these definitions

$$\cdots \longleftarrow \mathcal{S}_{q-1}(X) \xleftarrow{\delta_q} \mathcal{S}_q(X) \xleftarrow{\delta_{q+1}} \mathcal{S}_{q+1}(X) \longleftarrow \cdots$$

is a chain complex, which is also simply denoted by $\mathcal{S}_*(X)$. $\mathcal{S}_*(X)$ is called the **singular chain complex** (with coefficients in $\mathbf{P}$). Let X, Y be topological spaces and $f : X \to Y$ a continuous map. For all $q \in \mathbf{Z}$, let us denote by $\mathcal{S}_q(f; \mathbf{P})$, or simply $\mathcal{S}_q(f)$, the unique linear map

$$\mathcal{S}_q(f) : \mathcal{S}_q(X) \to \mathcal{S}_q(Y)$$

such that $\chi_s \mapsto \chi_{f \circ s}$. One can show that

$$\delta \circ S_q(f) = S_{q-1}(f) \circ \delta. \tag{2.5}$$

The family $\mathcal{S}_*(f) := \{\mathcal{S}_q(f) \mid q \in \mathbf{Z}\}$ is a chain morphism such that

1. $\mathcal{S}_*(id_X) = id_{\mathcal{S}_*(X)}$,
2. $\mathcal{S}_*(g \circ f) = \mathcal{S}_*(g) \circ \mathcal{S}_*(f), \ g \in Mor(Y, Z)$.

Thus, $\mathcal{S}_*(\cdot \,; \mathbf{P})$ is a covariant functor from the category of topological spaces to the category of chain complexes over $\mathbf{P}$. The qth homology module of the complex $\mathcal{S}_*(X)$ over $\mathbf{P}$ is called the qth **singular homology module** and is denoted by $H_q(X; \mathbf{P})$, or simply $H_q(X)$. An element of $H_q(X)$ is called a q-th homology class of X. In general, computing homology modules is a non-trivial task and requires the use of specialized tools. However, it is easy to show that $H_0(X; \mathbf{P})$ is a free $\mathbf{P}$-module on as many generators as there are path components of X. In particular, if X is path connected, then

$$H_0(X; \mathbf{P}) \cong \mathbf{P}.$$

Let X, Y be topological spaces and $f : X \to Y$ a continuous map. By passage to the quotient, $\mathcal{S}_q(f; \mathbf{P})$ induces the map

$$H_q(f; \mathbf{P}) : H_q(X) \to H_q(Y),$$

which we simply denote also by $H_q(f)$. $H_q(f)$ is a linear map such that

1. $H_q(id_X) = id_{H_q(X)}$,
2. $H_q(g \circ f) = H_q(g) \circ H_q(f), \ g \in Mor(Y, Z)$.

Thus, $H_q(\cdot \,; \mathbf{P})$ is a covariant functor from the category of topological spaces to the category of $\mathbf{P}$-modules. It follows that homeomorphic spaces have iso-morphic homology modules. This result is often expressed by saying that homology modules are topological invariants. In fact, one can show that homotopy equivalent spaces have isomorphic homology modules, or that homology modules are homotopy invariants.

Observe that singular 1-simplexes in a topological space X are paths in X. Thus, there exists a natural map

$$\phi : \pi_1(X, x_0) \to H_1(X; \mathbf{P})$$

such that, if γ is a loop at x_0, $\phi([\gamma])$ is the homology class of the singular 1-simplex γ. The precise connection between fundamental groups and homology groups of path connected topological spaces is given in the following theorem (see [160] for a proof):

Theorem 2.5 *Let X be a path connected topological space. The map ϕ defined above is a surjective homomorphism whose kernel is the commutator subgroup F of $\pi_1(X)$ (F is the subgroup of $\pi_1(X)$ generated by all the elements of the form $aba^{-1}b^{-1}$). Thus, $H_1(X; \mathbf{Z})$ is isomorphic to $\pi_1(X)/F$. In particular, $H_1(X)$ is isomorphic to $\pi_1(X)$ if and only if $\pi_1(X)$ is Abelian.*

In view of the above theorem the first homology group is sometimes referred to as the "Abelianization" of the fundamental group. For the relation between higher homology and homotopy groups an important result is the following Hurewicz isomorphism theorem, which gives sufficient conditions for isomorphisms between $H_q(X; \mathbf{Z})$ and $\pi_q(X)$ for $q > 1$.

Theorem 2.6 (Hurewicz) *Let X be a simply connected space. If there exists $j \in \mathbf{N}$ such that $\pi_j(X)$ is the first non-trivial higher homotopy group of X, then*

$$\pi_k(X) \cong H_k(X; \mathbf{Z}), \quad \forall k, \ 1 \le k \le j.$$

Thus for a simply connected space the first non-trivial homotopy and homology groups are in the same dimension and are equal.

Let A be a subspace of the topological space X. The pair (X, A) is called a **topological pair**. If (X', A') is another topological pair and $f : X \to X'$ is a continuous map such that $f(A) \subset A'$, then f is called a morphism of the topological pair (X, A) into (X', A') and is denoted by $f : (X, A) \to (X', A')$.

Let (X, A) be a topological pair. Then, $\forall q \in \mathbf{Z}$, $S_q(A)$ can be regarded as a submodule of $S_q(X)$ and $\delta_q(S_q(A)) \subset S_{q-1}(A)$. The quotient chain complex of $(S_*(X), \delta)$ by $S_*(A)$ is called the **relative singular chain complex** of X **mod** A and is denoted by $S_*(X, A)$. The qth homology module of this chain complex is denoted by $H_q(X, A)$, or $H_q(X, A; \mathbf{P})$ if one wants to stress the fact that the coefficients are in the principal ideal domain $\mathbf{P}$. $H_q(X, A)$ is called the qth **relative singular homology module** of X **mod** A. Let $Z_q(X, A)$, $B_q(X, A)$ be defined by

$$Z_q(X, A) := \{c \in S_q(X) \mid \delta c \in S_{q-1}(A)\},$$

$$B_q(X, A) := \{c \in S_q(X) \mid c = \delta w + c', \ w \in S_{q+1}(X), c' \in S_q(A)\}.$$

Then one can show that

$$H_q(X, A) \cong Z_q(X, A)/B_q(X, A).$$

Indeed, the above relation is sometimes taken as the definition of $H_q(X, A)$. The elements of $Z_q(X, A)$ (resp. $B_q(X, A)$) are called q-**cycles** (resp., q-**boundaries**) on X mod A. Let $f : (X, A) \to (X', A')$ be a morphism of topological pairs. Then the map $\mathcal{S}_q(f)$ sends $\mathcal{S}_q(A)$ into $\mathcal{S}_q(A')$ and hence $\mathcal{S}_q(f)$ induces, by passage to the quotient, the map $\tilde{\mathcal{S}}_q(f) : \mathcal{S}_q(X, A) \to \mathcal{S}_q(X', A')$. The family $\{\tilde{\mathcal{S}}_q(f) \mid q \in Z\}$ is denoted by $\tilde{\mathcal{S}}_*(f)$. It is customary to write simply $\mathcal{S}_q(f)$ and $\mathcal{S}_*(f)$ instead of $\tilde{\mathcal{S}}_q(f)$ and $\tilde{\mathcal{S}}_*(f)$, respectively. The map $\mathcal{S}_q(f)$ satisfies equation (2.5). From this it follows that it sends $Z_q(X, A)$ into $Z_q(X', A')$ and $B_q(X, A)$ into $B_q(X', A')$. Thus, $\mathcal{S}_q(f)$ induces, by passage to the quotient, a homomorphism

$$\tilde{H}_q(f; \mathbf{P}) : H_q(X, A) \to H_q(X', A').$$

It is customary to write $H_q(f; \mathbf{P})$, or simply $H_q(f)$, instead of $\tilde{H}_q(f; \mathbf{P})$. One can show that

$$H_q(id_{(X,A)}) = id_{H_q(X,A)}.$$

Moreover, if $g : (X', A') \to (X'', A'')$ is a morphism of topological pairs, then

$$H_q(g \circ f) = H_q(g) \circ H_q(f).$$

Thus, $\forall q \in \mathbf{Z}$, $\tilde{H}_q(\cdot; \mathbf{P})$ is a covariant functor from the category of topological pairs to the category of $\mathbf{P}$-modules.

Let (X, A) be a topological pair, $i : A \to X$ the natural inclusion map, and $j : (X, \emptyset) \to (X, A)$ the natural morphism of $(X, \emptyset)$ into (X, A). Then the sequence induced by these maps

$$0 \longrightarrow \mathcal{S}_*(A) \xrightarrow{\mathcal{S}_*(i)} \mathcal{S}_*(X) \xrightarrow{\mathcal{S}_*(j)} \mathcal{S}_*(X, A) \longrightarrow 0$$

is a short exact sequence of chain complexes. Moreover, one has the related connecting morphism h_* (see Appendix D)

$$h_* = \{h_q : H_q(X, A) \to H_{q-1}(A) \mid q \in \mathbf{Z}\}.$$

The corresponding long exact sequence

$$\cdots \longrightarrow H_{q+1}(X, A) \xrightarrow{h_{q+1}} H_q(A) \longrightarrow H_q(X) \longrightarrow$$

$$\longrightarrow H_q(X, A) \xrightarrow{h_q} H_{q-1}(A) \longrightarrow \cdots$$

is called the

Relative homology is useful in the evaluation of homology because of the following **excision property**. Let (X, A) be a topological pair and $U \subset A$.

Let $i : (X \setminus U, A \setminus U) \to (X, A)$ be the natural inclusion. We say that U can be **excised** and that i is an **excision** if

$$H_q(i) : H_q(X \setminus U, A \setminus U) \to H_q(X, A)$$

is an isomorphism. One can show that, if the closure $\bar{U} \subset A$, then U can be excised. If X is an n-dimensional topological manifold, then, using the excision property, one can show that, $\forall x \in X$,

$$H_n(X, X \setminus \{x\}) \cong \mathbf{P}.$$

Let U be a neighborhood of $x \in X$. If $j_x^U : (X, X \setminus U) \to (X, X \setminus \{x\})$ denotes the natural inclusion, then we have the homomorphism

$$H_n(j_x^U) : H_n(X, X \setminus U) \to H_n(X, X \setminus \{x\}).$$

One can show that, $\forall x \in X$, there exists an open neighborhood U of x and $\alpha \in H_n(X, X \setminus U)$ such that $\alpha_y := H_n(j_y^U)(\alpha)$ generates $H_n(X, X \setminus \{y\}), \forall y \in U$. Such an element α is called a **local P-orientation of X along U**. A **P-orientation system** of X is a set $\{(U_i, \alpha_i) \mid i \in I\}$ such that

1. $\bigcup_{i \in I} U_i = X$;
2. $\forall i \in I, \alpha_i$ is a local **P**-orientation of X along U_i;
3. $\alpha_{i,y} = \alpha_{j,y}, \ \forall y \in U_i \cap U_j$.

Given the **P**-orientation system $\{(U_i, \alpha_i) \mid i \in I\}$ of X, for each $x \in X$, $\exists i \in I$ such that $x \in U_i$ and hence we have a generator α_x of $H_n(X, X \setminus \{x\})$ given by $\alpha_x := \alpha_{i,x}$. Two **P**-orientation systems $\{(U_i, \alpha_i) \mid i \in I \}$, $\{(U_i', \alpha_i') \mid i \in I' \}$ are said to be **equivalent** if $\alpha_x = \alpha_x', \ \forall x \in X$. An equivalence class of **P**-orientation systems of X is denoted simply by α and is called a **P-orientation** of X. One can show that, if X is connected, then two **P**-orientations that are equal at one point are equal everywhere. A topological manifold is said to be **P-orientable** if it admits a **P**-orientation. A **P-oriented** manifold is a **P**-orientable manifold with the choice of a fixed **P**-orientation α. A manifold is said to be **orientable** (resp., **oriented**) when it is **Z**-orientable (resp., **Z**-oriented). We note that homology with integer (resp., rational, real) coefficients is often referred to as the integral (resp. rational, real) homology.

If X is a compact connected n-dimensional, **P**-oriented manifold, then

$$H_n(X) \cong \mathbf{P}.$$

This allows us to give the following definition of the fundamental class of a compact connected oriented manifold with orientation α. Let α_x be the local orientation at $x \in X$. Then there exists a unique generator of $H_n(X)$, whose image under the canonical map $H_n(X) \to H_n(X, X \setminus \{x\})$ is α_x. This generator of $H_n(X)$ is called the **fundamental class** of X with the orientation α and is denoted by $[X]$.

Using integral homology we can define the Betti numbers and Euler characteristic for certain topological spaces. They turn out to be integer-valued topological invariants. In order to define them, let us recall some results from algebra. Let V be a **P**-module. An element $v \in V$ is called a **torsion element** if there exists $a \in \mathbf{P} \setminus \{0\}$ such that $av = 0$. The set of torsion elements of V is a submodule of V denoted by V_t and called the **torsion submodule** of V. If $V_t = \{0\}$ then V is said to be **torsion free**. One can show (see Lang [244]) that if V is finitely generated then there exists a free submodule V_f of V such that

$$V = V_t \bigoplus V_f.$$

The dimension of V_f is called the **rank** of V. Let M be a topological manifold. If the homology modules $H_q(M; \mathbf{Z})$ are finitely generated, then the rank of $H_q(M; \mathbf{Z})$ is called the qth **Betti number** and is denoted by $b_q(M)$. In this case we define the **Euler** (or **Euler–Poincaré**) **characteristic** $\chi(M)$ of M by

$$\chi(M) := \sum_q (-1)^q b_q(M).$$

We observe that if M is compact then the homology modules are finitely generated. Roughly speaking, the Betti numbers count the number of holes of appropriate dimension in the manifold, whereas the torsion part indicates the twisting of these holes.

An example of this is the following. Recall that the **Klein bottle** K is obtained by identifying the two ends of the cylinder $[0, 1] \times S^1$ with an antipodal twist, i.e., by identifying $(0, \theta)$ with $(1, -\theta)$, $\theta \in S^1$. This twist is reflected in the torsion part of homology and we have $H_1(K; \mathbf{Z}) = \mathbf{Z} \oplus \mathbf{Z}_2$, whereas $H_1(K; \mathbf{R}) = \mathbf{R}$. Note that if we use homology with coefficient in $\mathbf{Z}_2$ then the torsion part also vanishes since $\mathbf{Z}_2$ has no non-trivial subgroups. If the integral domain **P** is taken to be the field $\mathbf{Q}$ or $\mathbf{R}$, then the Betti numbers remain the same but there is no torsion part in the homology modules.

By duality a homology theory gives a cohomology theory. As an example singular cohomology is defined as the dual of singular homology. The qth **singular cochain module** of a topological space X with coefficients in **P** is the dual of $\mathcal{S}_q(X; \mathbf{P})$ and is denoted by $\mathcal{S}^q(X; \mathbf{P})$ or simply $\mathcal{S}^q(X)$. If X, Y are topological spaces and $f : X \to Y$ is a continuous map, then we denote by $\mathcal{S}^q(f; \mathbf{P})$ or simply by $\mathcal{S}^q(f)$ the map

$$\mathcal{S}^q(f) := {}^t\mathcal{S}_q(f) : \mathcal{S}^q(Y) \to \mathcal{S}^q(X),$$

where we have used the notation ${}^tL$ for the transpose of the linear map L (here $L = \mathcal{S}^q(f)$). Then it is easy to verify that $\mathcal{S}^q(\cdot ; \mathbf{P})$ is a contravariant functor from the category of topological spaces to the category of **P**-modules. The qth **singular cohomology P-module** $H^q(X; \mathbf{P})$ or simply $H^q(X)$ is defined by

$$H^q(X) = \operatorname{Ker} {}^t\delta_{q+1}/\operatorname{Im} {}^t\delta_q,$$

where ${}^t\delta_{q+1} : \mathcal{S}^q(X) \to \mathcal{S}^{q+1}(X)$ is the qth **coboundary operator**. The module $Z^q(X) := \operatorname{Ker} {}^t\delta_{q+1}$ (resp., $B^q(X) := \operatorname{Im} {}^t\delta_q$) is called the qth singular cohomology module of **cocycles** (resp., **coboundaries**). In particular, the duality of $H^q(X, \mathbf{Z})$ with $H_q(X, \mathbf{Z})$ and the finite dimensionality of $H_q(X, \mathbf{Z})$ implies that $\dim H^q(X, \mathbf{Z}) = \dim H_q(X, \mathbf{Z}) = b_q(X)$, $\forall q \geq 0$, where $b_q(X)$ is the qth Betti number of X. If X, Y are topological spaces and $f : X \to Y$ is a continuous map, then $\mathcal{S}^q(f)$ sends $Z^q(Y)$ to $Z^q(X)$ and $B^q(Y)$ to $B^q(X)$. Hence, it induces, by passage to the quotient, the homomorphism

$$H^q(f; \mathbf{P}) \equiv H^q(f) : H^q(Y) \to H^q(X).$$

Then it is easy to verify that $H^q(\,\cdot\,; \mathbf{P})$ is a contravariant functor from the category of topological spaces to the category of $\mathbf{P}$-modules. With an analogous procedure one can define the qth relative singular cohomology modules for a topological pair (X, A), denoted by $H^q(X, A)$ (see Greenberg [160] for details). A comprehensive introduction to algebraic topology covering both homology and homotopy can be found in Tammo tam Dieck's book [98].

In dealing with noncompact spaces it is useful to consider singular cohomology with compact support that we now define. Let X be a topological manifold. The set $\mathcal{K}$ of compact subsets of X is a directed set with the partial order given by the inclusion relation. Let us consider the direct system

$$D = (\{H^q(X, X \setminus K) \mid K \in \mathcal{K}\}, \{f_K^{K'} \mid (K, K') \in \mathcal{K}_0^2\}),$$

where

$$\mathcal{K}_0^2 := \{(K, K') \in \mathcal{K}^2 \mid K \subset K'\}.$$

The map $f_K^{K'} : H^q(X, X \setminus K) \to H^q(X, X \setminus K')$ is the homomorphism induced by the inclusion. The qth **singular cohomology P-module with compact support** is the direct limit of the direct system D and is denoted by $H_c^q(X; \mathbf{P})$, or simply $H_c^q(X)$. Then, by definition

$$H_c^q(X) := \varinjlim H^q(X, X \setminus K).$$

We observe that if X is compact then X is the largest element of $\mathcal{K}$. Thus, if X is compact we have that $H_c^q(X) = H^q(X)$, $\forall q \in \mathbf{Z}$.

As with homology theories, there are several cohomology theories. An example is given by the **differentiable singular homology** (resp., cohomology) whose difference from singular homology (resp., cohomology) is essentially in the fact that its construction starts with **differentiable singular q-simplexes** instead of (continuous) singular q-simplexes. The differentiable singular homology (resp., cohomology) of X is denoted by ${}_\infty H_*(X; \mathbf{P})$ (resp., $H_\infty^*(X; \mathbf{P})$). Under very general conditions the various cohomology theories are isomorphic (see Warner [396]); for example, $H^*(X; \mathbf{P}) \cong H_\infty^*(X; \mathbf{P})$. In most physical applications we are interested in topological spaces that are differentiable manifolds. We now discuss the cohomology theory based on the cochain complex of differential forms on a manifold. This is the well known de Rham cohomology with real coefficients.

2.5 de Rham Cohomology

The **de Rham complex** of an m-dimensional manifold M is the cochain complex $(\Lambda(M), d)$ given by

$$0 \longrightarrow \Lambda^0(M) \stackrel{d}{\longrightarrow} \Lambda^1(M) \stackrel{d}{\longrightarrow} \cdots \stackrel{d}{\longrightarrow} \Lambda^n(M) \longrightarrow 0 \ . \qquad (2.6)$$

The cohomology $H^*(\Lambda(M), d)$ is called the **de Rham cohomology** of M and is denoted by $H^*_{deR}(M)$. The de Rham cohomology has a natural structure of graded algebra induced by the exterior product. The product on homogeneous elements is given by the map

$$\cup : H^i(M; \mathbf{P}) \times H^j(M; \mathbf{P}) \to H^{i+j}(M; \mathbf{P})$$

defined by

$$([\alpha], [\beta]) \mapsto [\alpha \wedge \beta].$$

This induced product in cohomology is in fact a special case of a cohomology operation called the cup product (see Spanier [357]).

If M, N are manifolds then we have

$$H^*_{\mathrm{deR}}(M \times N) = H^*_{\mathrm{deR}}(M) \hat{\otimes} H^*_{\mathrm{deR}}(N),$$

where $\hat{\otimes}$ denotes the graded tensor products. In particular, we can express the cohomology of $M \times N$ in terms of the cohomologies of M and N as follows:

$$H^k_{\mathrm{deR}}(M \times N) = \bigoplus_{k=i+j} H^i_{\mathrm{deR}}(M) \otimes H^j_{\mathrm{deR}}(N). \qquad (2.7)$$

In fact, the above formula holds more generally and is called the **Künneth formula**.

There is a canonical map ρ called the **de Rham homomorphism**

$$H^q_{\mathrm{deR}}(M) \to (_\infty H_q(M; \mathbf{R}))' \cong H^q_\infty(M; \mathbf{R})$$

given by the following pairing between de Rham cohomology classes $[\alpha]$ and real differentiable singular homology classes $[c]$

$$([\alpha], [c]) \mapsto \int_c \alpha.$$

One can show that this map ρ is independent of the choice of $\alpha \in [\alpha]$ and $c \in [c]$. The classical de Rham theorem says that the map ρ is an isomorphism. Thus, $\forall q$ we have

$$H_{\text{deR}}^q(M) \cong H_{\infty}^q(M; \mathbf{R}) \cong H^q(M; \mathbf{R}).$$

Let (M, g) be an oriented, closed (i.e., compact and without boundary) Riemannian manifold with metric volume form μ. Recall that the **Hodge–de Rham operator** $\Delta = d \circ \delta + \delta \circ d$ maps $\Lambda^k(M) \to \Lambda^k(M)$, $\forall k$ and that the Hodge star operator $*$ maps $\Lambda^k(M) \to \Lambda^{n-k}(M)$, $\forall k$. For further details, see Chapter 3. The map

$$\langle \, , \, \rangle : \Lambda^k(M) \times \Lambda^k(M) \to \mathbf{R}$$

defined by

$$\langle \alpha, \beta \rangle = \int_M \alpha \wedge *\beta = \int_M g(\alpha, \beta)\mu \tag{2.8}$$

is an inner product on $\Lambda^k(M)$. One can show that, for $\sigma \in \Lambda^{k+1}(M)$, we have

$$\langle d\alpha, \sigma \rangle = \langle \alpha, \delta\sigma \rangle \text{ and } \langle \Delta\alpha, \beta \rangle = \langle \alpha, \Delta\beta \rangle. \tag{2.9}$$

That is, δ is the adjoint of d and Δ is self-adjoint with respect to this inner product. Furthermore, $\Delta\alpha = 0$ if and only if $d\alpha = \delta\alpha = 0$. An element of the set

$$\mathcal{H}^k := \{\alpha \in \Lambda^k(M) \mid \Delta\alpha = 0\} \tag{2.10}$$

is called a **harmonic** k-form. It follows that a k-form is harmonic if and only if it is both closed and coclosed. The set $\mathcal{H}^k$ is a subspace of $\Lambda^k(M)$. Using these facts one can prove (see, for example, Warner [396]) the **Hodge decomposition theorem**, which asserts that $\mathcal{H}^k$ is finite-dimensional and $\Lambda^k(M)$ has a direct sum decomposition into the orthogonal subspaces $d(\Lambda^k(M))$, $\delta(\Lambda^k(M))$, and $\mathcal{H}^k$. Thus any k-form α can be expressed by the formula

$$\alpha = d\beta + \delta\gamma + \theta, \tag{2.11}$$

where $\beta \in \Lambda^{k-1}(M)$, $\gamma \in \Lambda^{k+1}(M)$ and θ is a harmonic k-form. For α in a given cohomology class, the harmonic form θ of equation (2.11) is uniquely determined. Thus, we have an isomorphism of the kth cohomology space $H^k(M; \mathbf{R})$ with the space of harmonic k-forms $\mathcal{H}^k$. Therefore, the kth Betti number b_k is equal to the $\dim \mathcal{H}^k$. This is an illustration of a relation between physical or analytic data (the solution space of a partial differential operator) on a manifold and its topology. A far-reaching, nonlinear generalization of this idea relating the solution space of Yang–Mills instantons to the topology of 4-manifolds appears in the work of Donaldson (see Chapter 9 for further details).

2.5.1 The Intersection Form

Let M be a closed (i.e., compact, without boundary), connected, oriented manifold of dimension $2n$. Let v denote the volume form on M defining the orientation. We shall use the de Rham cohomology to define ι_M, the intersection form of M as follows. Let $\alpha, \beta \in \Lambda^n(M)$ be two closed n-forms representing the cohomology classes $a, b \in H^n(M; \mathbf{Z}) \subset H^n(M; \mathbf{R})$ respectively, i.e., $a = [\alpha]$ and $b = [\beta]$. Now $\alpha \wedge \beta \in \Lambda^{2n}(M)$ and hence $\int \alpha \wedge \beta$ is well defined with respect to the volume form v. It can be shown that this integral is independent of the choice of forms α, β representing the cohomology classes a, b and takes values that are integral multiples of the volume of M. Thus, we can define the binary operator

$$\iota_M : H^n(M; \mathbf{Z}) \times H^n(M; \mathbf{Z}) \to \mathbf{Z}$$

by

$$\iota_M(a, b) = \int_M (\alpha \wedge \beta).$$

In what follows we shall use the same letter to denote the cohomology class and an n-form representing that class. It can be shown that ι_M is a symmetric, non-degenerate bilinear form on $H^n(M; \mathbf{Z})$. This symmetric, non-degenerate form ι_M is called the **intersection form** of M. The definition given above works only for smooth manifolds. However, as is the case with de Rham cohomology, the intersection form does not depend on the differential structure and is a topological invariant. In particular, it is defined for topological manifolds. In fact, the intersection form can also be defined for non-orientable manifolds by considering cohomology or homology with coefficients in $\mathbf{Z}_2$ instead of $\mathbf{Z}$. Now for a compact manifold M, $H^n(M; \mathbf{Z})$ is a finitely generated free Abelian group of rank b_n (the nth Betti number), i.e., an integral lattice of rank b_n. Thus, the intersection form gives us the map

$$\iota : M \mapsto \iota_M ,$$

which associates to each compact, connected, oriented topological manifold M of dimension $2n$ a symmetric, non-degenerate bilinear form ι_M on a lattice of rank b_n. Let (b^+, b^-) be the signature of the bilinear form ι_M. If $b_n > 0$ and c_j, $1 \leq j \leq b_n$, is a basis of the lattice $H^n(M; \mathbf{Z})$, then the intersection form is completely determined by the matrix of integers $\iota_M(c_j, c_k)$, $1 \leq j, k \leq b_n$. From Poincaré duality it follows that the intersection form is unimodular, i.e.,

$$|\det(\iota_M(c_j, c_k))| = 1.$$

If $b_n = 0$ then we take $\iota_M := \emptyset$ the empty form. Recall that on the abstract level two forms ι_1, ι_2 on lattices L_1, L_2, respectively, are said to be **equivalent** if there exists an isomorphism of lattices $f : L_1 \to L_2$ such that $f^* \iota_2 = \iota_1$.

The intersection form plays a fundamental role in Freedman's classification of topological 4-manifolds. We give a brief discussion of this result in the next section.

2.6 Topological Manifolds

In this section we discuss topological manifolds with special attention to low-dimensional manifolds. The following theorem gives some results on the existence of smooth structures on topological manifolds.

Theorem 2.7 *Let M be a closed topological manifold of dimension n. Then we have the following results:*

1. For $n \leq 3$ there is a unique compatible smooth structure on M.
2. For $n = 4$ there exist (infinitely many) simply connected manifolds that admit infinitely many distinct smooth structures. It is not known whether there are manifolds that admit only finitely many distinct smooth structures.
3. For $n \geq 5$ there are at most finitely many distinct compatible smooth structures.

Thus, dimension 4 seems very special. This is also true for open topological manifolds. There is a unique smooth structure on $\mathbf{R}^n$, $n \neq 4$, compatible with its standard topology. However, $\mathbf{R}^4$ admits uncountably many smooth structures. We do not know at this time if every open topological 4-manifold admits uncountably many smooth structures. For further results in the surprising world of 4-manifolds, see, for example, the book by Scorpan [343].

2.6.1 Topology of 2-Manifolds

The topology of 2-manifolds, or surfaces, was well known in the nineteenth century. Smooth, compact, connected and oriented 2-manifolds are called **Riemann surfaces**. An introduction to compact Riemann surfaces from various points of view and their associated geometric structures may be found in Jost [213]. They are classified by a single non-negative integer, the genus g. The genus counts the number of holes in the surface. There is a standard model Σ_g for a surface of genus g obtained by attaching g handles to a sphere (which has genus zero). Every smooth, compact, oriented surface is diffeomorphic to one and only one Σ_g. The classification is sometimes given in terms of the Euler characteristic of the surface. It is related to the genus by the formula $\chi(\Sigma_g) = 2 - 2g$.

In the classical theory of surfaces, homology classes $a, b \in H^1(M; \mathbf{Z})$ were represented by closed curves, which could be chosen to intersect transversally.

The intersection form was then defined by counting the algebraic number of intersections of these curves. Surfaces are completely classified by their intersection forms. In the orientable case we have the following well known result.

Theorem 2.8 *Let M, N be two closed, connected, oriented surfaces. Then $M \cong N$ (i.e., M is diffeomorphic to N) if and only if the intersection forms ι_M, ι_N are equivalent. Moreover, if $\iota_M = \emptyset$ then $M \cong S^2$ and if $\iota_M \neq \emptyset$ then there exists $k \in \mathbf{N}$ such that $\iota_M \cong k\sigma_1$, where*

$$\sigma_1 = \begin{pmatrix} 0 & 1 \\ 1 & 0 \end{pmatrix}$$

is a Pauli spin matrix, and $k\sigma_1$ is the block diagonal form with k entries of σ_1, and $M \cong kT^2$, where $T^2 = S^1 \times S^1$ is the standard torus and kT^2 is the connected sum of k copies of T^2.

The Riemann surface together with a fixed complex structure provides a classical model for one-dimensional algebraic varieties or complex curves. A compact surface corresponds to a projective curve. The genus of such a surface is equal to the dimension of the space of holomorphic one forms on the surface. This way of looking at a Riemann surface is crucial in the Gromov–Witten theory. We will not consider it in this book. The genus has also a topological interpretation as half the first Betti number of the surface. We note that the classification of orientable surfaces given by the above theorem can be extended to include non-orientable surfaces as well. We are interested in extending this theorem to the case of 4-manifolds. This was done by Freedman in 1981. Before discussing his theorem we consider the topology of 3-manifolds where no intersection form is defined.

2.6.2 Topology of 3-Manifolds

The classification of manifolds of dimension 3 or higher is far more difficult than that of surfaces. It was initiated by Poincaré in 1900. The year 1900 is famous for the Paris ICM and Hilbert's lecture on the major open problems in mathematics. The classification of 3-manifolds was not among Hilbert's problems. Armed with newly minted homology groups and his fundamental group Poincaré began his study by trying to characterize the simplest 3-manifold, the sphere S^3. His first conjecture was the following:

Let M be a compact connected 3-manifold with the same homology groups as the sphere S^3. Then M is homeomorphic to S^3.

In attempting to prove this conjecture Poincaré found a 3-manifold P that provided a counterexample to the conjecture. This 3-manifold P is now called the **Poincaré homology sphere**. It is denoted by $\Sigma(2,3,5)$ as it can be represented as a special case of the Brieskorn homology 3-spheres,

$$\Sigma(a_1, a_2, a_3) := \{(z_1, z_2, z_3) \mid z_1^{a_1} + z_2^{a_2} + z_3^{a_3} = 0\} \cap S^5, \ a_1, a_2, a_3 \in \mathbf{N}.$$

We will compute various new invariants of the Brieskorn homology 3-spheres in Chapter 10. Poincaré's original ingenious construction of P can be described via well known geometric figures. It is the space of all regular icosahedra inscribed in the standard unit 2-sphere. Note that each icosahedron is uniquely determined by giving one vertex on the sphere (2 parameters) and a direction to a neighboring vertex (1 parameter). It can be shown that the parameter (or moduli) space P of all icosahedra is a 3-manifold diffeomorphic to $\Sigma(2,3,5)$ and that it has the same homology groups as the sphere S^3. Poincaré showed that $\pi_1(P)$, the fundamental group of P, is non-trivial. Since $\pi_1(S^3)$ is trivial, P cannot be homeomorphic to S^3. Yet another description of P is obtained by observing that the rotation group $SO(3)$ maps S^2 to itself and the induced action on P is transitive. The isotropy group I_x of a fixed point $x \in P$ can be shown to be a finite group of order 60. Thus, P is homeomorphic to the coset space $SO(3)/I_x$. This fact can be used to show that $\pi_1(P)$ is a perfect group of order 120.

Icosahedron is one of the five **regular polyhedra** or solids known since antiquity. They are commonly referred to as **Platonic solids** (see Appendix B for some interesting properties of Platonic solids). Poincaré's counterexample showed that homology was not enough to characterize S^3 and that one has to take into account the fundamental group. He then made the following conjecture:

Let M be a closed simply connected 3-manifold. Then M is homeomorphic to S^3.

We give it in an alternative form, which is usrful for stating the generalized Poincaré conjecture in any dimension, with 3 replaced by a natural number n.

Poincaré Conjecture: *Let M be a closed connected 3-manifold with the same homotopy type as the sphere S^3. Then M is homeomorphic to S^3.*

Definition 2.2 *Let H^n denote the n-dimensional hyperbolic space. A 3-manifold X is said to be a* **Thurston 3-dimensional geometry** *if it is one of the following eight homogeneous manifolds (i.e., the group of isometries of X acts transitively on X).*

- *Three spaces of constant curvature $\mathbf{R}^3, S^3, H^3$.*
- *Two product spaces $\mathbf{R} \times S^2$, $\mathbf{R} \times H^2$.*
- *Three* **twisted product spaces**, *each of which is a Lie group with a left invariant metric. They are*

  - *the universal covering space of the group $SL(2, \mathbf{R})$;*
  - *the group of upper triangular matrices in $M_3(\mathbf{R})$ with all diagonal entries 1, called* **Nil**;
  - *the semidirect product of $\mathbf{R}$ and $\mathbf{R}^2$, where $\mathbf{R}$ acts on $\mathbf{R}^2$ through multiplication by the diagonal matrix $\mathrm{diag}(t, \ t^{-1})$, $t \in \mathbf{R}$, called* **Sol**.

We can now define a geometric 3-manifold in the sense of Thurston.

Definition 2.3 *A 3-manifold M is said to be* **geometric** *if it is diffeomorphic to X/Γ, where Γ is a discrete group of isometries of X (i.e. $\Gamma < \mathrm{Isom}(X)$) acting freely on X, and X is one of Thurston's eight 3-dimensional geometries.*

We can now state the Thurston geometrization conjecture.

Thurston Geometrization Conjecture: : Let M be a closed 3-manifold that does not contain two-sided projective planes. Then M admits a connected sum decomposition and a decomposition along disjoint incompressible tori and Klein bottles into a finite number of pieces each of which is a geometric manifold.

The Thurston geometrization conjecture was recently proved by Perelman using a generalized form of Hamilton's Ricci flow technique. This result implies the original Poincaré conjecture. We will comment on it in Chapter 6.

2.6.3 Topology of 4-manifolds

The importance of the intersection form for the study of 4-manifolds was already known since 1940 from the following theorem of Whitehead (see Milnor and Husemoller [285]).

Theorem 2.9 *Two closed, 1-connected, 4-manifolds are homotopy equivalent if and only if their intersection forms are equivalent.*

In the category of topological manifolds a complete classification of closed, 1-connected, oriented 4-manifolds has since been carried out by Freedman (see [136]). To state his results we begin by recalling the general scheme of classification of symmetric, non-degenerate, unimodular, bilinear forms (referred to simply as "forms" in the rest of this section) on lattices (see Milnor and Husemoller [285]). The classification of forms has a long history and is an important area of classical mathematics with applications to algebra, number theory, and more recently to topology and geometry. We have already defined two fundamental invariants of a form, namely its rank b_n and its signature (b^+, b^-). We note that sometimes the signature is defined to be the integer $b^+ - b^- = b_n - 2b^-$. We shall denote this integer by $\sigma(M)$, i.e., $\sigma(M) := b_n - 2b^-$. We say that a form ι on the lattice L is **even** or of **type II** if $\iota(a, a)$ is even for all $a \in L$. Otherwise we say that it is **odd** or of **type I**. It can be shown that for even (type II) forms, 8 divides the signature $\sigma(M)$. In particular, 8 divides the rank of a positive definite even form. The indefinite forms are completely classified by the rank, signature, and type. We have the following result.

Theorem 2.10 *Let ι be an indefinite form of rank r and signature (j, k), $j > 0, k > 0$. Then we have*

1. $\iota \cong j(1) \oplus k(-1)$ if it is odd (type I),
2. $\iota \cong m\sigma_1 \oplus pE_8$, $m > 0$ if it is even (type II),

where (1) and (−1) are 1 × 1 matrices representing the two possible forms of rank 1, σ_1 is the Pauli spin matrix defined earlier, and E_8 is the matrix associated to the exceptional Lie group E_8 in Cartan's classification of simple Lie groups, i.e.,

$$E_8 = \begin{pmatrix} 2 & -1 & 0 & 0 & 0 & 0 & 0 & 0 \\ -1 & 2 & -1 & 0 & 0 & 0 & 0 & 0 \\ 0 & -1 & 2 & -1 & 0 & 0 & 0 & 0 \\ 0 & 0 & -1 & 2 & -1 & 0 & 0 & 0 \\ 0 & 0 & 0 & -1 & 2 & -1 & 0 & -1 \\ 0 & 0 & 0 & 0 & -1 & 2 & -1 & 0 \\ 0 & 0 & 0 & 0 & 0 & -1 & 2 & 0 \\ 0 & 0 & 0 & 0 & -1 & 0 & 0 & 2 \end{pmatrix}.$$

The classification of definite forms is much more involved. The number $N(r)$ of equivalence classes of definite forms (which counts the inequivalent forms) grows very rapidly with the rank r of the form, as Table 2.3 illustrates.

Table 2.3 Number of inequivalent definite forms

r	8	16	24	32	40
$N(r)$	1	2	24	$\geq 10^9$	$\geq 10^{51}$

We now give some simple examples of computation of intersection forms that we will use later.

Example 2.8 *We denote $H^2(M; \mathbf{Z})$ by L in this example.*

1. *Let $M = S^4$; then $L = 0$ and hence $\iota_M = \emptyset$.*
2. *Let $M = S^2 \times S^2$; then L has a basis of cohomology classes α, β dual to the homology cycles represented by $S^2 \times \{(1, 0, 0)\}$ and $\{(1, 0, 0)\} \times S^2$, respectively. With respect to this basis the matrix of ι_M is the Pauli spin matrix*

$$\sigma_1 = \begin{pmatrix} 0 & 1 \\ 1 & 0 \end{pmatrix}.$$

3. *Let $M = \mathbf{CP}^2$; then $L = \mathbf{Z}$ and hence $\iota_M = (1)$.*
4. *Let $M = \overline{\mathbf{CP}}^2$, i.e., $\mathbf{CP}^2$ with the opposite complex structure and orientation. Then $L = \mathbf{Z}$ and hence $\iota_M = (-1)$.*

Whitehead's Theorem 2.9 stated above says that the map ι that associates to a closed, 1-connected topological 4-manifold its intersection form induces an injection from the homotopy equivalence classes of manifolds into the equivalence classes of forms. It is natural to study this map ι in greater detail. One can ask, for example, the following two questions.

1. Is the map ι surjective? I.e., given an intersection form μ, does there exist a manifold M such that $\iota_M = \mu$?
2. Does the injection from the homotopy equivalence classes of manifolds into the equivalence classes of forms extend to other equivalence classes of manifolds?

The first question is an existence question while the second is a uniqueness question. These questions can be restricted to different categories of manifolds such as topological or smooth manifolds. A complete answer to these questions in the topological category is given by the following theorem, proved by Freedman in 1981 [136].

Theorem 2.11 *Let $\mathcal{M}_{sp}$ (resp., $\mathcal{M}_{ns}$) denote the set of topological equivalence classes (i.e. homeomorphism classes) of closed, 1-connected, oriented, spin (resp., non-spin) 4-manifolds. Let $\mathcal{I}_{ev}$ (resp. $\mathcal{I}_{od}$) denote the set of equivalence classes of even (resp. odd) forms. Then we have the following:*

1. *the map $\iota : \mathcal{M}_{sp} \to \mathcal{I}_{ev}$ is bijective;*
2. *the map $\iota : \mathcal{M}_{ns} \to \mathcal{I}_{od}$ is surjective and is exactly two-to-one. The two classes in the preimage of a given form are distinguished by a cohomology class $\kappa(M) \in H^4(M; \mathbf{Z}_2)$ called the Kirby–Siebenmann invariant.*

We note that the **Kirby–Siebenmann invariant** represents the obstruction to the existence of a piecewise linear structure on a topological manifold of dimension ≥ 5. Applying the above theorem to the empty rank zero form provides a proof of the Poincaré conjecture for dimension 4. Freedman's theorem is regarded as one of the fundamental results of modern topology.

In the smooth category the situation is much more complicated. It is well known that the map ι is not surjective in this case. In fact, we have the following theorem:

Theorem 2.12 (Rochlin) *Let M be a smooth, closed, 1-connected, oriented, spin manifold of dimension 4. Then $\sigma(M)$, the signature of M is divisible by 16.*

Now, as we observed earlier, 8 always divides the signature of an even form, but 16 need not divide the form. Thus, we can define the **Rochlin invariant** $\rho(\mu)$ of an even form μ by

$$\rho(\mu) := \frac{1}{8}\sigma(\mu) \ (\text{mod } 2).$$

We note that the Rochlin invariant and the Kirby–Siebenmann invariant are equal in this case, but for non-spin manifolds the Kirby–Siebenmann invariant is not related to the intersection form and thus provides a further obstruction to smoothability. From Freedman's classification and Rochlin's theorem it follows that a topological manifold with nonzero Rochlin invariant is not smoothable. For example, the topological manifold $|E_8| := \iota^{-1}(E_8)$ corresponds to the equivalence class of the form E_8 and has signature 8 (Rochlin invariant 1) and hence is not smoothable.

For several years very little progress was made beyond the result of the above theorem in the smooth category. Then, in 1982, through his study of the topology and geometry of the moduli space of instantons on 4-manifolds Donaldson discovered the following, unexpected, result. The theorem has led to a number of important results including the existence of uncountably many exotic differentiable structures on the standard Euclidean topological space $\mathbf{R}^4$.

Theorem 2.13 (Donaldson) *Let M be a smooth closed 1-connected oriented manifold of dimension 4 with positive definite intersection form ι_M. Then $\iota_M \cong b_2(1)$, the diagonal form of rank b_2, the second Betti number of M.*

Donaldson's work uses in an essential way the solution space of the Yang–Mills field equations for $SU(2)$ gauge theories and has already had profound influence on the applications of physical theories to mathematical problems. In 1990 Donaldson obtained more invariants of 4-manifolds by using the topology of the moduli space of instantons. Donaldson theory led to a number of new results for the topology of 4-manifolds, but it was technically a difficult theory to work with. In fact, Atiyah announced Donaldson's new results at a conference at Duke University in 1987, but checking all the technical details delayed the publication of his paper until 1990. The matters simplified greatly when the Seiberg–Witten equations appeared in 1994. We discuss the Donaldson invariants of 4-manifolds in more detail in Chapter 9. It is reasonable to say that at that time a new branch of mathematics which may be called "Physical Mathematics" was created.

In spite of these impressive new developments, there is at present no analogue of the geometrization conjecture in the case of 4-manifolds. Here geometric topologists are studying the variational problems on the space of metrics on a closed oriented 4-manifold M for one of the classical curvature functionals such as the square of the L^2 norm of the Riemann curvature Rm, Weyl conformal curvature W, and its self-dual and anti-dual parts W_+ and W_-, respectively, and Ric, the Ricci curvature. The Hilbert–Einstein variational principle based on the scalar curvature functional and its variants are important in the study of gravitational field equations. Einstein metrics, i.e., metrics satisfying the equation

$$K := Ric - \frac{1}{4}Rg = 0$$

are critical points of all of the functionals listed above. Here K is the trace-free part of the Ricci tensor. In many cases the Einstein metrics are minimizers, but there are large classes of minimizers that are not Einstein metrics. A well-known obstruction to the existence of Einstein metrics is the Hitchin–Thorpe inequality $\chi(M) \geq \frac{3}{2}|\tau(M)|$, where $\chi(M)$ is the Euler characteristic and $\tau(M)$ is the signature of M. A number of new obstructions are now known. Some of these indicate that their existence may depend on the smooth structure of M as opposed to just the topological structure. These obstructions can be interpreted as implying a coupling of matter fields to gravity (see [269, 270, 94]). The basic problem is to understand the existence and moduli spaces of these metrics on a given manifold and perhaps to find a geometric decomposition of M with respect to a special functional. One of the most important tools for developing such a theory is the Chern–Gauss–Bonnet theorem which states that

$$\chi(M) = \frac{1}{8\pi^2} \int (|Rm|^2 - |K|^2)dv = \frac{1}{8\pi^2} \int \left(|W|^2 - \frac{1}{2}|K|^2 + \frac{1}{24}R^2 \right) dv.$$

This result allows one to control the full Riemann curvature in terms of the Ricci curvature Ric. It is interesting to note that in [242], Lanczos had arrived at the same result while searching for Lagrangians to generalize Einstein's gravitational field equations. He noted the curious property of the Euler class that it contains no dynamics (or is an invariant). He had thus obtained the first topological gravity invariant (without realizing it). Chern's fundamental paper [74] appeared in the same journal seven years later. Chern–Weil theory and Hirzebruch's signature theorem give the following expression for the signature $\tau(M)$:

$$\tau(M) = \frac{1}{12\pi^2} \int (|W_+|^2 - |W_-|^2)dv.$$

The Hitchin–Thorpe inequality follows from this result and the Chern–Gauss–Bonnet theorem. In dimension 4, all the classical functionals are conformally (or scale) invariant, so it is customary to work with the space of unit volume metrics on M.

2.7 The Hopf Invariant

As we remarked in the preface, mathematicians and physicists have often developed the same ideas from different perspectives. The Hopf fibration and Dirac's monopole construction provide an example of this. Each is based on the observation that S^2 (the base of the Hopf fibration) is a deformation retract of $\mathbf{R}^3 \setminus \{0\}$ (the base of the Dirac monopole field). Thus non-triviality of $\pi_3(S^2)$ can be interpreted as Dirac's monopole quantization condition. Other Hopf fibrations and Hopf invariants also arise in physical theories, as we indicate later in this section.

Let $f : S^{2n-1} \to S^n, n > 1$, be a continuous map. Let X denote the quotient space of the disjoint union $D^{2n} \bigsqcup S^n$ under the identification of

$x \in S^{2n-1} \subset D^{2n}$ with $f(x) \in S^n$. The map f is called the **attaching map**. The space X is called the **adjunction space** obtained by attaching the $2n$-cell $e^{2n} := (D^{2n}, S^{2n-1})$ to S^n by f and is denoted by $S^n \cup_f e^{2n}$. The cohomology $H^*(X; \mathbf{Z})$ of X is easy to calculate and is given by

$$H^i(X; \mathbf{Z}) = \begin{cases} \mathbf{Z} & \text{if } i = n \text{ or } 2n, \\ 0 & \text{otherwise} . \end{cases}$$

Let $\alpha \in H^n(X; \mathbf{Z})$ and $\beta \in H^{2n}(X; \mathbf{Z})$ be the generators of the respective cohomology groups. We note that in de Rham cohomology α, β can be identified with closed differential forms. It follows that α^2 is an integral multiple of β. The multiplier $h(f)$ is completely determined by the map f and is called the **Hopf invariant** of f. Thus we have

$$\alpha^2 = h(f)\beta.$$

It can be shown that $h(f)$ is, in fact, a homotopy invariant and hence defines a map (also denoted by h)

$$[f] \mapsto h(f) \text{ of } \pi_{2n-1}(S^n) \to \mathbf{Z}.$$

To include the case $n = 1$ we note that the double covering $c_2 : S^1 \to S^1$ has adjunction space $\mathbf{RP}^2$ and Hopf invariant 1. The following theorem is due to H. Hopf:

Theorem 2.14 *Let $\mathcal{F}_n := C(S^{2n-1}, S^n)$, $n > 0$, denote the space of continuous functions from S^{2n-1} to S^n. Then we have the following:*

1. *If $n > 1$ is odd, then $h(f) = 0$, $\forall f \in \mathcal{F}_n$.*
2. *If n is even, then for each $k \in \mathbf{Z}$ there exists a map $f_k \in \mathcal{F}_n$ such that $h(f_k) = 2k$.*
3. *If there exists $g \in \mathcal{F}_n$ such that $h(g)$ is odd, then $n = 2^m$, where m is a nonnegative integer.*
4. *Let $\pi \in \mathcal{F}_1$ (resp., $\mathcal{F}_2$) be the real (resp., complex) Hopf fibration. Then $h(\pi) = 1$.*

We note that the real Hopf fibration $\pi : S^1 \to \mathbf{RP}^1 \cong S^1$ occurs in the geometric quantization of the harmonic oscillator [271,272] while the complex Hopf fibration $\pi : S^3 \to \mathbf{CP}^1 \cong S^2$ occurs in the geometric construction of the Dirac monopole. It can be shown that the last result in the above theorem can be extended to include the quaternionic and octonionic Hopf fibrations (which arise in the solution of Yang–Mills equations on S^4 and S^8, respectively) and that this extended list exhausts all $\mathcal{F}_n$ that contain a map with Hopf invariant 1. This result is part of the following extraordinary theorem, which links several specific structures from algebra, topology, and geometry.

Theorem 2.15 *Let $S^{n-1} \subset \mathbf{R}^n$, $n > 0$ denote the standard $(n-1)$-sphere in the real Euclidean n-space with the convention that $S^0 := \{-1, 1\} \subset \mathbf{R}$ and $\mathbf{R}^0 := \{0\}$. Then the following statements are equivalent:*

1. *The integer $n \in \{1, 2, 4, 8\}$.*
2. *$\mathbf{R}^n$ has the structure of a normed algebra.*
3. *$\mathbf{R}^n$ has the structure of a division algebra.*
4. *$\mathbf{R}^{n-1}$ admits a cross product (or a vector product).*
5. *S^{n-1} is an H-space.*
6. *S^{n-1} is parallelizable (i.e., its tangent bundle is trivializable).*
7. *There exists a map $f : S^{2n-1} \to S^n$ with Hopf invariant $h(f) = 1$.*

The relation of conditions (1) and (3) with condition (4) and its generalizations have been considered in [126]. It is well known that complex numbers have applications to 2-dimensional geometry and its ring of **Gaussian integers** (i.e., numbers of the form $m + ni$, where $m, n \in \mathbf{Z}$) is used in many classical questions in arithmetic. Similarly, the quaternions are related to 3-dimensional and 4-dimensional geometry and they contain rings of integers with many properties similar to those of Gaussian integers. The octonions have applications to 7- and 8-dimensional geometry. In elementary algebra one encounters the construction of the complex numbers in terms of certain real matrices of order 2. This **doubling procedure** of constructing $\mathbf{C}$ from $\mathbf{R}$ was generalized by Dixon to construct $\mathbf{H}$ from $\mathbf{C}$ and the octonions $\mathbf{O}$ from $\mathbf{H}$. This procedure leads to identities expressing the product of two sums of 2^n squares as another such sum. The familiar identity from high school

$$(a^2 + b^2)(c^2 + d^2) = (ac + bd)^2 + (ad - bc)^2$$

is the special case $n = 1$. Many interesting further developments of these ideas can be found in the book [83] by Conway and Smith.

2.7.1 Kervaire invariant

In 1960 Kervaire defined a geometric topological invariant of a framed differential manifold M of dimension $m = 4n + 2$ generalizing the Arf invariant for surfaces. Ten years earlier Pontryagin had used the Arf invariant of surfaces embedded in S^{k+2} with trivialized normal bundle to compute the homotopy groups $\pi_{k+2}(S^k)$ for $k > 1$. This group can be identified with the cobordism group of such surfaces. The cobordism of manifolds and the corresponding cobordism groups were defined by Thom in 1952. Algebraically the Arf invariant is defined for any quadratic form over $\mathbf{Z}_2$. Kervaire defined a quadratic form q on the homology group $H_{2n+1}(M; \mathbf{Z}_2)$ by using the framing and the Steenrod squares. The **Kervaire invariant** is the Arf invariant of q. A general reference for this section is Snaith [356]. Kervaire used his invariant to obtain the first example of a non-smoothable 10-dimensional PL-manifold. In the smooth category the first three examples of manifolds with Kervaire invariant 1 are $S^1 \times S^1$, $S^3 \times S^3$, and $S^7 \times S^7$. In these three cases the Kervaire invariant is related to the Hopf invariant of certain maps of spheres. No further examples of manifolds with Kervaire invariant 1 were known for many years. The problem of finding the dimensions of framed manifolds for which the Kervaire invariant is 1 came to be know as the **Kervaire invariant 1 problem**.

In 1969 Bill Browder proved that the Kervaire invariant is 0 for a manifold M if its dimension is different from $2^{j+1} - 2$, $j \in \mathbf{N}$. By 1984, it was known that there exist manifolds of dimensions 30 and 62 with the Kervaire invariant 1. Then on April 21, 2009, during the Atiyah 80 conference at Edinburgh, Mike Hopkins announced that he, Mike Hill, and Doug Ravenel had proved that there are no framed manifolds of dimension greater than 126 with Kervaire invariant 1. The case $n = 126$ was open as of January 2010. Mike Hopkins gave a very nice review of the problem and indicated key steps in the proof at the Strings, Fields and Topology workshop at Oberwolfach (June, 2009). This section is based in part on that review. The proof makes essential use of ideas from a generalized cohomology theory, called **topological modular forms**, or tmf theory. It was developed by Hopkins and collaborators. Witten has introduced a homomorphism from the string bordism ring to the ring of modular forms, called the Witten genus. This can be interpreted in terms of the theory of **topological modular forms** or **tmf**. We have already seen in Theorem 2.15, how the spheres S^1, S^3, S^7 enter from various perspectives in it. We have also discussed their relation to the Hopf fibration and to different physical theories. The relation of the other manifolds with other parts of mathematics and with physics is unclear at this time.

The discussion of homotopy and cohomology given in this chapter forms a small part of an area of mathematics called algebraic topology, where these and other related concepts are developed for general topological spaces. Standard references for this material and other topics in algebraic topology are Bott and Tu [55], Massey [279], and Spanier [357]. A very readable introduction is given in Croom [90].

Chapter 3
Manifolds

3.1 Introduction

The mathematical background required for the study of modern physical theories and, in particular, gauge theories, is rather extensive and may be divided roughly into the following parts: elements of differential geometry, fiber bundles and connections, and algebraic topology of a manifold. The first two of these parts are nowadays fairly standard background for research workers in mathematical physics. In any case, physicists are familiar with classical differential geometry, which forms the cornerstone of Einstein's theory of gravitation (the general theory of relativity). Therefore, in this chapter we give only a summary of some results from differential geometry to establish notation and make the monograph essentially self-contained. Fiber bundles and connections are discussed in Chapter 4. Characteristic classes, which are fundamental in the algebraic topology of a manifold, are discussed in detail in Chapter 5. The study of various classical field theories is taken up in Chapter 6. The reader familiar with the mathematical material may want to start with Chapter 6 and refer back to the earlier chapters as needed.

There are several standard references available for further study of the material on differential geometry that are discussed here in Chapters 3 and in Chapters 4 and 5; see, for example, Greub, Halperin, and Vanstone [162,163,164], Kobayashi and Nomizu [225,226], Lang [245], and Spivak [358]. Some basic references for topology and related geometry and analysis are Palais [310], Porteous [319], and Booss and Bleecker [47], For references that also discuss some physical applications see Abraham and Marsden [1], Abraham, Marsden, and Ratiu [2], Choquet-Bruhat, DeWitt-Morette, and Dillard-Bleick [79], Choquet-Bruhat and DeWitt-Morette [78], Curtis and Miller [93], Felsager [124], Jost [214], Marsden [278], Nakahara [295], Sachs and Wu [331], Scorpan [343], and Trautman [378].

K. Marathe, *Topics in Physical Mathematics*, DOI 10.1007/978-1-84882-939-8_3,
© Springer-Verlag London Limited 2010

3.2 Differential Manifolds

The basic objects of study in differential geometry are manifolds and maps
between manifolds. Roughly speaking a manifold is a topological space obtained by patching together open sets in a Banach space. For many physical applications this space may be taken to be finite-dimensional. However,
infinite-dimensional manifolds arise naturally in field theories and, hence, in
this chapter our discussion of manifolds applies to arbitrary manifolds unless
otherwise stated.

Definition 3.1 *Let M be a topological space and F a Banach space. A* **chart**
(U, ϕ) is a pair consisting of an open set $U \subset M$ and a homeomorphism

$$\phi : U \to \phi(U) \subset F,$$

where $\phi(U)$ is an open subset of F. M is called a **topological manifold** *or
simply a* **manifold** *modeled on the Banach space F if M admits a family
$\mathcal{A} = \{(U_i, \phi_i)\}_{i \in I}$ of charts such that $\{U_i\}_{i \in I}$ covers M. This family $\mathcal{A}$ of
charts is said to be an* **atlas** *for M. If $(U_i, \phi_i), (U_j, \phi_j)$ are two charts and
$U_{ij} := U_i \cap U_j \neq \emptyset$, then*

$$\phi_{ij} := \phi_i \circ \phi_j^{-1} : \phi_j(U_{ij}) \to \phi_i(U_{ij})$$

is a homeomorphism. The maps ϕ_{ij} are called **transition functions** *of the
atlas $\mathcal{A}$. Various smoothness requirements on $\mathcal{A}$ are obtained by using the
transition functions. For example, if the ϕ_{ij} are C^p-diffeomorphisms (i.e.,
ϕ_{ij} and ϕ_{ij}^{-1} are of class C^p), $0 < p \leq +\infty$, then $\mathcal{A}$ is called a* **differential**
(or a **differentiable***) atlas of class C^p. Two differentiable atlases are said to
be compatible if their union is a differentiable atlas. Let $\mathcal{A}$ be a differentiable
atlas of class C^p on M. The maximal differentiable atlas of class C^p containing
$\mathcal{A}$ is called the* **differential structure** *on M of class C^p determined by $\mathcal{A}$.
A differential structure of class C^∞ is also called a* **smooth structure**. *The
corresponding atlas is called a* **smooth atlas**. *A manifold M together with a
differential structure is called a* **differential** *(or a* **differentiable***) manifold.
By definition the* **dimension** *of M, $\dim M$, is the dimension of F. If F is
$\mathbf{R}^m$ then M is called a* **real manifold** *of dimension m. If F is $\mathbf{C}^m$ and the
transition functions are holomorphic (complex analytic), then M is called a*
complex manifold *of complex dimension m.*

One often considers a class of manifolds subjected to additional topological restrictions such as compactness, paracompactness etc.. For example, the
result that locally compact Hausdorff topological vector spaces are finite-
dimensional, implies that locally compact Hausdorff manifolds are finite-
dimensional. We remark that a **finite atlas** (i.e., an atlas that consists of
a finite family of charts) always exists on a compact manifold.

The charts allow us to give an intrinsic formulation of various structures
associated with manifolds.

Definition 3.2 *Let M, N be real differential manifolds and $f : M \to N$. We say that f is* **differentiable** *(or* **smooth***) if, for each pair of charts, $(U, \phi), (V, \psi)$ of M and N, respectively, such that $f(U) \subset V$, the* **representative** *$\psi \circ f \circ \phi^{-1}$ of f in these charts is differentiable (or smooth). The set of all smooth functions from M to N is denoted by $\mathcal{F}(M, N)$. When $N = \mathbf{R}$ we write $\mathcal{F}(M)$ instead of $\mathcal{F}(M, \mathbf{R})$. A bijective differentiable $f \in \mathcal{F}(M, N)$ is called a* **diffeomorphism** *if f^{-1} is differentiable. The set of all diffeomorphisms of M with itself under composition is a group denoted by $\mathrm{Diff}(M)$. Diffeomorphism is an equivalence relation.*

The class of differential manifolds and differentiable maps forms a category (see Appendix C) that we denote by *DIFF*. The class of complex manifolds and complex analytic maps forms a subcategory of *DIFF*. We discuss some important complex manifolds but their physical applications are not emphasized in this book. An excellent introduction to this area may be found in Manin [257] and Wells [399, 398]. The class of topological manifolds and continuous maps forms a category that we denote by *TOP*. An important problem in the topology of manifolds is that of **smoothability**, i.e., to find when a given topological manifold admits a **compatible differential structure**. We note that a differential structure on a topological manifold M is said to be compatible if it is contained in the maximal atlas of M. It is well known that a connected topological manifold of dim < 4 admits a unique compatible smooth structure. We observe that a topological manifold of dim > 3 may admit inequivalent compatible differential structures or none at all.

Definition 3.3 *Let M be a differential manifold with differential structure $\mathcal{A}$ and let $U \subset M$ be open. The collection of all charts of $\mathcal{A}$ whose domain is a subset of U is an atlas for U, which makes U into a differential manifold. This manifold is called an* **open submanifold** *of M. More generally, a subset $S \subset M$ is said to be a* **submanifold** *of M if, $\forall x \in S$, there exists a chart (U, ϕ) with $x \in U$ such that*

1. $\phi(U) \subset G \oplus H$, where G, H are Banach spaces;
2. $\phi(U \cap S) = \phi(U) \cap (G \times \{b\})$, for some $b \in H$.

Then, denoting by π_1 the projection onto the first factor, $(U \cap S, \pi_1 \circ \phi)$ is a chart at x. The collection of all these charts is an atlas on S, which determines a differential structure on S. With this differential structure, S itself (with the relative topology) is a differential manifold. We observe that an open submanifold is a special case of a submanifold.

Let $\mathbf{R}_+^n := \{x = (x_1, \ldots, x_n) \in \mathbf{R}^n \mid x_n \geq 0\}$. The set $\mathbf{R}_0^n := \{x \in \mathbf{R}^n \mid x_n = 0\}$ is called the **boundary** of $\mathbf{R}_+^n$. Let U be an open subset of $\mathbf{R}_+^n$ in the relative topology. We denote by $\mathrm{bd}(U)$ the set $\mathrm{bd}(U) := U \cap \mathbf{R}_0^n$ and call $\mathrm{bd}(U)$ the boundary of U. We denote by $\mathrm{Int}(U)$ the set $\mathrm{Int}(U) := U \setminus \mathrm{bd}(U)$ and call it the **interior** of U. If U and V are subsets of $\mathbf{R}_+^n$ and $f : U \to V$, we say that f is of class C^p, $1 \leq p \leq +\infty$ (smooth if $p = +\infty$), if there exist open neighborhoods U_1 of U and V_1 of V, and a map $f_1 : U_1 \to V_1$ of class C^p such that f coincides with f_1 on U. The derivative $Df_1(x)$, $x \in U$, is independent of the choice of U_1, V_1, f_1. Thus we may define $Df(x) := Df_1(x)$, $x \in U$. If f is a smooth isomorphism (i.e., diffeomorphism), then f induces a diffeomorphism of $\operatorname{Int} U$ onto $\operatorname{Int} V$ and of δU onto δV. If M is a topological space, a **chart with boundary** for M is a pair (U, ϕ) where U is an open set of M and $\phi : U \to \phi(U) \subset \mathbf{R}_+^n$ is a homeomorphism onto the open subset $\phi(U)$ of $\mathbf{R}_+^n$. With obvious changes with respect to the definition of differentiable manifold, we have the notions of an **atlas with boundary** and of an n-**manifold with boundary**. Thus, an n-manifold with boundary is obtained by piecing together open subsets of the upper half-space $\mathbf{R}_+^n$. This construction can be extended to the infinite-dimensional case. In this case in place of $\mathbf{R}^n$ we have a Banach space F and in place of $\mathbf{R}_+^n$ we have the half-space

$$F_\lambda^+ := \{x \in F \mid \lambda(x) \geq 0\},$$

where λ is a continuous, linear functional on F (for details see Lang [245]). The **boundary** of an n-manifold M with boundary denoted by ∂M, is the subset of the points $x \in M$ such that there exists a chart with boundary (U, ϕ) with $x \in U$ and $\phi(x) \in \mathbf{R}_0^n$. The interior of M is the set $\operatorname{Int} M := M \setminus \partial M$. A differentiable manifold with empty boundary is called a differentiable manifold without boundary and, in this case, we recover our previous definition of a differentiable manifold. The differentiable structure of an n-manifold with boundary M induces, in a natural way, a differentiable structure on ∂M and $\operatorname{Int} M$ with which ∂M and $\operatorname{Int} M$ are manifolds without boundary of dimensions $n-1$ and n, respectively. Thus, the boundary $\partial(\partial M)$ of the boundary ∂M, of a manifold with boundary M, is empty. Moreover, if $f : M \to N$ is a diffeomorphism of manifolds with boundary, then the maps $f_{\operatorname{Int}} : \operatorname{Int} M \to \operatorname{Int} N$ and $f_\partial : \partial M \to \partial N$ induced by restriction are diffeomorphisms. We now give several examples of manifolds that appear in many applications.

Example 3.1 *The sphere S^n is the subset of $\mathbf{R}^{n+1}$ defined by*

$$S^n := \{(x_1, x_2, \ldots, x_{n+1}) \mid x_1^2 + x_2^2 + \cdots + x_{n+1}^2 = 1\}.$$

Let us consider the map (stereographic projection)

$$\phi : S^n \setminus \{(0, \ldots, 0, 1)\} \to \mathbf{R}^n$$

defined by

$$\phi(x_1, x_2, \ldots, x_{n+1}) := (x_1/(1 - x_{n+1}), \ldots, x_n/(1 - x_{n+1}))$$

and the map

$$\psi : S^n \setminus \{(0, \ldots, 0, -1)\} \to \mathbf{R}^n$$

defined by

$$\psi(x_1, x_2, \ldots, x_{n+1}) := (x_1/(1 + x_{n+1}), \ldots, x_n/(1 + x_{n+1})).$$

Then, for $\psi \circ \phi^{-1} : \mathbf{R}^n \setminus \{(0, \ldots, 0)\} \rightarrow \mathbf{R}^n \setminus \{(0, \ldots, 0)\}$ one has

$$\psi \circ \phi^{-1}(y_1, \ldots, y_n) = (y_1/(y_1^2 + \cdots + y_n^2), \ldots, y_n/(y_1^2 + \cdots + y_n^2)),$$

$\forall (y_1, \ldots, y_n) \in \mathbf{R}^n \setminus \{(0, \ldots, 0)\}$. Thus, $\psi \circ \phi^{-1}$ is smooth and $\{\phi, \psi\}$ is a smooth atlas, which makes S^n into a differential manifold. Let

$$D^n := \{(x_1, x_2, \ldots, x_n) \mid x_1^2 + x_2^2 + \cdots + x_n^2 \leq 1\}$$

be the unit disk in $\mathbf{R}^n$. D^n is a manifold with boundary and $\partial D^n = S^{n-1}$ is a manifold without boundary of dimension $n - 1$.

Example 3.2 Let $f : \mathbf{R}^n \rightarrow \mathbf{R}$ be a smooth map and let $S := f^{-1}(\{0\})$. Let us suppose that $S \neq \emptyset$ and that $\forall x \in S$, the Jacobian matrix of f at x has rank 1. Then for each $x \in S$ there exists k, $1 \leq k \leq n$, such that

$$\frac{\partial f}{\partial x_k}(x) \neq 0,$$

and hence, by the implicit function theorem, we have a smooth bijection of a neighborhood of x in S onto a neighborhood of $(x_1, \ldots, \hat{x}_k, \ldots, x_n) \in \mathbf{R}^{n-1}$ (the sign $\hat{\ }$ over a symbol denotes deletion). This bijection is a chart at x, and the collection of all the charts of this type is an atlas, which makes S into a differential manifold. We observe that, for the sphere S^n, we have

$$S^n = f^{-1}(\{0\})$$

with $f : \mathbf{R}^{n+1} \rightarrow \mathbf{R}$ defined by $f(x_1, \ldots, x_{n+1}) = x_1^2 + \cdots + x_{n+1}^2 - 1$. One can verify that the differential structure on S^n defined by the procedure of this example coincides with the one defined in the previous example.

Manifolds of the type discussed in the above example arise in many applications.

Example 3.3 Consider the equivalence relation in $\mathbf{R}^{n+1} \setminus \{0\}$ defined as follows. We say that two points x, y

$$x = (x_1, x_2, \ldots, x_{n+1}), y = (y_1, y_2, \ldots, y_{n+1}) \in \mathbf{R}^{n+1} \setminus \{0\}$$

are equivalent if there exists $\lambda \in \mathbf{R} \setminus \{0\}$ such that $x_i = \lambda y_i$, $\forall i \in \{1, 2, \ldots, n + 1\}$. Let us denote by $[x]$ the equivalence class containing x and by $\mathbf{RP}^n$ the set

$$\mathbf{RP}^n := \{[x] \mid x \in \mathbf{R}^{n+1} \setminus \{0\}\}.$$

Let U_i, $i \in \{1, 2, \ldots, n + 1\}$, be the subset of $\mathbf{RP}^n$ defined by

$$U_i := \{[(x_1, x_2, \ldots, x_{n+1})] \in \mathbf{RP}^n \mid x_i \neq 0\}.$$

The map $\phi_i : U_i \to \mathbf{R}^n$ such that

$$\phi_i([(x_1, x_2, \ldots, x_{n+1})]) = (x_1/x_i, \ldots, x_{n+1}/x_i)$$

is a chart, and one can show that the set $\{\phi_1, \ldots, \phi_{n+1}\}$ is an atlas for $\mathbf{RP}^n$. The set $\mathbf{RP}^n$ with the differential structure determined by this atlas is a differential manifold also denoted by $\mathbf{RP}^n$ and called the n-dimensional **real projective space**. *We observe that $\mathbf{RP}^n$ is the manifold of 1-dimensional vector subspaces (i.e., lines through the origin) in $\mathbf{R}^{n+1}$. Analogously, one defines the n-dimensional* **complex projective space** $\mathbf{CP}^n$. *Thus, $\mathbf{CP}^n$ is the set of complex lines in $\mathbf{C}^{n+1}$, i.e.,*

$$\mathbf{CP}^n := \{[x] \mid x \in \mathbf{C}^{n+1} \setminus \{0\}\},$$

where $[x] = \{\lambda x \mid \lambda \in \mathbf{C} \setminus \{0\}\}$. A similar construction extends to the quaternionic case to define the **quaternionic projective space** $\mathbf{HP}^n$. *Generalizing the above construction of $\mathbf{RP}^n$ one can define the manifold of p-dimensional vector subspaces of $\mathbf{R}^{n+p}$. It is called the* **Grassmann manifold** *of p-planes in $\mathbf{R}^{n+p}$ and is denoted by $G_p(\mathbf{R}^{n+p})$. The construction of Grassmann manifolds can be extended to the complex and quaternionic cases.*

The construction of Grassmann manifolds given in the above example generalizes to the case of p-dimensional subspaces of an infinite-dimensional Banach space F and gives examples of infinite-dimensional differential manifolds.

Example 3.4 *Let M, N be differential manifolds and let $(U, \phi), (V, \psi)$ be any two charts of M and N with ranges in the Banach spaces F and G, respectively. Let us denote by $\phi \times \psi$ the map*

$$\phi \times \psi : U \times V \to \phi(U) \times \psi(V) \subset F \oplus G$$

defined by $(x, y) \mapsto (\phi(x), \psi(y))$. The pair $(U \times V, \phi \times \psi)$ is a chart of $M \times N$ and the collection of all the charts of this type is an atlas for $M \times N$, which determines a differential structure on $M \times N$. With this differential structure $M \times N$ is a differential manifold, which is called the **product manifold** *of M and N and is also denoted by $M \times N$. Analogously, one defines the product manifold $M_1 \times M_2 \times \cdots \times M_n$ of any finite number of differential manifolds. An example of a product manifold is the n-dimensional* **torus** T^n *defined by*

$$T^n := \underbrace{S^1 \times \cdots \times S^1}_{n \text{ times}}.$$

Example 3.5 *Let $M(m, n; \mathbf{R})$ denote the set of all real $m \times n$ matrices. The map*

$$\phi : M(m, n; \mathbf{R}) \to \mathbf{R}^{mn}$$

*defined by $\phi(A) = (a_{11}, a_{12}, \ldots, a_{mn})$, for $A = (a_{ij}) \in M(m, n; \mathbf{R})$, is a chart on $M(m, n; \mathbf{R})$ that induces a differential structure on $M(m, n; \mathbf{R})$. With this differential structure $M(m, n; \mathbf{R})$ is a differential manifold of dimension mn. Let $M(n, \mathbf{R}) := M(n, n; \mathbf{R})$ and let us denote by $S(n, \mathbf{R})$ (resp., $A(n, \mathbf{R})$) the subset of $M(n, \mathbf{R})$ of the **symmetric** (resp., **anti-symmetric**) $n \times n$ matrices. It is easy to show that $S(n, \mathbf{R})$ and $A(n, \mathbf{R})$ are submanifolds of $M(n, \mathbf{R})$ of dimension $n(n + 1)/2$ and $n(n - 1)/2$, respectively. Analogously, the set $M(m, n; \mathbf{C})$ of complex $m \times n$ matrices is a differential manifold of dimension 2mn. Let $M(n, \mathbf{C}) := M(n, n; \mathbf{C})$ and let us denote by $S(n, \mathbf{C})$ (resp., $A(n, \mathbf{C})$) the subset of $M(n, \mathbf{C})$ of the **Hermitian** (resp., **anti-Hermitian**) $n \times n$ matrices. It is easy to show that $S(n, \mathbf{C})$ and $A(n, \mathbf{C})$ are submanifolds of $M(n, \mathbf{C})$ of dimension n^2. The map*

$$\det : M(n, \mathbf{R}) \to \mathbf{R},$$

*which maps $A \in M(n, \mathbf{R})$ to the determinant $\det A$ of A, is smooth. Hence $\det^{-1}(\mathbf{R} \setminus \{0\})$ is an open submanifold of $M(n, \mathbf{R})$ denoted by $GL(n, \mathbf{R})$. The set $GL(n, \mathbf{R})$ with matrix multiplication is a group. It is called the **real general linear group**. One can show that matrix multiplication induces a smooth map of $GL(n, \mathbf{R}) \times GL(n, \mathbf{R})$ into $GL(n, \mathbf{R})$. Similarly, one can define the **complex general linear group** $GL(n, \mathbf{C})$. The real and complex general linear groups are examples of an important class of groups called Lie groups, which are discussed in Section 3.6.*

Let M be a differential manifold and $(U, \phi), (V, \psi)$ be two charts of M at $p \in M$. The triples $(\phi, p, u), (\psi, p, v)$, for $u, v \in F$, are said to be equivalent if

$$D(\psi \circ \phi^{-1})(\phi(p)) \cdot u = v$$

where D is the derivative operator in a Banach space. This is an equivalence relation between such triples. A **tangent vector** to M at p may be defined as an equivalence class $[\phi, p, u]$ of such triples. Alternatively, one can define a tangent vector to M at p to be an equivalence class of smooth curves on M passing through p and touching one another at p (see, for example, Abraham et al. [2]). The set of tangent vectors at p is denoted by T_pM; this set is a vector space isomorphic to F, called the **tangent space** to M at p. The set

$$TM = \bigcup_{p \in M} T_pM$$

can be given the structure of a smooth manifold. This manifold TM is called the **tangent space** to M. A tangent vector $[\phi, p, u]$ at p may be identified with the directional derivative u_p^ϕ, also denoted by u_p, defined by

$$u_p : \mathcal{F}(U) \to \mathbf{R},$$

such that

$$u_p(f) = D(f \circ \phi^{-1})(\phi(p)) \cdot u.$$

If $\dim M = m$ and $\phi : q \mapsto (x^1, x^2, \ldots, x^m)$ is a chart at p, then the tangent vectors to the coordinate curves at p are denoted by $\partial/\partial x^i|_p$ or simply by $\partial/\partial x^i$ or $\partial_i, i = 1, \ldots, m$. The set of vectors $\partial_i, i = 1, \ldots, m$, form a basis of the tangent space at $p \in U$. A smooth map

$$X : M \to TM$$

is called a **vector field** on M if $X(p) \in T_pM, \forall p \in M$. The set of all vector fields on M is denoted by $\mathcal{X}(M)$. A vector field $X \in \mathcal{X}(M)$ can be identified with a derivation of the algebra $\mathcal{F}(M)$, i.e., with the linear map $\mathcal{F}(M) \to \mathcal{F}(M)$ defined by $f \mapsto Xf$ where $(Xf)(p) = X(p)f$ satisfying the Leibnitz property $X(fg) = fXg + gXf$. On a coordinate chart, the vector field X has the local expression $X = X^i\partial_i$ and hence Xf has the local expression $X = X^i\partial_i f$. In the above local expressions we have used **Einstein's summation convention** (summation over repeated indices) and in what follows we shall continue to use this convention. If $X, Y \in \mathcal{X}(M)$ then, regarding them as derivations of the algebra $\mathcal{F}(M)$, we can define the **commutator** or **bracket** of X, Y by

$$[X, \ Y] := X \circ Y - Y \circ X.$$

It is easy to check that $[X, \ Y]$ is in $\mathcal{X}(M)$. The local expression for the bracket is given by

$$[X, \ Y] = \left[\frac{\partial Y^k}{\partial x^j}X^j - \frac{\partial X^k}{\partial x^j}Y^j \right] \partial_k,$$

where $X = X^i\partial_i$ and $Y = Y^i\partial_i$. We note that $\mathcal{X}(M)$ under pointwise vector space operations and the bilinear operation defined by the bracket has the structure of a Lie algebra as defined in Chapter 1; i.e., the **Lie bracket**, $(X, Y) \mapsto [X, Y]$ satisfies the following conditions:

1. $[X, Y] = -[Y, X]$ (anticommutativity),
2. $[X, [Y, Z]] + [Y, [Z, X]] + [Z, [X, Y]] = 0$ (Jacobi identity).

If V and W are Lie algebras, a linear map $f : V \to W$ is said to be a **Lie algebra homomorphism** if

$$f([X, Y]) = [f(X), f(Y)], \quad \forall X, Y \in V.$$

The Lie algebra $\mathcal{X}(M)$ is, in general, infinite-dimensional.

If $f \in \mathcal{F}(M, N)$, the **tangent of** f **at** p, denoted by T_pf or $f_*(p)$, is the map

$$T_pf : T_pM \to T_{f(p)}N$$

such that

$$T_pf(u_p) \cdot g = u_p(g \circ f).$$

The **tangent** of f, denoted by Tf, is the map of TM to TN whose restriction to T_pM is T_pf.

Let E be a Banach space and F a closed subspace of E. We say that F is **complemented** in E if there exists a closed subspace G of E such that

1. $E = F + G$, i.e., $\forall u \in E$, there exist $v \in F$, $w \in G$ such that $u = v + w$;
2. $F \cap G = \{0\}$.

The space G in the above definition is called a **complement** of F in E. We say that $F \subset E$ **splits** in E, when F is a closed complemented subspace of E. We observe that if E is a Hilbert space, then any closed subspace of E splits in E and that if E is finite-dimensional, then every subspace of E splits in E. Let M, N be differential manifolds and $f \in \mathcal{F}(M, N)$. We say that f is a **submersion at** $p \in M$ if T_pf is surjective and $\operatorname{Ker} T_pf$ splits in T_pM. Equivalently, f is a submersion at $p \in M$, if there exist charts $\phi : U \rightarrow \phi(U) = A_1 \times A_2 \subset F \oplus G$ and $\psi : V \rightarrow \psi(V) = A_1$ at p and $f(p)$, respectively, such that $f(U) \subset V$ and the representative $\psi \circ f \circ \phi^{-1}$ of f in these charts is the projection onto the first factor. If M and N are finite-dimensional, then f is a submersion at $p \in M$ if and only if T_pf is surjective, i.e., the rank of the linear map T_pf is equal to $\dim N$. We say that f is a **submersion** if it is a submersion at p, $\forall p \in M$. One can show that a submersion is an open map. A point $q \in N$ is said to be a **regular value** of f if, $\forall p \in f^{-1}(\{q\})$, f is a submersion at p. By definition, points of $N \setminus f(M)$ are regular. An important result is given by the following theorem.

Theorem 3.1 *Let M, N be differential manifolds and $q \in N$ a regular value of $f \in \mathcal{F}(M, N)$. Then $f^{-1}(\{q\})$ is an $(m - n)$-dimensional submanifold of M and, $\forall p \in f^{-1}(\{q\})$, $T_p(f^{-1}(\{q\})) = \operatorname{Ker} T_pf$.*

An equivalence relation ρ on a differential manifold M is said to be **regular** if, for some differential structure on the set of equivalence classes M/ρ, the canonical projection $\pi : M \rightarrow M/\rho$ is a submersion. Such a differential structure, if it exists, is unique and, in this case, M/ρ with this differential structure is called the **quotient manifold** of M by ρ. A useful result is given by the next theorem.

Theorem 3.2 *Let M, N be differential manifolds and let $f \in \mathcal{F}(M, N)$ be a submersion. Let us denote by ρ the equivalence relation on M defined by f, i.e., $x\rho y$ if $f(x) = f(y)$. Then ρ is regular and $f(M)$ is an open submanifold of N diffeomorphic to the quotient manifold M/ρ.*

Example 3.6 *Let ρ be the equivalence relation of Example 3.3. Then, as a set, $\mathbf{RP}^n = (\mathbf{R}^{n+1} \setminus \{0\})/\rho$. One can easily verify that, with respect to the differential structure defined on $\mathbf{RP}^n$ in Example 3.3, the canonical projection $\pi : \mathbf{R}^{n+1} \setminus \{0\} \rightarrow \mathbf{RP}^n$ is a submersion, and $\mathbf{RP}^n$ is the quotient manifold $(\mathbf{R}^{n+1} \setminus \{0\})/\rho$. Let μ_n be the restriction to S^n of the map π given above and let ρ_n be the equivalence relation induced by μ_n on S^n. Observe that ρ_n identifies antipodal points x and $-x$. By Theorem 3.2, ρ_n is regular and*

S^n/ρ_n is diffeomorphic to $\mathbf{RP}^n$. Analogously, if ρ_n' (resp., ρ_n'') is the equivalence relation on S^{2n+1} (resp., S^{4n+3}) defining $\mathbf{CP}^n$ (resp., $\mathbf{HP}^n$), then
ρ_n' identifies all points of the circle $S_x^1 = \{\lambda x \mid |\lambda| = 1, \lambda \in \mathbf{C}\}$ (resp., ρ_n''
identifies all points of the sphere $S_x^3 = \{\lambda x \mid |\lambda| = 1, \lambda \in \mathbf{H}\}$). Thus, $\mathbf{CP}^n$
is diffeomorphic to S^{2n+1}/ρ_n' and $\mathbf{HP}^n$ is diffeomorphic to S^{4n+3}/ρ_n''. From
this it follows that these projective spaces are connected and compact.

Let M, N be differential manifolds and $f \in \mathcal{F}(M, N)$. We say that f is
an **immersion at** $p \in M$ if $T_p f$ is injective and $\mathrm{Im}\, T_p f$ splits in $T_{f(p)} N$.
Equivalently, f is an immersion at $p \in M$, if there exist charts $\phi : U \rightarrow$
$\phi(U) = A_1 \subset F$ and $\psi : V \rightarrow \psi(V) = A_1 \times A_2 \subset F \oplus G$ at p and $f(p)$
respectively, such that $f(U) \subset V$ and the representative $\psi \circ f \circ \phi^{-1}$ of f in
these charts is the natural injection $A_1 \rightarrow A_1 \times A_2$. If M and N are finite
dimensional then f is an immersion at $p \in M$ if and only if $T_p f$ is injective
and this is true if and only if the rank of the linear map $T_p f$ is equal to
$\dim M$. We say that f is an **immersion** if it is an immersion at p, $\forall p \in M$.
An immersion f is, locally, a diffeomorphism onto a submanifold of N, but
f need not be injective. Even if f is an injective immersion, $f(M)$ need not
be a submanifold of N. An injective immersion f is called an **embedding** if
$f(M)$ is a submanifold of N.

3.3 Tensors and Differential Forms

In this section we define various tensor spaces and spaces of differential forms
on a manifold and discuss some important operators acting on these spaces.
Let V denote a vector space and let V^* be the dual of V. We denote by
$T_s^r(V)$ (see Appendix C) the tensor space of type (r, s) over V, i.e.,

$$T_s^r(V) = \underbrace{V \otimes \ldots \otimes V}_{r \text{ times}} \otimes \underbrace{V^* \otimes \ldots \otimes V^*}_{s \text{ times}}.$$

Replacing $T_p M$ with various tensor spaces over $T_p M$, a construction similar
to that of the tangent space, defines the tensor spaces on M. We define $T_s^r M$
by

$$T_s^r M = \bigcup_{p \in M} T_s^r(T_p M),$$

where

$$T_s^r(T_p M) = \underbrace{T_p M \otimes \ldots \otimes T_p M}_{r \text{ times}} \otimes \underbrace{(T_p M)^* \otimes \ldots \otimes (T_p M)^*}_{s \text{ times}}.$$

The set $T_s^r M$ can be given the structure of a manifold and is called the
tensor space of type (r, s) of contravariant degree r and covariant degree s.
A smooth map

$$t : M \to T_s^r M$$

is called a **tensor field of type** (r, s) on M if $t(p) \in T_s^r(T_pM), \forall p \in M$. We note that, if M is finite-dimensional, then $T_s^r(T_pM)$ may be identified with a space of multilinear maps as follows. The element

$$u_1 \otimes \ldots \otimes u_r \otimes \alpha^1 \otimes \ldots \otimes \alpha^s \in T_s^r(T_pM)$$

is identified with the multilinear map

$$\underbrace{(T_pM)^* \times \ldots \times (T_pM)^*}_{r \text{ times}} \times \underbrace{(T_pM) \times \ldots \times (T_pM)}_{s \text{ times}} \to \mathbf{R}$$

defined by

$$(\beta^1, \ldots, \beta^r, v_1, \ldots, v_s) \mapsto \beta^1(u_1) \ldots \beta^r(u_r)\alpha^1(v_1) \ldots \alpha^s(v_s).$$

This map is extended by linearity to all of $T_s^r(T_pM)$. We note that, if $\dim M = m$, then $\dim T_s^r M = m^{r+s}$. We observe that $T_0^1 M = TM$. The space $T_1^0 M$ is denoted by T^*M and is called the **cotangent space** of M. The space $(T_pM)^*$ (also denoted by T_p^*M) is called the cotangent space at $p \in M$. We define $T_0^0(T_pM) := \mathbf{R}$. Hence, $T_0^0 M := M \times \mathbf{R}$. Thus, a tensor field of type $(0, 0)$ can be identified with a function on M, i.e., an element of $\mathcal{F}(M)$. A tensor field of type $(r, 0)$ (resp., $(0, r)$) is also called a **contravariant** (resp., **covariant**) tensor field of degree r. A contravariant tensor field t of degree r is said to be **symmetric** if, $\forall p \in M$,

$$t(p)(\beta^1, \ldots, \beta^r) = t(p)(\beta^{\sigma(1)}, \ldots, \beta^{\sigma(r)}), \quad \forall \beta^i \in (T_pM)^*, \forall \sigma \in S_r,$$

where S_r is the symmetric group on r letters. A contravariant tensor field t of degree r is said to be **skew-symmetric** if

$$t(p)(\beta^1, \ldots, \beta^r) = \text{sign}(\sigma)t(p)(\beta^{\sigma(1)}, \ldots, \beta^{\sigma(r)}),$$

$\forall p \in M, \forall \beta^i \in (T_pM)^*, \forall \sigma \in S_r$. Similar definitions of symmetry and skew symmetry apply to covariant tensor fields.

We define $A^0 M := M \times \mathbf{R}$ and denote by $A^k M$, $k \geq 1$, the manifold

$$A^k M = \bigcup_{p \in M} A^k(T_pM),$$

where

$$A^k(T_pM) = \underbrace{T_p^*M \wedge \ldots \wedge T_p^*M}_{k \text{ times}}$$

is the vector space of exterior k-forms on T_p^*M. The manifold $A^k M$ is called the **manifold of exterior k-forms** on M. We note that $A^1 M = T^*M$. A smooth map

$$\alpha : M \to A^k M$$

is called a k-form on M if $\alpha(p) \in A^k(T_p M)$, $\forall p \in M$. The space of k-forms on M is denoted by $\Lambda^k(M)$. We note, in particular, that $\Lambda^0(M)$ can be identified with $\mathcal{F}(M)$. We define the (graded) **exterior algebra** on M to be the (graded) vector space $\Lambda(M)$, defined by

$$\Lambda(M) = \bigoplus_{k=0}^{+\infty} \Lambda^k(M),$$

with the exterior product $\wedge$ as multiplication.

If $\dim M = m$ and $\{e_1, \ldots, e_m\}$ is a basis for $T_p^* M$, then

$$\{e_{i_1} \wedge \cdots \wedge e_{i_k}\}_{1 \le i_1 < \cdots < i_k \le m}$$

is a basis for $A^k(T_p M)$. Thus

$$\dim A^k M = \binom{m}{k}.$$

We note that $\Lambda^k(M) = \{0\}$ for $k > m$. An m-form ν is called a **volume form** or simply **volume** on M if, $\forall p \in M$, $\nu(p) \ne 0$. The manifold M is said to be **orientable** if it admits a volume. Two volumes ν, ω on M are said to be **equivalent** if $\omega = f\nu$ for some $f \in \mathcal{F}(M)$ such that $f(p) > 0$, $\forall p \in M$. An **orientation** of an orientable manifold M is an equivalence class $[\nu]$ of volumes on M. A pair $(M, [\nu])$, where $[\nu]$ is an orientation of the manifold M, is called an **oriented manifold**. Given an oriented manifold $(M, [\nu])$, a chart $\phi : q \mapsto (x_1(q), x_2(q), \ldots, x_m(q))$ on $U \subset M$ is said to be **positively oriented** if $dx_1 \wedge \cdots \wedge dx_m \in [\nu_{|U}]$. Let $(M, [\nu])$ be an oriented manifold with boundary and let $\mathcal{A} = \{\phi_i : U_i \to \phi_i(U_i) \subset \mathbf{R}_+^m \mid i \in I\}$ be an atlas on M of positively oriented charts. The orientation on δM induced by the forms $(-1)^m dx_1 \wedge \cdots \wedge dx_{m-1}$ relative to the charts of $\mathcal{A}$ is called the **induced orientation** on δM.

Let $f \in \mathcal{F}(M, N)$; then f induces the map

$$f^* : \Lambda(N) \to \Lambda(M),$$

called the **pull-back** map, defined as follows. If $\alpha \in \Lambda^0(N) = \mathcal{F}(N)$ then we define $f^*\alpha := \alpha \circ f \in \Lambda^0(M) = \mathcal{F}(M)$. If $\alpha \in \Lambda^k(N)$, $k \ge 1$, then $f^*\alpha \in \Lambda^k(M)$ is defined by

$$(f^*\alpha)(p)(u_1, \ldots, u_k) := \alpha(f(p))(T_p f(u_1), \ldots, T_p f(u_k)),$$

$\forall u_1, \ldots, u_k \in T_p M$. If $f : M \to N$ is a diffeomorphism and $X \in \mathcal{X}(N)$, the element $f^* X$ of $\mathcal{X}(M)$ defined by

$$f^* X = T f^{-1} \circ X \circ f$$

is called the pull-back of X by f. Thus, when f is a diffeomorphism one can extend the definition of the pull-back map f^* to the tensor fields of type (r, s). In particular if $X_1, \ldots, X_r \in \mathcal{X}(N)$ and $\alpha^1, \ldots, \alpha^s \in \Lambda^1(N)$, we have

$$f^*(X_1 \otimes \ldots X_r \otimes \alpha^1 \otimes \ldots \alpha^s) = f^* X_1 \otimes \ldots \otimes f^* X_r \otimes f^* \alpha^1 \otimes \ldots \otimes f^* \alpha^s.$$

Given $X \in \mathcal{X}(M)$ and $p \in M$, an **integral curve** of X through p is a smooth curve $c : I \to M$, where I is an open interval around $0 \in \mathbf{R}$, $c(0) = p$, and

$$\dot{c}(t) := Tc(t, 1) = X(c(t)), \qquad \forall t \in I.$$

A **local flow** of X at $p \in M$ is a map

$$F : I \times U \to M,$$

where U is an open neighborhood of p such that, $\forall q \in U$, the map $F_q : I \to M$ defined by

$$F_q : t \mapsto F(t, q)$$

is an integral curve of X through q. One can show that, $\forall X \in \mathcal{X}(M)$ and $\forall p \in M$, a local flow $F : I \times U \to M$ of X at p exists and the map F_t defined by

$$F_t(q) = F(t, q), \quad \forall q \in U$$

is a diffeomorphism of U onto some open subset U_t of M.

Definition 3.4 *Let $X \in \mathcal{X}(M)$ and let η be a tensor field of type (r, s) on M. The **Lie derivative** $L_X \eta$ of η with respect to X is the tensor field of type (r, s) defined by*

$$(L_X \eta)(p) = \frac{d}{dt}[(F_t^* \eta)(p)]|_{t=0}$$

$\forall p \in M$, where $F : I \times U \to M$ is a local flow of X at p. The above definition also applies to differential forms; then $\eta \in \Lambda^k(M)$ implies $L_X \eta \in \Lambda^k(M)$.

It can be shown that the definition of Lie derivative given above is independent of the choice of a local flow. We now give local expressions for the Lie derivative. For this it is enough to give such an expression on an open subset U of the Banach space F. These expressions will give the representative of the Lie derivative in every chart on M. We shall follow this practice whenever we give local expressions. We consider only the finite-dimensional case, i.e., $F = \mathbf{R}^n$; the infinite-dimensional expressions are analogous. Let $x = (x_1, x_2, \ldots, x_n) \in U$ and $X = X^k \partial_k$ (summation over repeated indices is understood). For $\eta = f \in \mathcal{F}(U)$, from the definition of Lie derivative it follows that

$$L_X f = \frac{\partial f}{\partial x^k} X^k. \tag{3.1}$$

For $\eta = Y \in \mathcal{X}(U)$ the expression of the Lie derivative $L_X Y$ is

$$L_X Y = \left[\frac{\partial Y^k}{\partial x^j} X^j - \frac{\partial X^k}{\partial x^j} Y^j \right] \partial_k, \qquad (3.2)$$

where $B = \{\partial_k := \partial/\partial x^k \mid k = 1, 2, \ldots, n\}$ is the natural basis of the tangent space at each point of U. From the definition of exterior differential operator given below, it follows that the dual base of B is $B' = \{dx^k \mid k = 1, 2, \ldots, n\}$. Then, for $\eta = \alpha = \alpha_k dx^k \in \Lambda^1(U)$ the expression of the Lie derivative $L_X \alpha$ becomes

$$L_X \alpha = \left[\frac{\partial \alpha_k}{\partial x^j} X^j + \frac{\partial X^j}{\partial x^k} \alpha_j \right] dx^k. \qquad (3.3)$$

The Lie derivative is a tensor derivation, i.e., it is linear and satisfies the Leibnitz product rule

$$L_X(\eta_1 \otimes \eta_2) = (L_X \eta_1) \otimes \eta_2 + \eta_1 \otimes L_X \eta_2, \qquad (3.4)$$

where η_1, η_2 are tensor fields. This fact and the local expressions given above allow us to write down the local expression for the Lie derivative of any tensor field of type (r, s).

Definition 3.5 *The **exterior differential operator** d of degree 1 on $\Lambda(M)$ is the map $d : \Lambda(M) \to \Lambda(M)$ defined as follows: If $f \in \Lambda^0(M)$, then $df \in \Lambda^1(M)$ is defined by*

$$df \cdot X = Xf.$$

If $\omega \in \Lambda^k(M)$, $k > 0$, then $d\omega$ is the $(k+1)$-form defined by

$$d\omega(X_0, X_1, \ldots, X_k) = \sum_{i=0}^{k} (-1)^i X_i(\omega(X_0, X_1, \ldots, \hat{X}_i, \ldots, X_k))$$

$$+ \sum_{0 \le i < j \le k} (-1)^{i+j} \omega(L_{X_i} X_j, X_0, \ldots, \hat{X}_i, \ldots, \hat{X}_j, \ldots, X_k), \qquad (3.5)$$

where $\hat{X}_h$ denotes suppression of X_h.

This definition implies that

$$d(\alpha \wedge \beta) = (d\alpha) \wedge \beta + (-1)^k \alpha \wedge d\beta, \qquad (3.6)$$

where $\alpha \in \Lambda^k(M)$, $\beta \in \Lambda(M)$, and

$$d^2 := d \circ d = 0. \qquad (3.7)$$

From the definition, it also follows that, if

$$\alpha = \sum_{i_1 < \cdots < i_k} \alpha_{i_1, \ldots, i_k} dx^{i_1} \wedge \cdots \wedge dx^{i_k}$$

is a local expression of the k-form α on a subset U of M, then the (local) expression of $d\alpha$ is given by

$$d\alpha = \sum_{j=1}^{n} \sum_{i_1 < \cdots < i_k} (\partial_j \alpha_{i_1,\ldots,i_k}) dx^j \wedge dx^{i_1} \wedge \cdots \wedge dx^{i_k}.$$

If $\alpha \in \Lambda(M)$, then α is said to be **closed** (resp., **exact**) if $d\alpha = 0$ (resp., $\alpha = d\beta$ for some $\beta \in \Lambda(M)$). As a consequence of equation (3.7) every exact form is closed. The converse of this statement is, in general, valid only locally, i.e., if $\alpha \in \Lambda(M)$ is closed then, $\forall p \in M$, there exists a neighborhood U of p such that $\alpha_{|U}$ is exact; i.e., there exists $\beta \in \Lambda(U)$ such that $\alpha = d\beta$. This statement is called the **Poincaré lemma**. In the physical literature β is referred to as a **potential** for α and one says that α is **derived from a potential**.

Definition 3.6 *Let $X \in \mathcal{X}(M)$ and $\alpha \in \Lambda(M)$. The (left)* **inner multiplication** $i_X \alpha$ *of α by X is defined as follows. If $\alpha \in \Lambda^0(M) = \mathcal{F}(M)$, then we define $i_X \alpha = 0$. If $\alpha \in \Lambda^k(M)$, $k \geq 1$, then $i_X \alpha \in \Lambda^{k-1}(M)$ is defined by*

$$i_X \alpha(X_1, \ldots, X_{k-1}) = \alpha(X, X_1, \ldots, X_{k-1}).$$

In the following theorem we collect some important properties of the operators L_X, d, and i_X.

Theorem 3.3 *Let M, N be differential manifolds, then we have*

1. *$d(f^*\alpha) = f^*(d\alpha)$, $f \in \mathcal{F}(M, N)$, $\alpha \in \Lambda(N)$.*
2. *$[X, Y] = L_X Y$, $X, Y \in \mathcal{X}(M)$.*
3. *$L_X = i_X \circ d + d \circ i_X$ on $\Lambda(M)$.*
4. *$d \circ L_X = L_X \circ d$ on $\Lambda(M)$.*
5. *$i_{[X,Y]} = L_X \circ i_Y - i_Y \circ L_X$ on $\Lambda(M)$.*

Let $(M, [\nu])$ be an oriented paracompact m-manifold. Let ω be an m-form on M with compact support $\operatorname{supp}\omega$. We now define the integral of ω with respect to the volume ν, denoted by $\int_M \omega d\nu$, or simply by $\int \omega$. First, let us suppose that there exists a positively oriented chart (U, ϕ) with $\operatorname{supp}\omega \subset U$. By the change of variables rule for integrals, we can show that the integral

$$\int \phi_* \omega \, dx^1 \ldots dx^m$$

does not depend on the choice of a positively oriented chart (U, ϕ) satisfying the condition that $\operatorname{supp}\omega \subset U$. Thus, in this case we can define

$$\int \omega := \int \phi_* \omega \, dx^1 \ldots dx^m.$$

For an arbitrary m-form ω with compact support, we choose an atlas $\mathcal{A} = \{(U_i, \phi_i) \mid i \in I\}$ of positively oriented charts and a smooth partition of unity $\mathcal{F} = \{f_i \mid i \in I\}$ subordinate to $\{U_i \mid i \in I\}$. We observe that the above definition of integral applies to $\int f_i\omega$, $\forall i \in I$. Furthermore, one can show that the number $\sum_{i \in I} \int f_i\omega$ does not depend on the choices made for $\mathcal{A}$ and $\mathcal{F}$. Thus, we define the **integral** of ω on M by

$$\int \omega = \sum_{i \in I} \int f_i\omega.$$

Using this definition of integration we now state the modern version of the classical theorem of Stokes.

Theorem 3.4 (Stokes' theorem) *If $(M, [\nu])$ is an oriented paracompact m-dimensional manifold with boundary ∂M and ω is an $(m-1)$-form on M with compact support, then*

$$\int_M d\omega = \int_{\partial M} i^*\omega,$$

where $i : \partial M \to M$ is the canonical injection.

One often writes simply $\int_{\partial M} \omega$ instead of $\int_{\partial M} i^*\omega$. The classical theorems of Green (relating the line and surface integrals) and Gauss (relating the surface and volume integrals) are special cases of the version of Stokes' theorem stated above. We observe that the definition of integral given above can be extended to forms with compact support that are only continuous (not necessarily smooth). Then, if f is a continuous function on M with compact support and ω is a volume on M, we may define the integral of f with respect to ω as $\int f\omega$. It can be shown that in a suitable topology this defines a continuous functional on the space of continuous functions on M with compact support. By the Riesz representation theorem, there exists a measure μ_ω on the σ-algebra of Borel sets in M such that

$$\int f\omega = \int f d\mu_\omega.$$

3.4 Pseudo-Riemannian Manifolds

In this section we take M to be a finite-dimensional manifold of dimension m. Let g be a tensor field of type $(0, 2)$; we say that g is **non-degenerate** if, for each $p \in M$, $g(p)$ is non-degenerate, i.e.

$$g(p)(u, v) = 0, \quad \forall v \in T_p M \Rightarrow u = 0.$$

A tensor field $g \in T_2^0 M$ which is symmetric and non-degenerate is called a **pseudo-metric** on M. Each $g(p)$ then defines an inner product on $T_p M$ of **signature** (r, s). The numbers r, s are invariants of the pseudo-metric.

The number r (resp., s) is the dimension of the positive eigenspace of $g(p)$ (resp., the negative eigenspace of $g(p)$) for each $p \in M$. The dimension of the negative eigenspace of g is called the **index** of g. It is denoted by i_g, i.e., the index $i_g = s$, where $r + s = m = \dim M$. If g is a pseudo-metric on M of index s then we say that (M, g) is a **pseudo-Riemannian** manifold of index s. In local orthonormal coordinates the metric is usually expressed as

$$ds^2 = \sum_{i=1}^{r} dx_i^2 - \sum_{i=1}^{s} dx_{r+i}^2.$$

If $s = 0$, i.e., $g(p)$ is positive definite $\forall p \in M$, then we say that (M, g) is a **Riemannian** manifold. If $s = 1$, i.e., $\forall p \in M$ the signature of $g(p)$ is $(m - 1, 1)$, then we say that (M, g) is a **Lorentz** manifold. We note that a pseudo-metric g induces an inner product on all tensor spaces, which we also denote by g. In particular, if U is an open subset of M and $\alpha = \sum_{i_1 < \cdots < i_k} \alpha_{i_1, \ldots, i_k} dx^{i_1} \wedge \cdots \wedge dx^{i_k}$, $\beta = \sum_{i_1 < \cdots < i_k} \beta_{i_1, \ldots, i_k} dx^{i_1} \wedge \cdots \wedge dx^{i_k}$ are k-forms on U, the expression of $g(\alpha, \beta)$ is the following: Let g^{-1} be the inverse matrix of $g = \{g_{ij}\}$ where $g_{ij} = g(\partial_i, \partial_j)$ and let $g^{i_1, \ldots, i_k; j_1, \ldots, j_k}$ denote the determinant of the $k \times k$-matrix obtained taking the rows $i_1, \ldots, i_k$ and the columns $j_1, \ldots, j_k$ of g^{-1}. Then

$$g(\alpha, \beta) = \sum_{i_1 < \cdots < i_k} \alpha_{i_1, \ldots, i_k} \beta^{i_1, \ldots, i_k},$$

where

$$\beta^{i_1, \ldots, i_k} = \sum_{j_1 < \cdots < j_k} g^{i_1, \ldots, i_k; j_1, \ldots, j_k} \beta_{j_1, \ldots, j_k}.$$

Local expressions in pseudo-Riemmanian geometry simplify greatly when they are referred to a coordinate chart with respect to which the matrix of g is diagonal. Such coordinate charts always exist at each point of M and the corresponding coordinates are called **orthogonal coordinates**.

There are several important differences in both the local and global properties of Riemannian and pseudo-Riemannian manifolds (see, for example, Beem and Ehrlich [34] and O'Neill [308]). Until recently, most physical applications involved pseudo-Riemannian (in particular, Lorentz) manifolds. However, the discovery of instantons and their possible role in quantum field theory and subsequent development of the so-called **Euclidean gauge theories** has led to extensive use of Riemannian geometry in physical applications. It is these Euclidean gauge theories that have been most useful in applications to geometric topology of low-dimensional manifolds.

If (M, g) is an oriented, pseudo-Riemannian manifold with orientation $[\nu]$, then we define the **metric volume form** μ by $\mu = \nu/|g(\nu, \nu)|^{1/2}$. On an oriented pseudo-Riemannian manifold (M, g) with metric volume μ, we define the **Hodge star operator**

$$* : \Lambda(M) \to \Lambda(M)$$

as follows. For $\beta \in \Lambda^k(M)$, $0 \le k \le m$, $*\beta \in \Lambda^{m-k}(M)$ is the unique form such that

$$\alpha \wedge *\beta = g(\alpha, \beta)\mu, \quad \forall \alpha \in \Lambda^k(M). \tag{3.8}$$

One can show that

$$* *\alpha = (-1)^{i_g + k(m-k)}\alpha , \quad \forall \alpha \in \Lambda^k(M), \tag{3.9}$$

where i_g is the index of the manifold M. In particular, we have

$$*\mu = (-1)^{i_g}, \quad *1 = \mu.$$

The star operator is linear. Thus, for the local expression, it is enough to give the expression for $*(dx^{i_1} \wedge \cdots \wedge dx^{i_k})$. Taking into account that locally

$$\mu = |\det g|^{1/2} dx^1 \wedge \cdots \wedge dx^n,$$

we have

$$*(dx^{i_1} \wedge \cdots \wedge dx^{i_k}) = |\det g|^{1/2} \sum_{j_1 < \cdots < j_k} (-1)^{j_1 + \cdots + j_k + k(k+1)/2}.$$

$$g^{j_1, \ldots, j_k; i_1, \ldots, i_k} dx^1 \wedge \cdots \wedge \widehat{dx^{j_1}} \wedge \cdots \wedge \widehat{dx^{j_k}} \wedge \cdots \wedge dx^n,$$

where $\widehat{dx^{j_r}}$ denotes suppression of dx^{j_r}.

Definition 3.7 *Let (M, g) be an m-dimensional oriented pseudo-Riemannian manifold of index i_g. The **codifferential** δ is the linear map $\delta : \Lambda(M) \to \Lambda(M)$ of degree -1, which is defined on $\Lambda^k(M)$ by*

$$\delta := (-1)^{i_g + mk + m + 1} * d*, \tag{3.10}$$

where $*$ is the Hodge star operator.

We observe that if $f \in \Lambda^0(M) = \mathcal{F}(M)$, then $\delta f = 0$. Furthermore, as a consequence of equations (3.7) and (3.9) we have

$$\delta^2 := \delta \circ \delta = 0. \tag{3.11}$$

If

$$\alpha = \sum_{i_1 < \cdots < i_k} \alpha_{i_1, \ldots, i_k} dx^{i_1} \wedge \cdots \wedge dx^{i_k}$$

is a k-form on a subset U of M, expressed in terms of orthogonal coordinates (x^i), then the expression for $\delta\alpha$ is given by

$$\delta\alpha = -|\det g|^{-1/2} \sum_{i_1 < \cdots < i_{k-1}} g_{i_1 i_1} \cdots g_{i_{k-1} i_{k-1}}$$

$$\cdot \partial_j(\alpha^{j,i_1,\ldots,i_{k-1}} |\det g|^{1/2})dx^{i_1} \wedge \cdots \wedge dx^{i_{k-1}}.$$

If $\alpha \in \Lambda(M)$, then α is said to be **coclosed** (resp., **coexact**) if $\delta\alpha = 0$ (resp. $\alpha = \delta\beta$ for some $\beta \in \Lambda(M)$). As a consequence of equation (3.11) every coexact form is coclosed. The converse of this statement is, in general, true only locally.

The exterior differential and codifferential operators on a pseudo-Riemannian manifold (M,g) are closely related to the classical differential operators gradient, curl, divergence, and Laplacian, as we now explain. Let V be a finite-dimensional, real vector space and $f : V \times V \to \mathbf{R}$ a bilinear map. Then f induces a linear map $f^\flat : V \to V^*$ defined by

$$f^\flat(u)(v) = f(u,v), \ \forall u,v \in V.$$

We note that f is non-degenerate if and only if $f^\flat$ is an isomorphism. In this case, we define $f^\sharp : V^* \to V$ by $f^\sharp := (f^\flat)^{-1}$. Applying this pointwise to the pseudo-Riemannian metric g we obtain the maps

$$g^\flat : \mathcal{X}(M) \to \Lambda^1(M), \quad g^\sharp : \Lambda^1(M) \to \mathcal{X}(M).$$

The map $g^\flat$ ($g^\sharp$) is said to **lower** (**raise**) indices. Locally, we have

$$g^\flat(X^i\partial_i) = X_j dx^j, \quad \text{where } X_j = X^i g_{ij},$$

and

$$g^\sharp(\alpha_j dx^j) = \alpha^i\partial_i, \quad \text{where } \alpha^i = \alpha_j g^{ij}.$$

The operation of lowering or raising indices has an obvious extension to tensors of arbitrary type. The **gradient** operator in a pseudo-Riemannian manifold (M,g) is the map grad $: \mathcal{F}(M) \to \mathcal{X}(M)$ defined by

$$\text{grad} := g^\sharp \circ d.$$

If f is a function on an open subset U of M, we have

$$\text{grad} f = g^{ij}\frac{\partial f}{\partial x^i}\partial_j.$$

The **divergence** operator in an oriented pseudo-Riemannian manifold (M,g) is the map div $: \mathcal{X}(M) \to \mathcal{F}(M)$ defined by

$$\text{div} := -\delta \circ g^\flat.$$

If X is a vector field on an open subset U of M, we have

$$\text{div } X = |\det g|^{-1/2}\partial_i(|\det g|^{1/2}X^i).$$

The **Hodge–de Rham** operator Δ (also called the **Laplacian**) in an oriented pseudo-Riemannian manifold (M, g) is the map $\Delta : \Lambda(M) \to \Lambda(M)$ defined by

$$\Delta := d\delta + \delta d. \tag{3.12}$$

The operator Δ is at the basis of Hodge–de Rham theory of harmonic forms. The classical **Laplace–Beltrami** operator or the Laplacian defined as $\mathrm{div} \circ \mathrm{grad}$ equals $-\Delta$ on functions. The **curl** operator in an oriented pseudo-Riemannian manifold (M, g) of dimension 3 is the map $\mathrm{curl} : \mathcal{X}(M) \to \mathcal{X}(M)$

$$\mathrm{curl} := g^{\sharp} \circ * \circ d \circ g^{\flat}.$$

If X is a vector field on an open subset U of M, we have

$$\begin{aligned}
\mathrm{curl}\, X = (-1)^{i_g} |\det g|^{-1/2} \{ &[\delta_2(g_{33}X^3) - \delta_3(g_{22}X^2)]\delta_1 \\
+ &[\delta_3(g_{11}X^1) - \delta_1(g_{33}X^3)]\delta_2 + [\delta_1(g_{22}X^2) - \delta_2(g_{11}X^1)]\delta_3 \}.
\end{aligned}$$

We observe that the well known classical results

$$A = \mathrm{grad}\, \phi \Rightarrow \mathrm{curl}\, A = 0 \quad \text{and} \quad B = \mathrm{curl}\, A \Rightarrow \mathrm{div}\, B = 0,$$

where ϕ is a function and A, B are vector fields on $\mathbf{R}^3$ are consequences of the relation $d^2 = 0$. Note that, in view of the Poincaré lemma, the converse implications frequently used in physical theories to define the scalar and vector potentials are valid only locally.

3.5 Symplectic Manifolds

Let M be an m-dimensional manifold and let $\omega \in \Lambda^2(M)$. We say that ω is non-degenerate if, $\forall p \in M$,

$$\omega(p)(u, v) = 0, \quad \forall v \in T_p M \quad \Rightarrow \quad u = 0. \tag{3.13}$$

If $\omega_{ij}(p)$ are the components of $\omega(p)$ in a local coordinate system at p, then condition (3.13) is equivalent to

$$\det \omega_{ij}(p) \neq 0, \qquad \forall p \in M. \tag{3.14}$$

Condition (3.14) together with the skew symmetry of ω implies that the dimension m must be even; i.e., $m = 2n$. Then the condition (3.13) is equivalent to the condition that $\omega^n := \omega \wedge \omega \wedge \cdots \wedge \omega$ be a volume form on M, i.e.,

$$\omega^n(p) \neq 0, \qquad \forall p \in M.$$

Recall that any 2-form α can be regarded as a bilinear map of T_pM and hence induces a linear map

$$\alpha^\flat(p) : T_pM \to T_p^*M$$

as follows:

$$\alpha^\flat(p)(u)(v) = \alpha(p)(u,v),$$

where $u, v \in T_pM$. The non-degeneracy of $\omega \in \Lambda^2(M)$ defined above is then equivalent to $\omega^\flat$ being an isomorphism. Its inverse is then denoted by $\omega^\sharp$. Thus, a non-degenerate 2-form sets up an isomorphism between vector fields and 1-forms. If $X \in \mathcal{X}(M)$ then we have $\omega^\flat(X) = i_X\omega$. Let α, β be 1-forms; the **bracket** of α and β is the 1-form

$$[\alpha, \beta] = \omega^\flat([\omega^\sharp(\alpha), \omega^\sharp(\beta)]).$$

We note that this form is well defined for any non-degenerate 2-form ω.

Definition 3.8 *A* **symplectic structure** *on a manifold M is a 2-form ω that is non-degenerate and closed. A* **symplectic manifold** *is a pair (M, ω), where ω is a symplectic structure on the manifold M.*

Example 3.7 *Let Q be an n-dimensional manifold. Let $P = T^*Q$ be the cotangent space of Q; then P carries a natural symplectic structure ω defined as follows: Let θ be the 1-form on P defined by*

$$\theta(\alpha_p)(X) = \alpha_p(\psi_*(X)), \qquad \forall \alpha_p \in T^*Q, \ X \in T_{\alpha_p}P,$$

where ψ is the canonical projection of $P = T^*Q$ to Q. We define $\omega = -d\theta$. The form θ is called the **canonical 1-form** *and ω the* **canonical symplectic structure** *on T^*Q. By definition, ω is exact and hence closed. Its non-degeneracy follows from a local expression for ω in a special coordinate system, called a* **canonical coordinate system**, *defined as follows: Let $\{q^i\}$ be local coordinates at $p \in Q$. Then $\alpha_p \in P$ can be expressed as $\alpha_p = p_idq^i$. We take $Q^i = q^i \circ \psi, P_i = p_i \circ \psi$ as the canonical coordinates of $\alpha_p \in P$. Using these coordinates, we can express the canonical 1-form θ as*

$$\theta = P_idQ^i.$$

It is customary to denote the canonical coordinates on P by the same letters q^i, p_i and from now on we follow this usage. The canonical symplectic structure ω is given by

$$\omega = -d(p_idq^i) = dq^i \wedge dp_i.$$

From this expression it follows that

$$\omega^n = dq^1 \wedge dp_1 \wedge \cdots \wedge dq^n \wedge dp_n \neq 0.$$

Further, the components of ω in this coordinate system are given by the matrix

$$(\omega_{ij}) = \begin{pmatrix} 0 & I \\ -I & 0 \end{pmatrix},$$

where I (resp., 0) denotes the $n \times n$ unit (resp., zero) matrix.

The above example is of fundamental importance in the theory of symplectic manifolds in view of the following theorem, which asserts that, at least locally, every symplectic manifold looks like T^*Q.

Theorem 3.5 (Darboux) Let ω be a non-degenerate 2-form on a $2n$-manifold M. Then ω is symplectic if and only if each $p \in M$ has a local coordinate neighborhood U with coordinates $(q^1, \ldots, q^n, p_1, \ldots, p_n)$ such that

$$\omega_{|U} = dq^i \wedge dp_i.$$

Example 3.7 is also associated with the geometrical formulation of classical Hamiltonian mechanics, where Q is the configuration space of the mechanical system and P is the corresponding phase space. We now explain this formulation.

If (M, ω) is a symplectic manifold, then the charts guaranteed by Darboux's theorem are called **symplectic charts** and the corresponding coordinates (q^i, p_i) are called **canonical coordinates**. If $M = T^*Q$, and ω is the canonical symplectic structure on it, then, in the physical literature, the q^i are called the **canonical coordinates** and the p_i the corresponding **conjugate momenta**. This terminology arises from the formulation of classical mechanics on T^*Q. We now indicate briefly the relation of the classical Hamilton's equations with symplectic manifolds.

Let (M, ω) be a symplectic manifold. A vector field $X \in \mathcal{X}(M)$ is called **Hamiltonian** (resp., **locally Hamiltonian**) if $\omega^\flat(X)$ is exact (resp., closed). The set of all Hamiltonian (resp., locally Hamiltonian) vector fields is denoted by $\mathcal{HX}(M)$ (resp. $\mathcal{LHX}(M)$). If $X \in \mathcal{HX}(M)$, then there exists an $H \in \mathcal{F}(M)$ such that

$$\omega^\flat(X) = dH. \tag{3.15}$$

The function H is called a Hamiltonian corresponding to X. If M is connected, then any two Hamiltonians corresponding to X differ by a constant. Conversely, given any $H \in \mathcal{F}(M)$ equation (3.15) defines the corresponding Hamiltonian vector field by $\omega^\sharp(dH)$, which is denoted by X_H. The integral curves of X_H are said to represent the evolution of the classical mechanical system specified by the Hamiltonian H. In a local canonical coordinate system, these integral curves appear as solutions of the following system of differential equations

$$\frac{dq^i}{dt} = \frac{\partial H}{\partial p_i}, \tag{3.16}$$

$$\frac{dp_i}{dt} = -\frac{\partial H}{\partial q^i}. \tag{3.17}$$

This is the form of the classical **Hamilton's equations**. Let $f, g \in \mathcal{F}(M)$; the **Poisson bracket** of f and g, denoted by $\{f, g\}$, is the function

$$\{f, g\} := \omega(X_f, X_g).$$

The Poisson bracket makes $\mathcal{F}(M)$ into a Lie algebra. If X is a Hamiltonian vector field with flow F_t, then Hamilton's equations can be expressed in the form

$$\frac{d}{dt}(f \circ F_t) = \{f \circ F_t, H\}.$$

A Poisson structure on a manifold is a Lie algebra structure on $\mathcal{F}(M)$ that is also a derivation in the first argument of the Lie bracket. The Lie bracket of f, g is also called the Poisson bracket and is denoted by $\{f, g\}$. We can define a Hamiltonian vector field on M corresponding to the Hamiltonian function H by $X_H(g) := \{g, H\}$ $\forall g \in \mathcal{F}(M)$. A manifold with a fixed Poisson structure is called a **Poisson manifold**. A symplectic manifold is a Poisson manifold but the converse is not true. A Poisson structure induces a map of T^*M to TM, but this map need not be invertible.

In view of Darboux's theorem, a symplectic manifold is locally standard (or rigid). Thus, topology of symplectic manifolds or symplectic topology is essentially global. This has been an active area of research with close ties to classical mechanics. Gromov's definition of pseudo-holomorphic curves in symplectic manifolds has provided a powerful tool for studying symplectic topology. For symplectic 4-manifolds this led to the Gromov–Witten, or GW, invariants. For a special class of 4-manifolds Taubes has shown that the GW and SW (Seiberg–Witten) invariants contain equivalent information. There is also a symplectic field theory introduced by Eliashberg and Hofer with connections to string theory and non-commutative geometry. We do not consider these topics in this book. A general reference for symplectic topology is the book by McDuff and Salamon [283]. For quantum cohomology see the books by Kock and Vainsencher [228], Manin [258], and McDuff and Salamon [282].

3.6 Lie Groups

A manifold sometimes carries an additional mathematical structure that is compatible with its differential structure. An important example of this is furnished by a Lie group.

Definition 3.9 *A **Lie group** G is a manifold that carries a compatible group structure, i.e., the operations of multiplication and taking the inverse are smooth. If G, H are Lie groups, a **Lie group homomorphism** $f : G \to H$ is a smooth group homomorphism of G into H, i.e., f is smooth as a map of manifolds and $f(ab) = f(a)f(b)$, $\forall a, b \in G$. **Isomorphism** and **automorphism** of Lie groups are defined similarly. Another useful concept is that* *of an anti-homomorphism. A smooth map $f : G \to H$ is called an* **anti-homomorphism** *of Lie groups if $f(ab) = f(b)f(a)$, $\forall a, b \in G$.*

We observe that it is enough to require that the multiplication be smooth because this implies that the operation of taking the inverse is smooth. Inversion, i.e., the map $\iota : G \to G$ defined by $a \mapsto a^{-1}$, $a \in G$, is an anti-automorphism of the Lie group G. Furthermore, one can show that if G, H are finite-dimensional Lie groups, then every continuous homomorphism of G into H is smooth and hence is a Lie group homomorphism.

Example 3.8 *Let F be a finite- (resp., infinite-) dimensional Banach space. The group of automorphisms (linear, continuous bijections) of F is a finite- (resp., infinite-) dimensional Lie group. Due to the fact that the locally compact Hausdorff manifolds are finite-dimensional, a locally compact Haudorff Lie group is finite-dimensional (see, for example Lang [245]).*

If G is a locally compact group, there exists a unique (up to a multiplicative constant factor) measure μ on the σ-algebra of Borel subsets of G that is left invariant, i.e.,

$$\mu(gU) = \mu(U), \quad \forall g \in G$$

and for all Borel subsets U of G. Such a measure is called a **Haar measure** for G. If G is compact, a Haar measure is also right invariant. Thus, finite-dimensional Lie groups have a Haar measure and this measure is also right invariant if the Lie group is compact.

A subgroup H of a Lie group is called a **Lie subgroup** if the natural injection $i : H \to G$ is an immersion. If H is a closed subgroup of G, then H is a submanifold of G and therefore is a Lie subgroup of G. It is easy to check that the groups $GL(n, \mathbf{R})$ and $GL(n, \mathbf{C})$ defined in Example 3.5 are Lie groups. We now give an example of Lie subgroups.

Example 3.9 *In Example 3.5, we introduced the smooth determinant map* $\det : M(n, \mathbf{R}) \to \mathbf{R}$. *Its restriction to $GL(n, \mathbf{R})$ is also smooth. In fact,*

$$det : GL(n, \mathbf{R}) \to \mathbf{R} \setminus 0 = GL(1, \mathbf{R})$$

is a Lie group homomorphism. It follows that the kernel of this homomorphism, $det^{-1}(\{1\})$ is a closed subgroup of $GL(n, \mathbf{R})$. It is called the **special real linear group** *and is denoted by $SL(n, \mathbf{R})$. Thus,*

$$SL(n, \mathbf{R}) := \{A \in GL(n, \mathbf{R}) \mid \det A = 1\}$$

is a Lie subgroup of $GL(n, \mathbf{R})$. Analogously, one can show that the **special complex linear group**

$$SL(n, \mathbf{C}) := \{A \in GL(n, \mathbf{C}) \mid \det A = 1\}$$

is a closed Lie subgroup of $GL(n, \mathbf{C})$.

The **orthogonal group** *$O(n)$ defined by*

$$O(n) := \{A \in GL(n, \mathbf{R}) \mid AA^t = 1\},$$

where A^t is the transpose of A, is the fixed point set of the automorphism of $GL(n, \mathbf{R})$ defined by $A \mapsto (A^t)^{-1}$. Hence, it is a closed Lie subgroup of $GL(n, \mathbf{R})$. The **special orthogonal group** $SO(n)$ is defined by

$$SO(n) := O(n) \cap SL(n, \mathbf{R}).$$

The group $SO(n)$ is usually referred to as the **rotation group** of the n-dimensional Euclidean space $\mathbf{R}^n$. It is the connected component of the identity in the orthogonal group. Similarly, the **unitary group** $U(n)$ defined by

$$U(n) := \{A \in GL(n, \mathbf{C}) \mid AA^\dagger = 1\},$$

where $A^\dagger$ denotes the conjugate transpose of A, is a closed Lie subgroup of $GL(n, \mathbf{C})$. The **special unitary group** $SU(n)$ is defined by

$$SU(n) := U(n) \cap SL(n, \mathbf{C}).$$

Another definition of these groups is given in Example 3.12.

We give other important examples of Lie groups later.

A **Lie group (left) action** (or a **G-action**) of a Lie group G on a manifold M is a smooth map

$$L : G \times M \to M, \text{ such that } L_g : M \to M$$

defined by $L_g(x) = L(g, x)$ (also denoted by gx) is a diffeomorphism of M $\forall g \in G$ and

$$\forall g_1, g_2 \in G, \ L_{g_1 g_2} = L_{g_1} \circ L_{g_2} \quad \text{and} \quad L_e = id_M, \tag{3.18}$$

where e is the identity element of G. The map L induces the map $\hat{L} : G \to \text{Diff}(M)$ defined by $g \mapsto L_g$. The conditions in equation (3.18) may be expressed by saying that the map $\hat{L} : G \to \text{Diff}(M)$ is a group homomorphism. In fact, this last statement is sometimes used as the definition of a Lie group action. The **orbit** of $x \in M$ under the G-action is the subset $\{gx \mid g \in G\}$ of M, also denoted by Gx. The set of the orbits of the G-action L on M is denoted by M/L or by M/G when L is understood. A G-action on M is said to be **transitive** if there is just one orbit; in this case we also say that G acts transitively on M. If $x \in M$, then the **isotropy group** H_x of the G-action is defined by

$$H_x = \{g \in G \mid gx = x\}.$$

If G acts transitively on M then $H_x \cong H_y$, $\forall x, y \in M$, and if H denotes the isotropy subgroup of some fixed point in M, then M is diffeomorphic to G/H. Such a differential manifold M is called a **homogeneous space** of G or a G-homogeneous space. In particular, if H is a closed subgroup of G, then

H is a Lie subgroup and the quotient G/H with the natural transitive action of G is a G-homogeneous space.

Example 3.10 *The rotation group $SO(n + 1)$ of $\mathbf{R}^{n+1}$ acts on the sphere S^n transitively. The isotropy group at the point $(1, 0, \ldots, 0) \in \mathbf{R}^{n+1}$ may be identified with $SO(n)$. Thus, the sphere S^n is a homogeneous space of the group $SO(n + 1)$, i.e.,*

$$S^n = SO(n + 1)/SO(n).$$

The **conformal group** *$SO(n, 1)$, defined in Example 3.12, acts transitively on the open unit ball $B^n \subset \mathbf{R}^{n+1}$ with isotropy group $SO(n)$ at the origin. Thus B^n is a homogeneous space of the conformal group, called the* **Poincaré model of hyperbolic space,** *i.e.*

$$B^n = SO(n, 1)/SO(n).$$

For $n = 5$ this construction occurs in the study of the moduli space of BPST instantons (see Section 9.2).

A G-action on M is said to be **free** if $gx = x$ for some $x \in M$ implies $g = e$, i.e., if $H_x = \{e\}$, $\forall x \in M$. A G-action on M is said to be **effective** if $gx = x$, $\forall x \in M$ implies $g = e$.

A vector field $X \in \mathcal{X}(M)$ is said to be **invariant under the action L of G on M** (or **G-invariant**) if

$$TL_g \circ X = X \circ L_g, \quad \forall g \in G;$$

i.e., the following diagram commutes:

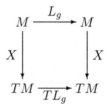

Equivalently, X is G-invariant if

$$(L_g)^* X = X, \quad \forall g \in G.$$

G-invariant tensor fields and **G-invariant differential forms** can be defined similarly. A **right G-action** on M may be defined similarly; then, with obvious changes, one has the related notions of orbit, transitive action, etc.

We now consider an important example of a natural action of G when the manifold is G itself and the action is by left multiplication. A vector field $X \in \mathcal{X}(G)$ is said to be **left invariant** if it is invariant under this action by left multiplication. The set of all left invariant vector fields is a finite-dimensional Lie subalgebra of the Lie algebra $\mathcal{X}(G)$ and is called the **Lie algebra of the group** G and is denoted by LG or $\mathfrak{g}$. The tangent space T_eG to G at the identity e is isomorphic, as a vector space, to $\mathbf{g}$. This isomorphism is used to make T_eG into a Lie algebra isomorphic to $\mathbf{g}$. Precisely, for $C \in T_eG$ we define the left invariant vector field X_C on G by

$$X_C(g) := TL_g(C), \quad \forall g \in G,$$

where $L_g : G \to G$ is left multiplication by g. Now for $A, B \in T_eG$ the Lie product $[A, B]$ is defined by

$$[A, B] := [X_A, X_B](e) \in T_eG,$$

where X_A, X_B are the left invariant vector fields corresponding, respectively, to A and B. Let G, H be Lie groups with Lie algebras $\mathfrak{g}, \mathfrak{h}$ respectively. Let $f : G \to H$ be a Lie group homomorphism and let $f_* : \mathfrak{g} \to \mathfrak{h}$ be the map defined by

$$f_*(X_A) := X_{\hat{A}}, \text{ where } \hat{A} := T_e f(A) \in T_eH;$$

i.e., $f_*(X_A)$ is the left invariant vector field on H determined by $\hat{A} \in T_eH$. Then one can show that f_* is a Lie algebra homomorphism of $\mathfrak{g}$ into $\mathfrak{h}$.

The element $A \in \mathfrak{g}$ generates a global one-parameter group ϕ_t of diffeomorphisms of G, determined by the global flow $\phi(t, g)$ of A. Thus $\phi_t(e)$ is the integral curve of the left invariant vector field A through e. We observe that $\phi_t(e)$ is the one-parameter subgroup of G generated by $A \in \mathfrak{g}$. Conversely, a one-parameter subgroup ϕ_t of G determines a unique element $A \in \mathfrak{g}$, namely the tangent to ϕ_t at $e \in G$. These observations play an important role in the construction of Lie algebras. We define the map $exp : \mathfrak{g} \to G$ by

$$exp : A \mapsto \phi_1(e).$$

This map is a homeomorphism on some neighborhood of $0 \in \mathfrak{g}$ and can be used to define a special coordinate chart on G called the **normal coordinate chart**. The map exp is called the **exponential map** and coincides with the usual exponential function for matrix groups and algebras. One can show that $t \mapsto exp(tA)$ is the integral curve of A through $e \in G$, i.e.,

$$exp(tA) = \phi_t(e).$$

The **adjoint action** Ad of G on itself is defined by the map

$$\mathrm{Ad} : h \mapsto \mathrm{Ad}(h) \text{ of } G \to \mathrm{Aut}\, G,$$

where $\mathrm{Ad}(h) : G \to G$ is defined by

$$\mathrm{Ad}(h)g = hgh^{-1}.$$

This action induces an action ad of G on $\mathfrak{g}$, which is a representation of G, called the **adjoint representation** of G on $\mathfrak{g}$. We observe that with the identification of T_eG with $\mathfrak{g}$,

$$\mathrm{ad}\, g = T(\mathrm{Ad}(g))|_{\mathfrak{g}}.$$

The set $GL(\mathfrak{g})$ of linear invertible transformations of $\mathfrak{g}$ is a Lie group with Lie algebra: the algebra $gl(\mathfrak{g})$ of linear transformations of $\mathfrak{g}$ with Lie product given by

$$[M, N] = M \circ N - N \circ M, \quad \forall M, N \in gl(\mathfrak{g}).$$

Thus the restriction of $(\mathrm{ad})_*$ to $\mathfrak{g}$, also denoted by ad_*, is a map of $\mathfrak{g}$ into $gl(\mathfrak{g})$. If V is a Lie algebra, let us denote by ad the map

$$\mathrm{ad} : V \to gl(V)$$

defined by

$$\mathrm{ad}(X)(Y) = [X, Y], \qquad \forall X, Y \in V. \tag{3.19}$$

The map ad is a Lie algebra homomorphism called the **adjoint representation** of V. In the case of $V = \mathfrak{g}$ one can show that ad_* coincides with ad defined in equation (3.19). Therefore, we use ad to also denote ad_*.

The contragredient of the representation ad of G on $\mathfrak{g}$, called the **coadjoint representation** of G on $\mathfrak{g}^*$, is denoted by ad^*. Thus, $\forall g \in G$,

$$ad^*(g) = (ad(g^{-1}))^t : \mathfrak{g}^* \to \mathfrak{g}^*,$$

is the transpose of the linear map $ad(g^{-1})$. It plays an important role in the theory of representations of nilpotent and solvable groups. The study of this representation is the starting point of the Kostant–Kirillov–Souriau theory of geometric quantization (see Abraham and Marsden [1] and [271, 272] and references therein). We now give a construction of the Lie algebras of the Lie groups $GL(n, \mathbf{R})$ and $GL(n, \mathbf{C})$ defined in Example 3.5 as well as those of $SL(n, \mathbf{R})$ and $SL(n, \mathbf{C})$ defined in Example 3.9.

Example 3.11 *We first observe that, when the differential manifold M is an open subset of a Banach space F, then $TM = M \times F$. The Lie group $G \equiv GL(n, \mathbf{R})$ is an open subset of the n^2-dimensional Banach space $M(n, \mathbf{R})$. Then $TG = GL(n, \mathbf{R}) \times M(n, \mathbf{R})$ and thus the Lie algebra of $GL(n, \mathbf{R})$ can be identified with $M(n, \mathbf{R})$ as a vector space. If $A \in M(n, \mathbf{R})$, then the left invariant vector field associated to A is*

$$X_A : GL(n, \mathbf{R}) \to GL(n, \mathbf{R}) \times M(n, \mathbf{R})$$

defined by

$$X_A(g) = (g, gA).$$

From this it follows that the Lie algebra structure of $M(n, \mathbf{R})$ is the one given by the commutator. Analogously, the Lie algebra of $GL(n, \mathbf{C})$ is the Lie algebra $M(n, \mathbf{C})$ with the Lie algebra structure given by the commutator. We note that the map exp defined above becomes in this case the usual exponential map defined by

$$\exp(A) = e^A = I + \frac{A}{1!} + \frac{A^2}{2!} + \cdots.$$

In view of the fact that $\det(e^A) = e^{\operatorname{tr}(A)}$, we conclude that

$$sl(n, \mathbf{R}) := \{A \in M(n, \mathbf{R}) \mid \operatorname{tr}(A) = 0\}$$

is the Lie algebra of $SL(n, \mathbf{R})$. Similarly,

$$sl(n, \mathbf{C}) := \{A \in M(n, \mathbf{C}) \mid \operatorname{tr}(A) = 0\}$$

is the Lie algebra of $SL(n, \mathbf{C})$.

We now give several examples of Lie groups and their Lie algebras, which appear in many applications.

Example 3.12 *Let g_k be the canonical pseudo-metric on $\mathbf{R}^n$ of signature $n - k$, i.e., g_k is the pseudo-metric on $\mathbf{R}^n$ whose matrix representation, also denoted by g_k, is*

$$g_k = \begin{pmatrix} I_k & 0 \\ 0 & -I_{n-k} \end{pmatrix},$$

where I_i denotes the $i \times i$ unit matrix, $i = k$, $n - k$, and the 0 denote suitable zero matrices. Let $O(k, n-k)$ be the group of linear transformations A of $\mathbf{R}^n$ such that $g_k(Ax, Ay) = g_k(x, y)$, $\forall x, y \in \mathbf{R}^n$. The group $O(k, n - k)$ can be identified with the group of $n \times n$ matrices A such that

$$A^t g_k A = g_k.$$

From this equation it follows that $\det A = \pm 1$, $\forall A \in O(k, n - k)$. We denote by $SO(k, n-k)$ the subgroup of $O(k, n-k)$ of matrices A such that $\det A = 1$. We write $O(n)$, $SO(n)$ in place, respectively, of $O(n, 0)$, $SO(n, 0)$. The group $O(n)$ (resp., $SO(n)$) is the orthogonal group (resp., special orthogonal group) in n dimensions defined in Example 3.9. Let $f : GL(n, \mathbf{R}) \to S(n, \mathbf{R})$ be the map defined by

$$f(A) = A^t g_k A.$$

The map f is smooth and then $f^{-1}(\{g_k\})$ is closed. Thus, $O(k, n - k) = f^{-1}(\{g_k\})$ is a closed subgroup of $GL(n, \mathbf{R})$ and thus is a Lie subgroup of $GL(n, \mathbf{R})$. For any $A \in GL(n, \mathbf{R})$, the linear map $Df(A)$ on $M(n, \mathbf{R})$ is given by

$$Df(A)B = B^t g_k A + A^t g_k B, \qquad \forall B \in M(n, \mathbf{R}).$$

Thus,

$$\text{Ker}\, Df(I_n) = \{B \in M(n, \mathbf{R}) \mid B^t g_k = -g_k B\}.$$

This implies that the rank of f at I_n is maximum (it is equal to $n(n+1)/2$) and that f is a submersion at I_n. By Theorem 3.1, the Lie algebra of $O(k, n-k)$, denoted by $o(k, n-k)$, can be identified with $\text{Ker}\, T_{I_n} f$ and thus is given by

$$o(k, n-k) = \{B \in M(n, \mathbf{R}) \mid B^t g_k = -g_k B\}.$$

This also implies that the dimension of $O(k, n-k)$ is $n(n-1)/2$. In particular, with $g_k = I_n$, the Lie algebra $o(n)$ of $O(n)$ is given by

$$o(n) = \{B \in M(n, \mathbf{R}) \mid B^t = -B\},$$

the Lie algebra of the antisymmetric $n \times n$ matrices. It is easy to see that $SO(k, n-k)$ is a Lie subgroup of $O(k, n-k)$ of dimension $n(n-1)/2$, given by the connected component of the identity element. Its Lie algebra denoted by $so(k, n-k)$ is given by

$$so(k, n-k) = \{B \in o(k, n-k) \mid \text{tr}(B) = 0\}.$$

The groups $O(n)$, $SO(n)$ are compact.

Example 3.13 *Let us denote by g_k the canonical non-degenerate Hermitian sesquilinear form on $\mathbf{C}^n$ of signature $n - k$, i.e. the sesquilinear form on $\mathbf{C}^n$ given by*

$$g_k(x, y) = \overline{x_1} y_1 + \cdots + \overline{x_k} y_k - \overline{x_{k+1}} y_{k+1} - \cdots - \overline{x_n} y_n$$

$\forall x, y \in \mathbf{C}^n$. Then the matrix representation of g_k is the same as for g_k of the previous example. Let $U(k, n-k)$ be the group of linear transformations A of $\mathbf{C}^n$ such that $g_k(Ax, Ay) = g_k(x, y)$, $\forall x, y \in \mathbf{C}^n$. The group $U(k, n-k)$ identifies with the group of $n \times n$ complex matrices A such that

$$A^\dagger g_k A \;=\; g_k,$$

where $A^\dagger$ denotes the conjugate transpose of A. From this equation it follows that $\det A = \pm 1$, $\forall A \in U(k, n-k)$. We denote by $SU(k, n-k)$ the subgroup of $U(k, n-k)$ of matrices A such that $\det A = 1$. We write $U(n)$, $SU(n)$ in place of $U(n, 0)$, $SU(n, 0)$, respectively. The group $U(n)$ (resp., $SU(n)$) is the unitary group (resp., special unitary group) in n dimensions defined in Example 3.9. Let $f : GL(n, \mathbf{C}) \to S(n, \mathbf{C})$ be the map defined by

$$f(A) = A^\dagger g_k A.$$

The map f is smooth and hence $f^{-1}(\{g_k\})$ is closed. Thus, $U(k, n-k) = f^{-1}(\{g_k\})$ is a closed subgroup of $GL(n, \mathbf{C})$ and thus a Lie subgroup of $GL(n, \mathbf{C})$. Proceeding as in the previous example, we find that the Lie algebra of $U(k, n-k)$, denoted by $u(k, n-k)$, is given by

$$u(k, n-k) = \{B \in M(n, \mathbf{C}) \mid B^\dagger g_k = -g_k B\}.$$

This also implies that the dimension of $U(k, n-k)$ is n^2. In particular, with $g_k = I_n$, the Lie algebra $u(n)$ of $U(n)$ is given by

$$u(n) = \{B \in M(n, \mathbf{C}) \mid B^\dagger = -B\},$$

the Lie algebra of the anti-Hermitian $n \times n$ matrices. One easily realizes that $SU(k, n-k)$ is a Lie subgroup of $U(k, n-k)$ of dimension $n^2 - 1$, with Lie algebra $su(n, n-k)$ given by

$$su(k, n-k) = \{B \in u(k, n-k) \mid tr(B) = 0\}.$$

The groups $U(n)$, $SU(n)$ are compact.

Example 3.14 *Let ω be the canonical symplectic structure on $\mathbf{R}^{2n}$ whose matrix representation, also denoted by ω, is*

$$\omega = \begin{pmatrix} 0 & I \\ -I & 0 \end{pmatrix},$$

where I (resp., 0) denotes the $n \times n$ unit (resp., zero) matrix. Let $SP(n, \mathbf{R})$ be the group of linear transformations A of $\mathbf{R}^{2n}$, such that $\omega(Ax, Ay) = \omega(x, y)$, $\forall x, y \in \mathbf{R}^{2n}$. The group $SP(n, \mathbf{R})$ can be identified with the group of $2n \times 2n$ matrices A such that

$$A^t \omega A = \omega.$$

*From this equation it follows that $\det A = \pm 1$, $\forall A \in SP(n, \mathbf{R})$. The group $SP(n, \mathbf{R})$ is called the **real symplectic group** in $2n$ dimensions. Let $f : GL(2n, \mathbf{R}) \to GL(2n, \mathbf{R})$ be the map defined by*

$$f(A) = A^t \omega A.$$

The map f is smooth and hence $f^{-1}(\{\omega\})$ is closed. Thus, $SP(n, \mathbf{R}) = f^{-1}(\{\omega\})$ is a closed subgroup of $GL(2n, \mathbf{R})$ and thus is a Lie subgroup of $GL(2n, \mathbf{R})$. Proceeding as in the previous example, we find that the Lie algebra of $SP(n, \mathbf{R})$, denoted by $sp(n, \mathbf{R})$, is given by

$$sp(n, \mathbf{R}) = \{B \in M(2n, \mathbf{R}) \mid B^t \omega = -\omega B\}.$$

*This also implies that the dimension of $SP(n, \mathbf{R})$ is $n(2n + 1)$. The group $SP(n, \mathbf{R})$ is non-compact. Analogously, one has the **complex symplectic group** $SP(n, \mathbf{C})$, which is a Lie group of dimension $2n(2n + 1)$.*

As a preparation for the next example we start by briefly recalling the notations and elementary properties of **quaternions**. The noncommutative ring of quaternions is denoted by $\mathbf{H}$ in honor of its discoverer, Hamilton (see

Appendix B for some historical remarks). It is obtained by adjoining to the field $\mathbf{R}$ the elements i, j, k satisfying the relations

$$i^2 = j^2 = k^2 = -1,$$

$$ij = -ji = k, \quad jk = -kj = i, \quad ki = -ik = j.$$

Thus, $\mathbf{H}$ is a 4-dimensional real vector space with basis $1, i, j, k$ and a quaternion $x \in \mathbf{H}$ has the following expression

$$x = x_0 + ix_1 + jx_2 + kx_3, \quad x_i \in \mathbf{R}, \quad 0 \le i \le 3.$$

The operation of conjugation on $\mathbf{H}$ is defined by

$$x = x_0 + ix_1 + jx_2 + kx_3 \mapsto \bar{x} := x_0 - ix_1 - jx_2 - kx_3.$$

Then $\overline{(xy)} = \bar{y}\bar{x}$ and each nonzero $x \in \mathbf{H}$ has a unique inverse x^{-1} given by

$$x^{-1} = \bar{x}/(x\bar{x}).$$

We denote by $\mathbf{H}^n$ the module over $\mathbf{H}$ of n-tuples of quaternions with the action given by right multiplication. It is customary to refer to a morphism of this module as an $\mathbf{H}$-linear, or simply a linear map.

Example 3.15 *We denote by $GL(n, \mathbf{H})$ the group of linear bijections of $\mathbf{H}^n$. $\mathbf{H}$ can be identified with $\mathbf{C}^2$ by the map $\phi : \mathbf{H} \to \mathbf{C}^2$ defined by*

$$\phi(x_0 + ix_1 + jx_2 + kx_3) = (x_0 + ix_3, x_2 + ix_1).$$

Then $GL(n, \mathbf{H})$ can be identified with the subgroup of $GL(2n, \mathbf{C})$ of the matrices A of the form

$$A = \begin{pmatrix} B & -\bar{C} \\ C & \bar{B} \end{pmatrix}, \tag{3.20}$$

where B, C are complex $n \times n$ matrices. The group $GL(n, \mathbf{H})$ is a Lie group. Let us denote by Q the quadratic form on $\mathbf{H}^n$ defined by

$$Q(x^1, x^2, \ldots, x^n) = x^1\overline{x^1} + x^2\overline{x^2} + \cdots + x^n\overline{x^n}, \qquad \forall x^1, x^2, \ldots, x^n \in \mathbf{H}.$$

We define the Lie subgroup $Sp(n)$ of $GL(n, \mathbf{H})$ by

$$Sp(n) := \{A \in GL(n, \mathbf{H}) \mid Q(Ax) = Q(x), \ \forall x \in \mathbf{H}^n\}.$$

*The group $Sp(n)$ is called the **quaternionic symplectic group** in n dimensions. This group can be identified with the group of the $2n \times 2n$ complex matrices A of the form given in equation (3.20) that satisfy the conditions*

$$A^\dagger A = 1, \quad A^t \omega A = \omega,$$

where ω is the matrix of the canonical symplectic form of the previous example. Thus, $Sp(n) \cong U(2n) \cap SP(n, \mathbf{C})$ and is therefore also called the **unitary symplectic group**. *In particular we have that $Sp(1) \cong SU(2)$.*

The classic reference for the theory of Lie groups is Chevalley [77]; a modern reference with geometric analysis is Helgason [188].

Chapter 4
Bundles and Connections

4.1 Introduction

In 1931 Hopf studied the set $[S^3,\ S^2]$ of homotopy classes of maps of spheres in his computation of $\pi_3(S^2)$. He showed that $\pi_3(S^2)$ is generated by the class of a certain map that is now well known as the **Hopf fibration** (see Example 4.5). This fibration decomposes S^3 into subspaces homeomorphic to S^1 and the space of these subsets is precisely the sphere S^2. In 1933 Seifert introduced the term **fiber space** to describe this general situation. The product of two topological spaces is trivially a fiber space, but the example of the Hopf fibration shows that a fiber space need not be a global topological product. It continues to be a local product and is now referred to as a **fiber bundle**.

Fiber spaces immediately acquired an important role in mathematics through the early work of Whitney, Serre, and Ehresmann. In 1935 Hopf generalized his early work to construct the fibration of S^7 over S^4 with fiber S^3 and, using octonions, also constructed a fibration of S^{15} over S^8 with fiber S^7. All of these examples have found remarkable application to fundamental solutions of gauge field equations (see Chapter 8 for details). For the applications we wish to consider, we mainly need vector bundles and principal and associated bundles with a fixed structure group. In Section 4.2 we give several alternative definitions of **principal bundle** and discuss the reduction and extension of the structure group of the bundle. In physical applications it is the **structure group** of the bundle, which serves as the **local symmetry group** or the **gauge group**, while the base is usually a space-time manifold. The **matter fields** interacting with gauge fields are sections of vector bundles associated to the given principal bundle. The associated bundles are considered in Section 4.3, where the bundles $\mathrm{Ad}(P)$ and $\mathrm{ad}(P)$, which play a fundamental role in gauge theory, are defined. In Section 4.4, we give several definitions of connection on a principal bundle and define the curvature 2-form. The structure equations and Bianchi identities satisfied by the curvature are also given there. Subsection 4.4.1 is devoted to a discussion of

K. Marathe, *Topics in Physical Mathematics*, DOI 10.1007/978-1-84882-939-8_4,

universal connections. The covariant derivative is introduced in Section 4.5, where the notion of parallel displacement is also discussed. The important class of linear connections, which include the widely used pseudo-Riemannian connections, is discussed in Section 4.6. The special case of dimension 4 leads to the definitions of various special types of manifolds that play a significant role in physical applications. A section on generalized connections concludes the chapter.

A standard reference for most of the material in this chapter is Kobayashi and Nomizu [225]. The theory of fiber bundles occupies a central place in modern mathematics. Recently, fiber bundles have provided the geometrical setting for studying gauge field theories. We discuss various basic aspects of field theories in Chapters 6 and 7. Standard references for the theory of fiber bundles are Husemöller [198] and Steenrod [359].

4.2 Principal Bundles

We begin with the definition of a differentiable fiber bundle.

Definition 4.1 *A **differentiable fiber bundle** E over B is a quadruple $\zeta = (E, B, \pi, F)$, where E, B, F are differentiable manifolds and the map $\pi : E \longrightarrow B$ is an open differentiable surjection satisfying the following* **local triviality (LT) property***:*

> *LT property—There exists an open covering $\{U_i\}_{i \in I}$ of B and a family ψ_i of diffeomorphisms $\psi_i : U_i \times F \longrightarrow \pi^{-1}(U_i)$, $\forall i \in I$, satisfying the condition $(\pi \circ \psi_i)(x, g) = x$, $\forall (x, g) \in U_i \times F$.*

The family $\{(U_i, \psi_i)\}_{i \in I}$ is called a **local coordinate representation** or a **local trivialization** of the bundle ζ. E is called the **total space** or the **bundle space**, B the **base space**, π the **bundle projection** of E on B, and F the **standard** or **typical fiber**. $E_x := \pi^{-1}(x)$ is called the **fiber** of E over $x \in B$.

The bundle $\tau = (B \times F, B, \pi, F)$, where π is the projection onto the first factor, is called the **trivial bundle** over B with fiber F. The total space of the trivial bundle τ is $B \times F$. For this reason a general fiber bundle is sometimes called a **twisted product** of B (the base) and F (the standard fiber). A fiber bundle $\zeta = (E, B, \pi, F)$ is sometimes indicated by a diagram as follows:

One can verify that, $\forall x \in U_i$, $\forall i \in I$, the map $\psi_{i,x} : F \to E_x$ defined by $g \mapsto \psi_i(x,g)$ is a diffeomorphism. If $U_i \cap U_j \neq \emptyset$ then we write $U_{ij} = U_i \cap U_j$ and define

$$\psi_{ij} : U_{ij} \to \text{Diff}(F) \qquad \text{by} \qquad \psi_{ij}(x) = \psi_{i,x}^{-1} \circ \psi_{j,x}.$$

The functions ψ_{ij} are called the **transition functions** for the local representation. They satisfy the condition

$$\psi_{ij}(x) \circ \psi_{jk}(x) \circ \psi_{ki}(x) = id_F, \qquad \forall x \in U_{ijk}, \qquad (4.1)$$

where we have written $U_{ijk} = U_i \cap U_j \cap U_k$. Condition (4.1) is referred to as the **cocycle condition** on the transition functions. We note that, given the base B, the typical fiber F and a family of transition functions $\{\psi_{ij}\}$ satisfying the cocycle condition (4.1), it is possible to construct the fiber bundle E. Given the bundles $\zeta = (E, B, \pi, F)$ and $\zeta' = (E', B', \pi', F')$, a **bundle morphism** f from ζ to ζ' is a differentiable map $f : E \to E'$ such that f maps the fibers of E smoothly to the fibers of E' and therefore induces a smooth map of B to B' denoted by f_0. Thus, we have the following commutative diagram:

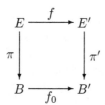

If $B = B'$, f is injective and $f_0 = id_B$, then we say that ζ is a **subbundle** of ζ'. If $\zeta = (E, B, \pi, F)$ is a fiber bundle, we frequently denote by E the fiber bundle ζ when B, π, F are understood. If $\zeta = (E, B, \pi, F)$ is a fiber bundle and $h \in \mathcal{F}(M, B)$, where M is a manifold, then we can define a bundle $h^*\zeta = (h^*E, M, h^*\pi, F)$ called the **pull-back** of the bundle E to M as follows. h^*E is the subset of $M \times E$ consisting of the pairs $(p, a) \in M \times E$ such that $h(p) = \pi(a)$, and one can show that it is a closed submanifold of $M \times E$. Thus, h^*E is obtained by attaching to each point $p \in M$ the fiber $\pi^{-1}(h(p))$ over $h(p)$. The map $h^*\pi$ is the restriction to h^*E of the natural projection of $M \times E$ onto M. The map h lifts to a unique bundle map $\hat{h} : h^*E \to E$ such that the following diagram commutes:

$$
\begin{array}{ccc}
h^*E & \xrightarrow{\ \hat{h}\ } & E \\
\downarrow{h^*\pi} & & \downarrow{\pi} \\
M & \xrightarrow{\ h\ } & B
\end{array}
$$

Let $\zeta = (E, B, \pi, F)$ be a fiber bundle. A smooth map $s : B \to E$ with $\pi \circ s = id_B$ is called a (smooth) **section** of the fiber bundle E over B. We denote by $\Gamma(B,\ E)$, or simply $\Gamma(E)$, the space of sections of E over B. If $U \subset B$ is open then we denote by $\Gamma(E_{|U})$ the set of sections of the bundle E restricted to U. If $p \in U$ and $s \in \Gamma(E_{|U})$ then we say that s is a **local section** of E at p.

Choosing local coordinates $\{x^i \mid 1 \leq i \leq m\}$ in a neighborhood of $p \in B$ and $\{(x^i, y^j)\mid 1 \leq i \leq m,\ 1 \leq j \leq n\}$ in a neighborhood of $s(p) \in E$, we may think of s as a function from an open subset of $\mathbf{R}^m$ to $\mathbf{R}^{m+n}$ such that

$$(x^i) \mapsto (x^i,\ y^j(x^1, x^2, \ldots, x^m)).$$

The Taylor expansion of s at p clearly depends on the local coordinates chosen. However, if two local sections s and t have the same kth order Taylor expansion at p in one coordinate system, then they have the same kth order expansion in any other coordinate system. This observation can be used to define an equivalence relation on local sections at p. An equivalence class determined at p by the section s is called the k-**jet** of s at p and is denoted by $j^k(s)_p$. We define the k-**jet of sections** of E over B, $J^k(E/B)$ (also denoted by $J^k(E)$) by

$$J^k(E/B) := \{j^k(s)_p \mid p \in B,\ s \text{ is a local section of } E \text{ at } p\}.$$

$J^k(E/B)$ is a fiber bundle over B with projection $\pi^k : J^k(E) \to B$ defined by $\pi^k(j^k(s)_p) = p$, and a fiber bundle over $J^l(E/B)$, $0 \leq l \leq k$, with projection $\pi^k_l : J^k(E) \to J^l(E)$ defined by $\pi^k_l(j^k(s)_p) = j^l(s)_p$. In particular, $J^k(E)$ is a fiber bundle over $J^0(E) = E$. A section $s \in \Gamma(E)$ induces a section $j^k(s) \in \Gamma(J^k(E))$ defined by $j^k(s)(p) = j^k(s)_p$. We call $j^k(s)$ the k-**jet extension** of s. The map $j^k : \Gamma(E) \to \Gamma(J^k(E))$, defined by

$$s \mapsto j^k(s),$$

is called the k-**jet extension map**. If M, N are manifolds then we define the space $J^k(M, N)$ of k-jets of maps of M to N by

$$J^k(M, N) = J^k((M \times N)/M),$$

where $M \times N$ is regarded as a trivial fiber bundle over M. It is possible to define the bundle $J^k(M, N)$ directly by considering the Taylor expansion of local maps up to order k. Jet bundles play a fundamental role in the geometrical formulation of variational problems and in particular of Lagrangian theories. We shall not make use of these jet bundle techniques in this book.

Let $\zeta = (E, B, \pi, F)$ be a fiber bundle. If a Lie group G is a subgroup of $\text{Diff}(F)$ such that for each transition function ψ_{ij} of ζ we have $\psi_{ij}(x) \in G$ $\forall x \in U_{ij}$, and ψ_{ij} is a smooth map of U_{ij} into G, then we say that ζ is a **fiber bundle with structure group** G. We now give the two most important cases of fiber bundles with structure group.

Definition 4.2 *A fiber bundle $\zeta = (E, B, \pi, F)$ with structure group G is called a* **vector bundle** *with fiber type F if F is a Banach space and G is the Lie group of the linear automorphisms (linear continuous bijections) of F. In particular, if F is a real (resp., complex) vector space of dimension n and $G = GL(n, \mathbf{R})$ (resp., $G = GL(n, \mathbf{C})$), then we call ζ a* **real** *(resp.,* **complex***) vector bundle of rank n.*

We note that in the case of a vector bundle of rank n the transition functions turn out to be automatically smooth. Let E, H be two vector bundles over B. The algebraic operations on vector spaces can be extended to define vector bundles such as $E \oplus H$, $E \otimes H$, and $\mathrm{Hom}(E, H)$ by using pointwise operations on the fibers over B. In particular, we can form the bundle $(\Lambda^k B) \otimes E$. The sections of this bundle are called k-**forms on B with values in the vector bundle** E, or simply, **vector bundle-valued** (E-valued) k-forms . We write $\Lambda^k(B, E)$ for the space of sections $\Gamma((\Lambda^k B) \otimes E)$. Thus, $\alpha \in \Lambda^k(B, E)$ can be regarded as defining for each $x \in B$ a k-linear, anti-symmetric map α_x of $(T_x B)^k$ into E_x. In particular, $\Lambda^0(B, E) = \Gamma(E)$. If E is a trivial vector bundle with fiber V, then we call $\Lambda^k(B, E)$ the space of k-forms with values in the vector space V, or vector valued (V-valued) k-forms, and denote it by $\Lambda^k(B, V)$. We observe that a Riemannian metric on M is a smooth section of the vector bundle $S^2(TM)$ on M, i.e., of the bundle whose fiber on $x \in M$ is the vector space $S^2(T_x M)$ of symmetric bilinear maps of $T_x M \times T_x M$ into $\mathbf{R}$. More generally, given the real (resp., complex) vector bundle E on M, a **Riemannian** (resp., **Hermitian**) **metric** on E is a smooth section s of the vector bundle $S^2(E)$ on M such that $s(x)$ is a bilinear (resp., sesquilinear), symmetric (resp., Hermitian) and positive definite map of $E_x \times E_x$ into $\mathbf{R}$ (resp., $\mathbf{C}$), $\forall x \in M$. By definition a Riemannian metric on M is a Riemannian metric on the vector bundle TM on M. A **Riemannian** (resp., **Hermitian**) **vector bundle** is a couple (E, s) where E is a real (resp., complex) vector bundle and s is a Riemannian (resp., Hermitian) metric on E. We note that if M is paracompact, then every real (resp., complex) vector bundle on M admits a Riemannian (resp., Hermitian) metric.

Let V_1, V_2, V_3 be vector spaces and $h : V_1 \times V_2 \to V_3$ be a bilinear form. Let $\alpha \in \Lambda^p(B, V_1)$, $\beta \in \Lambda^q(B, V_2)$; then we define $\alpha \wedge_h \beta \in \Lambda^{p+q}(B, V_3)$ as follows. Let $\{u_i\}$ be a basis of V_1 and $\{v_j\}$ a basis for V_2. Then we can write

$$\alpha = \alpha^i u_i, \quad \beta = \beta^j v_j,$$

where $\alpha^i \in \Lambda^p(B)$ and $\beta^j \in \Lambda^q(B)$ and we define

$$\alpha \wedge_h \beta := \alpha^i \wedge \beta^j h(u_i, v_j).$$

There are several important special cases of this operation. For example, if $V = V_1 = V_2 = V_3$ and if h is an inner product on the vector space V, then $\alpha \wedge_h \beta \in \Lambda^{p+q}(B)$. If V is a Lie algebra and h is the Lie bracket, then it is customary to denote $\alpha \wedge_h \beta$ by $[\alpha, \beta]$. Thus, we have

$$[\alpha, \beta] := \alpha^i \wedge \beta^j [u_i, u_j] = c_{ij}^k \alpha^i \wedge \beta^j u_k,$$

where c_{ij}^k are the structure constants of the Lie algebra V. If B is a pseudo-Riemannian manifold with metric g and h is an inner product on V, then for each $x \in B$, the space $\Lambda_x^k(B, V)$ becomes a pseudo-inner product space, with pseudo-inner product denoted by $\langle \ , \ \rangle_{(g,h)}$, defined by

$$\langle \alpha, \beta \rangle_{(g,h)} := \langle \alpha^i, \beta^j \rangle_g h(u_i, u_j), \qquad \alpha, \beta \in \Lambda_x^k(B, V),$$

where $< \ , \ >_g$ is the pseudo-inner product on $\Lambda_x^k(B)$ induced by g. If B is compact with volume form v_g, then we can integrate these local products over B to obtain a pseudo-inner product on $\Lambda^k(B, V)$ defined by

$$\langle\langle \alpha, \beta \rangle\rangle_{(g,h)} := \int_B \langle \alpha^i, \beta^j \rangle_g \, h(u_i, u_j) dv_g, \qquad \alpha, \beta \in \Lambda^k(B, V).$$

Recall that a natural symmetric bilinear form on a Lie algebra V is given by the **Killing form**

$$K(X, Y) := \text{Tr}(\text{ad}\, X \circ \text{ad}\, Y), \qquad \forall X, Y \in V,$$

where $\text{Tr}(L)$ denotes the trace of the linear transformation L of V. We note that, if c_{jk}^i are the structure constants of V with respect to a basis E_i, then

$$K(X, Y) = c_{ik}^r c_{jr}^k X^i Y^j.$$

Furthermore, K is non-degenerate if and only if the determinant of the matrix $\{c_{ik}^r c_{jr}^k\}$ is nonzero. By **Cartan's criterion** K is non-degenerate if and only if V is semisimple, i.e., if it has no nonzero Abelian ideals. If $V = \mathbf{g}$ is the Lie algebra of a connected Lie group G and $\mathbf{g}$ is semisimple, then, by a theorem of Weyl, K is negative definite if and only if G is compact. Thus, in this case $h := -K$ is an inner product on $\mathbf{g}$ which with h the Killing form and g a can be used to define the **norm** or **energy** of $\alpha \in \Lambda^k(B, V)$ by

$$\|\alpha\| := \sqrt{|\langle\langle \alpha, \alpha \rangle\rangle_{(g,h)}|}.$$

The constructions discussed above for vector-valued forms can be extended to apply to vector bundle-valued forms by using their pointwise vector space structures. We shall use them to define the Yang–Mills and other action functionals in gauge theories.

Example 4.1 *Let TM be the tangent space of the m-dimensional manifold M and $\pi : TM \to M$ the canonical projection. Then $(TM, M, \pi, \mathbf{R}^m)$ is a vector bundle of rank m, called the **tangent bundle** of M. A k-dimensional distribution on M is a vector bundle of rank k over M, which is a subbundle of the tangent bundle of M.*

Definition 4.3 *A fiber bundle* $\zeta = (P, M, \pi, F)$ *with structure group G is called a* **principal fiber bundle,** *or simply a* **principal bundle,** *over M with structure group G if F is a Lie group and G is the Lie group of the diffeomorphisms h of F such that*

$$h(g_1 g_2) = h(g_1) g_2.$$

The map $h \mapsto h(e)$, where e is the identity of F, is an isomorphism of G with F. A principal bundle over M with structure group G is denoted by $P(M, G)$.

Equivalently a principal bundle $P(M, G)$ with structure group G over M may be defined as follows.

Definition 4.4 *A principal bundle $P(M, G)$ with structure group G over M is a fiber bundle (P, M, π, G) with a free right action ρ of G on P, such that*

1. *the orbits of ρ are the fibers of $\pi : P \to M$, i.e., π may be identified with the canonical projection $P \to P/G$;*
2. *$\forall \psi : U \times G \to \pi^{-1}(U)$, such that ψ is a local trivialization of P, writing $u_x g$ in the place of $\rho(u_x, g)$, one has*

$$\psi_x^{-1}(u_x g) = \psi_x^{-1}(u_x) g, \quad \forall u_x \in P_x, \ g \in G.$$

We observe that, given the first definition one has the natural right action ρ of G on P defined by

$$\rho(u_x, g) = \psi_{i,x}(\psi_{i,x}^{-1}(u_x) g),$$

where ψ_i is a local trivialization of P at $x \in M$. This action ρ is a free right action and satisfies both conditions of Definition 4.4. Conversely, given the second definition, the $\psi_{ij}(x)$ are diffeomorphisms of G satisfying the defining condition

$$\psi_{ij}(x)(g_1 g_2) = \psi_{ij}(x)(g_1) g_2$$

of Definition 4.3. We give below an example of a principal bundle that is naturally associated with every manifold.

Example 4.2 (Bundle of frames) *Let M be an m-dimensional manifold. A* **frame** *$u = (u_1, \ldots, u_m)$ at a point $x \in M$ is an ordered basis of the tangent space $T_x M$. Let*

$$L_x(M) = \{u \mid u \text{ is a frame at } x \in M\},$$

$$L(M) = \bigcup_{x \in M} L_x(M).$$

Define the projection

$$\pi : L(M) \to M \text{ by } u \mapsto x, \text{ where } u \in L_x(M) \subset L(M).$$

The general linear group $GL(m, \mathbf{R})$ acts freely on $L(M)$ on the right by

$$(u, g) \mapsto ug = (u_i g_1^i, u_i g_2^i, \ldots, u_i g_m^i), \qquad g = (g_j^i) \in GL(m, \mathbf{R}).$$

It can be shown that $L(M)$ can be given the structure of a manifold, such that this $GL(m, \mathbf{R})$-action is smooth. Then $L(M)(M, GL(m, \mathbf{R}))$ is a principal bundle over M with structure group $GL(m, \mathbf{R})$. This principal bundle $L(M)$ is called the **bundle of frames** *or simply the* **frame bundle** *of M.*

Let $P(M, G)$ and $Q(N, H)$ be two principal fiber bundles. A **principal bundle homomorphism** of $Q(N, H)$ into $P(M, G)$ is a bundle homomorphism $f : Q \to P$ together with a Lie group homomorphism $\gamma : H \to G$, such that

$$f(uh) = f(u)\gamma(h), \ \forall u \in Q, h \in H.$$

When f satisfies the above condition we say that it is **equivariant** with respect to γ. A bundle isomorphism $f : P \to Pm$ which is equivariant with respect to the identity is called a **principal bundle automorphism**. The set of all principal bundle automorphisms of P is a subgroup of $\mathrm{Diff}(P)$, called the **automorphism group** of P. It is denoted by $\mathrm{Aut}(P)$. Returning to the general case, if $f : Q \to P$ is an embedding and γ is injective, then we say that Q is embedded in P. Note that in this case the induced morphism $f_0 : N \to M$ is also an embedding. If $M = N$ and $f_0 = id_M$, then Q is called a **reduced subbundle** of P or a **reduction of the structure group** G to H where H is regarded as a subgroup of G. The structure group G of the principal bundle $P(M, G)$ is said to be **reducible** to a subgroup $H \subset G$ if $P(M, G)$ admits a reduction of G to H. If H is a maximal compact subgroup of G, then it can be shown that the bundle $P(M, G)$ can be reduced to a bundle $Q(M, H)$. An application of this result to the bundle of frames $L(M)$ (Example (4.2)) shows that $L(M)(M, GL(m, \mathbf{R}))$ can be reduced to a subbundle $O(M)$ with structure group $O(m, \mathbf{R})$, the orthogonal group. The bundle $O(M)(M, O(m, \mathbf{R}))$ is called the **bundle of orthonormal frames** on M. Furthermore, this reduction is equivalent to the existence of a Riemannian structure on M. The structure group $O(m, \mathbf{R})$ of $O(M)$ can be reduced to the special orthogonal group $SO(m, \mathbf{R})$ if and only if the manifold M is orientable. The reduced subbundle of special orthonormal frames is denoted by $SO(M)(M, SO(m, \mathbf{R}))$.

On the other hand, in some situations one is interested in **extending**, or **lifting**, the structure group of a bundle to obtain a new principal bundle in the following sense. Let $P(M, G)$ be a principal bundle with structure group G. Recall that the center of a group H is the normal subgroup $Z(H)$ of H given by

$$Z(H) := \{x \in H \mid xh = hx, \ \forall h \in H\}.$$

Let H be a Lie group and let $f : H \to G$ be a surjective, covering homomorphism such that $K = \mathrm{Ker} f \subset Z(H)$, the center of H. We say that $P(M, G)$ has a **lift** to a principal bundle $Q(M, H)$ with structure group H if there exists a bundle map $\hat{f} : Q \to P$ such that

$$\hat{f}(uh) = \hat{f}(u)f(h), \qquad \forall u \in Q, h \in H.$$

Using Čech cohomology, Greub and Petry [165] have shown that there exists a **topological obstruction** $\eta(P) \in H^2(M, K)$ to the lifting of $P(M, G)$ to $Q(M, H)$ with the property that

$$h \in \mathcal{F}(N, M) \quad \text{implies that} \quad \eta(h^*(P)) = h^*(\eta(P)).$$

An important special case that appears in many applications is when f is the universal covering map. In this case K is isomorphic to $\pi_1(G)$, the fundamental group of G, and hence $\eta(P) \in H^2(M, \pi_1(G))$. When f is the universal covering map we say that $Q(M, H)$ is a **universal lift**, or a **universal extension**, of $P(M, G)$. The following theorem gives the obstruction $\eta(P)$ in terms of well known characteristic classes (see Chapter 5) for three frequently used groups.

Theorem 4.1 *In the following, H denotes the universal covering group of G.*

1. *Let $G = SO(n)$, then $H = Spin(n)$. In this case the obstruction $\eta(P) = w_2(P) \in H^2(M, \mathbf{Z}_2)$, where $w_2(P)$ is the second Stiefel–Whitney class of P.*
2. *Let $G = SO(3, 1)_+$, the connected component of the identity of the proper Lorentz group; then $H = Spin(3, 1) = SL(2, \mathbf{C})$. In this case $\eta(P) = w_2(P) \in H^2(M, \mathbf{Z}_2)$, where $w_2(P)$ is the second Stiefel–Whitney class of P.*
3. *Let $G = U(n)$, then $H = \mathbf{R} \times SU(n)$. In this case $\eta(P) = c_1(P) \in H^2(M, \mathbf{Z})$, where $c_1(P)$ is the first Chern class of P.*

The following example is a typical application of the above theorem.

Example 4.3 *Let M be an oriented Riemannian m-manifold. Then its bundle of special orthonormal frames $SO(M)$ is a principal $SO(m, \mathbf{R})$ bundle. We say that M admits a **spin structure** if the bundle $SO(M)$ can be extended to the group $Spin(m, \mathbf{R})$. By Theorem 4.1 it follows that M admits a spin structure if and only if the second Stiefel–Whitney class of M is zero. A **spin manifold** is a manifold M together with a fixed spin structure. The principal bundle obtained by this extension is called the **spin frame bundle** and is denoted by $Spin(M)(M, Spin(m, \mathbf{R}))$, or simply by $Spin(M)$.*

The definition of spin manifold can be extended to an oriented pseudo-Riemannian manifold. In the physical literature one usually deals with the case of a 4-dimensional Lorentz manifold. Item 2 of Theorem 4.1 refers to this case.

Let $P(M, G)$ be a principal bundle. The action ρ of G on P defines for each $u \in P$ the map $\rho_u : G \to P$ by $\rho_u(g) = \rho(u, g)$. This induces an injective homomorphism of the Lie algebra $\mathbf{g}$ of G into $\mathcal{X}(P)$ (the Lie algebra of the vector fields on P) as follows. Let $A \in \mathbf{g}$ and let $a_t = exp(tA)$ be the one-parameter subgroup of G generated by A. Restricting the action ρ of G to $exp(tA)$ we get a smooth curve

$$u \cdot exp(tA) := \rho_u(exp(tA)) = \rho(u, exp(tA))$$

through $u \in P$. The tangent vector to this curve at the point $u \in P$ is denoted by $\tilde{A}_u$. The **fundamental vector field** $\tilde{A} \in \mathcal{X}(P)$ of $A \in \mathbf{g}$ is defined by the map

$$u \mapsto \tilde{A}_u := T\rho_u(A).$$

One can verify that the map $\tilde{\rho}$ from $\mathbf{g}$ to $\mathcal{X}(P)$ defined by

$$A \mapsto \tilde{A}$$

is an injective Lie algebra homomorphism. We note that, since G acts freely on P, $A \neq 0$ implies $\tilde{A}_u \neq 0$, $\forall u \in P$. Moreover, $\tilde{A}$ has another important property, namely, that it is a vertical vector field in the sense of the following definition. A vector field $Y \in \mathcal{X}(P)$ is said to be a **vertical vector field** if Y_u is tangent to the fiber of P through u, $\forall u \in P$. We denote by $\mathcal{V}(P)$ the set of vertical vector fields on P. The set $\mathcal{V}(P)$ is called the **vertical bundle** of P. It is a Lie subalgebra of $\mathcal{X}(P)$. Since G preserves the fibers (i.e., acts vertically on P), $\tilde{A}$ is a vertical vector field and the map

$$\tilde{\rho} : \mathbf{g} \to \mathcal{V}(P) \text{ defined by } A \mapsto \tilde{A}$$

is a Lie algebra isomorphism into $\mathcal{V}(P)$ and induces a trivialization of the vertical bundle $\mathcal{V}(P) \cong P \times \mathbf{g}$. The inverse of this isomorphism associates to a vertical vector field an element of the Lie algebra $\mathbf{g}$. This result plays an important role in a definition of connection given later in this chapter.

4.3 Associated Bundles

Let $P(M, G)$ be a principal fiber bundle and F be a left G-manifold, i.e., G acts by diffeomorphisms from the left on F. We denote by r this action of G on F and by gf or $r(g)f$ the element $r(g, f)$. Then we have the following right action R of G on the product manifold $P \times F$

$$(u, f)g = (ug, r(g^{-1})f), \qquad \forall g \in G, (u, f) \in P \times F :$$

The action R is free and its **orbit space** $(P \times F)/R$ is denoted by $P \times_r F$ or by $E(M, F, r, P)$ (or simply E). We denote by

$$\mathcal{O} : P \times F \to E$$

the quotient map, called the **projection onto the orbits of** R, and by $[u, f]$ the **orbit** $\mathcal{O}(u, f)$. We have the following commutative diagram:

$$
\begin{array}{ccc}
P \times F & \xrightarrow{\ p_1\ } & P \\
{\scriptstyle \mathcal{O}}\big\downarrow & & \big\downarrow{\scriptstyle \pi} \\
E & \xrightarrow[\ \pi_E\ ]{} & M
\end{array}
$$

where $\pi_E : E \to M$ is defined by $[u, f] \mapsto \pi(u)$ and p_1 is the projection onto the first factor. (E, M, π_E, F) is a fiber bundle over M with fiber type F. We call $P \times_r F$ or $E(M, F, r, P)$ the **fiber bundle associated to** P with **fiber type** F. If F is a vector space then E is called the **vector bundle with fiber type** F **associated to** P.

Example 4.4 Let Ad *denote the adjoint action of* G *on itself. Then* $P \times_{\mathrm{Ad}} G$ *is a bundle of Lie groups associated to* P, *denoted by* $\mathrm{Ad}(P)$. *We note that* $\mathrm{Ad}(P)$ *is not a principal bundle, in general. Let* ad *denote the adjoint action of* G *on its Lie algebra* $\mathbf{g}$. *Then* $P \times_{\mathrm{ad}} \mathbf{g}$ *is a bundle of Lie algebras associated to* P, *denoted by* $\mathrm{ad}(P)$. *The bundles* $\mathrm{Ad}(P)$ *and* $\mathrm{ad}(P)$ *play a fundamental role in applications to gauge theory.*

Each $u \in P$ induces an isomorphism

$$
\tilde{u} : F \to E_{\pi(u)}, \quad \text{defined by } \tilde{u}(f) = \mathcal{O}(u, f),
$$

where $\mathcal{O}$ is the orbit projection. We note that the map $\tilde{u}$ satisfies the relation

$$
\widetilde{(ug)}(f) = \tilde{u}(gf), \qquad \forall g \in G.
$$

We denote by $\mathcal{F}_G(P, F)$ the space of G-**equivariant maps** of P to F, i.e.,

$$
\mathcal{F}_G(P, F) := \{f : P \to F \mid f(ug) = g^{-1}(f(u)), \ \forall g \in G\}.
$$

Recall that $\Gamma(E) = \Gamma_M(E(M, F, r, P))$ is the space of smooth sections of E over M. There exists a one-to-one correspondence between $\mathcal{F}_G(P, F)$ and $\Gamma(E)$ which is defined as follows: For $f \in \mathcal{F}_G(P, F)$ we define $s_f \in \Gamma(E)$ by

$$
s_f(x) = [u, f(u)], \quad \text{where } u \in \pi^{-1}(x).
$$

It is easy to verify that s_f is well defined and that $f \mapsto s_f$ is a one-to-one correspondence from $\mathcal{F}_G(P, F)$ to $\Gamma(E)$ with the inverse defined as follows: If $s \in \Gamma(E)$, then $f_s \in \mathcal{F}_G(P, F)$ is defined by

$$
f_s(u) = \tilde{u}^{-1}(s(\pi(u))).
$$

Example 4.5 Let M *be an* m-*dimensional manifold. The tangent bundle* TM *of* M *is an associated bundle of* $L(M)$ *with fiber type* $\mathbf{R}^m$ *and left action* *r given by the defining representation of $GL(m, \mathbf{R})$ on $\mathbf{R}^m$, i.e.,*

$$TM = E(M, \mathbf{R}^m, r, L(M)).$$

For $u \in L_x(M)$ we have the map

$$\tilde{u} : \mathbf{R}^m \to T_x(M),$$

which is a linear isomorphism. Thus, we could define a frame by the map $\tilde{u}$ using the vector bundle structure of TM with the action of $GL(m, \mathbf{R})$ given by the composition

$$\tilde{u} \cdot g = \tilde{u} \circ g, \qquad \forall g \in GL(m, \mathbf{R}),$$

where g is regarded as a map from $\mathbf{R}^m$ to $\mathbf{R}^m$. Similarly it can be shown that the various tensor bundles $T_s^r(M)$ and form bundles $\Lambda^k(M)$ are associated bundles of $L(M)$.

Example 4.6 *Let M be a spin manifold with spin bundle $Spin(M)$. Let $Spin(M) \times_\rho V$ be the bundle associated to $Spin(M)$ by the representation ρ of $Spin(m, \mathbf{R})$ on the complex vector space V. Then $S_\rho = \Gamma(Spin(M) \times_\rho V)$ is called the space of **spinors of type** ρ. By the first part of the Theorem 4.1 we can conclude that M admits a spin structure if and only if $w_2(M) := w_2(SO(M))$, the second Stiefel–Whitney class of M, is zero. Topological classification of spin structures is attributed to Milnor. A detailed account of spin geometry may be found in Lawson and Michelsohn [248]. The mathematical foundations of the theory of spinors over Lorentz manifolds were laid by Cartan in [71], where the Dirac operator on spinors was introduced to study Dirac's equation for the electron. This operator and its various extensions play a fundamental role in the study of the topology and the geometry of manifolds arising in gauge theory.*

Associated bundles can be used to give the following formulation of the concept of reduction of the structure group of a principal bundle.

Theorem 4.2 *If H is a closed subgroup of G, then the structure group G of $P(M, G)$ is reducible to H if and only if the associated bundle $E(M, G/H, q, P(M, G))$ admits a section (where q is the canonical action of G on the quotient G/H).*

This theorem is useful in studying the existence of pseudo-Riemannian structures on a manifold and their topological implications.

4.4 Connections and Curvature

The idea of Riemannian manifold was introduced in Riemann's famous lecture "Über die Hypothesen, welche der Geometrie zugrunde liegen" (On the hypotheses that lie at the foundations of geometry). Riemann's ideas were extended by Ricci and Levi-Civita, who gave a systematic account of Riemannian geometry and also introduced the notion of covariant differentiation. Their work influenced Einstein (via Grossmann), who used it to formulate his general theory of relativity. Weyl and Cartan introduced the idea of affine, projective, and conformal connections and found interesting applications of these to physical theories. The notion of connection in a fiber bundle was introduced by Ehresmann and this influenced all subsequent developments of the theory of connections. In particular his notion of a Cartan connection includes affine, projective, and conformal connections as special cases.

In this section we give several definitions of connection on a principal bundle and define the curvature 2-form. The structure equations and Bianchi identities satisfied by the curvature are also given there. A subsection is devoted to a detailed discussion of universal connections.

Let $P(M, G)$ be a principal bundle with structure group G and canonical projection π over a manifold M of dimension m. Recall that a k-dimensional distribution on P is a smooth map $L : P \to TP$ such that $L(u)$ is a k-dimensional subspace of $T_u P$, for all $u \in P$.

Definition 4.5 *A **connection** Γ in $P(M, G)$ is an m-dimensional distribution $H : u \mapsto H_u P$, on P such that the following conditions are satisfied for all $u \in P$:*

1. *$T_u P = V_u P \oplus H_u P$, where $V_u P = \mathrm{Ker}(\pi_{*u})$ is the **vertical subspace** of the tangent space $T_u P$;*
2. *$H_{\rho_a(u)} P = (\rho_a)_* H_u P$, $\forall a \in G$, where ρ_a is the right action of G on P determined by a.*

*The first condition in the above definition can be taken as a definition of a **horizontal distribution** H. The second condition may be rephrased as follows:*

2'. The distribution H is invariant under the action of G on P.

We call $H_u P$ (also denoted simply by H_u) the Γ-**horizontal subspace**, or simply the horizontal subspace of $T_u P$. The union HP of the horizontal subspaces is a manifold, called the **horizontal bundle** of P. The union VP of the vertical subspaces is a manifold called the **vertical bundle** of P. We will also write simply V_u instead of $V_u P$. A vector field $X \in \mathcal{X}(P)$ is called **vertical** if $X(u) \in V_u P$, $\forall u \in P$. We note that while the definition of HP depends on the connection on P, the definition of the vertical bundle VP is independent of the connection on P. Condition (1) allows us to decompose each $X \in T_u P$ into its **vertical part** $v(X) \in V_u$ and the **horizontal part** $h(X) \in H_u$. If $Y \in \mathcal{X}(P)$ is a vector field on P then

$$v(Y) : P \rightarrow TP \text{ defined by } u \mapsto v(Y_u)$$

and

$$h(Y) : P \rightarrow TP \text{ defined by } u \mapsto h(Y_u)$$

are also in $\mathcal{X}(P)$. We observe that

$$\pi_{*u} : H_u \rightarrow T_{\pi(u)} M$$

is an isomorphism. The vector field $Y \in \mathcal{X}(P)$ is said to be Γ-**horizontal** (or simply horizontal) if $Y_u \in H_u$, $\forall u \in P$. The set of horizontal vector fields is a vector subspace but not a Lie subalgebra of $\mathcal{X}(P)$. For $X \in \mathcal{X}(M)$ the Γ-**horizontal lift** (or simply the **lift**) of X to P is the unique horizontal field $X^h \in \mathcal{X}(P)$ such that $\pi_* X^h = X$. Note that $X^h \in \mathcal{X}(P)$ is invariant under the action of G on P and every horizontal vector field $Y \in \mathcal{X}(P)$ invariant under the action of G is the lift of some $X \in \mathcal{X}(M)$, i.e., $Y = X^h$ for some $X \in \mathcal{X}(M)$. In the following proposition we collect some important properties of the horizontal lift of vector fields.

Proposition 4.3 *Let $P(M, G)$ be a principal bundle with connection Γ. Let $X, Y \in \mathcal{X}(M)$ and $f \in \mathcal{F}(M)$; then we have the following properties:*

1. $X^h + Y^h = (X + Y)^h$,
2. $(\pi^* f) X^h = (fX)^h$,
3. $h([X^h , Y^h]) = [X , Y]^h$.

A smooth curve c in P (i.e. c is a smooth function from some open interval $I \subset \mathbf{R}$ into P) is called a **horizontal curve** if $\dot{c}(t) \in H_{c(t)}$, $\forall t \in I$. A section $s \in \Gamma(P)$ is called **parallel** if

$$s_*(T_x M) \subset H_{s(x)}, \qquad \forall x \in M;$$

i.e., if $s \circ c$ is a horizontal curve for all curves c in M. Given a curve $x : [0, 1] \rightarrow M$ and $w_0 \in P$ such that $\pi(w_0) = x(0)$, there is a unique horizontal curve $w : [0, 1] \rightarrow P$ such that $w(0) = w_0$ and

$$\pi(w(t)) = x(t), \qquad \forall t \in [0, 1].$$

The curve w in P is called the **horizontal lift of the curve** x starting from $w_0 \in P$. If P_0 (resp., P_1) denotes the fiber of P over $x(0)$ (resp., $x(1)$) then the horizontal lift of x to P induces a diffeomorphism of P_0 with P_1 called the **parallel displacement** along x. In particular, given a loop (closed curve) x at $p \in M$, i.e., $p = x_0 = x_1$, the horizontal lift of x to P induces an automorphism of the fiber $\pi^{-1}(p)$. The set of all such automorphisms forms a group called the **holonomy group** of the connection Γ at $p \in M$. It is denoted by $\Phi(p)$. The subset of $\Phi(p)$ corresponding to parallel displacement along loops homotopic to the constant loop at p turns out to be a subgroup of $\Phi(p)$. It is called the **restricted holonomy group** of the connection Γ

at $p \in M$ and is denoted by $\varPhi^0(p)$. Given a point $u \in \pi^{-1}(p)$, each $\alpha \in \varPhi(p)$ (resp., $\varPhi^0(p)$) determines a unique $g \in G$ such that $\alpha(u) = ug$. The map $\alpha \mapsto g$ of $\varPhi(p)$ (resp., $\varPhi^0(p)$) to G is an isomorphism onto a subgroup $\varPhi_u$ (resp., $\varPhi_u^0$) called the **holonomy group** (resp., **restricted holonomy group**) of the connection $\varGamma$ at $u \in P$. It can be shown that if M is paracompact, then the holonomy group $\varPhi_u$ is a Lie subgroup of the structure group G, $\varPhi_u^0$ is a normal subgroup of H_u and $\varPhi_u/\varPhi_u^0$ is countable. Now we let $P(u)$ denote the set of points of P that can be joined to u by a horizontal curve. Then we have the following theorem.

Theorem 4.4 *Let M be a connected, paracompact manifold and $\varGamma$ a connection in the principal bundle $P(M, G)$. Then*

1. *$P(u)$, $u \in P$, is a reduced subbundle of P with structure group $\varPhi_u$. It is called the **holonomy bundle** (of $\varGamma$) through u.*
2. *The holonomy bundles $P(u)$, $u \in P$, are isomorphic with one another and partition P into a disjoint union of reduced subbundles.*
3. *The connection $\varGamma$ is reducible to a connection in $P(u)$.*

We now give two important characterizations of a connection that are often used as definitions of a connection. Recall first that a k-form $\alpha \in \varLambda^k(P, V)$ (the space of k-forms on P with values in the vector space V) can be identified with the map

$$u \mapsto \alpha_u, \qquad u \in P,$$

where

$$\alpha_u : \underbrace{T_u P \times \cdots \times T_u P}_{k \text{ times}} \to V$$

is a multilinear anti-symmetric map. Given a basis $\{v_i\}_{1 \le i \le n}$ in V, we can express α as the formal sum

$$\alpha = \sum_{i=1}^{n} \alpha_i v_i,$$

where $\alpha_i \in \varLambda^k(P)$, $\forall i$. We define a 1-form $\omega \in \varLambda^1(P, \mathbf{g})$ on P with values in the Lie algebra $\mathbf{g}$ by using the connection $\varGamma$ as follows:

$$\omega_u(X_u) = \tilde{\rho}_u^{-1}(v(X_u)), \qquad \forall u \in P, \quad \forall X_u \in T_u P, \tag{4.2}$$

where $\tilde{\rho}_u : \mathbf{g} \to \mathcal{V}(P)_u$ is the isomorphism induced by the action of G on P. It is customary to write (4.2) in the form

$$\omega(X) = \tilde{\rho}^{-1}(v(X)). \tag{4.3}$$

The 1-form ω is called the **connection 1-form** of the connection $\varGamma$. It can be shown that the connection 1-form ω satisfies the following conditions:

$$\omega(\tilde{A}) = A, \qquad \forall A \in \mathbf{g}, \tag{4.4}$$

$$(\rho_a)^*\omega = \text{ad}(a^{-1})\omega, \qquad \forall a \in G. \tag{4.5}$$

Condition (4.5) means that ω is G-equivariant, i.e., the following diagram commutes:

$$
\begin{array}{ccc}
T_u P & \xrightarrow{\;T\rho_a\;} & T_{ua} P \\[2pt]
\Big\downarrow{\scriptstyle \omega_u} & & \Big\downarrow{\scriptstyle \omega_{ua}} \\[6pt]
\mathfrak{g} & \xrightarrow[\text{ad}(a^{-1})]{} & \mathfrak{g}
\end{array}
$$

Condition (4.5) can also be expressed as

$$\omega(ua)(T\rho_a(X(u))) = \text{ad}(a^{-1})(\omega(u)X(u)), \qquad \forall X \in \mathcal{X}(P),$$

where we have written $X(u)$ for X_u etc. Conditions (4.4) and (4.5) characterize the connection 1-form ω on P. In fact, one may give the following alternative definition of a connection.

Definition 4.6 *A **connection** in $P(M,G)$ is a 1-form ω on P with values in the Lie algebra $\mathfrak{g}$) (i.e., $\omega \in \Lambda^1(P,\mathfrak{g})$), which satisfies conditions (4.4) and (4.5) given above.*

Note that given a 1-form ω satisfying conditions (4.4) and (4.5), we can define the distribution $H : u \mapsto H_u$ on P by

$$H_u := \{Y \in T_u P \mid \omega_u(Y) = 0\}. \tag{4.6}$$

One can then verify that the distribution H is m-dimensional and defines a connection Γ according to the Definition 4.5 and that the connection 1-form associated to Γ is ω. Let $\mathfrak{h}$ be a Lie subalgebra of $\mathfrak{g}$. We say that the connection ω is **reducible** if $\omega(X) \in \mathfrak{h}, \forall X \in \mathcal{X}(P)$. A connection ω that is not reducible is called **irreducible**. These concepts play an important role in studying the space of all connections on P. These spaces are fundamental in the construction of various gauge fields that we study in later chapters.

We now motivate another characterization of a connection in terms of a local representation of the bundle $P(M,G)$. Let ω be a connection on $P(M,G)$. Let $\{(U_i, \psi_i)\}_{i \in I}$ be a local representation of $P(M,G)$ with transition functions

$$\psi_{ij} : U_{ij} \to G.$$

Let $e \in G$ be the identity element of G and let $s_i : U_i \to P$ be a local section defined by

$$s_i(x) = \psi_i(x, e).$$

Define the family $\{\omega_i\}_{i \in I}$ of 1-forms

$$\omega_i \in \Lambda^1(U_i, \mathbf{g}) \quad \text{by} \quad \omega_i = s_i^*(\omega),$$

where ω is a connection on P. Let $\Theta \in \Lambda^1(G, \mathbf{g})$ be the canonical left invariant 1-form on G defined by $\Theta(A) = A$, $\forall A \in \mathbf{g}$. Alternatively, Θ can be defined by

$$\Theta(g) := TL_{g^{-1}} : T_g G \to T_e G \equiv \mathbf{g}, \ \forall g \in G.$$

Then writing $\Theta_{ij} = \psi_{ij}^* \Theta$, we obtain the following relations

$$\omega_j(x) = \mathrm{ad}(\psi_{ij}(x)^{-1})\omega_i(x) + \Theta_{ij}(x), \qquad \forall x \in U_{ij} \text{ and } \forall i, j \in I. \quad (4.7)$$

Thus, given a connection ω on P, we have a family $\{(U_i, \psi_i, \omega_i)\}_{i \in I}$ where $\{(U_i, \psi_i)\}_{i \in I}$ is a local representation of P and $\{\omega_i\}_{i \in I}$ is a family of 1-forms satisfying conditions (4.7). Conversely, given such a family $\{(U_i, \psi_i, \omega_i)\}_{i \in I}$ satisfying conditions (4.7), a connection is determined in the following way. Let $\psi_i^{-1} : \pi^{-1}(U_i) \to U_i \times G$ and let p_1, p_2 be the canonical projections of $U_i \times G$ on the first and second factors, respectively. Let

$$\alpha_i = (p_1 \circ \psi_i^{-1})^* \omega_i + (p_2 \circ \psi_i^{-1})^* \Theta.$$

Then $\alpha_i \in \Lambda^1(\pi^{-1}(U_i), \mathbf{g})$. Define $\omega \in \Lambda^1(P, \mathbf{g})$ by $\omega = \alpha_i$ on $\pi^{-1}(U_i)$. This is well defined because on the intersection $\pi^{-1}(U_{ij})$ the conditions guarantee that $\alpha_i = \alpha_j$. Then one can show that ω is a connection according to Definition 4.6. Thus we may give the following third definition of a connection equivalent to the two definitions given above.

Definition 4.7 *A* **connection** *in the bundle $P(M, G)$ is a family of triples $\{(U_i, \psi_i, \omega_i)\}_{i \in I}$, where $\{(U_i, \psi_i)\}_{i \in I}$ is a local representation of P and $\{\omega_i \in \Lambda^1(U_i, \mathbf{g})\}_{i \in I}$ is a family of 1-forms satisfying the relations (4.7).*

Frequently, it is this local definition that is used to construct a connection via a suitable local representation of P.

Example 4.7 *Let M be a paracompact manifold and let $P = L(M)$ be the bundle of frames on M. Recall that M admits a Riemannian metric so that the bundle of frames $L(M)$ can be reduced to a bundle of orthonormal frames. Using the local isomorphism with $\mathbf{R}^m$ we can pull back the flat connection on the frame bundle of $\mathbf{R}^m$ to $L(M)$, locally. These local connections can be pieced together by a partition of unity argument to obtain the Riemannian connection in $L(M)$.*

Let $\phi \in \Lambda^k(P, V)$ be a k-form in P with values in a finite-dimensional vector space V. Let $r : G \to GL(V)$ be a representation of G on V. We say that ϕ is **pseudo-tensorial** of type (r, V) if

$$\rho_a^* \phi = r(a^{-1}) \cdot \phi, \ \forall a \in G.$$

A connection 1-form ω on P is pseudo-tensorial of type $(\mathrm{ad}, \mathbf{g})$. The form $\phi \in \Lambda^k(P, V)$ is called **horizontal** if

$$\phi(X_1, X_2, \ldots, X_k) = 0$$

whenever some X_i, $1 \leq i \leq k$, is vertical. The form $\phi \in \Lambda^k(P, V)$ is called **tensorial** of type (r, V) if it is horizontal and pseudo-tensorial of type (r, V). If the k-form $\phi \in \Lambda^k(P, V)$ is tensorial of type (r, V), then there exists a unique k-form s_ϕ on M with values in the vector bundle $E = P \times_r V$ defined as follows:

$$s_\phi(x)(X_1, X_2, \ldots, X_k) = \tilde{u}\phi(u)(Y_1, Y_2, \ldots, Y_k), \qquad \forall x \in M,$$

where $u \in \pi^{-1}(x)$ and $Y_i \in T_u P$ such that $T\pi(Y_i) = X_i$, $1 \leq i \leq k$. Due to tensoriality of ϕ this definition of s_ϕ is independent of the choice of u and Y_i. We call $s_\phi \in \Lambda^k(M, E)$ the k-**form associated to** ϕ. We note that the one-to-one correspondence $f \mapsto s_f$ between $\mathcal{F}_G(P, F)$ and $\Gamma(E)$ extends to a one-to-one correspondence $\phi \mapsto s_\phi$ of tensorial forms of type (r, V) and forms with values in the vector bundle E defined above.

Given a connection 1-form ω on P we define the **exterior covariant differential**

$$d^\omega : \Lambda^k(P, \mathbf{g}) \to \Lambda^{k+1}(P, \mathbf{g})$$

on (k-forms on) P (with values in $\mathbf{g}$) by

$$d^\omega \alpha(X_0, \ldots, X_k) = d\alpha(h(X_0), \ldots, h(X_k)), \qquad \forall \alpha \in \Lambda^k(P, \mathbf{g}).$$

We define the **curvature** 2-form $\Omega \in \Lambda^2(P, \mathbf{g})$ of the connection 1-form ω by

$$\Omega := d^\omega \omega. \tag{4.8}$$

It is easy to verify that Ω is a tensorial 2-form of type (ad, $\mathbf{g}$) and that it satisfies the condition

$$d\omega(X, Y) = \Omega(X, Y) - [\omega(X), \ \omega(Y)]. \tag{4.9}$$

Equation (4.9) is called the **structure equation** and is often written in the form

$$d\omega = \Omega - \omega \wedge \omega.$$

Using definition (4.8) we obtain the **Bianchi identity** on the bundle P,

$$d^\omega \Omega = 0. \tag{4.10}$$

The one-to-one correspondence $\phi \mapsto s_\phi$ of tensorial forms of type (ad, $\mathbf{g}$) and forms with values in the vector bundle $\mathrm{ad}(P)$ defined above associates to the curvature form Ω a unique 2-form $F_\omega \in \Lambda^2(M, \mathrm{ad}(P))$ so that

$$F_\omega = s_\Omega. \tag{4.11}$$

The 2-form F_ω is called the **curvature 2-form** on M with values in the **adjoint bundle** $\mathrm{ad}(P)$ corresponding to the connection ω on P.

4.4.1 Universal Connections

We now discuss an important class of principal bundles with natural connections, which play a fundamental role in the classification of principal bundles with connections. Let F stand for $\mathbf{R}, \mathbf{C}$, or $\mathbf{H}$. On F^n we have the standard inner product defined by

$$\langle u, v \rangle := \sum_{i=1}^{n} \bar{u}_i v_i,$$

where the bar denotes the conjugation on F, which is the identity on $\mathbf{R}$ and complex (resp., quaternionic) conjugation when F is $\mathbf{C}$ (resp., $\mathbf{H}$). Let $U_F(n)$ denote the connected component of the identity of the Lie group of isometries of F^n (i.e., linear automorphisms of F^n preserving the above inner product). Then we have

$$U_F(n) = \begin{cases} SO(n), & \text{if } F = \mathbf{R}, \\ U(n), & \text{if } F = \mathbf{C}, \\ Sp(n), & \text{if } F = \mathbf{H}. \end{cases}$$

The real dimension of $U_F(n)$ is $\frac{1}{2}n(n+1)\dim_{\mathbf{R}} F - n$.

A k-**frame** in F^n is an ordered orthonormal set $(u_1, u_2, \ldots, u_k)$ of k vectors in F^n. The set of k-frames can be given a natural structure of smooth manifold. The connected component of the k-frame $(e_1, e_2, \ldots, e_k)$, where $\{e_i, \ 1 \le i \le n\}$ is the set of standard orthonormal basis vectors in F^n, is called the **Stiefel manifold** of k-frames in F^n. It is denoted by $V_k(F^n)$. The group $U_F(n)$ acts transitively on $V_k(F^n)$. The stability group of this action at the k-frame $(e_1, e_2, \ldots, e_k)$ can be identified with the subgroup $U_F(l)$ of the group $U_F(n)$, where $l = n - k$. In block matrix form we can write an element of the group $U_F(l)$ as

$$\begin{pmatrix} I_k & O_{(k,n-k)} \\ O_{(n-k,k)} & T \end{pmatrix},$$

where I_k is the $k \times k$ unit matrix, $O_{(i,j)}$ is the $i \times j$ zero matrix, and T is the $l \times l$ matrix in $U_F(l)$ (regarded as the group of isometries of F^l). Hence, we can identify the Stiefel manifold $V_k(F^n)$ as the left coset space $U_F(n)/U_F(l)$. Thus, $U_F(n)$ is a principal $U_F(l)$-bundle over the Stiefel manifold $V_k(F^n)$. We use the following notation for indices:

1. lower case Latin letters $i, j, \ldots$ take values from 1 to k;
2. upper case Latin letters $A, B, \ldots$ take values from 1 to $l = n - k$;
3. Greek letters $\alpha, \beta, \ldots$ take values from 1 to n.

Then the canonical projection $\pi_0 : U_F(n) \to V_k(F^n)$ is given by

$$\begin{pmatrix} a_{ij} & a_{iB} \\ a_{Aj} & a_{AB} \end{pmatrix} \to (u_1, u_2, \ldots, u_k) \in V_k(F^n),$$

where $u_j := \sum_{\alpha=1}^{n} e_\alpha a_{\alpha j}$. We note that the subgroup $U_F(k)$ of the group $U_F(n)$ formed by elements of the type

$$\begin{pmatrix} T & O_{(k,n-k)} \\ O_{(n-k,k)} & I_{n-k} \end{pmatrix}$$

acts on F^n, sending a k-frame to a k-frame and inducing a free action of $U_F(k)$ on the Stiefel manifold $V_k(F^n)$ on the right. The quotient manifold of $V_k(F^n)$ under this action is called the **Grassmann manifold** and is denoted by $G_k(F^n)$. It can be defined directly as the set of k-planes through the origin in F^n. Two elements $u, v \in V_k(F^n)$ determine the same k-plane in F^n if and only if there exists $t \in U_F(k)$ such that $ut = v$. Thus, $V_k(F^n)$ can be regarded as a principal $U_F(k)$-bundle over the Grassmann manifold $G_k(F^n)$ with canonical projection π. In particular, $G_1(\mathbf{C}^n) = \mathbf{CP}^n$ and each point is a complex line. The assignment of this line to the corresponding point defines a complex line bundle which is called the **tautological** complex line bundle. Similar definition can be give for the real and quaternionic case.

Let Θ denote the canonical Cartan 1-form on the Lie group $U_F(n)$ with values in the corresponding Lie algebra $u_F(n)$. It is often written in the form

$$\Theta = g^{-1}dg = (\Theta_{\alpha\beta}), \tag{4.12}$$

where $(\Theta_{\alpha\beta})$ are F-valued 1-forms on $U_F(n)$ which satisfy the relations

$$\bar{\Theta}_{\alpha\beta} + \Theta_{\alpha\beta} = 0. \tag{4.13}$$

The form Θ is left invariant and right equivariant under the adjoint action and satisfies the **Maurer–Cartan equations**

$$d\Theta + \Theta \wedge \Theta = 0, \quad \text{i.e., } d\Theta_{\alpha\beta} + \Theta_{\alpha\gamma} \wedge \Theta_{\gamma\beta} = 0. \tag{4.14}$$

The forms $\Theta_{\alpha j}$ are horizontal relative to the projection π_0. The forms Θ_{ij} are invariant under the action of $U_F(l)$ on $U_F(n)$ and hence project to forms on the Stiefel manifold $V_k(F^n)$, which we continue to denote by the same symbols. The 1-form $\omega = (\Theta_{ij})$ with values in the Lie algebra $u_F(k)$ satisfies the conditions for it to be a connection form on the bundle $V_k(F^n)$ regarded as a principal $U_F(k)$-bundle over the Grassmann manifold $G_k(F^n)$. The connection defined by this form ω is called a **universal connection** in view of the fact that the principal bundle $V_k(F^n)(G_k(F^n), U_F(k))$ with connection ω is m-classifying for $m < kl$, i.e. given a principal $U_F(k)$-bundle P over a compact manifold M of dim $M \leq m$ with connection α there exists a map $f : M \to G_k(F^n)$ such that $P = f^*(V_k(F^n))$ and $\alpha = f^*\omega$ (see Chapter 5 for further discussion).

The Stiefel and Grassmann manifolds carry a natural Riemannian structure, which can be described as follows. The quadratic form $\sum_{\alpha,j} \bar{\Theta}_{\alpha j} \otimes \Theta_{\alpha j}$ on $U_F(n)$ is invariant under the action of $U_F(l)$ and hence descends to the quotient to define the metric $d\sigma^2$ on $V_k(F^n)$ such that

$$\sum_{\alpha,j} \bar{\Theta}_{\alpha j} \otimes \Theta_{\alpha j} = \pi_0^*(d\sigma^2).$$

The quadratic form $\sum_{A,j} \bar{\Theta}_{Aj} \otimes \Theta_{Aj}$ on $U_F(n)$ is invariant under the left $U_F(l)$ and the right $U_F(k)$ action and therefore descends to define a Riemannian metric ds^2 on $G_k(F^n)$. We have the following relations

$$\sum_{A,j} \bar{\Theta}_{Aj} \otimes \Theta_{Aj} = (\pi \circ \pi_0)^*(ds^2), \qquad (4.15)$$

$$d\sigma^2 = \pi^*(ds^2) + \sum_{i,j} \bar{\Theta}_{ij} \otimes \Theta_{ij}. \qquad (4.16)$$

We note that this metric is useful in generalized Kaluza–Klein theories. In the special case of $k = 1$ and $F = \mathbf{C}$ (resp., $\mathbf{H}$), the metric $d\sigma^2$ reduces to the natural Riemannian metric on a sphere S^{2n-1} (resp., S^{4n-1}), and the metric ds^2 reduces to the **Fubini–Study metric** on the corresponding projective space $\mathbf{CP}^{n-1}$ (resp., $\mathbf{HP}^{n-1}$). These are the classical **Hopf fibrations**, also called the **Hopf fiberings**. In the case $n = 2$, the complex fibration appears in consideration of the Dirac monopole whereas the quaternionic fibration corresponds to the well-known BPST instanton solution of the Yang–Mills equations (see Chapter 8 for further details).

We observe that $U_F(n)$ is a $U_F(k) \times U_F(l)$-bundle over $G_k(F^n)$ and hence we may view $U_F(n)$ as a restriction of the bundle of orthonormal frames of $G_k(F^n)$ corresponding to the canonical injection $U_F(k) \times U_F(l) \hookrightarrow U_F(kl)$. The form $\delta_{ij}\Theta_{AB} + \delta_{AB}\Theta_{ij}$ on $U_F(n)$ defines the **Levi-Civita connection** for $G_k(F^n)$.

4.5 Covariant Derivative

Let $P(M, G)$ be a principal bundle and $E(M, F, r, P)$ be the associated fiber bundle over M with fiber type F and action r. A connection Γ in P allows us to define the notion of a **horizontal vector field** on E. Let $w \in E$ and $(u, a) \in \mathcal{O}^{-1}(w)$, where $\mathcal{O} : P \times F \to E$ is the canonical orbit projection. Define

$$f_a : P \to E \quad \text{by } u \mapsto \mathcal{O}(u, a),$$

where $\mathcal{O}(u, a)$ is the orbit of (u, a) in E. Now define $H_w = H_w E \subset T_w E$ by

$$H_w := (f_a)_*(H_u P),$$

where $H_u P$ is the horizontal subspace of $T_u P$. It can be shown that $H_w E$ is independent of the choice of $(u, a) \in \mathcal{O}^{-1}(w)$ and is thus well-defined. The assignment $w \mapsto H_w$, $w \in E$, defines a **connection** on E. Thus, one has the notions of horizontal vector field on E, horizontal lift of a vector field on M, horizontal curve, and horizontal lift of a curve to E. Given a curve $x : [0,\ 1] \rightarrow M$ and $w_0 \in E$ such that $\pi_E(w_0) = x(0)$, there is a unique horizontal lift $w : [0,1] \rightarrow E$ of the curve x such that $w(0) = w_0$ and

$$\pi_E(w(t)) = x(t), \qquad \forall t \in [0,\ 1],$$

i.e., the curve w in E is horizontal and $\pi_E(w(t)) = x(t)$. Given the curve x joining x_0 and x_1, the horizontal lift induces a diffeomorphism of $\pi_E^{-1}(x_0)$ and $\pi_E^{-1}(x_1)$ called **parallel translation** or **parallel displacement** of fibers of E. In particular, if E is a vector bundle (i.e., the fiber type F is a vector space), then parallel displacement is an isomorphism and we may define the **covariant derivative** $\nabla_{\dot{x}(t)}s$ of a section s at $x(t)$ along the vector $\dot{x}(t)$ by the formula

$$\nabla_{\dot{x}(t)}s = \lim_{h \to 0} \frac{1}{h}[c_{t,t+h}^{-1}(s(x(t+h))) - s(x(t))],$$

where

$$c_{t,t+h} : \pi_E^{-1}(x(t)) \rightarrow \pi_E^{-1}(x(t+h))$$

is the parallel displacement along x from $x(t)$ to $x(t+h)$. If $X \in \mathcal{X}(M)$ and x is the integral curve of X through x_0, so that $X_{x(t)} = \dot{x}(t)$, then the above definition may be used to define the **covariant derivative** $\nabla_X s$.

Theorem 4.5 *The covariant derivative operator* $\nabla_X : \Gamma(E) \rightarrow \Gamma(E)$ *satisfies the following relations,* $\forall X, Y \in \mathcal{X}(M)$, $\forall f \in \mathcal{F}(M)$ *and* $\forall t, s \in \Gamma(E)$:

1. $\nabla_{X+Y}s = \nabla_X s + \nabla_Y s$,
2. $\nabla_X(s+t) = \nabla_X s + \nabla_X t$,
3. $\nabla_{fX}s = f\nabla_X s$,
4. $\nabla_X(fs) = f\nabla_X s + (Xf)s$.

We note that the covariant derivative may be defined in terms of the Lie derivative as follows. Recall first that there is a one-to-one correspondence between G-equivariant functions from P to F and sections in $\Gamma(E)$, which is defined as follows. Let $s \in \Gamma(E)$; then we define

$$f_s : P \rightarrow F \quad \text{by } u \mapsto \tilde{u}^{-1}(s(x)),$$

where $\pi(u) = x$ and $\tilde{u} : F \rightarrow E_x$ is the isomorphism defined by $u \in P$. Conversely, given a G-equivariant map f from P to F, define $s_f \in \Gamma(E)$ by

$$s_f(x) = \mathcal{O}(u, f(u)),$$

where $u \in \pi^{-1}(x)$. This is well-defined because of equivariance. The covariant derivative $\nabla_X s$ corresponds to $L_{\hat{X}}(f_s)$ where f_s is defined above and $\hat{X}$ is the horizontal lift of X to P. The covariant derivative defines a map ∇ from $\Gamma(E)$ to $\Gamma(T^*(M) \otimes E)$ as follows. Recall that a section $s \in \Gamma(T^*(M) \otimes E)$ may be regarded as defining, for each $x \in M$, a map $s(x) : T_x M \to E_x$. For $s \in \Gamma(E)$, let $\nabla s : M \to T^*(M) \otimes E$ be the map defined by

$$\nabla s(x) \cdot X_x = (\nabla_X s)(x).$$

Then we have

$$\nabla(fs) = df \otimes s + f\nabla s, \qquad \forall f \in \mathcal{F}(M) \quad \text{and} \quad \forall s \in \Gamma(E). \tag{4.17}$$

To indicate the dependence of ∇ on the connection 1-form ω, it is customary to denote it by ∇^ω, and the same notation is used for its natural extension to E-valued tensors on M. We note that the operator ∇^ω maps $\Lambda^0(M, E)$ to $\Lambda^1(M, E)$. We denote the extension of the operator ∇^ω to $\Lambda^p(M, E)$ by d^ω, in agreement with the notation used for vector-valued forms. It is defined as follows:

$$d^\omega \alpha(X_0, \ldots, X_p) := \sum_{j=0}^{p} (-1)^j \nabla_{X_j} (\alpha(X_0, \ldots, \hat{X}_j, \ldots, X_p))$$
$$+ \sum_{i<j} (-1)^{i+j} \alpha([X_i, X_j], X_0, \ldots, \hat{X}_i, \ldots, \hat{X}_j, \ldots, X_p),$$

where the hat sign $\hat{\ }$ on a vector field denotes deletion of that vector field. The operator d^ω is called the **exterior covariant differential** or simply the **covariant differential** on M. Applying the above equation to the curvature 2-form $F_\omega \in \Lambda^2(M, \mathrm{ad}(P))$ we obtain the Bianchi identity on M,

$$d^\omega F_\omega = 0. \tag{4.18}$$

The **classical Bianchi identity** is a special case of the above identity when ω is the Levi-Civita connection on the principal bundle of orthonormal frames. The operator d^ω satisfies the following relation:

$$d^\omega(\beta \wedge \alpha) = (d^\omega \beta) \wedge \alpha + (-1)^{\deg(\beta)} \beta \wedge (d^\omega \alpha), \tag{4.19}$$

where $\alpha \in \Lambda^*(M, E)$ and $\beta \in \Lambda^*(M, \mathbf{R})$ is a homogeneous form. In fact, this relation can be used to define the operator d^ω. It is customary to write d^ω_p for the above operator if we want to emphasize its action on p-forms, otherwise we write d^ω to denote any one of these operators. Thus our earlier definition of $d^\omega : \Lambda^0(M, E) \to \Lambda^1(M, E)$, combined with the above definition, gives us the sequence

$$0 \longrightarrow \Lambda^0(M, E) \xrightarrow{d^\omega} \Lambda^1(M, E) \xrightarrow{d^\omega} \Lambda^2(M, E) \xrightarrow{d^\omega} \cdots,$$

which is called the **generalized de Rham sequence**. We note that if $\alpha \in \Lambda^0(M, E)$, then we have

$$d^\omega \circ d^\omega(\alpha) = \alpha \Omega. \tag{4.20}$$

Thus, we can consider the curvature Ω as a zeroth order operator, which acts as an **obstruction** to the vanishing of $(d^\omega)^2 := d^\omega \circ d^\omega$. We will find these concepts useful in the study of gauge fields and their associated fields.

4.6 Linear Connections

Let $L(M)$ be the bundle of frames of M. Then we have seen that $L(M)$ is a principal bundle with structure group $GL(m, \mathbf{R})$. A connection on this principal bundle is called a **linear connection**. If Γ is a linear connection on M, the 1-form ω of Γ is a 1-form on $L(M)$ with values in $gl(m, \mathbf{R})$. Let us denote by $\{u_j^i \mid i, j = 1, 2, \ldots, m\}$ the natural basis of $gl(m, \mathbf{R})$, where u_j^i is the $n \times n$ matrix such that the only non-zero element is 1 at the ith column and jth row. Then locally we have

$$\omega = \Gamma^i_{jk} dx^j u^k_i. \tag{4.21}$$

The functions Γ^i_{jk} are called the **Christoffel symbols** of the linear connection Γ. One can show that

$$\nabla_{\partial_i} \partial_j = \Gamma^k_{ij} \partial_k.$$

Recall that the tangent bundle $TM = E(M, \mathbf{R}^m, GL(m, \mathbf{R}), L(M))$ is a vector bundle associated to the bundle of frames $L(M)$. In particular, a frame $u \in L(M)$ induces an isomorphism

$$\tilde{u} : \mathbf{R}^m \to T_{\pi(u)}M.$$

Let $X \in T_u L(M)$ and define the 1-form $\theta \in \Lambda^1(L(M), \mathbf{R}^m)$ by

$$\theta_u(X) = \tilde{u}^{-1}(\pi_*(X)). \tag{4.22}$$

Then θ is a tensorial 1-form on $L(M)$ of type $(r, \mathbf{R}^m)$, where r is the defining representation of $GL(m, \mathbf{R})$. In the physics literature the form θ is frequently called the **soldering form**. If we choose a basis for the Lie algebra $gl(m, \mathbf{R})$ and a basis for $\mathbf{R}^m$, then we can express the connection 1-form ω as m^2 1-forms ω_{ij} and the form θ as m 1-forms θ_k, $1 \le k \le m$. These $m^2 + m$ forms are globally defined on $L(M)$ and make $L(M)$ into a **parallelizable manifold**. In particular, we can define a Riemannian metric on $L(M)$ by

$$ds^2 = \sum_{i=1}^{m}\sum_{j=1}^{m}\omega_{ij}^2 + \sum_{k=1}^{m}\theta_k^2. \tag{4.23}$$

Using this Riemannian metric we can make $L(M)$ into a metric space. This metric can be used to show that a manifold M that admits a linear connection (in particular, a Lorentz connection) must be a metric space. Thus, in considering manifolds of interest in physical applications one may restrict attention to manifolds that are metric spaces. The metric on $L(M)$ is called the **bundle metric** or **b-metric** and is used in defining the so-called b-boundaries for space-time manifolds. The bundle metric was introduced by Marathe in [265] to study the topology of spaces admitting a linear connection and independently by Schmidt in [338] in his study of singular points in general relativity.

We observe that the linear connections are distinguished from connections in other principal bundles by the existence of the soldering form. Thus, in addition to the curvature 2-form $\Omega = d^{\omega}\omega$, we have the **torsion 2-form** $\vartheta = d^{\omega}\theta \in \Lambda^2(L(M), \mathbf{R}^m)$. A linear connection ω is called **torsion free** if the torsion 2-form $\vartheta = d^{\omega}\theta = 0$.

Let ω be a linear connection on the m-manifold M, i.e., a connection on the principal $GL(m, \mathbf{R})$-bundle of frames of M. If $x \in M$, $X_x, Y_x \in T_x M$, let us consider the element

$$\tilde{u}(\vartheta(X_u^*, Y_u^*)) \in T_x M,$$

where $u \in L(M)_x$ and X_u^*, Y_u^* are elements of $T_u L(M)$ such that $\pi_* X_u^* = X_x$, $\pi_* Y_u^* = Y_x$. One can show that $\tilde{u}(\vartheta(X_u^*, Y_u^*))$ does not depend on the choices of u, X_u^*, Y_u^*. Thus, it is easy to see that

$$T(X_x, Y_x) := \tilde{u}(\vartheta(X_u^*, Y_u^*)), \qquad X_x, Y_x \in T_x M$$

defines a tensor field $T \in \Gamma(T_2^1(M))$, which is called the **torsion tensor field**, or simply the **torsion**. Analogously, the **curvature tensor field** or simply the **curvature** is the tensor field $R \in \Gamma(T_3^1(M))$ such that

$$R(X_x, Y_x)Z_x := \tilde{u}[\Omega(X_u^*, Y_u^*)(\tilde{u}^{-1}(Z_x))],$$

$\forall x \in M$, $X_x, Y_x, Z_x \in T_x M$. We observe that $\Omega(X_u^*, Y_u^*)$ is in the Lie algebra $gl(m, \mathbf{R})$ of $GL(m, \mathbf{R})$, while $\tilde{u}^{-1}(Z_x) \in \mathbf{R}^m$ and thus $\Omega(X_u^*, Y_u^*)(\tilde{u}^{-1}(Z_x)) \in \mathbf{R}^m$. One can show that the torsion T and the curvature R can be expressed in terms of the covariant derivative $\nabla \equiv \nabla^{\omega}$ as follows:

$$T(X, Y) = \nabla_X Y - \nabla_Y X - [X, Y], \tag{4.24}$$

$$R(X, Y)Z = [\nabla_X, \nabla_Y]Z - \nabla_{[X,Y]}Z. \tag{4.25}$$

Locally, with $X = X^i \partial_i$, $Y = Y^j \partial_j$

$$T(X,Y) = (\Gamma_{ij}^k - \Gamma_{ji}^k)X^i Y^j \partial_k \tag{4.26}$$

and thus

$$T_{ij}^k \equiv dx^k(T(\partial_i, \partial_j)) = \Gamma_{ij}^k - \Gamma_{ji}^k. \tag{4.27}$$

From the above local expression it follows that a linear connection is torsion-free if and only if its Christoffel symbols are symmetric in the lower indices. Hence, a torsion-free connection is sometimes called a **symmetric** connection. Analogously, defining

$$R_{ijk}^l := dx^l(R(\partial_j, \partial_k)\partial_i), \tag{4.28}$$

we have

$$R_{ijk}^l = \partial_j \Gamma_{ki}^l - \partial_k \Gamma_{ji}^l + \Gamma_{jr}^l \Gamma_{ki}^r - \Gamma_{kr}^l \Gamma_{ji}^r. \tag{4.29}$$

We observe that

$$R_{ijk}^l = -R_{ikj}^l \tag{4.30}$$

and, for a torsion-free connection,

$$R_{[ijk]}^l \equiv R_{ijk}^l + R_{kij}^l + R_{jki}^l = 0. \tag{4.31}$$

By contraction, from the curvature one obtains the following important tensor field. The **Ricci tensor field** is the tensor field Ric $\in \Gamma(T_2^0(M))$ defined by

$$\mathrm{Ric}(X_x, Y_x) = \alpha^k(R(e_k, X_x)Y_x),$$

$\forall x \in M$, $X_x, Y_x \in T_x M$, where $\{e_1, \ldots, e_m\}$ is a basis of $T_x M$ with dual basis $\{\alpha_1, \ldots, \alpha_m\}$. Thus locally, the components R_{ij} of the **Ricci tensor** are given by

$$R_{ij} := \mathrm{Ric}(\partial_i, \partial_j) = R_{jki}^k. \tag{4.32}$$

If M is a Riemannian manifold, then the structure grou $GL(m, \mathbf{R})$ can be reduced to the orthogonal group $O(m, \mathbf{R})$. A connection on the reduced bundle is called a **Riemannian connection**. Similarly if M is a Lorentz manifold, then the structure group $GL(m, \mathbf{R})$ can be reduced to the Lorentz group. The frames in the reduced bundle are called **local inertial frames** and the connection on the reduced bundle is called a **Lorentz connection**. If $m = 4$ then the Lorentz manifold M is called a **space-time manifold**. The connection and curvature can be interpreted as representing a **gravitational potential** and **gravitational field** when they satisfy **gravitational field equations**. We will discuss them in Chapter 6. The Levi-Civita connection defined in the next paragraph is determined by the metric tensor g and hence in this case g is interpreted as representing the **gravitational potential**. Einstein's field equations are then expressed as non-linear partial diferential equations for the components of g.

A linear connection ω on a pseudo-Riemannian manifold (M, g) is called a **metric connection** if the metric g is **covariantly constant**, i.e., $\nabla^\omega g = 0$.

Among all metric connections on a pseudo-Riemannian manifold there exists a unique torsion-free, metric connection λ called the **Levi-Civita connection**. This fact is sometimes referred to as the fundamental theorem of pseudo-Riemannian geometry. The Levi-Civita connection is also referred to as the **symmetric connection**. The curvature 2-form F_λ of the Levi-Civita connection λ is denoted by R and is called the **Riemann curvature** of M. Locally, the **Christoffel symbols** of the Levi-Civita connection are given, in terms of the metric g, by

$$\Gamma_{ij}^k = \frac{1}{2}g^{hk}(\partial_j g_{ih} + \partial_i g_{jh} - \partial_h g_{ij}). \tag{4.33}$$

The Ricci tensor of a pseudo-Riemannian manifold (M, g) is symmetric. Other important quantities for (M, g) are the scalar curvature and the sectional curvature which we now define. The **scalar curvature** is the function $S \in \mathcal{F}(M)$ defined by

$$S = g^{ij}R_{ij}. \tag{4.34}$$

Using this definition we obtain the following expression for the **trace-free part** K of the Ricci tensor

$$K_{ij} = R_{ij} - \frac{1}{m}Sg_{ij} . \tag{4.35}$$

Let p be a 2-dimensional subspace of T_xM and let $\{e_1, e_2\}$ be an orthonormal basis for p. The quantity

$$\kappa(p) := e_1^\flat(R(e_1, e_2)e_2) \tag{4.36}$$

does not depend on the choice of the orthonormal basis $\{e_1, e_2\}$ and is called the **sectional curvature** along p. The local expression for the covariant derivative

$$\nabla_X Y = (\delta_j Y^i X^j + \Gamma_{jk}^i X^j Y^k)\delta_i, \quad X, Y \in \mathcal{X}(M) \tag{4.37}$$

shows that $(\nabla_X Y)(x)$ depends on the value of X only in x. This allows us to define the covariant derivative of a vector field X along a curve $c : I \to M$ as the curve

$$\frac{DX}{dt} : I \to TM$$

in TM given by

$$\frac{DX}{dt}(t) = (\nabla_{\dot{c}}X)(c(t)),$$

where $\dot{c}(t) = Tc(t, 1)$. The vector field X is said to be **autoparallel** along c if $\nabla_{\dot{c}}X = 0$. A **geodesic** is a curve γ in M that is **autoparallel** along itself, i.e., it satisfies the equation

$$\nabla_{\dot{\gamma}}\dot{\gamma} = 0. \tag{4.38}$$

In local coordinates this gives the system of differential equations

$$\ddot{\gamma}^i + \Gamma^i_{jk}\dot{\gamma}^j\dot{\gamma}^k = 0, \tag{4.39}$$

where i, j, k go from 1 to $\dim M$ and the Einstein summation convention is used. These equations always admit a local solution giving a geodesic starting at a given point in a given direction. On a complete manifold there exists a piecewise smooth geodesic joining any two points. The classical Hopf–Rinow theorem states that on a complete manifold every geodesic can be extended to a geodesic defined for all time $t \in \mathbf{R}$. In classical mechanics a particle of unit mass moving with velocity v has kinetic energy $\frac{1}{2}v^2$. For a curve $c : I \to M$ the tangent vector $\dot{c}(t)$ is the velocity vector at time t. Let P_I denote the set of all smooth curves from I to M. Then it can be shown that the critical points of the energy functional

$$E(c) = \frac{1}{2}\int_I g(\dot{c}, \dot{c})dt$$

when c varies over P_I are the geodesics.

Let M be an oriented Riemannian manifold. The identification of the Lie algebra $so(m)$ with $\Lambda^2(\mathbf{R}^m)$ allows us to identify $\operatorname{ad} L(M)$ with $\Lambda^2(M)$. Thus for each $x \in M$, R defines a symmetric, linear transformation of $\Lambda^2_x(M)$. The dimension 4 is further distinguished by the fact that $so(4) = so(3) \oplus so(3)$ and that this decomposition corresponds to the decomposition $\Lambda^2(M) = \Lambda^2_+(M) \oplus \Lambda^2_-(M)$ into ± 1 eigenspaces of the Hodge star operator. The Riemann curvature R also decomposes into $SO(4)$-invariant components induced by the above direct sum decomposition of $\Lambda^2(M)$. These components are the **self-dual** (resp., **anti-dual**) **Weyl tensor** $\mathcal{W}^+$ (resp., $\mathcal{W}^-$), the **trace-free part** K of the Ricci tensor, and the scalar curvature S. Thus

$$R = \mathcal{W}^+ \oplus \mathcal{W}^- \oplus (K \times_c g) \oplus (g \times_c g)S \tag{4.40}$$

where $\times_c$ is the **curvature product** defined in Section 6.7. These components are used to define several important classes of 4-manifolds. Thus, M is called a **self-dual manifold** (resp., **anti-dual manifold**) if $\mathcal{W}^- = 0$ (resp., $\mathcal{W}^+ = 0$). It is called **conformally flat** if it is both self-dual and anti-dual or if the full **Weyl tensor** is zero, i.e., $\mathcal{W} := \mathcal{W}^+ + \mathcal{W}^- = 0$. A pseudo-Riemannian manifold (M, g) is said to be an **Einstein manifold** if

$$\operatorname{Ric} = \Lambda g , \tag{4.41}$$

where Λ is a constant. By contraction of both sides of equation (4.41) we get $\Lambda = S/m$. Hence equation (4.41) is equivalent to the vanishing of the trace-free part K of the Ricci tensor, i.e., $K = 0$. Einstein manifolds correspond to a class of gravitational instantons ([269]). Einstein manifolds were characterized in [353] by the commutation condition $[R, *] = 0$, where the Riemann curvature R and the Hodge star operator $*$ are both regarded as linear transformations of $\Lambda^2(M)$. This condition was generalized in [268, 267] to obtain a new formulation of the gravitational field equations, which are discussed in in Section 6.7. For the study of Riemannian manifolds in dimension 4 and manifolds of differentiable mappings see, for example, Donaldson and Kronheimer [112].

4.7 Generalized Connections

The various definitions of connection in principal and associated bundles discussed in this chapter are adequate for most of the applications discussed in this book. However, it is possible to define the notion of connection on an arbitrary fiber bundle. We call this a generalized connection. It can be used to give an alternative formulation of some aspects of gauge theories. There is extensive work in this area but we will not use this approach in our work.

Let E be a fiber bundle over B with projection $p : E \to B$. Let VE denote the vertical vector bundle over E. VE is a subbundle of the tangent bundle TE, the fiber $V_u E$, $u \in E$ being the tangent space to the fiber $E_{p(u)}$ of E passing through u. Let $p^*(TB)$ be the pull-back of the tangent bundle of B to E. Then we have the following short exact sequence of vector bundles and morphisms over E:

$$0 \longrightarrow VE \xrightarrow{i} TE \xrightarrow{a} \pi^*(TB) \longrightarrow 0 , \qquad (4.42)$$

where i is the injection of the vertical bundle VE into the tangent bundle, and a is defined by $(e, X_e) \mapsto (e, p_*(X_e))$. We define a **generalized connection** on E to be a splitting of the above exact sequence, i.e., a vector bundle morphism $c : \pi^*(TB) \to TE$ such that $a \circ c = id_{\pi^*(TB)}$. Let $b \in B$ and let $e \in p^{-1}(b)$, then the splitting c induces an injection $\hat{c}_e : T_b(B) \to T_e(E)$, $X \mapsto c(e, X)$. We call $\hat{c}_e(X)$ the **horizontal lift** of the tangent vector X to $e \in E$. We call $\hat{c}_e(T_b(B))$ the **horizontal space at** $e \in E$ and denote it by $H_e E$. The spaces $H_e E$ are the fibers of the horizontal bundle HE and we have the decomposition

$$TE = HE \oplus VE.$$

We note that this decomposition corresponds to the first condition of Definition 4.5. Alternatively, a connection may be defined as a section of the first jet bundle $J^1(E)$ over E. The definitions of covariant derivative, covariant differential, and curvature can be formulated in this general context. An introduction to this approach and to its physical applications is given in [153].

If $E = P(M, G)$, then we can recover the usual definition of connection as follows. The action of G on P extends to all the vector bundles in the exact sequence (4.42) to give the following short exact sequence of vector bundles over M:

$$0 \longrightarrow VP/G \xrightarrow{i} TP/G \xrightarrow{\hat{a}} \pi^*(TM)/G \longrightarrow 0.$$

We can rewrite the above sequence as follows:

$$0 \longrightarrow \operatorname{ad}(P) \overset{i}{\longrightarrow} TP/G \overset{\hat{a}}{\longrightarrow} TM \longrightarrow 0. \tag{4.43}$$

A connection on P is then defined as a splitting of the short exact sequence (4.43). This splitting induces a splitting of the sequence (4.42) and we recover Definition 4.5 of the connection given earlier. This approach helps to clarify the role of the structure group in the usual definition of connection in a principal bundle.

Chapter 5
Characteristic Classes

5.1 Introduction

In 1827 Gauss published his classic book *Disquisitiones generales circa su-perficies curvas*. He defined the **total curvature** (now called the **Gaussian curvature**) κ as a function on the surface. In his famous *theorema egregium* Gauss proved that the total curvature κ of a surface S depends only on the **first fundamental form** (i.e., the metric) of S. Gauss defined the **integral curvature** $\kappa(\Sigma)$ of a bounded surface Σ to be $\int_\Sigma \kappa \, d\sigma$. He computed $\kappa(\Sigma)$ when Σ is a geodesic triangle to prove his celebrated theorem

$$\kappa(\Sigma) := \int_\Sigma \kappa \, d\sigma = A + B + C - \pi, \tag{5.1}$$

where A, B, C are the angles of the geodesic triangle Σ. Gauss was aware of the significance of equation (5.1) in the investigation of the **Euclidean parallel postulate** (see Appendix B for more information). He was interested in surfaces of constant curvature and mentions a surface of revolution of constant negative curvature, namely, a **pseudosphere**. The geometry of the pseudosphere turns out to be the non-Euclidean geometry of Lobačevski–Bolyai.

Equation (5.1) is a special case of the well-known **Gauss–Bonnet theorem**. When applied to a compact, connected surface Σ the Gauss-Bonnet theorem states that

$$\int_\Sigma \kappa \, d\sigma = 2\pi\chi(\Sigma), \tag{5.2}$$

where $\chi(\Sigma)$ is the Euler characteristic of Σ. The left hand side of equation (5.2) is arrived at through using the differential structure of Σ, while the right hand side depends only on the topology of Σ. Thus, equation (5.2) is a relation between geometric (or analytic) and topological invariants of Σ. This result admits far-reaching generalizations. In particular, it can be regarded as a prototype of an index theorem. The Gauss–Bonnet theorem was

K. Marathe, *Topics in Physical Mathematics*, DOI 10.1007/978-1-84882-939-8_5, 137
© Springer-Verlag London Limited 2010

generalized to Riemannian polyhedra by Allendoerffer and Weil and to arbitrary manifolds by Chern. In this latter generalization the Gaussian integral curvature is replaced by an invariant formed from the Riemann curvature. It forms the starting point of the theory of characteristic classes.

Gauss' idea of studying the geometry of a surface intrinsically, without leaving it (i.e., by means of measurements made on the surface itself), is of fundamental importance in modern differential geometry and its applications to physical theories. We are similarly compelled to study the geometry of the three-dimensional physical world by the **intrinsic method**, i.e., without leaving it. This idea was already implicit in Riemann's work, which extended Gauss' intrinsic method to the study of manifolds of arbitrary dimension. This work together with the work of Ricci and Levi-Civita provided the foundation for Einstein's theory of general relativity. The constructions discussed in this chapter extend these ideas and provide important tools for modern mathematical physics.

5.2 Classifying Spaces

Let G be a Lie group. The classification of principal G-bundles over a manifold M is achieved by the use of **classifying spaces**. A topological space $B_k(G)$ is said to be k-**classifying** for G if the following conditions hold:

1. There exists a contractible space $E_k(G)$ on which G acts freely and $B_k(G)$ is the quotient of $E_k(G)$ under this G-action such that

$$E_k(G) \to B_k(G)$$

   is a principal fiber bundle with structure group G.
2. Given a manifold M of dim $\leq k$ and a principal bundle $P(M, G)$, there exists a continuous map $f : M \to B_k(G)$ such that the pull-back $f^*(E_k(G))$ to M is a principal bundle with structure group G that is isomorphic to P.

It can be shown that homotopic maps give rise to equivalent bundles and that all principal G-bundles over M arise in this way. Let $[M, B_k(G)]$ denote the set of equivalence classes under homotopy, of maps from M to $B_k(G)$. Then the **classifying property** may be stated as follows:

Theorem 5.1 (Classifying property) *Let M be a compact, connected manifold and G a compact, connected Lie group. Then there exists a one-to-one correspondence between the set $[M, B_k(G)]$ of homotopy classes of maps and the set of isomorphism classes of principal G-bundles over M.*

The spaces $E_k(G)$ and $B_k(G)$ may be taken to be manifolds for a fixed k. However, classifying spaces can be constructed for arbitrary finite-dimensional manifolds. They are denoted by $E(G)$ and $B(G)$ and are in general infinite-dimensional. The spaces $E(G)$ and $B(G)$ are called **universal classifying spaces** for principal G-bundles.

Example 5.1 *The* **complex Hopf fibration** $S^{2n+1} \to \mathbf{CP}^n$ *is a principal $U(1)$-bundle and is n-classifying. By forming a tower of these fibrations by inclusion, i.e., by considering two series of inclusions*

$$S^3 \subset S^5 \subset S^7 \subset \cdots$$

$$\mathbf{CP}^1 \subset \mathbf{CP}^2 \subset \mathbf{CP}^3 \subset \cdots,$$

we obtain a direct system of principal $U(1)$-bundles. Taking the direct limit of this system we get the spaces S^∞ and $\mathbf{CP}^\infty$ such that S^∞ is a principal $U(1)$-fibration over $\mathbf{CP}^\infty$. Thus $B(U(1)) = \mathbf{CP}^\infty$ and $E(U(1)) = S^\infty$. This classification is closely related to the electromagnetic field as a $U(1)$-gauge field and, in particular, to the construction of the Dirac monopole and the monopole quantization condition.

A similar argument applied to the quaternionic Hopf fibration gives the classifying spaces for principal $SU(2)$-bundles as indicated in the following example.

Example 5.2 *The* **quaternionic Hopf fibration** $S^{4n+3} \to \mathbf{HP}^n$ *is a principal $SU(2)$-bundle and is n-classifying. By considering the towers*

$$S^7 \subset S^{11} \subset S^{15} \subset \cdots$$

$$\mathbf{HP}^1 \subset \mathbf{HP}^2 \subset \mathbf{HP}^3 \subset \cdots,$$

we obtain the universal classifying spaces $B(SU(2)) = \mathbf{HP}^\infty$ and $E(SU(2)) = S^\infty$. This example is closely related to the classification of Yang–Mills fields with gauge group $SU(2)$, and, in particular, to the construction and classification of instantons.

5.3 Characteristic Classes

The beginning of the theory of **characteristic classes** was made in the 1930s by Stiefel and Whitney. Stiefel studied certain homology classes of the tangent bundle of a smooth manifold, while Whitney considered the case of an arbitrary sphere bundle and introduced the concept of a characteristic cohomology class. In the next decade, Pontryagin constructed important new characteristic classes by studying the homology of real Grassmann manifolds, and Chern defined characteristic classes for complex vector bundles. Chern's study of the cohomology of complex Grassmann manifolds also led to a better understanding of Pontryagin's real characteristic classes.

We begin with a brief discussion of the **Stiefel–Whitney classes** in terms of certain cohomology operations. Recall first that the de Rham cohomology has a graded algebra structure induced by the exterior product. This induced product in cohomology is in fact a special case of a **cohomology operation** in algebraic topology called the **cup product**:

$$\cup : H^i(M;\mathbf{P}) \times H^j(M;\mathbf{P}) \to H^{i+j}(M;\mathbf{P}), \quad (\alpha, \beta) \mapsto \alpha \cup \beta.$$

The cup product induces the structure of a graded ring on the cohomology space $H^*(M;\mathbf{P})$. The ring $H^*(M;\mathbf{P})$ is called the **cohomology ring** of M. If P is a field, then the ring $H^*(M;\mathbf{P})$ is a P-algebra called the **cohomology algebra** of M. In the rest of this paragraph we take $\mathbf{P} = \mathbf{Z}_2$ and omit its explicit indication. We also use the **Steenrod squaring operations** Sq^i, $i \geq 0$, which are characterized by the following four properties.

1. For each topological pair (X, A) and $\forall n \geq 0$ the map

$$Sq^i : H^n(X, A) \to H^{n+i}(X, A)$$

   is an **additive homomorphism**.
2. (**Naturality**) If $f : (X, A) \to (Y, B)$ is a morphism of topological pairs, then

$$Sq^i \circ f^* = f^* \circ Sq^i,$$

   where $f^* : H^*(Y, B) \to H^*(X, A)$ is the homomorphism induced by f.
3. If $\alpha \in H^n(X, A)$, then

$$Sq^0(\alpha) = \alpha, \quad Sq^n(\alpha) = \alpha \cup \alpha \text{ and } Sq^i(\alpha) = 0, \quad \forall i > n.$$

4. (**Cartan formula**) Let α, β be such that $\alpha \cup \beta$ is defined. Then

$$Sq^k(\alpha \cup \beta) = \sum_{i+j=k} Sq^i(\alpha) \cup Sq^j(\beta).$$

Theorem 5.2 *Let*

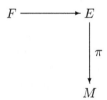

be a real vector bundle of rank n (i.e., $\dim F = n$) over the base manifold M and let E^0 be the complement of the image of the zero section θ of E over M. Let $E_x^0 = E^0 \cap E_x$, $x \in M$, then we have the following results.

1.

$$H^i(E_x,\ E_x^0) = \begin{cases} 0 & i \neq n, \\ \mathbf{Z}_2 & i = n. \end{cases}$$

2.

$$H^i(E,\ E^0) = \begin{cases} 0 & i < n, \\ H^{i-n}(M) & i \geq n. \end{cases}$$

3. *There exists a unique class $u \in H^n(E, E^0)$ such that $\forall x \in M$ the class u restricted to (F_x, F_x^0) is the unique non-zero class in $H^n(F_x, F_x^0)$. Furthermore, the cup product by u defines an isomorphism c_u by*

$$c_u : \alpha \mapsto \alpha \cup u \text{ of } H^k(E) \to H^{k+n}(E, E^0), \qquad \forall k. \qquad (5.3)$$

The class u introduced in the above theorem is called the **Thom class**. Now the zero section θ embeds M as a deformation retract of E with retraction map $\pi : E \to M$. Thus, $\pi^* : H^k(M) \to H^k(E)$ is an isomorphism for all k. The **Thom isomorphism** $\phi : H^k(M) \to H^{k+n}(E, E_0)$ is defined as the composition of the two isomorphisms π^* and c_u,

$$H^k(M) \xrightarrow{\pi^*} H^k(E) \xrightarrow{\cup u} H^{k+n}(E, E_0), \quad \phi(\alpha) = \pi^*(\alpha) \cup u.$$

Composing the Steenrod squaring operation Sq^k with ϕ^{-1} results in a homomorphism

$$H^n(E, E_0) \xrightarrow{Sq^k} H^{n+k}(E, E_0) \xrightarrow{\phi^{-1}} H^k(M), \qquad \forall k \geq 0.$$

We define the kth Stiefel–Whitney class $w_k(\xi) \in H^k(M)$ of the vector bundle $\xi = (E, \pi, M)$ as the image of the Thom class u under the above homomorphism, i.e.,

$$w_k(\xi) = \phi^{-1}(Sq^k(u)). \qquad (5.4)$$

We remark that the definitions of the Thom class and the Thom isomorphism extend to the case of oriented vector bundles and their integral cohomology. Theorem 5.2 also extends to this case with obvious modifications. The class $\chi \in H^n(M; \mathbf{Z})$, which corresponds to the Thom class under the canonical isomorphism $\pi^* : H^n(M; \mathbf{Z}) \to H^n(E; \mathbf{Z})$, is called the **Euler class** of the oriented bundle ξ. Moreover, we have the following proposition.

Proposition 5.3 *Let M be an n-dimensional manifold. The natural homomorphism $H^n(M; \mathbf{Z}) \to H^n(M; \mathbf{Z}_2)$, induced by the reduction of coefficients mod 2, maps the Euler class $\chi(\xi)$ to the Stiefel–Whitney class $w_n(\xi)$ in the top dimension.*

The Stiefel–Whitney classes are characterized by the following four properties.

1.
$$w_i(\xi) \in H^i(M; \mathbf{Z}_2), \qquad i \geq 0.$$

In particular, $w_0(\xi) = 1$ and $w_i(\xi) = 0$ for $i > \text{rank}(\xi)$.

2. (**Naturality**) If $\xi' = (E', \pi', M')$ is another vector bundle and $f : M \to M'$ is covered by a bundle map from ξ to ξ', then

$$w_i(\xi) = f^*(w_i(\xi')).$$

3. (**Whitney product formula**) With the above notation, if $M = M'$ then

$$w_k(\xi \oplus \xi') = \sum_{i=0}^{k} w_i(\xi) \cup w_{k-i}(\xi'). \tag{5.5}$$

4. If γ is the canonical line bundle over the real projective space $\mathbf{RP}^1 \cong S^1$, then $w_1(\gamma) \neq 0$.

We recall that the ring of **formal series** over $H^*(M; \mathbf{Z}_2)$ is the set

$$\left\{ \sum_{i=0}^{\infty} a_i \mid a_i \in H^i(M; \mathbf{Z}_2) \right\}$$

with termwise addition and with multiplication induced by the cup product. One can show that the set of invertible elements in this ring consists of elements with $a_0 = 1$. The **total Stiefel–Whitney class** is defined as the formal series

$$w(\xi) = 1 + w_1(\xi) + \cdots + w_n(\xi) + \cdots . \tag{5.6}$$

Its inverse

$$\bar{w} = 1 + \sum_{i=1}^{\infty} \bar{w}_i$$

can be computed by formal power series expansion. Each $\bar{w}_i$ is then expressed in terms of w_j, $j \leq i$. For example,

$$\bar{w}_1 = w_1, \quad \bar{w}_2 = w_1^2 + w_2.$$

We define the ith **Stiefel–Whitney class of a manifold** M by

$$w_i(M) := w_i(TM).$$

If M can be immersed in $\mathbf{R}^{m+k}$ then we have the following theorem.

Theorem 5.4 (Whitney duality) *Let ν denote the normal bundle of M in $\mathbf{R}^{m+k}$. Then*

$$w_i(\nu) = \bar{w}_i(M), \ \forall i \in \mathbf{N}.$$

In particular, $\bar{w}_i(M) = 0$ for $i > k$.

Applying this theorem to projective spaces, we get the following best estimate result for immersions of compact manifolds in Euclidean spaces.

Theorem 5.5 *Let* $m = 2^r$, $r \in \mathbf{N}$. *If* $\mathbf{RP}^m$ *can be immersed in* $\mathbf{R}^{m+k}$, *then* $k \geq m - 1$.

Let M be a compact m-dimensional manifold and let

$$\underline{r} = (r_1, \ldots, r_m), \text{ such that } \sum_{i=1}^{m} ir_i = m,$$

where r_i, $1 \leq i \leq m$, are non-negative integers. Define the monomial $w_{\underline{r}}(M)$ by

$$w_{\underline{r}}(M) := w_1(M)^{r_1} \cdots w_m(M)^{r_m}.$$

The value of $w_{\underline{r}}(M)$ on the fundamental homology class $[M]$ is called the **Stiefe–Whitney number** of M associated to the monomial $w_{\underline{r}}(M)$. Any M has a finite set of Stiefel–Whitney numbers. One important application of these numbers is found in the following theorem of Pontryagin and Thom.

Theorem 5.6 *Let* M *be a compact* m-*dimensional manifold. Then* M *can be realized as the boundary of a compact* $(m+1)$-*dimensional manifold if and only if all the Stiefel–Whitney numbers of* M *are zero.*

This theorem is the starting point of **Thom's theory of cobordism.**

Definition 5.1 *Two closed* m-*dimensional manifolds* M_1, M_2 *are said to be* **cobordant** *($M_1 \sim_{cb} M_2$) if their disjoint union* $M_1 \sqcup M_2$ *is the boundary of a compact* $(m+1)$-*dimensional manifold. The relation* $\sim_{cb}$ *is an equivalence relation, and the equivalence class* $[M]$ *of* M *under this relation is called an* **unoriented cobordism class** *of* M. *The class of closed* m-*dimensional manifolds together with unoriented cobordism as morphism is called the* m-*dimensional* **cobordism category**.

Theorem 5.6 and the above definition lead to the following result.

Theorem 5.7 $[M_1] = [M_2]$ *if and only if all Stiefel–Whitney numbers of* M_1 *and* M_2 *are the same.*

Define the set

$$\mathcal{T}_m = \{[M] \mid M \text{ is a closed } m\text{-dimensional manifold}\}. \tag{5.7}$$

Disjoint union induces an operation on $\mathcal{T}_m$ (denoted by $+$) with which it is an Abelian group. In view of Theorem 5.7 the group $(\mathcal{T}_m, +)$ is finite. It is called the m-dimensional **unoriented cobordism group**. Thom proved that the unoriented cobordism group $\mathcal{T}_m$ is canonically isomorphic to the homotopy group π_{m+k} of a certain universal space for a sufficiently large k. We note that $\{\mathcal{T}_m\}$ can be made into a **graded ring** with the product induced by Cartesian product of manifolds. The zero element of this ring is

represented by the class of the empty manifold and the unity by the class of
the one-point manifold. It is called the **unoriented cobordism ring**. There
are several theories of cobordism for manifolds carrying additional structure
such as orientation, group action, etc. (see Stong [360]). The 2-dimensional
cobordism category was used by Segal in his axioms for conformal field theory.
They form the starting point for the Atiyah–Segal axioms for (topological)
quantum field theories. These axioms are discussed in Chapter 7.

Let $\xi = (E, B, \pi, F)$ be a vector bundle with a Riemannian metric. Let

$$D_x^n := \{v \in E_x \mid \|v\| \leq 1\},$$

$$S_x^{n-1} := \{v \in E_x \mid \|v\| = 1\}.$$

The space

$$D(\xi) = \bigcup_{x \in B} D_x^n \quad (\text{resp.,} \ S(\xi) = \bigcup_{x \in B} S_x^{n-1})$$

has a natural structure of a manifold with which it becomes a bundle over
B, called the **disk bundle** of ξ (resp., the **sphere bundle** of ξ). The bundles $D(\xi)$ (resp., $S(\xi)$) defined by a different choice of the metric on ξ are
equivalent (isomorphic as bundles). The quotient bundle $T(\xi) = D(\xi)/S(\xi)$
is called the **Thom space** of ξ. The cobordism groups $\mathcal{T}_m$ are isomorphic
to $\pi_{m+k}(T(\gamma^k))$, where γ^k is the universal k-plane bundle. Thom has shown
that these isomorphisms induce an isomorphism $\theta : \{\mathcal{T}_m\} \to \pi_m(T)$ of graded
rings. Let $P(M, G)$ be a principal bundle over a manifold M with structure
group G. We will define a set of cohomology classes in $H^*(M; \mathbf{R})$ (the cohomology ring of M with real coefficients) associated to the principal bundle,
which characterize the bundle up to bundle isomorphism. They are called the
real characteristic classes of $P(M, G)$. There are several different ways of
defining these characteristic classes (see Husemoller [198], Kamber and Tondeur [215], and Milnor and Stasheff [286]). We define them by using connections in the bundle P and the **Weil homomorphism** defined below.

Let $\rho : G \to GL(V)$ be a representation of G on the real vector space V.
Let $S^k(V)$ denote the set of k-linear symmetric real-valued functions on V
and let

$$S(V) = \oplus_{k=0}^{\infty} S^k(V)$$

be the **symmetric algebra** of V with product of $f \in S^h(V)$, $g \in S^k(V)$
defined by

$$(f \cdot g)(v_1, \ldots, v_{h+k}) =$$

$$\frac{1}{(h+k)!} \sum_{\sigma \in S_{h+k}} f(v_{\sigma(1)}, \ldots, v_{\sigma(h)}) g(v_{\sigma(h+1)}, \ldots, v_{\sigma(h+k)}),$$

where S_n, $n \in \mathbf{N}$, is the symmetric group of permutations on n numbers.
The group G acts on $S^k(V)$ by an action induced by the representation ρ,
which we also denote by ρ, given by

$$(\rho(a)f)(v_1, \ldots, v_k) = f(\rho(a^{-1})v_1, \ldots \rho(a^{-1})v_k).$$

We note that from this definition it follows that

$$\rho(a)(\rho(b)f) = \rho(ab)f.$$

Let $(e_1, \ldots, e_n)$ be a basis of V. The relation

$$P_f(x_1, \ldots, x_n) = f(v, v, \ldots, v), \quad v = \sum_{i=1}^{n} x_i e_i$$

allows us to establish a correspondence between functions in $S^k(V)$ and the space $\mathcal{P}^k$ of **homogeneous polynomials** of degree k in n variables with the coefficient of $x_{i_1}^{m_1} \cdots x_{i_r}^{m_r}$, $m_1 + \cdots + m_r = k$, given by

$$\frac{k!}{m_1! \cdots m_r!} f(\underbrace{e_{i_1}, \ldots, e_{i_1}}_{m_1 \text{ times}}, \ldots, \underbrace{e_{i_r}, \ldots, e_{i_r}}_{m_r \text{ times}}).$$

This correspondence can be extended to $S(V)$ and turns out to be an algebra homomorphism. The inverse of the mapping $f \mapsto P_f$ is called **polarization**. If $P \in \mathcal{P}^k$, we denote by f_P the element of $S^k(V)$ obtained by polarization of P. For example, if $P \in \mathcal{P}^2$, then

$$f_P(v_1, v_2) = \frac{1}{2}\left[P(x_1 + y_1, \ldots, x_n + y_n) - P(x_1, \ldots, x_n) - P(y_1, \ldots, y_n)\right],$$

where $v_1 = \sum_{i=1}^{n} x_i e_i$, $v_2 = \sum_{i=1}^{n} y_i e_i$. If $P \in \mathcal{P}^3$, then

$$\begin{aligned}
f_P(v_1, v_2, v_3) = \frac{1}{6}\big[& P(x_1 + y_1 + z_1, \ldots, x_n + y_n + z_n) \\
& - P(x_1 + y_1, \ldots, x_n + y_n) - P(x_1 + z_1, \ldots, x_n + z_n) \\
& - P(y_1 + z_1, \ldots, y_n + z_n) + P(x_1, \ldots, x_n) \\
& + P(y_1, \ldots, y_n) + P(z_1, \ldots, z_n)\big],
\end{aligned}$$

where $v_1 = \sum_{i=1}^{n} x_i e_i$, $v_2 = \sum_{i=1}^{n} y_i e_i$, $v_3 = \sum_{i=1}^{n} z_i e_i$. We define

$$I^k(V, \rho) := \{f \in S^k(V) \mid \rho(a)(f) = f, \quad \forall a \in G\}. \tag{5.8}$$

In view of the isomorphism between **symmetric functions** and **homogeneous polynomials** described above, we call the space $I^k(V, \rho)$ the space of **G-invariant symmetric polynomials of degree k on** V, the action of G by ρ being understood. We define the space $I(V, \rho)$ of all G-invariant symmetric polynomials on V by

$$I(V, \rho) := \bigoplus_{k=0}^{\infty} I^k(V, \rho). \tag{5.9}$$

It is easy to verify that $I(V, \rho)$ is a subalgebra of $S(V)$. In what follows we will take V to be the Lie algebra $\mathbf{g}$ of the Lie group G and ρ to be the adjoint representation of G on $\mathbf{g}$, and we denote $I^k(V, \rho)$ by $I^k(G)$ and $I(V, \rho)$ by $I(G)$.

Let ω be the connection 1-form of a connection on $P(M, G)$ and Ω the curvature 2-form of ω. For $f \in I^k(G)$, we define the $2k$-form f_Ω on P by

$$f_\Omega(X_1, \ldots, X_{2k}) = \frac{1}{(2k)!} \sum_{\sigma \in S_{2k}} \operatorname{sgn}(\sigma) f(u_1, \ldots, u_k), \qquad (5.10)$$

where $\quad u_i = \Omega(X_{\sigma(2i-1)}, X_{\sigma(2i)}) \in \mathbf{g}, \quad 1 \leq i \leq k.$

Since the curvature form Ω is tensorial with respect to the adjoint action of G on $\mathbf{g}$ and f is G-invariant, the $2k$-form f_Ω on P descends to a $2k$-form $\hat{f}_\Omega$ on M; i.e., there exists a form $\hat{f}_\Omega \in \Lambda^{2k}(M)$ such that

$$\pi^*(\hat{f}_\Omega) = f_\Omega,$$

where π is the canonical bundle projection of P on M. It can be shown that the form $\hat{f}_\Omega$ is closed and hence defines a cohomology class $[\hat{f}_\Omega] \in H^{2k}(M; \mathbf{R})$. Furthermore, this cohomology class turns out to be independent of the choice of a particular connection ω on P, i.e., if ω_1, ω_2 are two connections on P with respective curvature forms Ω_1, Ω_2 then $[\hat{f}_{\Omega_1}] = [\hat{f}_{\Omega_2}] \in H^{2k}(M; \mathbf{R})$. In view of this result we can define a map

$$w_k : I^k(G) \to H^{2k}(M; \mathbf{R}) \quad \text{by} \quad w_k(f) := [\hat{f}_\Omega]. \qquad (5.11)$$

The family of maps $\{w_k\}$ defines the map

$$w : I(G) \to H^*(M; \mathbf{R}). \qquad (5.12)$$

The map w is called the **Weil homomorphism**. We note that w is an algebra homomorphism of the algebra $I(G)$ into the **cohomology algebra** $H^*(M; \mathbf{R})$. The image of w is an algebra called the **real characteristic algebra** of the bundle P. The construction discussed above extends to the algebra $I_\mathbf{C}(G)$ of the complex valued G-invariant polynomials on $\mathbf{g}$ to give us the complex algebra homomorphism

$$w_\mathbf{C} : I_\mathbf{C}(G) \to H^*(M; \mathbf{C}). \qquad (5.13)$$

This is called the **Chern–Weil homomorphism**. The image of $w_\mathbf{C}$ is an algebra called the **complex characteristic algebra** of the bundle P.

An element $p \in w(I(G)) \subset H^*(M; \mathbf{R})$ is called a **real characteristic class** of the bundle P. Similarly, an element $c \in w_\mathbf{C}(I_\mathbf{C}(G)) \subset H^*(M; \mathbf{C})$ is called a **complex characteristic class** of the bundle P.

Characteristic classes are topological invariants of the principal bundle P and characterize P up to isomorphism. They can also be viewed as **topological invariants** of the vector bundle associated to P by the fundamental or defining representation of the structure group G. It is possible to give an axiomatic formulation of characteristic classes of vector bundles directly (see Kobayashi and Nomizu [226] and Milnor and Stasheff [286]). In studying the properties of characteristic classes we will use either of these formulations as is convenient.

For the Lie groups that are commonly encountered in physical theories we can show that the algebra of characteristic classes is finitely generated. We also exhibits a basis for each such algebra. Before giving several examples of this, we recall the correspondence between homogeneous polynomials of degree k and symmetric functions in $S^k(V)$. In the following examples V is a vector space of matrices and the polynomial variables are the entries of these matrices. In view of this remark, in the examples discussed below, we give only the homogeneous polynomials p_k of degree k for the construction of the characteristic classes.

Example 5.3 *Let $G = GL(n, \mathbf{R})$; then the **characteristic algebra** $I(G)$ is generated by $p_0, p_1, \ldots, p_n$, where the p_i are defined by*

$$\det\left(\lambda I_n - \frac{1}{2\pi}A\right) = \sum_{i=0}^{n} p_i \lambda^{n-i}, \qquad A \in \mathbf{gl}(n, \mathbf{R}).$$

*If Ω is the curvature form of some connection ω on P, the kth **Pontryagin class** of P is defined to be the cohomology class $\mathbf{p}_k$ represented by the unique closed $4k$-form β_k on M such that $\pi^*(\beta_k) = (p_{2k})_\Omega$; i.e., the kth Pontryagin class is given by*

$$\mathbf{p}_k := w(p_{2k}) = [\beta_k].$$

The $4k$-form $(p_{2k})_\Omega$ can be expressed as

$$(p_{2k})_\Omega = \frac{1}{(2\pi)^{2k}(2k)!} \sum \delta^{j_1, \ldots, j_{2k}}_{i_1, \ldots, i_{2k}} \Omega^{i_1}_{j_1} \wedge \cdots \wedge \Omega^{i_{2k}}_{j_{2k}}, \qquad (5.14)$$

where the sum is over all ordered subsets $(i_1, \ldots, i_{2k})$ of $2k$ different elements in $(1, \ldots, n)$ and all permutations $(j_1, \ldots, j_{2k})$ of $(i_1, \ldots, i_{2k})$, $\delta^{j_1, \ldots, j_{2k}}_{i_1, \ldots, i_{2k}}$ denotes the sign of the permutation and the Ω^i_j are the components of Ω in $\mathbf{gl}(n, \mathbf{R})$.

The reason for considering only p_i with even index i is given in the following example.

Example 5.4 *Let $G = O(n, \mathbf{R})$. The characteristic algebra $I(G)$ is generated by $p_0, p_2, p_4, \ldots$ defined as in Example 5.3. Observe that for $A \in o(n, \mathbf{R}) = so(n, \mathbf{R})$ one has*

$$\det\left(\lambda I_n - \frac{1}{2\pi}A\right) = \det\left(\lambda I_n + \frac{1}{2\pi}A\right).$$

Therefore $p_1 = p_3 = \ldots = 0$ and p_{2k} again corresponds to the kth Pontryagin class. In Example 5.3, $p_{2k+1} \neq 0$ in general but $w(p_{2k+1}) = 0$, since every $GL(n, \mathbf{R})$-connection is reducible to an $O(n, \mathbf{R})$-connection.

Example 5.5 *With the notation of Example 5.4, we construct an $SO(2m)$-invariant polynomial, called the Pfaffian, which is not invariant under the action of $O(2m)$. The **Pfaffian** Pf is the homogeneous polynomial of degree m such that, for $A \in so(2m)$,*

$$Pf(A) = \frac{1}{2^{2m}\pi^m m!}\sum_\sigma \text{sgn}(\sigma)A_{\sigma(1)\sigma(2)}\cdots A_{\sigma(2m-1)\sigma(2m)}, \qquad (5.15)$$

*where the sum is over all permutations of $(1, 2, \ldots, 2m)$. The class $w(Pf)$ is called the **Euler class** of $P(M, SO(2m))$ and is the class $[\gamma]$ such that*

$$\pi^*(\gamma) = \frac{1}{2^{2m}\pi^m m!}\sum_\sigma \text{sgn}(\sigma)\Omega^{\sigma(1)}_{\sigma(2)} \wedge \cdots \wedge \Omega^{\sigma(2m-1)}_{\sigma(2m)}. \qquad (5.16)$$

The Euler class occurs as the generalized curvature in the Chern–Gauss–Bonnet theorem, which states that

$$\int_M Pf(\hat{\Omega}) = \chi(M). \qquad (5.17)$$

This theorem generalizes the classical Gauss–Bonnet theorem for compact, connected surfaces in $\mathbf{R}^3$. For such a surface the $Pf(\hat{\Omega})$ is a multiple of the volume form on M. The multiplier κ is the Gaussian curvature and the above theorem reduces to

$$\int_M \kappa = \chi(M).$$

The expression for the Euler class of a 4-manifold as an integral of a polynomial in the Riemann curvature was obtained by Lanczos [242] in his study of Lagrangians for generalized gravitational field equations. He observed that the integral was invariant and so did not contain any dynamics, but he did not recognize its topological significance. The general formula for the Euler class of an arbitrary oriented manifold was obtained by Chern and it was in studying this generalization that Chern was led to his famous characteristic classes, now called the **Chern classes**.

Example 5.6 *Let $G = GL(n, \mathbf{C})$. $I_\mathbf{C}(G)$ is generated by the characteristic classes $c_0, c_1, \ldots, c_n$ defined by*

$$\det\left(\lambda I_n - \frac{1}{2\pi i}A\right) = \sum_{i=0}^n c_i \lambda^{n-i}.$$

The classes $w(c_k) \in H^{2k}(M; \mathbf{C})$, $k = 1, 2, \ldots, n$, are called the *Chern classes* of P.

Example 5.7 $G = U(n)$. $I_{\mathbf{C}}(G)$ *is generated by the classes* c_k *defined in Example 5.6. However, for* $A \in u(n)$ *one has the following complex conjugation condition:*

$$\det\left(\lambda I_n - \frac{1}{2\pi i} A\right) = \overline{\det\left(\lambda I_n - \frac{1}{2\pi i} A\right)}.$$

Therefore the class $w(c_k)$ *turns out to be a real cohomology class, i.e.,* $w(c_k) \in H^{2k}(M; \mathbf{R}) \subset H^{2k}(M; \mathbf{C})$, *where* $H^{2k}(M; \mathbf{R})$ *is regarded as a subset of* $H^{2k}(M; \mathbf{C})$ *by the isomorphism induced by the inclusion of* $\mathbf{R}$ *into* $\mathbf{C}$. *In view of the fact that the* $GL(n, \mathbf{C})$-*connection is reducible to a* $U(n)$-*connection we find that the classes* $w(c_k)$ *of Example 5.6 are in fact real.*

As we have indicated above, the characteristic classes turn out to be real cohomology classes. Indeed, the normalizing factors that we have used in defining them make them integral cohomology classes.

The **total Chern class** $c(P)$ of P is defined by

$$c(P) := 1 + c_1(P) + \cdots + c_n(P), \tag{5.18}$$

and the **Chern polynomial** $c(t)$ is defined by

$$c(t) := \sum_{i=0}^{n} t^{n-i} c_i(P) = t^n + t^{n-1} c_1(P) + \cdots + t c_{n-1}(P) + c_n(P). \tag{5.19}$$

We factorize the Chern polynomial formally as follows:

$$c(t) = (t + x_1)(t + x_2) \cdots (t + x_n). \tag{5.20}$$

The Chern classes are then expressed in terms of the **formal generators** $x_1, \ldots, x_n$ by elementary symmetric polynomials. For example,

$$c_1(P) = x_1 + x_2 + \cdots + x_n$$
$$c_2(P) = \sum_{1 \leq i < j \leq n} x_i x_j$$
$$\vdots$$
$$c_n(P) = x_1 x_2 \cdots x_n.$$

It is well known that every symmetric polynomial in $x_1, \ldots, x_n$ can be expressed as a polynomial in elementary symmetric polynomials. In particular, the homogeneous polynomial $\sum_{i=1}^{n} x_i^k$ of degree k can be expressed in terms of the first k Chern classes. These expressions are useful in many characteristic classes.

The **Chern character** $ch(P)$ is the polynomial in x_i defined by

$$ch(P) = \sum_{i=1}^{n} e^{x_i} \in H^*(M; \mathbf{Q}), \tag{5.21}$$

where the exponential function on the right hand side is interpreted as a formal power series, which in fact is finite and terminates after $\frac{1}{2}(\dim M)$ terms. It is customary to say that the Chern character is defined by the **generating function** e^z. The first few terms on the right hand side, expressed in terms of the Chern classes, give us the following formula

$$ch(P) = n + c_1(P) + \left[\frac{1}{2}c_1^2(P) - c_2(P)\right] + \cdots.$$

We now give examples of some other important characteristic classes that can be defined by generating functions and then expressed as polynomials in x_i.

- The **Todd class** $\tau(P) \in H^*(M; \mathbf{Q})$ defined by the generating function $z/(1 - e^{-z})$ is given by

$$\tau(P) = \prod_{i=1}^{n} \frac{x_i}{1 - \exp(-x_i)}. \tag{5.22}$$

  Its expression in terms of the Chern classes is given by

$$\tau(P) = 1 + \frac{1}{2}c_1 + \frac{1}{12}(c_1^2 + c_2) + \frac{1}{24}c_1 c_2 + \cdots. \tag{5.23}$$

  The Todd class appears in the statement of a version of the Riemann–Roch theorem.
- The **Hirzebruch L-polynomial** defined by the generating function $z/\tanh(z)$ is given by

$$L(P) = \prod_{i=1}^{n} \frac{x_i}{\tanh(x_i)}. \tag{5.24}$$

  Its expression in terms of the Pontryagin classes is given by

$$L(P) = 1 + \frac{1}{3}p_1 + \frac{1}{45}(7p_2 - p_1^2) + \cdots. \tag{5.25}$$

  The Hirzebruch L-polynomial appears in the Hirzebruch signature theorem, which is discussed later in this chapter.
- The $\hat{A}$ **genus** defined by the generating function $z/(2\sinh(z/2))$ is given by

$$\hat{A}(P) = \prod_{i=1}^{n} \frac{x_i}{2\sinh(x_i/2)}. \tag{5.26}$$

Its expression in terms of the Pontryagin classes is given by

$$\hat{A}(P) = 1 - \frac{1}{3}p_1 + \frac{1}{5760}(7p_1^2 - 4p_2) + \cdots. \tag{5.27}$$

The $\hat{A}$ genus is a topological invariant of spin manifolds. It is in establishing the divisibility properties of this invariant that Atiyah and Singer were led to their famous index theorem.

The characteristic classes satisfy the following properties.
(i) **Naturality** with respect to pull-back of bundles:
If $f : N \to M$ and f^*P is the pull-back of the principal bundle $P(M, G)$ to N, then we have

$$b(f^*(P)) = f^*(b(P)), \tag{5.28}$$

where $b(P)$ is a characteristic class of the bundle P.
(ii) The **Whitney sum rule**:

$$b(E_1 \oplus E_2) = b(E_1)b(E_2), \tag{5.29}$$

where E_1, E_2 are vector bundles and b is a characteristic class.

If E is a complex vector bundle with fiber $\mathbf{C}^n$, which is associated to $P(M, GL(n, \mathbf{C}))$, then the kth Chern class $c_k(E)$ of E is represented by $w(c_k) \in H^{2k}(M; \mathbf{C})$. If V is a real vector bundle over M and V^c its complexification, then the kth Pontryagin class $p_k(V)$ of V is given by

$$p_k(V) = (-1)^k c_{2k}(V^c) \in H^{4k}(M; \mathbf{R}). \tag{5.30}$$

We now work out in detail the Chern classes for an $SU(2)$-bundle P over a compact manifold M. They are used in defining invariants of isospin gauge fields and in particular the instanton numbers.

Let

$$A = \begin{pmatrix} a_{11} & a_{12} \\ a_{21} & a_{22} \end{pmatrix}$$

be a matrix in the Lie algebra $su(2)$. The algebra of characteristic classes is generated by the images of the invariant symmetric polynomials $\mathrm{Tr}\, A = a_{11} + a_{22}$ and $\det A = a_{11}a_{22} - a_{12}a_{21}$. Let ω be a connection on P with curvature Ω. Writing $a_{lk} = -\Omega_{lk}/2\pi i$ we get the polynomials in curvature representing the Chern classes

$$c_1(P) = -\frac{1}{2\pi i}(\Omega_{11} + \Omega_{22}) = 0 \qquad \text{(by the Lie algebra property)},$$

$$c_2(P) = -\frac{1}{4\pi^2}(\Omega_{11} \wedge \Omega_{22} - \Omega_{12} \wedge \Omega_{21})$$

$$= \frac{1}{8\pi^2}\, \mathrm{Tr}(\Omega \wedge \Omega).$$

The 2-form $\hat{\Omega}$ induced by the curvature on the base M will be denoted by F_ω. We will see later that F_ω corresponds to a gauge field when ω is identified with a gauge potential. Evaluating the second Chern class on the fundamental cycle of M, we get

$$c_2(P)[M] = \frac{1}{8\pi^2} \int_M \text{Tr}(F_\omega \wedge F_\omega). \tag{5.31}$$

In view of the integrality of the Chern classes, the number k defined by

$$k = -c_2(P)[M] \tag{5.32}$$

is an integer called the **topological charge** or **topological quantum number** or the **instanton number** of the principal $SU(2)$-bundle P. The topological charge may also be defined as $p_1(P)[M]$, where p_1 is the first Pontryagin class. If M is orientable and $\dim M = 4$, then we have

$$k = p_1(P) = -c_2(P). \tag{5.33}$$

The characteristic classes described above are often referred to as the **primary characteristic classes**. These classes do not depend on the choice of connection used to define them and are in fact topological invariants of the principal bundle $P(M, G)$. They can be obtained by pulling back suitable universal classes, as can be shown by using the following theorem due to Narasimhan and Ramanan [296, 297].

Theorem 5.8 *Let G be a Lie group with $\pi_0(G) < \infty$ and k a positive integer. Then there exists a principal G-bundle $E^k(B^k, G)$ with connection θ^k, which is k-universal for principal G-bundles with connection; i.e., for any compact manifold M with $\dim M < k$ and principal bundle $P(M, G)$ with connection ω there exists a map $f : M \to B^k$, defined up to homotopy such that P is the pull-back of E^k to M by f. Then we have the following commutative diagram:*

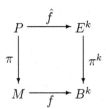

where π, π^k are the bundle projections, $P = f^* E^k$, and $\hat{f}$ is the lift of f. Furthermore, we have

$$\omega = (\hat{f})^* \theta^k.$$

We note that the spaces B^k and E^k form two inductive sets and their direct limits define topological spaces BG and EG, respectively, such that EG is a principal G-bundle over BG. The spaces BG, EG are called the **universal classifying spaces** for principal G-bundles. They are not finite-dimensional manifolds. However, in most applications it is sufficient to choose a large enough k so that the classifying spaces can be taken as finite-dimensional manifolds.

Example 5.8 *The* **Stiefel manifold** $V_{\mathbf{R}}(n+k,n)$ *of orthonormal n-tuples of vectors in $\mathbf{R}^{n+k}$ is a principal $O(n)$-bundle over the* **Grassmann manifold** $G_{\mathbf{R}}(n+k,n)$ *of n-planes in $\mathbf{R}^{n+k}$. This bundle has a natural connection θ^k. The bundle $V_{\mathbf{R}}(n+k,n)$ with connection θ^k is k-universal for principal $O(n)$- bundles. Similarly, the principal bundle $V_{\mathbf{C}}(n+k,n)(G_{\mathbf{C}}(n+k,n),U(n))$ with connection $\theta^k_{\mathbf{C}}$ is k-universal for principal $U(n)$-bundles.*

This example plays a fundamental role in the proof of Theorem 5.8.

5.3.1 Secondary Characteristic Classes

Primary characteristic classes of principal bundles have become important tools for defining and analyzing topological invariants in physical theories and in particular in gauge theories. Recently, another set of characteristic classes, called the **secondary characteristic classes**, have appeared in the Lagrangian formulation of quantum field theories. The most well-known of these classes are the **Chern–Simons classes**. We now discuss a construction which leads to another proof of the independence of the primary characteristic classes of the choice of connections and at the same time prepares the way for defining secondary characteristic classes. Recall that a standard r-simplex is defined by

$$\Delta^r = \left\{ (t_0,\ldots,t_r) \in \mathbf{R}^{r+1} \mid t_i \geq 0,\ \sum_{i=0}^{r} t_i = 1 \right\}. \tag{5.34}$$

Let ω_i, $0 \leq i \leq r$, be a set of connections on $P(M,G)$. Let $\omega = \sum_{i=0}^{r} t_i \omega_i$. Then ω is a connection on P with curvature $\Omega = d^\omega \omega$. Let $f \in I^k(G)$; define

$$\Delta_f(\omega_0,\ldots,\omega_r) := (-1)^{[\frac{r+1}{2}]} \int_{\Delta^r} f(\Omega,\ldots,\Omega), \tag{5.35}$$

where $[\frac{r+1}{2}]$ is the integer part of $\frac{r+1}{2}$ and the integration is along the fiber Δ^r of the bundle $P \times \Delta^r \to P$ with respect to the standard volume form $dt_1 \wedge \ldots \wedge dt_r$ of Δ^r. Thus, $\Delta_f(\omega_0,\ldots,\omega_r)$ is a $(2k-r)$-form on P, which descends to M in view of the fact that f is an invariant polynomial. We continue to denote this form on M by the same notation. Thus we have a map

$$\Delta(\omega_0,\ldots,\omega_r) : I^k(G) \to \Lambda^{2k-r}(M) \quad \text{defined by } f \mapsto \Delta_f(\omega_0,\ldots,\omega_r). \tag{5.36}$$

This map extends to an algebra homomorphism

$$\Delta(\omega_0, \ldots, \omega_r) : I(G) \to \Lambda(M), \tag{5.37}$$

which is called the **Bott homomorphism** relative to $(\omega_0, \ldots, \omega_r)$ [50]. The following theorem gives an important relation among the forms Δ_f.

Theorem 5.9 *Let ω_i, $0 \le i \le r$, be a set of connections on $P(M, G)$ and $f \in I^k(G)$; then*

$$d(\Delta_f(\omega_0, \ldots, \omega_r)) = \sum_{i=0}^{r} (-1)^i \Delta_f(\omega_0, \ldots, \hat{\omega}_i, \ldots, \omega_r), \tag{5.38}$$

where $\hat{}$ over a variable denotes that this variable is deleted.

Applying the above theorem to the case of a 0-simplex and a 1-simplex we obtain the following corollaries.

Corollary 5.10 *In the case $r = 0$ we have*

$$d(\Delta_f(\omega_0)) = 0.$$

Hence, $\Delta_f(\omega_0)$ is a closed form and defines an element of the cohomology space $H^{2k}(M; \mathbf{R})$. The Bott homomorphism induces the homomorphism

$$[\Delta(\omega_0)] : I^k(G) \to H^{2k}(M; \mathbf{R}), \quad \text{defined by } f \mapsto [\Delta_f(\omega_0)]. \tag{5.39}$$

Corollary 5.11 *In the case $r = 1$ we have*

$$d(\Delta_f(\omega_0, \omega_1)) = \Delta_f(\omega_0) - \Delta_f(\omega_1) \in \Lambda^{2k}(M, \mathbf{R}).$$

Thus,

$$[\Delta_f(\omega_0)] = [\Delta_f(\omega_1)] \in H^{2k}(M; \mathbf{R}),$$

and hence the homomorphism $[\Delta(\omega_0)]$ of Corollary 5.10 is independent of the connection ω_0 and is, in fact, the Chern–Weil homomorphism w defined earlier.

Corollary 5.11 allows us to define secondary characteristic classes in the following way. Let

$$I^k(G)_{(\omega)} := \{ f \in I^k(G) \mid \Delta_f(\omega) = 0 \},$$

and let

$$I^k(G)_{(\omega_1, \omega_2)} := I^k(G)_{(\omega_1)} \cap I^k(G)_{(\omega_2)}.$$

Let $f \in I^k(G)_{(\omega_1, \omega_2)}$; then by Corollary 5.11, $d(\Delta_f(\omega_1, \omega_2)) = 0$. Hence, $\Delta_f(\omega_1, \omega_2)$ defines a cohomology class in $H^{2k-1}(M; \mathbf{R})$. We call the class $[\Delta_f(\omega_1, \omega_2)] \in H^{2k-1}(M; \mathbf{R})$ a **simple secondary characteristic class** of the triple (P, ω_1, ω_2). We note that in geometrical mechanics and geometric quantization theory, an important role is played by the **Maslov class**, which can be interpreted as a secondary characteristic class. A general discussion of symplectic geometry and its relation to the secondary characteristic classes can be found in Vaisman [390].

We now proceed to define the Chern–Simons classes of $P(M, G)$ with connection ω (see [75, 76]). The basic tool is the **transgression form** $T_f(\omega) \in H^{2k-1}(P; \mathbf{R})$ defined by

$$T_f(\omega) := - \int_{\Delta^1} f(\Omega_t, \ldots, \Omega_t) = k \int_0^1 f(\omega, \Omega_t, \ldots, \Omega_t)dt, \ f \in I^k(G),$$

where $\int_{\Delta^1}$ denotes integration along the fiber and

$$\Omega_t := dt \wedge \omega + t\Omega + t(1+t)d\omega \in \Lambda^2(P, \mathbf{g}), \qquad 0 \le t \le 1.$$

Roughly speaking Ω_t can be thought of as the "curvature " form corresponding to the "connection form" $t\omega$ on $P \times \Delta^1$. We note that, in general, the transgression forms do not descend to the base M. However, we have the following theorem.

Theorem 5.12 *Let ω be a connection form on $P(M, G)$. Then*

$$d(T_f(\omega)) = \pi^*(\Delta_f(\omega)), \qquad \forall f \in I^k(G).$$

From this theorem it follows that, if $f \in I^k(G)_{(\omega)}$ then $T_f(\omega)$ defines a cohomology class $[T_f(\omega)] \in H^{2k-1}(P; \mathbf{R})$, which is called the Chern–Simons class of (P, ω) related to $f \in I^k(G)_{(\omega)}$. The relation between the secondary characteristic classes and the Chern–Simons classes is given by the following proposition.

Proposition 5.13 *Let ω_1, ω_2 be two connection forms on the principal bundle $P(M, G)$ and let $f \in I^k(G)_{(\omega_1, \omega_2)}$. Then*

$$\pi^*[\Delta_f(\omega_1, \omega_2)] = [T_f(\omega_2)] - [T_f(\omega_1)].$$

An interesting application of the transgression forms is the following description of the de Rham cohomology ring of an important class of principle bundles.

Theorem 5.14 (Chevalley) *Let G be a compact, connected, semi-simple Lie group of rank r (dimension of maximal torus in G) and M a compact manifold. Let ω be a connection form on the principal bundle $P(M, G)$. Then the ring $I(G)$ of invariant polynomials is generated by a set of r elements $f_1, f_2, \ldots, f_r$ and the de Rham cohomology $H^*(P; \mathbf{R})$ is given by the quotient ring*

$$H^*(P; \mathbf{R}) = A/dA,$$

where

$$A = \Lambda(M)[T_{f_1}(\omega), T_{f_2}(\omega), \ldots, T_{f_r}(\omega)]$$

is the ring of polynomials in $[T_{f_1}(\omega), T_{f_2}(\omega), \ldots, T_{f_r}(\omega)]$ *with coefficients in* $\Lambda(M)$, *the ring of forms on* M *(pulled back to* P *by the bundle projection).*

If (M, g) is a Riemannian manifold, $P = L(M)$ the bundle of frames of M with structure group $GL(m, \mathbf{R})$ and λ the Levi-Civita connection on P, then we have the following result.

Theorem 5.15 *Let* $f \in I^{2k}(GL(m, \mathbf{R}))$. *Let* g' *be a Riemannian metric conformal to* g *and let* λ' *the corresponding Levi-Civita connection on* P. *Then there exists a form* $W \in \Lambda^{4k-1}(P)$ *such that*

$$T_f(\lambda') - T_f(\lambda) = dW. \tag{5.40}$$

It can be shown that the form W in equation (5.40) can be expressed in terms of the **Weyl conformal curvature tensor**.

We now consider a special case in which we can associate to $f \in I^k(G)_{(\omega)}$ a form $\alpha_f \in H^{2k-1}(M; \mathbf{R}/\mathbf{Z})$. Recall that the short exact sequence

$$0 \longrightarrow \mathbf{Z} \xrightarrow{\iota} \mathbf{R} \xrightarrow{p} \mathbf{R}/\mathbf{Z} \longrightarrow 0$$

where ι is the inclusion and p is the canonical projection, induces the long exact sequence in the cohomology of M

$$\cdots \longrightarrow H^j(M; \mathbf{Z}) \xrightarrow{\iota} H^j(M; \mathbf{R}) \xrightarrow{p} H^j(M; \mathbf{R}/\mathbf{Z}) \longrightarrow \cdots$$

and a similar sequence in the cohomology of P. Using these sequences we obtain the following theorem.

Theorem 5.16 *Let* $f \in I^k(G)_{(\omega)}$ *be such that* $w(f)$ *is an integral cohomology class, i.e.,* $w(f) \in \iota(H^j(M; \mathbf{Z})) \subset H^j(M; \mathbf{R})$. *Then there exists a form* $\alpha_f \in H^{2k-1}(M; \mathbf{R}/\mathbf{Z})$ *such that*

$$p[T_f(\omega)] = \pi^*(\alpha_f) \in H^{2k-1}(P; \mathbf{R}/\mathbf{Z}).$$

It is this form $\alpha_f \in H^{2k-1}(M; \mathbf{R}/\mathbf{Z})$ that appears as the **Chern–Simons term** in the Lagrangian of field theories on manifolds with boundary. If A denotes the pull-back to M of the connection form ω by a local section of P, and F the corresponding curvature form on M (in the physics literature, A is called the gauge potential and F the gauge field on M), then the first three Chern–Simons terms have the following local expressions:

$$\alpha_1 = \frac{i}{2\pi} \operatorname{tr} A,$$

$$\alpha_3 = \frac{1}{2} \left(\frac{i}{2\pi} \right)^2 \operatorname{tr}(A \wedge F + \frac{2}{3} A^3),$$

$$\alpha_5 = \frac{1}{6}\left(\frac{i}{2\pi}\right)^3 \mathrm{tr}(A \wedge F \wedge F + \frac{3}{2}A^3 \wedge F + \frac{3}{5}A^5),$$

where

$$A^n := \underbrace{A \wedge A \wedge \ldots \wedge A}_{n \text{ terms}}.$$

The secondary characteristic classes and the Chern–Simons classes are also used in describing anomalies such as chiral and gravitational anomalies in field theories. Using quantum field theory on a 3-manifold with the Lagrangian function given by the Chern–Simons term α_3, Witten obtained a physical interpretation of the Jones polynomial of a link by expressing it as the expectation value of a quantum observable. This work ushered in topological quantum field theory (TQFT) as a new area of research in physical mathematics. We discuss this work in Chapter 11.

5.4 *K*-theory

The foundations of *K*-theory were laid by A. Grothendieck in the framework of algebraic geometry, in his formulation of the Riemann–Roch theorem, providing a new and powerful tool that can be regarded as a generalized cohomology theory. Grothendieck's ideas have led to other K-theories, notably algebraic and topological K-theories. Grothendieck (b. 1928) was awarded a Fields Medal at the ICM 1966, in Moscow, for work that gave a new unifying perspective in the study of geometry, number theory, topology, and complex analysis. The Grothendieck Festschrift, celebrating his 60th birthday, contains articles on his incredible achievements, which opened up many new fields of research in mathematics. Volume 3 (1989) of the journal *K-Theory* contains two articles on Grothendieck's work in *K*-theory, one by Serre and the other by his mentor Dieudonné. (For a recent update on topological and bivariant *K*-theories see the book by Cuntz et al. [92].) Grothendieck unveiled his work in the first lecture at the first Arbeitstagung, organized by Friedrich Hirzebruch[1] in Bonn in 1957. At the 2007 Arbeitstagung, celebrating 50 years of these extraordinary meetings, Hirzebruch gave the first lecture, offering his point of view on the first Arbeitstagungen, with special emphasis on 1957, 1958, and 1962. He explained that Grothendieck lectured for twelve hours spread out over four days on "Kohärente Garben und verallgemeinerte Riemann–Roch–Hirzebruch-Formel auf algebraischen Mannigfaltigkeiten"[2]. In these lectures he proved a far reaching generalization of the

---

[1] Hirzebruch is the founding director of the Max Planck Institute for Mathematics in Bonn. This institute and the Arbeitstagung have had a very strong and wide ranging impact on mathematical research since their inception.

[2] Coherent sheaves and generalized Riemann–Roch–Hirzebruch formula for algebraic manifolds.

Riemann–Roch–Hirzebruch (or RRH) theorem, which was itself an important extension of the classical **Riemann–Roch** (or **RR**) theorem. This theorem is now known as the **Grothendieck–Riemann–Roch (GRR)** theorem. Grothendieck proved this theorem by using his new theory that he called K-**theory**. When Hirzebruch asked Grothendieck to explain the significance of "K" in K-theory, Grothendiek replied: "'H' was already used for homology and I did not like 'I, J', so I decided to call my theory K-theory."

The starting point of Grothendieck's proof is the construction of the Grothendieck ring of the projective algebraic variety X. The **Grothendieck ring** $K(X)$ is constructed with complex vector bundles and coherent sheaves on X. Recall that a vector bundle corresponds to the locally free sheaf of its sections. A coherent sheaf has a resolution in terms of vector bundles. This allows a coherent sheaf S to be considered as an element of the ring $K(X)$, as an alternating sum of vector bundles. The definition of $K(X)$ is based on the notion of **ring completion** of a **semi-ring**. Recall that a semi-ring S satisfies all the axioms of a ring except for the existence of the additive inverse. If S has a unity then it is unique and is denoted by 1_S or simply by 1. The non-negative integers $\{0, 1, \dots\}$ with the usual addition and multiplication form a commutative semi-ring with unity (i.e., multiplicative identity). The ring completion of the semi-ring $\mathbf{N}$ is the ring $\mathbf{Z}$ of all integers with the usual addition and multiplication. In general, the **ring completion** of a semi-ring S is a pair $(\widetilde{S}, f)$, where $\widetilde{S}$ is a ring and $f : S \to \widetilde{S}$ is a morphism of semi-rings such that the following universal property is satisfied: If $h : S \to R$ is any morphism of S into a ring R then there exists a unique ring morphism $\widetilde{h} : \widetilde{S} \to R$ such that $\widetilde{h} \circ f = h$, i.e., the following diagram commutes:

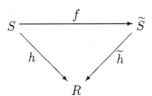

We recall the construction of $\widetilde{S}$. Define the relation $\sim$ on $S \times S$ by $(a, b) \sim (c, d)$ if there exists $e \in S$ such that $a + d + e = c + b + e$. It is easy to verify that this is an equivalence relation. We define $\widetilde{S} := (S \times S)/\sim$ and denote by $[a, b] \in \widetilde{S}$ the equivalence class of (a, b). Addition and multiplication in $\widetilde{S}$ are defined by

$$[a, b] + [c, d] = [a + c, b + d],$$
$$[a, b] \times [c, d] = [ac + bd, bc + ad].$$

We denote the class $[0, 0]$ simply by 0 and define $-[a, b] := [b, a]$. We define the map
$$f : S \to \widetilde{S} \quad \text{by } f(a) = [a, 0], \qquad \forall a \in S.$$

It is customary to identify S with the image of f and to denote the class $[a, 0]$ simply by a and to denote $[a, b]$ by $a - b$.

The following example is fundamental in our considerations.

Example 5.9 (The semi-ring $Vect_F(X)$) *Let X be any compact manifold and F denote either of the fields $\mathbf{R}$ or $\mathbf{C}$. Let $Vect_F(X)$ be the set of isomorphism classes of F-vector bundles over X. Then the set $Vect_F(X)$ has a natural* **commutative semi-ring** *structure with unity, defined by*

1. *the Whitney sum $(\alpha, \beta) \mapsto \alpha \oplus \beta$ as addition,*
2. *the tensor product $(\alpha, \beta) \mapsto \alpha \otimes \beta$ as multiplication,*
3. *θ^1, the class of the trivial line bundle as unity,*

where $\theta^n \in Vect_F(X)$ denotes the isomorphism class of the trivial n-plane bundle $X \times F^n$ over X.

The Grothendieck ring $K_F(X)$ is defined to be the ring completion of the semi-ring $Vect_F(X)$. Thus elements of $K_F(X)$ can be written as $A - B$, $A, B \in Vect_F(X)$. If V, W are vector bundles over X, then it is customary to call $[V] - [W]$ a **virtual vector bundle** and to denote it simply by $V - W$. Now we recall that the Chern character Ch satisfies the following relations:

$$Ch(V \oplus W) = Ch(V) + Ch(W), \ Ch(V \otimes W) = Ch(V) \, Ch(W), \quad (5.41)$$

where $V, W \in Vect_{\mathbf{C}}(X)$. Hence, $Ch : Vect_{\mathbf{C}}(X) \to H^*(X; \mathbf{R})$ is a semi-ring morphism of $Vect_{\mathbf{C}}(X)$ into the cohomology ring $H^*(X; \mathbf{R})$ and hence lifts to a unique ring morphism (also denoted by Ch) $Ch : K_{\mathbf{C}}(X) \to H^*(X; \mathbf{R})$ defined by

$$Ch([V] - [W]) = Ch(V) - Ch(W). \quad (5.42)$$

This morphism allows us to extend the definition of Chern character to virtual vector bundles. In fact, the image of this morphism lies in the even cohomology with rational coefficients, i.e.,

$$Ch : K_{\mathbf{C}}(X) \to H^{\text{even}}(X; \mathbf{Q}) = \bigoplus_{i=0}^{\infty} H^{2i}(X; \mathbf{Q}) \subset H^*(X; \mathbf{R}). \quad (5.43)$$

In particular, if X is compact then the Chern character induces an isomorphism

$$K_{\mathbf{C}}(X) \otimes \mathbf{Q} \cong H^{\text{even}}(X; \mathbf{Q}). \quad (5.44)$$

In the case of spheres we can say more, namely,

$$K_{\mathbf{C}}(S^{2n}) \cong H^*(S^{2n}; \mathbf{Z}) \subset H^*(S^{2n}; \mathbf{R}). \quad (5.45)$$

We note that in the original definition of $K(X)$, the space X is a projective algebraic variety. Thus a coherent sheaf S can be regarded as an element of $K(X)$ and the Chern character $Ch(S)$ is well-defined. Grothendieck then defines the **push forward** of the sheaf S by an algebraic map $f : X \to Y$ of

algebraic varieties, as a sheaf on Y. It is denoted by $f_!S$. The GRR theorem
is then expressed by the formula;

$$f_*(Ch(S)\tau(TX)) = Ch(f_!S)\tau(TY) , \qquad (5.46)$$

where f_* is the homomorphism on cohomology induced by the map f. If the
space Y is a point then the GRR formula (5.46) reduces to the RRH formula.
The original RRH formula is in terms of a holomorphic vector bundle E over
a compact complex manifold X of complex dimension n. The **RRH formula**
can be expressed as follows:

$$(Ch(E)\tau(TX))[X] = \chi(X, E), \qquad (5.47)$$

where TX is the holomorphic tangent bundle of X, $[X]$ is the fundamental
class of X, and $\chi(X, E)$ is the **holomorphic Euler characteristic** of E in
sheaf cohomology defined by

$$\chi(X, E) := \sum_{i=0}^{\infty} (-1)^i \dim H^i(X, E). \qquad (5.48)$$

The Chern character and the Todd class lie in the cohomology ring of X,
and evaluation on the fundamental homology class $[X]$ is obtained from the
pairing of homology and cohomology (i.e., integration over X of the total class
in $H^{2n}(X, E)$ in the expansion of $Ch(E)\tau(TX)$). All the sheaf cohomology
spaces $H^i(X, E)$ are finite-dimensional. They equal zero for $i > 2n + 1$ so the
sum is finite. The RRH theorem, proved in 1954, provides the long sought
after generalization of the RR theorem from Riemann surfaces (i.e.r, complex
curves) to complex manifolds of arbitrary finite dimension. In fact, the RR
theorem in the current form was proved by Riemann's student Gustav Roch,
improving on Riemann's inequality in the 1850s. It provides an important tool
in the computation of the dimension of the space of meromorphic functions
on a compact connected Riemann surface Σ satisfying certain conditions. It
relates the complex analytic properties of Σ to its global topological properties, namely its Euler characteristic or, equivalently, its genus. The RRH
theorem is very much in the spirit of the RR theorem relating holomorphic
and topological data related to a fixed variety. GRR theorem changes these
statements to a statement about morphism of varieties inaugurating the categorical approach which paved the way for K-theories for other structures.
These observations are useful in the study of index theorems for families of
elliptic operators such as the Dirac operator coupled to gauge fields.

We note that $X \to Vect_F(X)$ defines a **contravariant functor** $Vect_F$
from the **category of manifolds** to the **category of semi-rings**. Similarly, $X \to K_F(X)$ defines a contravariant functor K_F from the category of
manifolds to the category of rings. We recall that the F-**rank** of a vector
bundle over X is well defined by the requirement of connectedness of X.
Therefore, the map

$$\rho : Vect_F(X) \to \mathbf{Z}, \quad [\alpha] \mapsto \mathrm{rank}_F(\alpha),$$

is well defined. It is easy to see that ρ is a morphism of semi-rings and hence lifts to a ring morphism $\widetilde{\rho} : K_F(X) \to \mathbf{Z}$. We define the **reduced Grothendieck ring** $\widetilde{K}_F(X)$ to be the kernel of $\widetilde{\rho}$, i.e.,

$$\widetilde{K}_F(X) := \ker(\widetilde{\rho}).$$

We denote by $\epsilon : \mathbf{Z} \to K_F(X)$ the map defined by $\epsilon(1) := \theta^1$ and note that $\widetilde{\rho} \circ \epsilon = id_{\mathbf{Z}}$. For a positive integer n, $\epsilon(n) = \theta^n = [X \times F^n]$, the class of the trivial n-plane bundle over X. One can show that the generic element of $K_F(X)$ can be written in the form $[\alpha] - \theta^n$, $[\alpha] \in Vect_F(X)$, and that

$$K_F(X) \cong \widetilde{K}_F(X) \oplus \mathbf{Z}. \qquad (5.49)$$

The contravariant functor K_F can be used to give alternative definitions of the maps $\widetilde{\rho}$ and ϵ as follows: For $a \in X$, let $\iota : \{a\} \to X$ be the natural injection. This induces a ring morphism $K_F(\iota) : K_F(X) \to K_F(\{a\})$. Observing that $K_F(\{a\}) \cong \mathbf{Z}$, we can show that $K_F(\iota)$ can be identified with $\widetilde{\rho}$. Similarly, we may identify ϵ with $K_F(\pi)$, where $\pi : X \to \{a\}$ is the natural projection.

It is possible to define the ring $\widetilde{K}_F(X)$ directly by using the relation of **stable equivalence** of vector bundles. We say that two vector bundles α, β over X are **stably equivalent** or **s-equivalent** if there exist natural numbers k, n such that

$$[\alpha \oplus \theta^k] = [\beta \oplus \theta^n] \in Vect_F(X).$$

It can be shown that stable equivalence is an equivalence relation on $Vect_F(X)$. Let us denote by $E_s(X)$ the set of **stable equivalence classes**. Under the operation induced by direct sum the set $E_s(X)$ is a group. We observe that a generic element of $\widetilde{K}_F(X)$ can be written in the form $[\alpha] - \theta^{r(\alpha)}$, $[\alpha] \in Vect_F(X)$, where we have written $r(\alpha)$ for $\mathrm{rank}_F(\alpha)$. Let $[\alpha]_s$ denote the stable equivalence class of α. The map

$$\phi : \widetilde{K}_F(X) \to E_s(X) \qquad \text{defined by } [\alpha] - \theta^{r(\alpha)} \mapsto [\alpha]_s$$

is an isomorphism with inverse $\psi : E_s(X) \to \widetilde{K}_F(X)$ defined by $[\alpha]_s \mapsto [\alpha] - \theta^{r(\alpha)}$.

If Y is a closed subspace of X, we denote by X/Y the topological space obtained from X by identifying Y to a point denoted by $\{y\}$. If $Y = \emptyset$, we consider X/Y as obtained from X by adjoining a disjoint point, which we also denote by $\{y\}$. Thus, in any case $(X/Y, \{y\})$ is a pointed topological space. We define the **relative K-group** of X with respect to Y, denoted by $K_F(X, Y)$, to be the reduced K-group of X/Y, i.e.,

$$K_F(X, Y) := \widetilde{K}_F(X/Y).$$

If $Y = \emptyset$ it follows that $K_F(X, \emptyset) \cong K_F(X)$. If X is locally compact we may identify $X/\emptyset$ with the one-point compactification, denoted by $X \cup \{\infty\}$, of X; in this case we define

$$K_F(X) := K_F(X/\emptyset, \{\infty\}) = \widetilde{K}_F(X/\emptyset).$$

Thus, by definition,

$$K_F(\mathbf{R}^2) = \widetilde{K}_F(S^2).$$

We shall use the relative K-groups in the K-theoretic formulation of index theorems later in this chapter.

If we consider only the group structure of $K_F(X)$ and $\widetilde{K}_F(X)$ then the above considerations can be applied also to the case when $F = \mathbf{H}$, the division ring of quaternions. The functors $\widetilde{K}_\mathbf{R}, \widetilde{K}_\mathbf{C}, \widetilde{K}_\mathbf{H}$ are usually denoted by $\widetilde{KO}, \widetilde{KU}$, and $\widetilde{KSp}$ and are called the **real, complex**, and **quaternionic** K-groups respectively. The reduced Grothendieck groups of S^n, $n > 1$, are given by the following theorem.

Theorem 5.17 *For $n > 1$ we have the following group isomorphisms:*

$$\widetilde{KO}(S^n) = \pi_{n-1}(SO(\infty)),$$
$$\widetilde{KU}(S^n) = \pi_{n-1}(SU(\infty)),$$
$$\widetilde{KSp}(S^n) = \pi_{n-1}(Sp(\infty)).$$

The following theorem is of fundamental importance in the K-theory treatment of periodicity theorems.

Theorem 5.18 *Let X be a compact manifold. Then in the complex case we have the following isomorphism*

$$KU(X) \otimes KU(S^2) \cong KU(X \times S^2),$$

where $KU(S^2)$ is the free Abelian group on two generators 1 and η (the class of the complex Hopf fibration of S^3 over S^2, which is the tautological complex line bundle over $\mathbf{CP}^1 = S^2$). In the real case we have the isomorphism

$$KO(X) \otimes KO(S^8) \cong KO(X \times S^8),$$

where $KO(S^8)$ is the free Abelian group on two generators 1 and η_8, the class of the real 8-dimensional Hopf bundle (for further details see [198]).

The K-theory interpretation of the Bott periodicity theorem given below is a direct consequence of the above two theorems.

Corollary 5.19
$$\widetilde{KO}(S^n) = \widetilde{KO}(S^{n+8}),$$
$$\widetilde{KU}(S^n) = \widetilde{KU}(S^{n+2}).$$

It can be shown that the quaternionic K-groups are related to real K-groups of spheres by the following relations:

$$\widetilde{KO}(S^n) = \widetilde{KSp}(S^{n+4}), \tag{5.50}$$

$$\widetilde{KO}(S^{n+4}) = \widetilde{KSp}(S^n). \tag{5.51}$$

These relations and Corollary 5.19 imply the following periodicity relation for the quaternionic K-groups of spheres:

$$\widetilde{KSp}(S^n) = \widetilde{KSp}(S^{n+8}),$$

thus allowing us to calculate all the K-groups of spheres using the following table.

Table 5.1 Reduced Grothendieck groups of spheres

n	$\widetilde{KO}(S^n)$	$\widetilde{KU}(S^n)$	$\widetilde{KSp}(S^n)$
1	$\mathbf{Z}_2$	0	0
2	$\mathbf{Z}_2$	$\mathbf{Z}$	0
3	0	0	0
4	$\mathbf{Z}$	$\mathbf{Z}$	$\mathbf{Z}$
5	0	0	$\mathbf{Z}_2$

We observe that these groups correspond to the stable homotopy groups given in the Bott periodicity table in Chapter 2.

Grothendieck groups and rings have many interesting properties that parallel those of classical cohomology theories. For this reason K-theory is sometimes referred to as a generalized cohomology theory. Grothendieck's ideas were extended to the domain of topology, differential geometry, and algebra by Atiyah, Hirzebruch, and other mathematicians. In **topological K-theory** one associates to any compact topological space X, a group $K(X)$ constructed from the category of vector bundles on X. In **algebraic K-theory** one associates to a ring R with unity the group $K(R)$ constructed from the category of finitely generated projective right R-modules. One of the main problems in algebraic K-theory is the computation of $K(R)$ for special classes of rings. In differential geometry it is reasonable to think of K-theory as a generalized cohomology theory that deals with classes of stable vector bundles. In this case the set $K(X)$ can be given the structure of a ring.

Applications of K-theory have provided new links and simpler proofs of several important results in geometry, topology, and algebra. In particular, as we discussed earlier, the Bott periodicity theorem can be interpreted as a theorem in K-theory. J. F. Adams solved the long outstanding problem of the existence of vector fields on spheres by using K-theory. Today K-theory has developed into an important discipline in its own right with its own aforementioned journal K-Theory. A very readable account of K-theory may be found in Karoubi [216].

5.5 Index Theorems

The **Atiyah–Singer index theorem** is one of the most important results of modern mathematics. The theorem—or rather a set of theorems collectively referred to as index theorems—were developed in a series of papers by Atiyah, Singer and their collaborators. A good introduction to this and other related results may be found in Gilkey [154], Lawson and Michelsohn [248], Palais [311], and Shanahan [348]. Its application to gauge theories is discussed in the book by Booss and Bleecker [47]. We give below a statement of some versions of the index theorem and also consider special cases. Index theorems relate analytic data of an operator on bundles over a manifold to the topological data of these bundles. We discuss the relevant operator theory in Appendix D. We also introduce below some additional machinery needed for the statements of various index theorems.

1. If X and Y are compact spaces, one can define an **outer product** $\dot{\otimes}$ of X and Y's respective Grothendieck groups $K(X)$ and $K(Y)$:

$$\dot{\otimes} : K(X) \times K(Y) \to K(X \times Y)$$

   such that $[E]\dot{\otimes}[F]$ is the class in $K(X \times Y)$ determined by the vector bundle $E\dot{\otimes}F$ over $X \times Y$ with fiber $E_x \otimes F_y$ over (x, y). This product can be extended to the case of locally compact spaces (see, for example, Booss and Bleecker [47]).

2. Let E^1 and E^2 be vector bundles over the compact spaces X_1 and X_2, respectively, with $A = X_1 \cap X_2 \neq \emptyset$ and let $X = X_1 \cup X_2$. Let f be a vector bundle isomorphism of $E^1_{|A}$ onto $E^2_{|A}$. We denote by $E^1 \cup_f E^2$ the vector bundle over X obtained by identifying the fiber E^1_x, $x \in A$, with the fiber E^2_x through the isomorphism f_x. Let B^+ (resp., B^-) denote the upper (resp., lower) closed hemisphere of S^2. Thus $B^+ \cap B^- = S^1$. Consider the above construction with $X_1 = B^+$, $X_2 = B^-$, $E^1 = B^+ \times \mathbf{C}$ (resp., $E^2 = B^- \times \mathbf{C}$) the complex, trivial line bundle over B^+ (resp., B^-), $f : S^1 \times \mathbf{C} \to S^1 \times \mathbf{C}$ the map defined by $f(z, z_1) = z_1/z$. Let $E_{-1} := (B^+ \times \mathbf{C}) \cup_f (B^- \times \mathbf{C})$; then E_{-1} is a vector bundle over S^2. If $\theta^1 := [S^2 \times \mathbf{C}]$ denotes the class of the trivial line bundle, the element $b := [E_{-1}] - \theta^1$ is in the kernel of $K(S^2) \to \mathbf{Z}$ and thus $b \in K(\mathbf{R}^2)$. The class b is called the **Bott class**. Let X be a locally compact space; the **Bott periodicity theorem** (complex case) asserts that the map $m_b : K(X) \to K(X \times S^2)$ defined by outer product by b, i.e., $m_b(u) := u\dot{\otimes}b$, $u \in K(X)$, is an isomorphism.

3. Let E^1 and E^2 be vector bundles over X and let A be a closed subspace of X. Let f be a vector bundle isomorphism of $E^1_{|A}$ onto $E^2_{|A}$. Associated with the triple $(E_1, E_2; f)$ there is a unique, canonically constructed element in $K(X, A)$. Let $X_1 = X \times \{1\}$, $X_2 = X \times \{2\}$, and Z be the union of X_1 and X_2 with $(x, 1)$ and $(x, 2)$ identified for all $x \in A$. Let $\pi_r : Z \to X_r$, $r = 1, 2$ be the natural maps, and $W = \pi_1^* E_1 \cup_f \pi_2^* E_2$. Let $j : X_2 \to Z$ be the natural injection. One can show that $[W] - [\pi_2^* E_2]$ is in the kernel of $K(j)$ and thus can be considered an element of $K(Z, X_2)$. We now observe that Z with X_2 reduced to a point can be identified with X in which A is reduced to a point. Thus, $K(Z, X_2) \cong K(X, A)$ and hence $[W] - [\pi_2^* E_2]$ can be identified with an element of $K(X, A)$, which we denote $[f]_K$.

4. Let E, F be two vector bundles over a compact Riemannian manifold M and let $P \in El_k(E, F)$ be an elliptic operator of order k from $\Gamma(E)$ to $\Gamma(F)$. Define the disk bundle $DM := \{u \in T^*M \mid \|u\| \leq 1\}$. Applying the above construction to the triple $(\pi^*(E), \pi^*(F); \sigma_k(P))$, we have $[\sigma_k(P)]_K \in K(DM, \partial DM)$. But $DM/\partial DM$ is naturally homeomorphic to the one-point compactification of T^*M and by means of the Riemannian metric we may identify T^*M with TM. Thus we may consider $[\sigma_k(P)]_K \in K(TM)$.

5. Recall that every m-dimensional compact manifold M can be trivially embedded in some $\mathbf{R}^{m+n}$ in the following sense. The restriction of $T\mathbf{R}^{m+n}$ to M allows one to define the normal bundle of M with fibers N_x such that $T_x \mathbf{R}^{m+n} = T_x M \oplus N_x$. For large enough n the normal bundle is trivial and in this case we say that M is trivially embedded in $\mathbf{R}^{m+n}$. Furthermore, by choosing a trivialization we can write $N = M \times \mathbf{R}^n$. Thus N becomes a tubular neighborhood of M in $\mathbf{R}^{m+n}$ and $TN = TM \times \mathbf{R}^{2n}$. Thus, an element $a \in K(TM \times \mathbf{R}^{2n})$ may be identified with an element in $K(\mathbf{R}^{2m+2n})$ also denoted by a. Thus applying $m+n$ times the inverse m_b^{-1} of m_b to a gives an element of $\mathbf{Z}$, i.e., $m_b^{-(m+n)}(a) \in \mathbf{Z}$ if $a \in K(TM \times \mathbf{R}^{2n})$.

We can now state the K-theoretic version of the index theorem.

Theorem 5.20 (Atiyah–Singer) *Let M be a closed, oriented, Riemannian manifold of dimension m, which is trivially embedded in $\mathbf{R}^{m+n}$. Let E and F be Hermitian vector bundles over M and $P \in El_k(E, F)$ be an elliptic operator of order k from $\Gamma(E)$ to $\Gamma(F)$. Then*

$$\mathrm{Ind}(P) = (-1)^m m_b^{-(m+n)}([\sigma_k(P)]_K \dot{\otimes} b^n). \tag{5.52}$$

The cohomological versions of the Atiyah–Singer index theorem are obtained by lifting to $K(M)$ certain characteristic classes. One form is the following.

Theorem 5.21 (Atiyah–Singer) *Let M be a compact manifold of dimension m and let $P \in El_k(E, F)$ be an elliptic operator of order k from $\Gamma(E)$ to $\Gamma(F)$. Then the index of P is given by*

$$\text{Ind}(P) = (-1)^m \{ch[\sigma(P)] \cdot \tau(TM \otimes \mathbf{C})\}[TM], \qquad (5.53)$$

where τ is the Todd class and $[TM]$ is the fundamental class of the tangent bundle.

We note that one can always roll up an elliptic complex to obtain a unique elliptic operator with the same index. However, sometimes it is advantageous to consider the full complex. A formulation of the index theorem for differential complexes is the following.

Theorem 5.22 (Atiyah-Singer) *Let M be a compact manifold of dimension m and let (E, L) be an elliptic differential complex over M. Then there exists a compactly supported cohomology class $a(E, L) \in H_c^*(TM, \mathbf{Q})$ such that*

$$\text{Ind}(E, L) = (a(E, L) \cdot \tau(TM \otimes \mathbf{C}))([TM]), \qquad (5.54)$$

where τ is the Todd class and $[TM]$ is the fundamental class of the tangent bundle.

The statement of the index theorem takes a much simpler form in the special case when all the Laplacians of the elliptic complex are second order operators. In this case it can be shown that there exists a cohomology class $b(E, L) \in H^m(M)$ with the property that

$$\text{Ind}(E, L) = (b(E, L))([M]).$$

Furthermore, $b(E, L) = 0$ for odd m. Hence, m odd implies that $\text{Ind}(E, L) = 0$. The index theorems for classical elliptic complexes are special cases of this formula, as indicated below.

As we discussed earlier, for the de Rham complex we have $E_i = \Lambda^i$, $L = d$, and $b(E, L) =$ Euler class of M, and the index theorem takes the form $\text{Ind}(E, L) = \chi(M)$, the Euler characteristic of M.

We now consider in detail the Hirzebruch signature operator D^+ and state the relation of its index to the Hirzebruch signature of M. Let M be a compact, oriented Riemannian manifold of dimension $4n$. We define the involution operator $j : \Lambda^k \to \Lambda^{4n-k}$ by

$$\alpha \mapsto j(\alpha) = i^{k(k-1)+2n} * \alpha = (-1)^n i^{k(k-1)} * \alpha.$$

The operator j extends to Λ and satisfies $j^2 = 1$. We denote by Λ_+ (resp., Λ_-) the eigenspace of j for the eigenvalue $+1$ (resp., -1). Define $D^+ = d + \delta|_{\Lambda_+}$. Then

$$D^+ : \Lambda_+ \to \Lambda_-$$

is an elliptic operator called the **Hirzebruch signature operator**. We now define the Hirzebruch signature $\sigma(M)$ of M. Consider the bilinear operator

$$h : H^{2n} \times H^{2n} \to R$$

defined by

$$(\alpha, \beta) \mapsto \int_M (\alpha \wedge \beta).$$

In the above formula we have used $(\alpha \wedge \beta)$ to denote the cohomology class and a $2n$-form representing that class. We note that h is the intersection form on M as defined in Chapter 2. Let (e^+, e^-) be the signature of the intersection form h. Then the **Hirzebruch signature** is defined by

$$\sigma(M) = e^+ - e^-.$$

Let us denote by Δ_+ (resp., Δ_-) the Laplacian on Λ_+ (resp., Λ_-). If $\mathcal{H}^{2n}_+$ (resp., $\mathcal{H}^{2n}_-$) denotes the space of harmonic $2n$-forms in Λ_+ (resp. Λ_-), then one can show that

$$\sigma(M) = \dim \mathcal{H}^{2k}_+ - \dim \mathcal{H}^{2k}_-,$$

$$\mathrm{Ind}(D^+) = \dim \ker \Delta_+ - \dim \ker \Delta_- = \dim \mathcal{H}^{2k}_+ - \dim \mathcal{H}^{2k}_-.$$

Hence,

$$\mathrm{Ind}(D^+) = \sigma(M).$$

We observe that $\sigma(M)$ could be defined in a purely topological way through the cup product, because

$$\int_M (\alpha \wedge \beta) = ([\alpha] \cup [\beta])([M]).$$

A deeper result is the **Hirzebruch signature theorem**, which gives

$$Ind(D^+) = \int_M L_k = L_k[M],$$

where L_k is the top degree form of the Hirzebruch L-polynomial. This result can be obtained as a special case of the cohomological version of the Atiyah–Singer index theorem.

A similar formulation can be given for the Dolbeault complex of a complex manifold and the spin complex of a spin manifold.

Chapter 6
Theory of Fields, I: Classical

6.1 Introduction

In recent years gauge theories have emerged as primary tools for research in elementary particle physics. Experimental as well as theoretical evidence of their utility has grown tremendously in the last two decades. The isospin gauge group $SU(2)$ of Yang–Mills theory combined with the $U(1)$ gauge group of electromagnetic theory has lead to a unified theory of weak interactions and electromagnetism. We give an account of this unified electroweak theory in Chapter 8. In this chapter we give a mathematical formulation of several important concepts and constructions used in classical field theories. We begin with a brief account of the physical background in Section 6.2. Gauge potential and gauge field on an arbitrary pseudo-Riemannian manifold are defined in Section 6.3. Three different ways of defining the group of gauge transformations and their natural equivalence is also considered there. The geometric structure of the space of gauge potentials is discussed in Section 6.4 and is then applied to the study of Gribov ambiguity in Section 6.5. A geometric formulation of matter fields is given in Section 6.6. Gravitational field equations and their generalization is discussed in Section 6.7. Finally, Section 6.8 gives a brief indication of Perelman's work on the geometrization conjecture and its relation to gravity.

The literature in the area covered in Chapter 6 is rather vast. We cite here only a few references that may be consulted for additional information about the material of this chapter and related aspects of gauge theories from the physicist's point of view. They are Cheng and Li [73], Faddeev and Slavnov [120], and Quigg [322]. Classical field theories need to be quantized before they can be applied to study elementary particle physics. While there is no mathematically satisfactory theory of quantization of fields, physicists have developed several workable methods of field quantization. The most notable among these is Feynman's method of path integration. For the path integral method of quantization of field theories and related topics, see, for

K. Marathe, *Topics in Physical Mathematics*, DOI 10.1007/978-1-84882-939-8_6,   169
© Springer-Verlag London Limited 2010

example, Feynman and Hibbs [125] and Schulman [340]. Standard references for a general discussion of the mathematical aspects of gauge theories are the books by Bleecker [43], Booss and Bleecker [47], Freed and Uhlenbeck [135], and the papers [116, 273]. For discussion of modern geometry and its relation to physics see, for example, the books by Frankel [132] and Novikov and Taimanov [301].

6.2 Physical Background

In this section we give a brief account of some ideas and results from modern physics which are important in **elementary particle physics**. We use the physical terminology found in standard texts in modern physics. Elementary particle physics is also referred to as **high energy physics**. The term "high energy" refers to the kinetic energy of particles used in accelerators and colliders to produce particle reactions. The principal experimental method of producing elementary particles and studying their size and structure is to accelerate them to high energies and shoot them against a given target, which may be another accelerated beam. The distance and the time scale of the resulting reaction is extremely small. The dispariy of the extremely small scale (distances of less than 10^{-13} cm and duration of the order of 10^{-10} sec or less) of elementary particles and the extremely high energies required to produce and study them may be explained by two fundamental principles of modern physics, now discussed.

First, Einstein's well-known **mass–energy relation**

$$E = mc^2,$$

where c is a universal constant (the velocity of light in a vacuum $\cong 3 \cdot 10^{10}$ cm/sec), allows us to calculate the energy E required for the creation (or released with the annihilation) of mass m. It also allows us to express both energy and mass in term of the same physical unit. In elementary particle physics this unit is taken to be the electron-volt (eV), the kinetic energy acquired by an electron in passing through an electric potential of 1 volt. In terms of this unit the electron rest-mass is $m_e \cong 0.5$ MeV (1 MeV $= 10^6$ eV). The proton rest mass is $m_p \cong 1$ GeV (1 GeV $= 10^9$ eV), while the mass of the heaviest known elementary particle, the Z^0, was first confirmed by observations made in the early 1980s to be approximately 91 GeV.

Einstein is considered one of the greatest physicists of all time. His special theory of relativity provides the foundation for both classical and quantum field theories. The general theory of relativity revealed the intimate connection between space-time geometry and gravity. Einstein received the 1921 Nobel Prize in physics for his many contributions to theoretical physics and especially for his discovery of the law of the photoelectric effect. The photoelectric effect provides conclusive evidence for the quantum nature of light. Combined with the known wave properties of light, it gave an example of the wave–particle duality.

Second, **Heisenberg's uncertainty principle** states that in a simultaneous measurement of two conjugate observables A, B of a quantum system in a given state ψ, one has

$$\Delta_\psi A \Delta_\psi B \geq \hbar/2,$$

where $\Delta_\psi C$ denotes the standard deviation in the measurement of the observable C in the state ψ and $\hbar \cong 10^{-27}$ erg sec is a universal constant related to the Planck constant h by the relation $\hbar = h/2\pi$.

Werner Heisenberg[1] (1901–1976) was one of the greatest physicists of the twentieth century. He is considered one of the founders of quantum mechanics. Heisenberg received the Nobel Prize in physics in 1932. The uncertainty principle is a cornerstone of quantum physics. In particular, applying the uncertainty relation to the conjugate observables position Q and momentum P yields that quantum states in which the precision of the measurement of the position is high are also states with large uncertainty in the measurement of momentum. Thus, the more precisely the position of a particle is determined, the less precisely its momentum is known. A similar situation exists with respect to the simultaneous measurement of the conjugate observables time and energy. This explains in part the requirements as well as the limitations of high energy experiments. Indeed, the energies required for the production of some conjectured elementary particles such as the Higgs boson are so high that they do not seem to be reachable in the near future. In the remaining part of this section we touch upon some basic aspects of the development of elementary particle physics.

Elementary particle physics is the science that studies the ultimate constituents of matter, the interactions among them, and the fundamental forces acting on them. In this broad sense elementary particle physics deals with questions that have been asked since ancient times. For example, it is well known that the classical Greek philosophers discussed such questions. Indeed the concept of atom as an indivisible fundamental constituent of matter can be traced to early Greek writings. Of the four fundamental forces known today, the force of gravity and some of its properties have been recongnized since ancient times. However, the beginning of a theory of the gravitational field was made only in the sixteenth century by Galileo, and the first complete theory of gravity was worked out by Newton in the following century. Newton's main motivation was to obtain a mathematical explanation of Kepler's laws of planetary motion, one of the most important instances in which experimental physics required and led to the development of a new area of

---

[1] I had the opportunity to visit Prof. Heisenberg at the MPI in Munich in 1970 and cherish the autographed copy of his autobiography *Der Teil und das Ganze*, which he gave me at that time.

mathematics. Newton developed the calculus to provide an appropriate tool for studying Kepler's laws. The second law is a conversion law, which provides a first integral of motion. The equation of an ellipse in polar coordinates with origin at one of the focal points was already well-known. Using this and the second law Newton could easily calculate the radial accaleration of the planet. This calculation led him to his law of gravitation. If one starts with this law then one sees that orbits depend on initial conditions and are, in general, conic sections. The elliptic orbits arise from special initial conditions.

Some properties of electric and magnetic forces were also known for a long time, but a systematic experimental and theoretical study of these forces started only in the eighteenth century, culminating in the unified theory of the electromagnetic field developed by Maxwell. The discovery of the periodic table and radioactivity led scientists to suspect that in spite of their name, atoms may not be the indivisible, ultimate constituents of matter. This suspicion was confirmed with the discovery of the electron and the nucleons (proton and neutron) as subatomic particles.

The foundations of quantum theory were established during the first quarter of the twentieth century, The theory immediately met with great success in explaining a wide range of phenomena in molecular, atomic, and nuclear physics. This, coupled with the discovery of subatomic particles, created a new and rapidly growing branch of physics, namely, elementary particle physics. The principal theoretical tools for studying elementary particles and their interactions are provided by (quantum) gauge field theories. We now discuss two well-known examples of classical gauge theories. The problem of their quantization is discussed from a physical perspective in Chapter 7.

Maxwell's electromagnetic theory provides the simplest example of a gauge theory, with the field equations being given by **Maxwell's equations**. We therefore begin with a brief review of Maxwell's equations, which in their classical form are given by:

$$\operatorname{div} B = 0$$
$$\operatorname{curl} E = -\partial B / \partial t$$
$$\operatorname{div} E = \rho$$
$$\operatorname{curl} B = J + \partial E / \partial t,$$

where the electric field E and the magnetic field B are time-dependent vector fields on some subset of $\mathbf{R}^3$ and ρ and J are the charge and current densities, respectively. These equations unified the separate theories of electricity and magnetism and paved the way for important advances in both experimental and theoretical physics. With the introduction of the 4-dimensional Minkowski space-time M^4 it became possible to describe both the electric and magnetic fields as parts of a skew-symmetric tensor field, or a differential 2-form F, on M^4 as follows. Using the standard chart on M^4 and the induced bases of the tensor spaces, the tensor F has components given by:

$$F_{k4} = E_k, 1 \le k \le 3 \quad F_{12} = B_3, \quad F_{23} = B_1, \quad F_{31} = B_2.$$

Maxwell's equations written in terms of F (regarded as a 2-form) are

$$dF = 0, \quad \delta F = j, \tag{6.1}$$

where $j = (J, \rho)$ is the current density 1-form and $\delta = *d*$ is the codifferential operator on 2-forms of M^4. The source-free field equations are obtained by setting $j = 0$ and can be expressed as

$$dF = 0, \quad d * F = 0. \tag{6.2}$$

This is the modern form of Maxwell's source free field equations. It is well-known that Maxwell's equations are globally invariant under the conformal group and in particular, under the Lorentz group. Their Lorentz invariance is the starting point of Einstein's theory of special relativity. They are also invariant under a local group of transformations known as **gauge trans-formations**. This invariance arises as follows. On M^4 the equation $dF = 0$ implies that the field F is derivable from a 4-potential or 1-form A, i.e., $F = dA$. However, the potential A is not uniquely determined. If B is another potential such that $F = dB$, then $d(B - A) = 0$. Every closed form on M^4 is exact and this implies that $B - A = d\psi$, $\psi \in \mathcal{F}(M^4)$. Thus, we may think of the potential B as obtained by a gauge transformation of A by ψ. We can write this transformation as

$$A \mapsto A^\psi := A + d\psi, \qquad \psi \in \mathcal{F}(M^4). \tag{6.3}$$

Since ψ is real-valued Weyl considered $\psi(x)$ to be the choice of the scale at x. The infinitesimal change $d\psi = \psi(x + dx) - \psi(x)$ corresponds up to first order to a change of local scale. This is not related to the conformal invariance of the Maxwell equations. In Einstein's theory of gravitation the base manifold M is a 4-dimensional Lorentz manifold, usually referred to as a space-time manifold, and its pseudo-metric or the corresponding Levi-Civita connection plays the role of "gravitational potential." Now Maxwell's equations written by using F admit immediate generalization to the case of the Minkowski space replaced by an arbitrary 4-dimensional Lorentz manifold. However, in contrast to the geometric description of a gravitational field on M, the electromagnetic field is put in "by hand" as a 2-form on M satisfying the generalized Maxwell's equations. Thus, it was natural to look for a unified geometric theory of gravity and electromagnetism. Weyl sought to incorporate the electromagnetic field into the geometric structures associated to the space-time manifold as arising from **local scale invariance**. He referred to this scale invariance as **"eich-invarianz"** and this is the origin of the modern term **gauge invariance**. In fact, with slight modification, replacing local scale by **local phase** taking values in the unitary group $U(1)$, one obtains a formulation of Maxwell's equations as gauge field equations[2] with the gauge group $U(1)$. The gauge transformation with the real-valued function ψ of equation (6.3) replaced by the gauge transformation $g := e^{i\psi} \in U(1)$ allows us to rewrite equation (6.3) as

$$iA \mapsto iA^g := g^{-1}(iA)g + g^{-1}dg, \qquad g = e^{i\psi} \in \mathcal{F}(M^4, U(1)). \qquad (6.4)$$

Note that the gauge potential iA takes values in the Lie algebra $u(1) = i\mathbf{R}$ of the gauge group $U(1)$. We study gauge transformations with arbitrary gauge group G later in this chapter. Equation (6.4) is a special case of the equation (6.18) of a general gauge transformation. We discuss this formulation of Maxwell's equations and the interpretation of the form F as a $U(1)$ gauge field on M^4 in Chapter 8.

Another approach to a unified treatment of electromagnetic and gravitational fields leads to the **Kaluza–Klein theory**, which uses a 5-dimensional pseudo-Riemannian manifold and a suitable $(4 + 1)$-dimensional decomposition to obtain Einstein's and Maxwell's equations from the 5-dimensional metric. The ideas of Kaluza–Klein theory have been applied to higher dimensional manifolds for studying coupled field equations and the problem of dimensional reduction. For a modern treatment of Kaluza–Klein theories, see Hermann [190] and Coquereaux, Jadczyk [85].

In 1954 Yang and Mills [413, 412] obtained the following now well-known non-Abelian gauge field equations for the vector potential b_μ of isotopic spin in interaction with a field ψ of isotopic spin $1/2$. These are the original **Yang–Mills equations** for the $SU(2)$ gauge group

$$\partial f_{\mu\nu}/\partial x_\nu + 2\epsilon(b_\nu \times f_{\mu\nu}) + J_\mu = 0 , \qquad (6.5)$$

where the quantities

$$f_{\mu\nu} = \partial b_\mu/\partial x_\nu - \partial b_\nu/\partial x_\mu - 2\epsilon b_\mu \times b_\nu, \qquad (6.6)$$

are the components of an $SU(2)$-gauge field and J_μ is the current density of the source field ψ. There was no immediate physical application of these equations since they seemed to predict massless gauge particles as in Maxwell's theory. In Maxwell's theory the predicted massless particle is identified as photon, the massless carrier of electromagnetic field. No such identification could be made for the massless particles predicted by pure (i.e., source-free) Yang–Mills equations. This difficulty is overcome by the introduction of the **Higgs mechanism** [192], which shows how spontaneous symmetry-breaking can give rise to massive gauge vector bosons by a gauge transformation to a

---

[2] I discussed this in my talk at the Geometry and Physics Workshop organized by Prof. Raoul Bott at MSRI, Berkeley in 1994. After my talk Bott remarked: "We teach Harvard students to think of functions as Lie algebra valued 0-forms so that they know the distinction between scale and phase." When one of his students said he never learned this in his courses, Bott gave a heary laugh.

particular local gauge. This paved the way for a gauge theoretic formulation of the standard model of electroweak theory and subsequent development of the general framework for a unified treatment of strong, weak and electromagnetic interactions.

The Yang–Mills equations may be thought of as a matrix-valued generalization of the equations for the classical vector potential of Maxwell's theory. The gauge field that they obtained turns out to be the curvature of a connection in a principal fiber bundle with gauge group $SU(2)$. The general theory of such connections was developed in 1950s by Ehresman [117]. However, physicists continued to use the classical theory of connections and curvature that was the cornerstone of Einstein's general relativity theory. In fact, using this classical theory, Ikeda and Miyachi[3] [199, 200] essentially linked the Yang–Mills theory with the theory of connections. However, this work does not seem to be well-known in the mathematical physics community, and it was not until the early 1970s that the identification between curvature of a connection in a principal bundle and the gauge field was made. This identification unleashed a flurry of activity among both physicists and mathematicians and has already had great successes, some of which were indicated in the preface. Since the physical and mathematical theories have developed independently each has its well-established terminology. We have used the notation that is primarily used in the mathematical literature but we have also taken into account the terminology that is most frequently used in physics. To help the reader we have given Appendix A, a table indicating the correspondence between the terminologies of physics and mathematics, prepared along the lines of Trautman [378]. The physical literature on gauge theory is vast and is, in general, aimed at applications to elementary particle physics and quantum field theories. The most successful quantum field theory is quantum electrodynamics (QED), which deals with quantization of electromagnetic fields. Its predictions have been verified to a very high degree of accuracy. However, there is as yet no generally accepted mathematical theory of quantization of gauge fields.

After this brief look at two classical gauge theories, we now turn to a discussion of some major developments in elementary particle physics in the last 30 years. As we observed earlier, the aim of elementary particle physics is to study the structure of matter in terms of some fundamental system of constituents. The last three decades have seen a dramatic increase in our knowledge of both the theoretical and the experimental aspects of elementary particle physics. A host of new experimental results have come out especially from a new generation of particle accelerators. We have identified two types of elementary particles, namely the **leptons** and the **hadrons**. In addition to these there are the carriers of the forces of interaction. Each particle also has a corresponding **anti-particle** with the same rest mass and spin, but with opposite charge. All particles are subject to two kinds of statis-

---

[3] We would like to thank Prof. Akira Asada of Shinshu University for introducing us to Prof. Miyachi and his work.

tics. Those subject to **Fermi–Dirac statistics** have half-integer spin and are called **fermions**, while those subject to **Bose–Einstein statistics** are called **bosons**. Leptons come in three pairs: the **electron**, the **muon**, and the **tau** particle, each with its corresponding massless **neutrino**. The hadrons are divided into two groups called baryons (with half-integer spin) and mesons (with integer spin). The well-known nuclear particles proton and neutron are baryons, while pion and kaon are examples of mesons. A large number of other hadrons have been discovered. We now have strong evidence that the properties of hadrons can be explained by considering them as composed of **quarks** u (**up**), d (**down**), c (**charm**), s (**strange**), t (**top** or **truth**), and b (**bottom** or **beauty**). Each quark also carries a **color** index corresponding to the **color gauge group** $SU_c(3)$. Each quark has its **anti-quark** partner. The hadrons are made up of combinations of quarks and anti-quarks. However, they are supposed to remain confined inside hadrons and hence are unobservable. This phenomenon is called **quark confinement**. The fact that all searches for free quarks since 1977 have had negative results strongly supports the hypothesis of quark confinement. In view of this we will not call quarks fundamental particles. In view of the indirect yet rather strong experimental evidence now available we can consider quarks as virtual particles that form the fundamental constituents of hadrons.

Table 6.1 displays what we consider to be the fundamental constituents of matter at this time. Each constituent is identified by its name, symbol, charge (in units of the proton charge), and mass or range of mass (in units of GeV), in that order. The information in the table is a summary of our knowledge of the fundamental constituents of matter at this time. These particles are subject to the various fundamental forces, which act via their carrier particles. Three of them have interpretation as gauge fields. They are combined to obtain the standard model of fundamental particles and forces described later. The data in Table 6.1 and Table 6.2 are taken from [11], with 2009 web updates by the Particle Data Group. They are not needed for the mathematical formulation discussed in detail in Chapter 8.

All matter is subjected to one or more of the four **fundamental forces**. They are the well known classical, long range forces of **electromagnetic** and **gravitational** fields and the more recently discovered short range forces of **weak** and **strong** interactions. Leptons are subject to all but the strong interaction while hadrons participate in all of them. In classical field theory it was assumed that each particle generates a set of fields that extend over entire space. A neutral, massive particle generates gravitational field while a charged particle also generates electromagnetic field. Thus, for example, two particles with charge of the same sign exert a repulsive force on each other as a result of the interaction of their fields. In quantum field theory it is postulated that all forces act by exchange of **carrier particles** or **quanta**. For the gravitational field this carrier particle is called the **graviton**. It has spin 2 related to the representation of the Lorentz group on symmetric tensors of order 2. This is consistent with the observation that gravitation is always

Table 6.1 Fundamental constituents of matter

Leptons		Quarks	
electron	electron neutrino	up	down
e	ν_e	u	d
-1	0	$2/3$	$-1/3$
0.0005	$0\ (<10^{-17})$	$0.0015-0.0033$	$0.0035-0.006$
muon	muon neutrino	charmed	strange
μ	ν_μ	c	s
-1	0	$2/3$	$-1/3$
0.105	$0\ (<10^{-3})$	$1.2-1.3$	$0.6-1.3$
tau	tau neutrino	top (or truth)	bottom (or beauty)
τ	ν_τ	t	b
-1	0	$2/3$	$-1/3$
1.78	$0\ (<0.25)$	? 171	$4.1-4.4$

an "attractive force." There is no experimental or theoretical evidence for the graviton. For the electromagnetic field, the prototype of Abelian gauge fields, this particle has been identified as the **photon**. The beta decay of unstable, radioactive nuclei provided an early example of weak interaction. Its analysis led Pauli and Fermi to conjecture the existence of the **neutrino** (Italian for "tiny neutron"). The prediction and detection of the carrier particles W^+, W^-, Z^0 of the weak force is one of the triumphs of the electroweak theory (a $U(1) \times SU(2)$-gauge theory), which provides a unified treatment of electromagnetic and weak interactions. The particles W^+, W^-, Z^0 are called **weak intermediate vector bosons**. The $U(1)$ factor of the **electroweak gauge group** is called the **weak hypercharge gauge group** and is denoted by $U_Y(1)$. The $SU(2)$ factor of the electroweak gauge group is called the **weak isospin gauge group** and is denoted by $SU_L(2)$. Hence, the **electroweak gauge group** is often denoted by $U_Y(1) \times SU_L(2)$. The subscript L in the term $SU_L(2)$ denotes the action on left-handed fermions. In fact, the electroweak theory is **left-right asymmetric**. For further details see [11].

The energy of the carrier particles and the range of corresponding interaction are related by **Heisenberg's uncertainty principle**. The infinite range of electromagnetic interaction is consistent with the zero rest mass of its carrier particle, the photon. The observed short range of the weak interaction requires that its carrier particles be massive. A very readable account of the problems and triumphs of electroweak theory is given in [330]. The strong force corresponds to the color gauge group $SU_c(3)$ and has eight exchange particles, called **gluons**. The corresponding quantum field theory is called **quantum chromodynamics** or QCD for short. The success of the electroweak theory has led scientists to a unified theory of strong, weak, and

electromagnetic fields based on the gauge group $SU_c(3) \times SU_L(2) \times U_Y(1)$.
This theory is usually referred to as the **standard model** of strong, weak,
and electromagnetic interactions or SM for short. We discuss a model of the
electroweak theory and the related Higgs mechanism of symmetry-breaking
in Chapter 8, where a brief account of the standard model will also be given.
Our current knowledge of the fundamental forces and their carrier particles
is summarized in Table 6.2.

Table 6.2 Fundamental forces and carrier particles

Force		Carrier Particles			
Name	Range	Name	Symbol	Mass	Spin
gravitation	∞	graviton	Γ	0	2
electromagetism	∞	photon	γ	0	1
weak	10^{-16}	weak intermediate vector bosons	W^+	80	1
			W^-	80	1
			Z^0	91	1
strong	10^{-14}	gluons	G^a	0	1

From the experimental point of view the main difficulty in studying strong
interaction is that the energy required for the creation and detection of the
gluons is very high. The Large Hadron Collider (LHC) at CERN, which has
recently gone on line, is expected to bring protons into head-on collision
at extremely high energies. Now the well-known de Broglie **wave-particle
duality** principle tells us that particles are also waves. This principle is used
in electron microscopes, which exploit the short wavelength of an electron to
reveal details unseen with visible light. The higher the energy, the greater
is the probability for creating more massive particles out of a collision and
the shorter the wavelength corresponding to these particles. This will allow
scientists to penetrate still further into the fine structure of matter. When
fully operational, the LHC is expected to re-create conditions prevailing in
the very early universe (about 10^{-10} seconds after the so-called "Big Bang")
causing nuclear matter to transform into quark-gluon plasma. Studies of such
a state of matter are expected to shed new light on a number of unexplained
phenomena such as quark confinement and the masses of elementary particles.

From the theoretical side, there are several other proposals beyond the
standard model for a unified treatment of electromagnetic, weak, and strong
interactions. The most extensively studied models are those of the grand
unified theories (also known as GUTs), the technicolor models, string and
superstring theories, and supersymmetric theories. However, none of these
has been found to be completely satisfactory. The known theories of gravitational forces differ substantially from those of the other three forces. The
quantum theories of gravitational forces as well as unified theories of all the

known forces should at present be considered to be speculative at best. Work
to understand and explain all these theories in precise mathematical terms
should remain on the agenda of mathematical physics of the twenty-first century.

6.3 Gauge Fields

The theory of gauge fields and their associated fields, such as the Yang–Mills–
Higgs fields, was developed by physicists to explain and unify the fundamental
forces of nature. The theory of connections in a principal fiber bundle was
developed by mathematicians during approximately the same period, but as
we pointed out in the preface, the fact that they are closely related was not
noticed for many years. Since then substantial progress has been made in
understanding this relationship and in applying it successfully to problems
in both physics and mathematics. In physical applications one is usually
interested in a fixed Lie group G called the **gauge group**, which represents an
internal or local symmetry of the field. The base manifold M of the principal
bundle $P(M, G)$ is usually the space-time manifold or its Euclidean version,
i.e., a Riemannian manifold of dimension 4. But in some physical applications,
such as superspace, Kaluza–Klein and string theories, the base manifold can
be an essentially arbitrary manifold. In this section we formulate the theory
of gauge fields on an arbitrary pseudo-Riemannian base manifold.

Let (M, g) be an m-dimensional pseudo-Riemannian manifold and G a
Lie group which we take as the gauge group of our theory. Let $P(M, G)$
be a principal bundle with the gauge group G as its structure group. A
connection in P is called a **gauge connection**. The connection 1-form ω is
called the **gauge connection form** or simply the **gauge connection**. A
global gauge or simply a **gauge** is a section $s \in \Gamma(P)$. The gauge potential
A on M in gauge s is obtained by pull-back of the gauge connection ω on P
to M by s, i.e., $A = s^*(\omega)$. A global gauge and hence the gauge potential on
M exists if and only if the bundle P is trivial. A **local gauge** is defined as a
section of the bundle $P(M, G)$ restricted to some open subset $U \subset M$. Local
gauges defined for the local representations $(U_i, \psi_i)_{i \in I}$ of P always exist. Let
$t \in \Gamma(U_i, P)$ be a local gaugel then the 1-form $t^*\omega \in \Lambda^1(U_i, \mathbf{g})$ is called the
gauge potential in the local gauge t and is denoted by A_t. If the local gauge
t is given, we often denote A_t by A and call it a **local gauge potential**. In
electromagnetism, the gauge group $G = U(1)$, the circle group. An element
$e^{i\theta} \in U(1)$ is determined by the phase θ. Thus, in this case, a local gauge
over an open set $V \subset M$ can be regarded as a choice of a phase in the bundle
$P_{|V} = V \times G$ at each point of V. For this reason, the total space of the bundle
P is sometimes called the **space of phase factors** in the physics literature.

Let $\Omega = d^\omega \omega$ be the curvature 2-form of ω with values in the Lie algebra
$\mathbf{g}$. We call Ω the **gauge field on** P. Although this terminology is fairly

standard, we would like to warn the reader that sometimes, in the physics literature, our gauge potential is called the gauge field and our gauge field is called the field strength tensor. As we have seen in Chapter 4, there exists a unique 2-form F_ω on M with values in the Lie algebra bundle $\mathrm{ad}\, P$ associated to the curvature 2-form Ω such that

$$F_\omega = s_\Omega. \tag{6.7}$$

The 2-form $F_\omega \in \Lambda^2(M, \mathrm{ad}\, P)$ is called the **gauge field on** M corresponding to the gauge connection ω. The gauge field F_ω is globally defined on M, even though, in general, there is no corresponding globally defined gauge potential on M. If we are given a local gauge potential $A_t \in \Lambda^1(U_i, \mathbf{g})$, then on U_i we have the relations

$$t^*\omega = A_t \quad \text{and} \quad F_\omega = d^\omega A_t . \tag{6.8}$$

Example 6.1 (The Dirac Monopole) *Let* $S^3(S^2, U(1))$ *be the principal* $U(1)$-*bundle over* S^2 *determined by the Hopf fibration of* S^3. *Let* μ *denote the connection 1-form of the canonical connection on this bundle and let* F_μ *be the corresponding gauge field on* S^2. *In this case there is no globally defined gauge potential on* S^2. *We need at least two charts to cover* S^2 *and therefore at least two local potentials, which give rise to a single globally defined gauge field. This field can be shown to be equivalent to the Dirac monopole field. The Dirac monopole quantization condition corresponds to the classification of principal* $U(1)$-*bundles over* S^2. *These are classified by* $\pi_1(U(1)) \cong \mathbf{Z}$. *In general, the principal* G-*bundles over* S^2 *are classified by* $\pi_1(G)$. *Thus* $\pi_1(SU(2)) = \mathrm{id}$ *implies that there is a unique* $SU(2)$-*monopole on* S^2 *and* $\pi_1(SO(3)) = \mathbf{Z}_2$ *implies that there are two inequivalent* $SO(3)$-*monopoles on* S^2 *(see [161, 411, 410] for further details).*

Gauge potentials and gauge fields acquire physical significance only after one postulates the field equations to be satisfied by them. These equations and their consequences must then be subjected to suitably devised experiments for verification. On more than one occasion a theory was abandoned when its predictions seemed to contradict an experimental result, but later this experiment or its conclusions turned out to be incorrect and the abandoned theory turned out to be correct. In any case there is no natural mathematical method for assigning field equations to gauge fields. Thus the Riemann curvature of a space-time manifold M is the gauge field corresponding to the gauge potential given by the Levi-Civita connection on the orthonormal frame bundle of M, but it does not describe the gravitational field until it is subjected to Einstein's field equations. If instead it satisfies Yang–Mills equations, then it describes a classical Yang–Mills field. This aspect of gauge fields is already evident in the following remark of Yang:

> The electromagnetic field is a gauge field. Einstein's gravitational theory is intimately related to the concept of gauge fields, although to *identify*

the gravitational field as a gauge field is not an absolutely straightforward matter.

However, a study of physically interesting field equations such as Maxwell's equations of electromagnetic field and their quantization indicates some desirable features for the gauge field equations. One of these features is gauge invariance of the field equations. This requirement is formulated in terms of the group of gauge transformations, which acts on the various fields involved. In the following we give a mathematical formulation of this group.

The group $\mathrm{Diff}(P)$ of the diffeomorphisms of P is too large to serve as a group of gauge transformations, since it mixes up the fibers of P. The requirement that fibers map to fibers may be expressed by the condition that the following diagram commutes:

$$\begin{array}{ccc} P & \overset{f}{\longrightarrow} & P \\ {\scriptstyle \pi}\downarrow & & \downarrow{\scriptstyle \pi} \\ M & \underset{f_M}{\longrightarrow} & M \end{array}$$

i.e.,

$$\pi \circ f = f_M \circ \pi. \tag{6.9}$$

We say that the map f is **fiber-preserving** if condition (6.9) is satisfied. We note that condition (6.9) does not depend on the principal bundle structure of P and hence can be imposed on any fiber bundle. The pair (f, f_M) satisfying the condition (6.9) is called a **fiber bundle automorphism**. In this case we call f a **projectable diffeomorphism** or **transformation** of P covering the diffeomorphism f_M. The projectable diffeomorphisms form a group $\mathrm{Diff}_M(P)$, defined by

$$\mathrm{Diff}_M(P) := \{f \in \mathrm{Diff}(P) \mid f \text{ is projectable}\}.$$

Let ϕ^t be a one-parameter group of projectable transformations of P with the associated vector field $X \in \mathcal{X}(P)$ and let ϕ_M^t be the corresponding one parameter group in $\mathrm{Diff}(M)$ with the associated vector field $X_M \in \mathcal{X}(M)$. Then X is a **projectable vector field** on P, i.e., the pair of vector fields (X, X_M) satisfies the condition

$$\pi_*(X(u)) = X_M(\pi(u)), \qquad \forall u \in P.$$

The set $\mathcal{X}_M(P)$ of projectable vector fields forms a Lie subalgebra of the Lie algebra $\mathcal{X}(P)$. The group $\mathrm{Diff}_M(P)$ (resp., the algebra $\mathcal{X}_M(P)$) and its subgroups (resp., subalgebras) arise in many applications. They are usually obtained by requiring that the transformations occurring in them preserve some additional structure on the fiber bundle. For example, condition (6.9) is satisfied if $f \in \mathrm{Diff}(P)$ is G-equivariant, i.e., $f(ug) = f(u)g$, $\forall u \in P$, $g \in G$. We are thus led to define the set $\mathrm{Aut}(P)$ by

$$\mathrm{Aut}(P) := \{f \in \mathrm{Diff}(P) \mid f \text{ is } G\text{-equivariant}\}. \tag{6.10}$$

The set $\mathrm{Aut}(P)$ is a group called the **group of generalized gauge transformations**. From the point of view of differential geometry, the group $\mathrm{Aut}(P)$ is just the group of principal bundle automorphisms of P. We note that the fiber-preserving property of the generalized gauge transformation f completely determines the diffeomorphism f_M. We define the **group of gauge transformations** $\mathcal{G}(P)$ to be the subgroup $\mathrm{Aut}_0(P)$ of the group $\mathrm{Aut}(P)$ of generalized gauge transformations. Thus

$$\mathcal{G}(P) := \mathrm{Aut}_0(P) = \{f \in \mathrm{Aut}(P) \mid f_M = id_M\}. \tag{6.11}$$

Then $\mathcal{G}(P)$ (also denoted simply by $\mathcal{G}$) is a normal subgroup of $\mathrm{Aut}(P)$. From definition (6.11) it is clear that $f \in \mathcal{G}$ if and only if it is a smooth fiber-preserving map of P into itself commuting with the action of the gauge group G on P, i.e., f satisfies the conditions

$$\pi \circ f = \pi \tag{6.12}$$

and

$$f(p \cdot g) = f(p) \cdot g, \qquad \forall p \in P, \ \forall g \in G. \tag{6.13}$$

From definitions (6.10) and (6.11) we obtain the following exact sequence of groups:

$$\mathcal{G} \xrightarrow{\ i\ } \mathrm{Aut}(P) \xrightarrow{\ j\ } \mathrm{Diff}(M),$$

where i denotes the inclusion map and j is defined by

$$j(f) = f_M, \quad \forall f \in \mathrm{Aut}(P).$$

Note that the exactness means here that $\mathrm{Im}(i) = \mathrm{Ker}(j)$. The map i is injective, so by adding the identity at the beginning of the above sequence we get a 4-term exact sequence. However, the map j is not, in general, surjective. Thus, the 4-term exact sequence cannot be extended to a 5-term or short exact sequence. The following example[4] illustrates this.

Example 6.2 *Recall that the principal $U(1)$ bundles over S^2 are classified by the integers. For each $n \in \mathbf{Z}$ there exists a unique equivalence class P_n of $U(1)$ bundles over S^2. As we observed in Example 6.1, this corresponds to Dirac's monopole quantization condition. The Hopf fibration of S^3 discussed there is in the class P_1. Now let $\alpha : S^2 \to S^2$ be the antipode map, i.e., $\alpha(x) = -x, \forall x \in S^2$. Then the pull-back bundle of the Hopf fibration $\alpha^*(S^3)$*

---

[4] This example was suggested by Stefan Wagner, a doctoral student of Prof. Neeb at TU Darmstadt.

is also a principal $U(1)$ bundle. But this bundle is in the class P_{-1}. Thus α cannot be lifted to an automorphism of the Hopf fibration.

This example leads to the following proposition.

Proposition 6.1 *Let* $\mathrm{Diff}_P(M)$ *denote the subgroup of* $\mathrm{Diff}(M)$ *defined by*

$$\mathrm{Diff}_P(M) := \{\alpha \in \mathrm{Diff}(M) \mid \alpha^*(P) \equiv P\}.$$

Then we have the short exact sequence

$$1 \longrightarrow \mathcal{G} \xrightarrow{i} \mathrm{Aut}(P) \xrightarrow{j} \mathrm{Diff}_P(M) \longrightarrow 1.$$

In several applications one is interested in splitting the above exact sequence or in finding conditions that imply the equality $\mathrm{Diff}_P(M) = \mathrm{Diff}(M)$ so that one may try to construct an extension of $\mathrm{Diff}(M)$ by $\mathcal{G}$. Additional geometric structures may also be involved in this process. For example, if $P = L(M)$, the bundle of frames of M, then it is a principal bundle but carries the additional structure given by the soldering form θ and we have the following proposition.

Proposition 6.2 *Let M be a manifold with a linear connection. Then there exists a natural lift $\lambda : \mathrm{Diff}(M) \to \mathrm{Diff}(L(M))$, which splits the exact sequence of groups*

$$1 \longrightarrow \mathcal{G} \xrightarrow{i} \mathrm{Diff}_M(L(M)) \xrightarrow{j} \mathrm{Diff}(M) \longrightarrow 1.$$

Furthermore, the map $f \in \mathrm{Diff}(L(M))$ is the natural lift of a diffeomorphism $f_M \in \mathrm{Diff}(M)$, i.e., $f = \lambda(f_M)$ if and only if f leaves the soldering form invariant (i.e., $f^*\theta = \theta$).

When M is a 4-dimensional Lorentz manifold, connections on the frame bundle $L(M)$ play the role of gravitational potentials. Action functionals involving connections and metrics on M form the starting point of gauge theories of gravitation.

A physical interpretation of a gauge transformation $f \in \mathcal{G}$ is that f is a **local** (i.e., **pointwise**) **change of gauge** For this reason, G is sometimes called a local symmetry group and $\mathcal{G}$ is called the local gauge group; but we will not use this terminology. Let t be a local gauge over U. Then t is a section of the bundle $P_{|U}$, i.e., for $x \in U$, $t(x)$ is in $P_x = \pi^{-1}(x)$, the fiber of P over x and $f(t(x))$ is also in P_x. Therefore, there exists a unique element $\hat{f}(x) \in G$ such that

$$f(t(x)) = (t(x)).(\hat{f}(x)), \qquad \forall x \in U.$$

The map $\hat{f} : U \to G$ is a local representation of the gauge transformation f. If the bundle P is trivial, then we can take $U = M$ and in this case a gauge transformation can be identified with a map of M to G.

We now consider two alternative definitions of the group of gauge transformations. Note first that the space $\mathcal{F}(P, G)$ of all smooth functions $f : P \to G$ with pointwise multiplication is a group. Let $\mathcal{F}_G(P, G)$ denote the subset of all G-equivariant functions (with respect to the adjoint action), i.e.,

$$\mathcal{F}_G(P, G) := \{f : P \to G \mid f(u\alpha) = \alpha^{-1}f(u)\alpha, \ \forall u \in P, \ \forall \alpha \in G\}.$$

Then it is easy to verify that $\mathcal{F}_G(P, G)$ is a group. Let $\mathrm{Ad}(P)$ denote the bundle $(P \times_{\mathrm{Ad}} G)$ over M associated to P by the adjoint action of G on itself. It is a bundle of Lie groups with fiber G. The set $\Gamma(\mathrm{Ad}(P)) := \Gamma(P \times_{\mathrm{Ad}} G)$ of sections of the associated bundle $\mathrm{Ad}(P)$ with pointwise multiplication is a group. The relation between these groups is established in the following theorem.

Theorem 6.3 *There exists an isomorphism between each pair of the following three groups:*

1. *the group of gauge transformations $\mathcal{G}$;*
2. *the group $\mathcal{F}_G(P, G)$ of all functions $f : P \to G$ such that f is G-equivariant, with respect to the adjoint action of G on itself;*
3. *the group $\Gamma(\mathrm{Ad}(P))$ of sections of the associated bundle $\mathrm{Ad}(P)$ over M.*

Proof: For $g \in \mathcal{G}$ we define $\bar{g} : P \to G$ by

$$\bar{g}(u) = a,$$

where $a \in G$ is the unique element such that $g(u) = ua$. It can be verified that the map $T : g \mapsto \bar{g}$ is a one-to-one correspondence from $\mathcal{G}$ to $\mathcal{F}_G(P, G)$ with inverse given by the map from $\mathcal{F}_G(P, G)$ to $\mathcal{G}$ such that $f \mapsto g_f$ where $g_f(u) = uf(u)$. Using the definition of T and of the G action it is easy to verify that

$$T(gh) = \overline{gh} = \bar{g}\bar{h} = T(g)T(h).$$

It follows that T is an isomorphism of groups.

The correspondence between $\mathcal{F}_G(P, G)$ and $\Gamma(P \times_{\mathrm{Ad}} G)$ is a special case of the correspondence between $\mathcal{F}_G(P, F)$ and $\Gamma(E(M, F, r, P))$ (see Chapter 4) with $F = G$ and $r = \mathrm{Ad}$, the adjoint action of G on itself. Thus, $f \in \mathcal{F}_G(P, G)$ corresponds to a section $s_f \in \Gamma(P \times_{\mathrm{Ad}} G)$ defined by

$$s_f(x) = f(u), \ u \in \pi^{-1}(x).$$

We note that s_f is well defined in view of the G-equivariance of f. On the other hand a section $s \in \Gamma(P \times_{\mathrm{Ad}} G)$ defines an element $f_s \in \mathcal{F}_G(P, G)$ by

$$f_s(u) = s(x), \qquad u \in \pi^{-1}(x).$$

One can verify that the map S defined by $S : f \mapsto s_f$ is an isomorphism of the group $\mathcal{F}_G(P, G)$ with the group $\Gamma(P \times_{\mathrm{Ad}} G)$.   □

In view of the above theorem we use any one of the three representations above for the group of gauge transformations as needed. For example, regarding $\mathcal{G}$ as the space of sections of $(P \times_{\mathrm{Ad}} G)$, the bundle of groups (not a principal G-bundle), we can show that a suitable Sobolev completion (see Appendix D) of $\mathcal{G}$ (also denoted by $\mathcal{G}$) is a Hilbert Lie group (i.e., $\mathcal{G}$ is a Hilbert manifold with smooth group operations). Let ad denote the adjoint action of the Lie group G on its Lie algebra $\mathfrak{g}$. Let $E(M, \mathfrak{g}, \mathrm{ad}, P)$ be the associated vector bundle with fiber type $\mathfrak{g}$ and action ad, the adjoint action of G on $\mathfrak{g}$. Recall that this bundle is a bundle of Lie algebras denoted by $P \times_{\mathrm{ad}} \mathfrak{g}$ or $\mathrm{ad}\, P$. We denote $\Gamma(\mathrm{ad}\, P)$ by $\mathcal{LG}$; it is a Lie algebra under the pointwise bracket operation. The algebra $\mathcal{LG}$ is called the **gauge algebra** of P. It can be shown that a suitable Sobolev completion of $\mathcal{LG}$ is a Banach Lie algebra with well-defined exponential map to $\mathcal{G}$. It is the Lie algebra of the infinite-dimensional Banach Lie group $\mathcal{G}$. An alternative characterization of the gauge algebra is given by the following theorem.

Theorem 6.4 *The set $\mathcal{F}_G(P, \mathfrak{g})$ of all G-equivariant (with respect to the adjoint action of G on its Lie algebra $\mathfrak{g}$) functions with the pointwise bracket operation is a Lie algebra isomorphic to the gauge algebra $\mathcal{LG}$.*

6.4 The Space of Gauge Potentials

Without any assumption of compactness for M or G it can be shown that $\mathcal{G}$ is a Schwartz Lie group (i.e., a Lie group modeled on a Schwartz space) with Lie algebra consisting of sections of $\mathrm{ad}\, P$ of compact support. While this approach has the advantage of working in full generality, the technical difficulties of working with spaces modelled on an arbitrary locally convex vector space can be avoided by considering Sobolev completions of the relevant objects as follows. In this section we consider a fixed principal bundle $P(M, G)$ over a compact, connected, oriented, m-dimensional Riemannian base manifold M with compact, semisimple gauge group G. These assumptions are satisfied by most Euclidean gauge theories that arise in physical or mathematical applications. The base manifold is typically a sphere S^n or a torus T^n or their products such as $S^n \times T^m$. Thus, for $n = 4$ one frequently considers as a base $S^4, T^4, S^3 \times S^1, or S^2 \times S^2$. With appropriate boundary conditions on gauge fields one may also include non-compact bases such as $\mathbf{R}^4$ or $\mathbf{R}^3 \times S^1$. The gauge group G is generally one of the following: $U(n)$, $SU(n)$, $O(n)$, $SO(n)$, or one of their products. For example, the gauge group of electroweak theory is $SU(2) \times U(1)$. As we discussed above, the gauge connections (gauge potentials) and the gauge fields acquire physical significance only after field equations, to be satisfied by them, are postulated. However, the topology and geometry of the space of gauge connections has significance for all physical theories and especially for the problem of quantization of gauge theories. They are also fundamental in studying low-dimensional topology. Various aspects of the topology and geometry of the space of gauge connections and its orbit spaces have been studied in [22, 153, 233, 234]. We denote by $\mathcal{A}(P)$ the **space of gauge potentials** or **connections** on P defined by

$$\mathcal{A}(P) := \{\omega \in \Lambda^1(P, \mathbf{g}) \mid \omega \text{ is a connection on } P\}. \qquad (6.14)$$

If P is fixed we will denote $\mathcal{A}(P)$ simply by $\mathcal{A}$ and a similar notation will be followed for other related spaces. From the definition of connection it follows that $\omega_1, \omega_2 \in \mathcal{A}$ implies that $\omega_1 - \omega_2$ is horizontal and of type $(rmad, \mathbf{g})$ and, therefore, defines a unique 1-form on M with values in the associated bundle $\operatorname{ad} P := P \times_{\operatorname{ad}} \mathbf{g}$. Then we have that, for a fixed connection α,

$$\mathcal{A} \cong \{\alpha + \pi^* A \mid A \in \Lambda^1(M, \operatorname{ad} P)\}. \qquad (6.15)$$

From the above isomorphism it follows that the space $\mathcal{A}$ is an affine space with the underlying vector space $\Lambda^1(M, \operatorname{ad} P)$. Thus, the tangent space $T_\alpha \mathcal{A}$ is isomorphic to $\Lambda^1(M, \operatorname{ad} P)$ and we identify these two spaces. If $\langle \ , \ \rangle_{\mathbf{g}}$ is a G-invariant inner product on $\mathbf{g}$, we have a natural inner product defined on $T_\alpha \mathcal{A}$ as follows. First, for $A, B \in T_\alpha \mathcal{A}$, we define

$$\langle A, B \rangle \in \mathcal{F}(M) \text{ by } \langle A, B \rangle := g^{ij} \langle A_i, B_j \rangle_{\mathbf{g}},$$

where g^{ij} are the components of the metric tensor g on M with respect to the base $\{dx^i\}$ of $T_x^* M$ and $A = A_i(x) dx^i$, $B = B_i(x) dx^i$. We note that $A_i(x)$, $B_i(x)$ are elements of the fiber of the Lie algebra bundle $\operatorname{ad} P$ over $x \in M$. Then we define the inner product $\langle\langle A, B \rangle\rangle_\alpha$, or simply $\langle\langle A, B \rangle\rangle$, by

$$\langle\langle A, B \rangle\rangle_\alpha = \int_M g^{ij} \langle A_i, B_j \rangle_{\mathbf{g}}, \quad \forall A, B \in T_\alpha \mathcal{A} . \qquad (6.16)$$

The map $\alpha \mapsto \langle\langle A, B \rangle\rangle_\alpha$ defines a **weak Riemannian metric** or the L^2 **Riemannian metric** on $\mathcal{A}$.

We observe that an invariant inner product always exists for semisimple Lie algebras and is given by a multiple of the Killing form K on $\mathbf{g}$ defined by

$$K(X, Y) = \operatorname{Tr}(\operatorname{ad} X \operatorname{ad} Y).$$

The inner product defined in (6.16) can be extended to $\Lambda^k(M, \operatorname{ad} P)$. A connection ω on P defines a covariant derivative

$$\nabla^\omega : \Lambda^0(M, \operatorname{ad} P) \to \Lambda^1(M, \operatorname{ad} P)$$

which is compatible with the metric on $\operatorname{ad} P$, i.e.,

$$\langle \nabla_X^\omega \psi, \ \phi \rangle + \langle \nabla_X^\omega \phi, \ \psi \rangle = X(\langle \phi, \ \psi \rangle),$$

for all $\phi,\ \psi \in \Lambda^0(M, \mathrm{ad}\, P)$ and $X \in \mathcal{X}(M)$. The covariant derivative has a natural extension to $(\mathrm{ad}\, P)$-valued tensors that is also denoted by ∇^ω. The corresponding covariant exterior derivative is denoted by d^ω. We now give definitions of several terms that occur frequently in physical applications.

Definition 6.1 *Let (M, g) be a compact, connected, m-dimensional, Riemannian manifold and let $P(M, G)$ be a principal bundle over M with compact, semisimple, n-dimensional gauge group G. Let $\langle\ ,\ \rangle_{\mathfrak{g}}$ be a G-invariant inner product on its Lie algebra $\mathfrak{g}$. For $x \in M$, the metric g_x induces inner products on the tensor spaces and spaces of differential forms that we also denote by g_x. Let $\{e^i(x)\}_{1 \leq i \leq n}$ be a basis for the fiber $(\mathrm{ad}\, P)_x$ of the Lie algebra bundle $\mathrm{ad}\, P$. Let $\alpha, \beta \in \Lambda^p(M, \mathrm{ad}\, P)$ and $\gamma \in \Lambda^q(M, \mathrm{ad}\, P)$. Locally, we can write*

$$\alpha(x) = \alpha_i(x) \otimes e^i(x),\ \ where\ \alpha_i(x) \in \Lambda_x^p(M)\ and\ e^i(x) \in (\mathrm{ad}\, P)_x,\ \forall i,$$

with similar expressions for β and γ. Then we have the following definitions:

1. *The product $\langle \alpha, \beta \rangle \in \mathcal{F}(M)$ is defined by $x \mapsto \langle \alpha, \beta \rangle_x$, where*

$$\langle \alpha, \beta \rangle_x := g_x(\alpha_i(x), \beta_j(x)) \langle e^i(x), e^j(x) \rangle_{\mathfrak{g}}.$$

The corresponding local norm $|\alpha| \in \mathcal{F}(M)$ is defined by

$$x \mapsto |\alpha|_x := \sqrt{\langle \alpha, \alpha \rangle_x}\ ,\ \ \forall x \in M.$$

2. *The inner product $\langle\langle \alpha, \beta \rangle\rangle \in \mathbf{R}$ and the corresponding norm are defined by*

$$\langle\langle \alpha, \beta \rangle\rangle := \int_M \langle \alpha, \beta \rangle dv_g\ and\ \|\alpha\| := \sqrt{\langle\langle \alpha, \alpha \rangle\rangle}.$$

3. *The formal adjoint of $d^\omega : \Lambda^p(M, \mathrm{ad}\, P) \to \Lambda^{p+1}(M, \mathrm{ad}\, P)$, denoted by δ^ω, is defined by*

$$\langle\langle d^\omega \alpha, \sigma \rangle\rangle = \langle\langle \alpha, \delta^\omega \sigma \rangle\rangle,\ \ \ \ \ \forall \sigma \in \Lambda^{p+1}(M, \mathrm{ad}\, P).$$

4. *The product $\alpha \dot\wedge \gamma \in \Lambda^{p+q}(M)$ is defined by*

$$x \mapsto (\alpha \dot\wedge \gamma)_x = (\alpha_i(x) \wedge \gamma_j(x)) \langle e^i(x), e^j(x) \rangle_{\mathfrak{g}} \in \Lambda_x^{p+q}(M).$$

5. *The bracket $[\alpha,\ \gamma]_\wedge$, or simply $[\alpha,\ \gamma]$ of bundle-valued forms, is defined by*

$$x \mapsto [\alpha,\ \gamma]_x = (\alpha_i(x) \wedge \gamma_j(x))[e^i(x), e^j(x)] \in \Lambda_x^{p+q}(M, \mathrm{ad}\, P).$$

We note that this product is also denoted by $\alpha \wedge \gamma$.

The group $\mathcal{G}$ acts on the space of gauge connections $\mathcal{A}(P)$. We can describe this action in two different ways. Let $f \in \mathcal{G}$. Then the first is the right action $R_{f^{-1}}$ obtained by pulling back the connection form, i.e.,

$$f \cdot \omega := R_{f^{-1}}(\omega) = (f^{-1})^* \omega, \quad f \in \mathcal{G}, \ \omega \in \mathcal{A}(P). \tag{6.17}$$

The second is the left action L_f obtained by pushing forward the horizontal distribution defined by the given connection by f_*. It can be shown that the two are related by $L_f = R_{f^{-1}}$. Thus we see that there is essentially only one natural action of $\mathcal{G}$ on $\mathcal{A}$. We shall use both forms of this action as convenient. We now derive a local expression for this action when G is one of the classical matrix groups. Let A_i (resp., A_j) be a local gauge potential in local gauge t_i (resp., t_j) over U_i (resp., U_j), corresponding to the gauge connection ω. Let

$$t_i = \psi_{ij} t_j, \ \psi_{ij} : U_i \cap U_j \to G$$

be the local expression of the gauge transformation f. Then equation (6.17) becomes

$$A_j = \psi_{ij}^{-1} A_i \psi_{ij} + \psi_{ij}^{-1} d\psi_{ij}.$$

It is customary to write g for ψ_{ij} and A, A^g for A_i, A_j, respectively. Then the local expression of equation (6.17) becomes

$$A^g = g^{-1} A g + g^{-1} dg. \tag{6.18}$$

Equation (6.4) for the gauge transformation of an electromagnetic potential is a special case of equation (6.18) corresponding to $G = U(1)$. In many applications it is the local form (6.18) of the equation (6.17) that is used for calculating the gauge transforms of potentials and fields. For example, using equation (6.18) we obtain the following expression for the gauge transform F_ω^g of the gauge field F_ω

$$F_\omega^g = g^{-1} F_\omega g. \tag{6.19}$$

From equation (6.19) it follows that $|F_\omega|$ is a gauge invariant function on M, i.e.,

$$|F_\omega^g| = |F_\omega| \in \mathcal{F}(M).$$

It is this gauge invariant function that is used in defining the Yang–Mills action functional (see Chapter 8).

We say that connections $\alpha, \beta \in \mathcal{A}$ are **gauge equivalent** if there exists a gauge transformation $f \in \mathcal{G}$ such that $\beta = f \cdot \alpha$. From the definition of the action of $\mathcal{G}$ on $\mathcal{A}$ given above, it follows that each equivalence class of gauge equivalent connections is an orbit of $\mathcal{G}$ in $\mathcal{A}$. The orbit space $\mathcal{O} = \mathcal{A}/\mathcal{G}$ thus represents gauge inequivalent connections and is called the **moduli space of gauge potentials** on $P(M, G)$.

Gauge field equations that arise in physical applications are partial differential equations on manifolds. Their analysis is greatly facilitated by considering Sobolev completions of the various infinite-dimensional spaces involved in the formulation and study of these equations. Now we define Sobolev norms and completions of some of the structures used in this chapter. We use the assumptions and notation of Definition 6.1. Let E be a Riemannian vector bundle over the manifold M associated to $P(M, G)$. Recall that $\Lambda^p(M, E) = \Gamma(A^p(M) \otimes E)$ is the space of p-forms on M with values in E. Fixing a connection α on P we get the induced covariant derivative ∇^α on $\Lambda^p(M, E)$. If λ is the Levi-Civita connection on (M, g), then we can define the covariant derivative

$$\nabla^{(\omega, \lambda)} : \Lambda^k(M, \operatorname{ad} P) \to \Gamma(T^*M \otimes A^k T^*M \otimes \operatorname{ad} P)$$

or simply ∇ by

$$\nabla \equiv \nabla^{(\omega, \lambda)} := 1 \otimes \nabla^\omega + \nabla^\lambda \otimes 1.$$

Using the covariant derivative ∇ we define a Sobolev k-norm on $\Lambda^p(M, E)$ by

$$\|\phi\|_k = \left(\sum_{j=0}^{k} \int_M |(\nabla)^j \phi|^2 \right)^{1/2}, \quad \phi \in \Lambda^p(M, E).$$

The completion of $\Lambda^p(M, E)$ in this norm is a Hilbert space (under the associated bilinear form) denoted by $H_k(\Lambda^p(M, E))$. A different choice of the connection on P and metrics on M and E gives an equivalent norm. The map $d^\alpha : \Lambda^p(M, E) \to \Lambda^{p+1}(M, E)$ extends to a smooth bounded map of Hilbert spaces (also denoted by d^α)

$$d^\alpha : H_k(\Lambda^p(M, E)) \to H_{k-1}(\Lambda^{p+1}(M, E)).$$

For $p = 0$, this map has finite dimensional kernel and closed range. In general, the sequence

$$0 \longrightarrow H_k(\Lambda^0(M, E)) \xrightarrow{d^\alpha} H_{k-1}(\Lambda^1(M, E)) \xrightarrow{d^\alpha} \cdots$$

fails to be a complex, the obstruction being provided by the curvature of α. In particular, fixing a connection α on P gives an identification of $\mathcal{A}$ with $\Lambda^1(M, \operatorname{ad} P)$ and we denote the corresponding Sobolev completion of $\mathcal{A}$ in the k-norm by $H_k(\mathcal{A})$. This Sobolev k-norm defines a **strong Riemannian metric** on $\mathcal{A}$. Strong Riemannian metrics and the weak or L^2 metric on $\mathcal{A}$ defined above are used in studying the geometry of $\mathcal{A}$ and other related spaces in Chapter 9. The curvature map

$$F : \omega \mapsto F_\omega \quad \text{of } \mathcal{A} \to \Lambda^2(M, \operatorname{ad} P)$$

extends to a smooth bounded Hilbert space map from $H_{k+1}(\mathcal{A})$ into $H_k(\Lambda^2(M, \operatorname{ad} P))$ for $k \geq 1$. The Sobolev completion of the gauge group $\mathcal{G}$ is obtained by considering $\mathcal{G}$ as a subset of $\Lambda^0(M, P \times_\rho \operatorname{End}(V))$, where $\rho : G \to \operatorname{End}(V)$ is a faithful representation of the gauge group G. We define $H_k(\mathcal{G})$ to be the closure of $\mathcal{G}$ in $H_k(\Lambda^0(M, P \times_\rho \operatorname{End}(V)))$. The Lie algebra structure of $\mathcal{LG} = \Lambda^0(M, \operatorname{ad} P)$ extends to a Lie algebra structure on its Sobolev completion $H_k(\Lambda^0(M, \mathrm{ad}\, P))$. The above discussion leads to the following theorem.

Theorem 6.5 *Let $k > \frac{1}{2}(m+1)$; then we have the following:*

1. *The Sobolev completion $H_k(\mathcal{G})$, is an infinite-dimensional Lie group modeled on a separable Hilbert space with Lie algebra $H_k(\mathcal{LG})$.*
2. *The action of $\mathcal{G}$ on $\mathcal{A}$ defined by equation (6.17) extends to a smooth action of $H_{k+2}(\mathcal{G})$ on $H_{k+1}(\mathcal{A})$.*
3. *The curvature map $F : \mathcal{A} \rightarrow \Lambda^2(M, \mathrm{ad}\, P)$ extends to a smooth, bounded, $H_{k+2}(\mathcal{G})$-equivariant, Hilbert space map from $H_{k+1}(\mathcal{A})$ into $H_k(\Lambda^2(M, \mathrm{ad}\, P))$, i.e.,*

$$F(g.\omega) = g^{-1}F(\omega)g, \qquad \forall g \in H_{k+2}(\mathcal{G}).$$

From now on we consider that all objects requiring Sobolev completions have been completed in appropriate norms and drop the H_k from $H_k(object)$.

In many applications one is interested in the orbit space $\mathcal{O} = \mathcal{A}/\mathcal{G}$ whose points correspond to equivalence classes of gauge connections. The orbit space is given the quotient topology and is a Hausdorff topological space. However, in general, the action of $\mathcal{G}$ on $\mathcal{A}$ is not free and $\mathcal{O}$ fails to be a manifold. We now discuss two methods of suitably modifying $\mathcal{A}$ or $\mathcal{G}$ to obtain orbit spaces with nice mathematical structure.

1. Let $\mathcal{G}_0 \subset \mathcal{G}$ denote the group of **based gauge transformations**, defined by

$$\mathcal{G}_0 = \{f \in \mathcal{G}|\ f(u_0) = u_0 \ \text{for some fixed } u_0 \in P\}.$$

The group $\mathcal{G}_0$ is also called the **restricted group of gauge transformations**. A based gauge transformation that fixes a connection is the identity. Therefore $\mathcal{G}_0$ acts freely on $\mathcal{A}$ and the orbit space $\mathcal{O}_0 = \mathcal{A}/\mathcal{G}_0$ is an infinite dimensional Hilbert manifold. $\mathcal{A}(\mathcal{O}_0, \mathcal{G}_0)$ is a principal fiber bundle with canonical projection

$$\pi_0 : \mathcal{A} \rightarrow \mathcal{A}/\mathcal{G}_0 = \mathcal{O}_0.$$

The relation between $\mathcal{G}, \mathcal{G}_0$ and the gauge group G is given by the following proposition.

Proposition 6.6 *The group $\mathcal{G}_0$ is a normal subgroup of $\mathcal{G}$ and the quotient $\mathcal{G}/\mathcal{G}_0$ is isomorphic to the gauge group G.*

Proof: We note that if $\pi(u_0) = x_0$ then a based gauge transformation f fixes every point of the fiber $\pi^{-1}(x_0)$. If $h \in \mathcal{G}$, then $h(u_0) \in \pi^{-1}(x_0)$ and hence $f(h(u_0)) = h(u_0)$; i.e., $(h^{-1}fh)(u_0) = u_0$. This proves that $\mathcal{G}_0$ is a normal subgroup of $\mathcal{G}$. Now define $\hat{h}$ to be the unique element of G such that $h(u_0) = (u_0)\hat{h}$. It is now easy to verify that the map

$$T_{u_0} : h \mapsto \hat{h} \quad \text{of } \mathcal{G} \rightarrow G$$

is a surjective, homomorphism with kernel $\mathcal{G}_0$. From this it follows that the map

$$h\mathcal{G}_0 \mapsto \hat{h} \quad \text{of} \quad \mathcal{G}/\mathcal{G}_0 \to G$$

is well defined and is an isomorphism of groups.   □

2. Let $Z(G)$ denote the center of the gauge group G. Denote by $\mathcal{Z}$ the group $\Gamma(P \times_{\mathrm{Ad}} Z(G)) \cong Z(\mathcal{G})$. Then $\mathcal{G}_c = \mathcal{G}/\mathcal{Z}$ is called the **group of effective gauge transformations**. By an argument similar to that in the above proposition we can show that $\mathcal{Z} \cong Z(G)$. Let $\mathcal{A}_{ir} \subset \mathcal{A}$ denote the space of **irreducible connections**. Then $\mathcal{G}_c$ acts freely on $\mathcal{A}_{ir}$ and we denote by $\mathcal{O}_{ir}$ the orbit space $\mathcal{A}_{ir}/\mathcal{G}_c$. $\mathcal{O}_{ir}$ is an infinite-dimensional Hilbert manifold and $\mathcal{A}_{ir}(\mathcal{O}_{ir}, \mathcal{G}_c)$ is a principal fiber bundle with canonical projection

$$\pi_c : \mathcal{A}_{ir} \to \mathcal{A}_{ir}/\mathcal{G}_c = \mathcal{O}_{ir}.$$

We now discuss the topology of the various spaces and fibrations introduced above and use it to study the problem of global gauge fixing. We begin by considering a fiber bundle E over base B and fiber F as follows:

Given a point $x_0 \in B$ and a point $u_0 \in \pi^{-1}(x_0) \cong F$, there exists a natural group homomorphism $\partial : \pi_n(B, x_0) \to \pi_{n-1}(F, u_0)$. This homomorphism together with the homomorphisms induced by the inclusion $\iota : F \to E$ and by the bundle projection p leads to the following long exact sequence of homotopy groups

$$\cdots \to \pi_{k+1}(E, u_0) \to \pi_{k+1}(B, x_0) \to \pi_k(F, u_0) \to \pi_k(E, u_0) \to \cdots.$$

Applying the above homotopy sequence to the various fiber bundles related to the based and effective gauge transformations we obtain the following theorem.

Theorem 6.7 *Let G be a compact, simply connected, non-trivial Lie group and let $P(M, G)$ be a principal bundle over a closed, simply connected base manifold M. Then we have*

1. $\pi_j(\mathcal{G}) \cong \pi_j(\mathcal{G}_0)$, for $j = 0, 1$;
2. $\pi_j(\mathcal{G}) \cong \pi_j(\mathcal{G}_c)$, for $j \geq 2$;
3. $\pi_j(\mathcal{G}_0) \cong \pi_j(\mathcal{O}_0)$, $\forall j$;
4. $\pi_j(\mathcal{G}_c) \cong \pi_j(\mathcal{O}_{ir})$, $\forall j$.

Furthermore, if M is orientable (resp., spin) then there exists a non-negative integer j such that $\pi_j(\mathcal{G}_0)$ (resp., $\pi_j(\mathcal{G}_c)$) is non-trivial.

Now we use these results and related constructions to study the **Gribov ambiguity**.

6.5 Gribov Ambiguity

Recall that the orbit space $\mathcal{O}$ whose points represent gauge equivalent connections (gauge potentials), is defined by

$$\mathcal{O} = \mathcal{A}/\mathcal{G}.$$

We denote by

$$p : \mathcal{A} \to \mathcal{O} = \mathcal{A}/\mathcal{G}$$

the infinite-dimensional principal $\mathcal{G}$-bundle with natural projection p. The group of gauge transformations $\mathcal{G}$ acts on $\mathcal{A}$ by

$$(g, \omega) \mapsto g \cdot \omega, \qquad g \in \mathcal{G}, \ \omega \in \mathcal{A} \, .$$

The induced action on the curvature Ω is given by $g.\Omega = g\Omega g^{-1}$. These transformation properties are used to construct gauge invariant functionals on $\mathcal{A}$. In the Feynman integral approach to quantization one must integrate these gauge invariant functions over the orbit space $\mathcal{O}$ to avoid the infinite contribution coming from gauge equivalent fields. However, the mathematical nature of this space is essentially unknown. Physicists often try to get around this difficulty by choosing a section $s : \mathcal{O} \to \mathcal{A}$ and integrating over its image $s(\mathcal{O}) \subset \mathcal{A}$ with a suitable weight factor such as the Faddeev–Popov determinant, which may be thought of as the Jacobian of the change of variables effected by $p_{|s(\mathcal{O})} : s(\mathcal{O}) \to \mathcal{O}$. This procedure amounts to a choice of one connection in $\mathcal{A}$ from each equivalence class in $\mathcal{O}$ and is referred to as **gauge fixing**. The question of the existence of such sections is thus crucial for this approach. For the trivial $SU(2)$-bundle over $\mathbf{R}^4$, Gribov showed that the so called Coulomb gauge fails to be a section, i.e., the Coulomb gauge is not a true global gauge. Gribov showed that the local section corresponding to the Coulomb gauge at the zero connection if extended (under some boundary conditions) intersects the orbit through zero at large distances and thus fails to be a section. The boundary conditions imposed by Gribov amount to the gauge potential being defined over the compactification of $\mathbf{R}^4$ to S^4. He also discussed a similar problem for $\mathbf{R}^3$. This non-existence of a global gauge is referred to as the **Gribov ambiguity**. In view of this negative result, it is natural to ask if any true gauge exists under these boundary conditions. Without any boundary conditions it is possible to exhibit a global gauge, but it does not seem to have any physical meaning. We show that, in fact, the Gribov ambiguity is present in all physically relevant cases, so that no global gauge exists. The existence of Gribov ambiguity can be proven for a number of different base manifolds and gauge groups. For examples and further details see [166, 167, 220, 351].

The Gribov ambiguity is a consequence of the topology of the principal bundle $\mathcal{A}$ over $\mathcal{O}$ as we now explain. Let $P(S^4, SU(2))$ be a principal bundle. Recall that $\mathcal{A}$ is isomorphic to the vector space of 1-forms on S^4 with values in the vector bundle $\operatorname{ad} P = P \times_{\operatorname{ad}} \mathfrak{g}$ once a connection is fixed. If $\alpha \in \mathcal{A}$ is a fixed connection we can write

$$\mathcal{A} \cong \{\alpha + A \mid A \in \Lambda^1(S^4, \operatorname{ad} P)\},$$

where the pull-back of A to P is understood. Consider the slice S_α defined by

$$S_\alpha := \{\alpha + A \mid \delta^\alpha A = 0\} \subset \mathcal{A}.$$

We call this the **generalized Coulomb gauge**. In particular, if $\alpha = 0$ then

$$S_0 = \{A \mid \delta^0 A = 0\}.$$

Locally the condition $\delta^0 A = 0$ can be written as

$$\sum_i \frac{\partial A^i}{\partial x^i} = 0.$$

Locally (or on $\mathbf{R}^4$ as a base) one can find a connection with zero time component that is gauge equivalent to the given connection. The gauge condition then reduces to the classical Coulomb gauge condition

$$\operatorname{div} A = 0.$$

It is convenient to reformulate the definition of the group $\mathcal{G}$ of gauge transformations as follows. Let

$$E_2 = E(M, \mathbf{C}^2, r, P)$$

be the vector bundle associated to P with fiber $\mathbf{C}^2$, where r is the defining or standard representation of $SU(2)$ on $\mathbf{C}^2$. Then

$$\mathcal{G} \cong \{h \in \Gamma(\operatorname{Hom}(E_2, E_2)) \mid h(x) \in SU(2), \forall x \in M\}.$$

Define the isotropy group $\mathcal{G}_\alpha$ of a fixed connection α by

$$\mathcal{G}_\alpha = \{g \in \mathcal{G} \mid g \cdot \alpha = \alpha\}.$$

It is easy to see that $g \in \mathcal{G}_\alpha$ if and only if

$$d^\alpha g = 0.$$

In particular, g is completely determined by specifying its value at a single point, say $x_0 \in M$. To study the question of the irreducibility of α, we consider the holonomy group $H_\alpha \subset SU(2)$ of the connection α at x_0. We observe that $g \in \mathcal{G}_\alpha$ if and only if $g(x_0) \in SU(2)$ and $[g(x_0), H_\alpha] = 0$. If α is irreducible then this is equivalent to requiring that $g(x_0) \in Z(SU(2))$ (the center of $SU(2)$, which is isomorphic to $\mathbf{Z}_2$). Recall that the group $\mathcal{G}_c$ of **effective gauge transformations** is the quotient of $\mathcal{G}$ by the center $Z(\mathcal{G})$, which in this case is isomorphic to $\mathbf{Z}_2$ and hence

$$\mathcal{G}_c = \mathcal{G}/\mathbf{Z}_2 \ .$$

Let $\mathcal{A}_{ir} \subset \mathcal{A}$ be the set of **irreducible connections**. $\mathcal{G}_c$ acts on $\mathcal{A}_{ir}$ freely and the quotient $\mathcal{O}_{ir}$ is the **orbit space** of irreducible connections. $\mathcal{A}_{ir}$ is a principal $\mathcal{G}_c$-bundle over $\mathcal{O}_{ir}$, i.e.,

$$\mathcal{A}_{ir} = P(\mathcal{O}_{ir}, \mathcal{G}_c).$$

We now compute the homotopy groups of the space $\mathcal{A}_{ir}$. Let $f : S^k \to \mathcal{A}_{ir} \subset \mathcal{A}$ be a continuous map. Regarding f as a map of the boundary of a $(k+1)$-simplex Δ_{k+1} we can extend f linearly to a map of Δ_{k+1} to $\mathcal{A}$. It can be shown the extended map is homotopic to a map g, which actually lies in $\mathcal{A}_{ir}$. The construction uses the fact that the set $\mathcal{A}_r$ of reducible connections is a closed, nowhere dense, stratified subset of $\mathcal{A}$. A simple argument then shows that $f \sim g \sim c_\alpha$, where c_α is the constant map $c_\alpha(x) = \alpha$, $\forall x \in S^k$. Thus

$$\pi_k(\mathcal{A}_{ir}) = 0. \tag{6.20}$$

Under certain topological conditions it can be shown that $\mathcal{A}_{ir} = \mathcal{A}$. For example, if $P(M, SU(2))$ is a non-trivial bundle and the second cohomology group $H^2(M, \mathbf{Z}) = 0$, then $\mathcal{A}_{ir} = \mathcal{A}$. In particular, this second condition is satisfied by $M = S^4$ and we have the following proposition.

Proposition 6.8 Let P be a fixed non-trivial $SU(2)$-bundle over S^4, then there exists some k such that

$$\pi_k(\mathcal{G}_c) \neq 0. \tag{6.21}$$

Proof: By definition of $\mathcal{G}_c$ we have the following short exact sequence of groups

$$0 \to \mathbf{Z}_2 \to \mathcal{G} \to \mathcal{G}_c \to 0.$$

Therefore, if $\mathcal{G}$ is connected then $\pi_1(\mathcal{G}_c)$ is non-trivial. For $k > 1$ we have $\pi_k(\mathcal{G}) = \pi_k(\mathcal{G}_c)$. Recall that the group $\mathcal{G}_0$ is a group of based gauge transformations over some fixed point of the manifold, which we may take to be the point at infinity (i.e., the north pole) on S^4. We have the following short exact sequence of groups

$$0 \to \mathcal{G}_0 \to \mathcal{G} \to SU(2) \to 0.$$

It induces the following long exact sequence in homotopy

$$\cdots \to \pi_{k+1}(SU(2)) \to \pi_k(\mathcal{G}_0) \to \pi_k(\mathcal{G}) \to \pi_k(SU(2)) \to \cdots$$

We shall use the following part of the above sequence

$$\pi_3(\mathcal{G}) \to \pi_3(SU(2)) \to \pi_2(\mathcal{G}_0) \to \pi_2(\mathcal{G}) \to \pi_2(SU(2)). \tag{6.22}$$

In the present case of an $SU(2)$-bundle over S^4 it can be shown that

$$\pi_k(\mathcal{G}_0) \cong \pi_{k+4}(SU(2)).$$

In particular,

$$\pi_2(\mathcal{G}_0) \cong \pi_6(SU(2)) = \mathbf{Z}_{12}.$$

We also know that

$$\pi_3(SU(2)) = \mathbf{Z} \text{ and } \pi_2(SU(2)) = 0.$$

Thus, (6.22) becomes

$$\cdots \to \pi_3(\mathcal{G}) \to \mathbf{Z} \to \mathbf{Z}_{12} \to \pi_2(\mathcal{G}) \to 0.$$

Hence, if $\pi_2(\mathcal{G}) = 0$ then $\pi_3(\mathcal{G}) \neq 0$. Thus either

$$\pi_2(\mathcal{G}_c) = \pi_2(\mathcal{G}) \neq 0 \quad \text{or} \quad \pi_3(\mathcal{G}_c) = \pi_3(\mathcal{G}) \neq 0.$$

Thus, there exists some k such that $\pi_k(\mathcal{G}_c) \neq 0$. □

We note that equation (6.21) is a consequence of Theorem 6.7 applied to the case when $M = S^4$ and $G = SU(2)$. However, we have given the above proof to illustrate the kind of computations involved in establishing such results. Using the result (6.21) we can prove that no continuous global gauge $s : \mathcal{O} \to \mathcal{A}$ exists in this case. For if such an s exists then it induces a map s_{ir}, that is,

$$s_{ir} : \mathcal{O}_{ir} \to \mathcal{A}_{ir},$$

which is a section of the principal $\mathcal{G}_c$-bundle $\mathcal{A}_{ir}$ and we have a corresponding trivialization of the bundle $\mathcal{A}_{ir} = \mathcal{O}_{ir} \times \mathcal{G}_c$. Now choose some k such that $\pi_k(\mathcal{G}_c) \neq 0$. Then we have

$$0 = \pi_k(\mathcal{A}_{ir}) = \pi_k(\mathcal{O}_{ir}) \times \pi_k(\mathcal{G}_c) \neq 0.$$

This contradiction shows that a continuous global gauge s does not exist in this case.

The non-existence of a global gauge fixing need not prevent an application of path-integral methods. For example, one may use the fact that the orbit space $\mathcal{O}_{ir}$ and the space $\mathcal{A}_{ir}$ are paracompact. Thus we may be able to find a suitable locally finite covering and a subordinate partition of unity for $\mathcal{O}_{ir}$ and construct local gauges and local weights to define the path integrals. However, for this procedure to work we need an explicit description of the locally finite covering and local weights for the Faddeev–Popov approach. More generally, one can consider physical configuration spaces that are subspaces or quotient spaces of the space of gauge potentials $\mathcal{A}$. This observation is the starting point of the development of a geometric setting for field theories in [276], on which part of our treatment of classical and quantum field theories and coupled fields is based.

In [167] Gribov proposed another interesting approach that we now discuss. It is based on the assumption that the physically relevant part of the configuration space can be defined by the following two conditions:

1. local transversality to the gauge orbits,
2. positivity of the Faddeev–Popov determinant.

The second condition insures that the effective contributions at various orders in perturbation theory do not tend to infinity at finite distances. The region of the configuration space satisfying the above two conditions is called the **Gribov region**. It is denoted by Ω. The boundary $\partial\Omega$ of the Gribov region is called the **first Gribov horizon**. It has been conjectured by Gribov that the problem of confinement of gluons may be due to the restriction of the configuration space to the physically allowable region Ω. It can be shown that the two conditions defining the Gribov region Ω are the conditions for a local minimum of the L^2 norm on each gauge orbit. In [97] it is shown that the L^2 norm attains its absolute minimum on each gauge orbit and hence each gauge orbit intersects the Gribov region at least once at the absolute minimum. It would be interesting to study the structure of the set of absolute minima and to check if this set is the appropriate region of integration for the Feynman path integral method of quantization.

6.6 Matter Fields

Let $E(M, F, r, P)$ be a vector bundle associated to the principal bundle $P(M, G)$. We call a section $\phi \in \Gamma(E)$ an E-**field** (or a **generalized Higgs field**) on M. If $E = \mathrm{ad}(P) := P \times_{\mathrm{ad}} \mathfrak{g}$ then ϕ is called the Higgs field (in the adjoint representation) on M. In general there are several fields that can be defined on bundles associated with a given manifold. For example, on a Lorentz 4-manifold the Levi-Civita connection is interpreted as representing a gravitational potential. Recall that the Levi-Civita connection is the unique torsion-free, metric connection defined on the bundle of orthonormal frames $\mathcal{O}(M)$ of M. In general if M is a pseudo-Riemannian manifold, we can define the space of linear connections $\mathcal{A}(\mathcal{O}(M))$ on M by

$$\mathcal{A}(\mathcal{O}(M)) := \{\alpha \in \Lambda^1(\mathcal{O}(M), so(m)) \mid \alpha \text{ is a connection on } \mathcal{O}(M)\}.$$

Generalized theories of gravitation often use this space as their configuration space. If M is a spin manifold and $S(M)(M, Spin(m))$ is the $Spin(m)$-principal bundle, then one can consider the space $\mathcal{A}(S(M))$ of spin connections on the bundle $S(M)$, defined by

$$\mathcal{A}(S(M)) := \{\beta \in \Lambda^1(S(M), spin(m)) \mid \beta \text{ is a connection on } S(M)\}.$$

For a given signature (p, q) we may consider the space $\mathcal{RM}_{(p,q)}(M)$ of all pseudo-Riemannian metrics on M of signature (p, q). The space $\mathcal{RM}_{(m,0)}(M)$ of Riemannian metrics on M is denoted simply by $\mathcal{RM}(M)$. Recall that there is a canonical principal $GL(m, \mathbf{R})$-bundle over M, namely $L(M)$ the bundle of frames of M. If $\rho : GL(m, \mathbf{R}) \to \operatorname{End} V$ is a representation of $GL(m, \mathbf{R})$ on V and $E(M, V, \rho, L(M))$ is the corresponding associated bundle of $L(M)$, then we denote by $\mathcal{W}$ the space of E-fields $\Gamma(E)$, i.e.,

$$\mathcal{W} = \Gamma(E(M, V, \rho, L(M))).$$

Thus we see that we have an array of fields on a given base manifold M and we must specify the equations governing the evolution and interactions of these fields and study their physical meaning. There is no standard procedure for doing these things. In many physical applications one obtains the coupled field equations of interacting fields as the Euler–Lagrange equations of a variational problem with the Lagrangian constructed from the fields. For any given problem the Lagrangian is chosen subject to certain invariance or covariance requirements related to the symmetries of the fields involved. We now discuss three general conditions that are frequently imposed on the Lagrangians in physical theories. In what follows we restrict ourselves to a fixed, compact 4-manifold M as the base manifold, but the discussion can be easily extended to apply to an arbitrary base manifold. Let $P(M, G)$ be a principal bundle over M whose structure group G carries a bi-invariant metric h. For example if G is a semisimple Lie group, then a suitable multiple of the Killing form on $\mathfrak{g}$ provides a bi-invariant metric on G.

We want to consider coupled field equations for a metric $g \in \mathcal{RM}(M)$, a field $\phi \in \mathcal{W}(M)$ (a section of the bundle associated to the frame bundle $L(M)$), a connection $\omega \in \mathcal{A}(P)$ and a generalized Higgs field $\psi \in \mathcal{H} = \Gamma E(M, V_r, r, P)$, where $r : G \to \operatorname{End}(V_r)$ is a representation of the gauge group G. Thus, our **configuration space** is defined by

$$\mathcal{C} := \mathcal{RM} \times \mathcal{W} \times \mathcal{A} \times \mathcal{H}.$$

We assume that the field equations are the variational equations of an action integral defined by a Lagrangian L on the configuration space with values in $\Lambda^4(M)$. When a fixed volume form such as the metric volume form is given, we may regard L as a real-valued function. We shall use any one of these conventions without comment. The action $\mathcal{E}$ is given by

$$\mathcal{E}(g, \phi, \omega, \psi) = \int_M L(g, \phi, \omega, \psi).$$

There are various groups associated with the geometric structures involved in the construction of these fields, groups that have natural actions on them. We shall require the Lagrangian to satisfy one or more of the following conditions: i) naturality, ii) local regularity, iii) conformal invariance.

i) **Naturality**: Naturality with respect to the group of generalized gauge transformations is defined as follows. Let $F \in \mathrm{Diff}_M(P)$ be a generalized transformation covering $f \in \mathrm{Diff}(M)$. Then by naturality with respect to $Diff_M(P)$ we mean that

$$L(f^*g, f_\rho^*\phi, F^*\omega, F_r^*\psi) = f^*L(g, \phi, \omega, \psi), \qquad (6.23)$$

where f_ρ^* is the induced action of f on $\mathcal{W}$, and F_r^* is the action induced by the generalized gauge transformation F on $\mathcal{H}$. In the absence of the principal bundle P this condition reduces to naturality with respect to $\mathrm{Diff}(M)$ and is Einstein's condition of general covariance of physical laws derived from the Lagrangian formalism. Further, in the absence of $\mathcal{W}(M)$ this condition corresponds to the covariance of gravitational field equations when the Lagrangian is taken to be the standard Einstein–Hilbert Lagrangian. If we require naturality with respect to the group $\mathcal{G}$, then condition (6.23) is precisely the principle of gauge invariance introduced by Weyl. Since in this case $f = id$, condition (6.23) becomes

$$L(g, \phi, F^*\omega, F_r^*\psi) = L(g, \phi, \omega, \psi).$$

The concept of natural tensors on a Riemannian manifold was introduced by Epstein. It was extended to oriented Riemannian manifolds in [361], where a functorial formulation of naturality is given and a complete classification of natural tensor fields is given under some regularity conditions.

ii) **Local regularity**: Given any coordinate chart on M and a local gauge we can express the various potentials and fields with respect to induced bases. We require that in this system the Lagrangian be expressible as a universal polynomial in

$$(\det g)^{-1/2}, \ (\det h)^{-1/2}, \ g_{ij}, \ \partial^{|\alpha|}g_{ij}/\partial x^\alpha, \ \phi_{|\beta|}, \ \omega_{|\gamma|}, \ \psi_{|\delta|}, \dots$$

where α, β, γ, $\delta, \dots$, are suitable multi-indices (i.e., in the coefficients and derivatives of the potentials and fields in the induced bases). In physical applications one often restricts the order of derivatives that can occur in this local expression to at most two. For example in gravitation one considers natural tensors satisfying the conditions that they contain derivatives up to order 2 and depend linearly on the second order derivatives. Then it is well-known that such tensors can be expressed as

$$c_1 R^{ij} + c_2 g^{ij} S + c_3 g^{ij},$$

where R^{ij} are the components of the Ricci tensor Ric and S is the scalar curvature. Einstein's equations with or without the cosmological constant involve the above combination with suitable values of the constants c_1, c_2, c_3. Einstein's field equations and their generalization are discussed in the next section.

Applying the classification theorem of [361] to $SO(4)$-actions on the metric and gauge fields, we get the following general form for the Lagrangian:

$$
\begin{aligned}
L(g, \omega) &= c_1 S^2 + c_2 \|K\|^2 + c_3 \|\mathcal{W}^+\|^2 + c_4 \|\mathcal{W}^-\|^2 \\
&\quad + c_5 \|F_\omega \wedge F_\omega\| + c_6 \|F_\omega \wedge (*F_\omega)\|,
\end{aligned} \tag{6.24}
$$

where S, K, $\mathcal{W}^+$, $\mathcal{W}^-$ are the $SO(4)$-invariant components of the Riemannian curvature and F_ω is the gauge field of the gauge potential ω. For a suitable choice of constants in the above Lagrangian we obtain various topological invariants of M and P as well as the pure Yang–Mills action. For example, the first Pontryagin class of M is given by

$$
p_1(M) = \frac{1}{4\pi^2} \int_M (|\mathcal{W}^+|^2 - |\mathcal{W}^-|^2).
$$

The first Pontryagin class of P is given by

$$
p_1(P) = \frac{1}{8\pi^2} \int_M (|F_\omega^+|^2 - |F_\omega^-|^2),
$$

which turns out to be the instanton number of P.

To satisfy the conditions of naturality and local regularity for fields coupled to gauge fields physicists often start with ordinary derivatives of associated fields and the coupling is achieved by replacing these by gauge covariant derivatives in the Lagrangian. This is called the principle of **minimal coupling** (or **minimal interaction**). These two requirements can also be formulated by taking the Lagrangian to be defined on sections of suitable jet bundles on the space of connections. Using this approach a generalization of the classical theorem of Utiyama has been obtained [153].

iii) **Conformal invariance**: A **conformal transformation** of a manifold M is a diffeomorphism $f : M \to M$ such that

$$
f^* g = e^{2\sigma} g.
$$

The condition of conformal invariance of the Lagrangian may be expressed as follows

$$
L(e^{2\sigma} g, \phi, \omega, \psi) = L(g, \phi, \omega, \psi), \qquad \forall \sigma \in \mathcal{F}(M).
$$

In general, Lagrangians satisfying the conditions of naturality and regularity need not satisfy the condition of conformal invariance. This condition is often used to select parameters such as the dimension of the base space and the rank of the representation. A particular case of (6.24) is the Yang–Mills Lagrangian with action

$$\mathcal{E} = \frac{1}{8\pi^2} \int_M F_\omega \wedge *F_\omega.$$

It is an example of a Lagrangian that is conformally invariant only if $\dim M = 4$.

A large number of Lagrangians satisfying the naturality and regularity requirements are used in the physics literature. They are broadly classified into Lagrangians where gauge and other force fields are coupled to fields of bosonic matter, which has integral spin, and to fields of fermionic matter, which has half-integral spin. For example, a bosonic Lagrangian is given by

$$L_{\text{boson}}(g, \omega, \psi) = \|F_\omega\|^2 + \|\nabla\psi\|^2 + \frac{1}{6} S\|\psi\|^2 - V(\psi),$$

where V is the potential function taken to be a gauge-invariant polynomial of degree ≤ 4 on the fibers of E. If M is a spin manifold and if Σ is a bundle associated to the spin bundle, then we define an E-valued spinor to be a section of $\Sigma \otimes E$. An example of a fermion Lagrangian is given by

$$L_{\text{fermion}}(g, \omega, \xi) = \|F_\omega\|^2 + \langle \mathcal{D}(\xi), \xi \rangle,$$

where $\xi \in \Gamma(\Sigma \otimes E)$ and $\mathcal{D}$ is the Dirac operator on E-valued spinors. Several important properties of coupled field equations are studied in [314, 58].

6.7 Gravitational Field Equations

There are several ways of deriving Einstein's gravitational field equations. For example, as we observed in the previous section, we can consider natural tensors satisfying the conditions that they contain derivatives of the fundamental (pseudo-metric) tensor up to order 2 and depend linearly on the second order derivatives. Then we obtain the tensor

$$c_1 R^{ij} + c_2 g^{ij} S + c_3 g^{ij},$$

where R^{ij} are the components of the Ricci tensor Ric and S is the scalar curvature. Requiring this tensor to be divergenceless and using the Bianchi identities leads to the relation $c_1 + 2c_2 = 0$ between the constants c_1, c_2, c_3. Choosing $c_1 = 1$ and $c_3 = 0$ we obtain Einstein's equations (without the cosmological constant), which may be expressed as

$$E = -T, \tag{6.25}$$

where $E := \text{Ric} - \frac{1}{2}Sg$ is the **Einstein tensor** and T is an energy-momentum tensor on the space-time manifold which acts as the source term. Now the

Bianchi identities satisfied by the curvature tensor imply that

$$\text{div } E := \delta(g^b E) = 0.$$

Hence, if Einstein's equations (6.25) are satisfied, then for consistency we must have

$$\text{div } T = 0. \tag{6.26}$$

Equation (6.26) is called the differential (or local) law of conservation of energy and momentum. However, integral (or global) conservation laws can be obtained by integrating equation (6.26) only if the space-time manifold admits Killing vectors. Thus, equation (6.26) has no clear physical meaning, except in special cases. An interesting discussion of this point is given by Sachs and Wu [331]. Einstein was aware of the tentative nature of the right hand side of equation (6.25), but he believed strongly in the expression on the left hand side. By taking the trace of both sides of equation (6.25) we are led to the condition

$$S = t \tag{6.27}$$

where t denotes the trace of the energy-momentum tensor. The physical meaning of this condition seems even more obscure than that of condition (6.26). If we modify equation (6.25) by adding the cosmological term cg (c is called the **cosmological constant**) to the left hand side of equation (6.25), we obtain Einstein's equation with cosmological constant

$$E + cg = -T. \tag{6.28}$$

This equation also leads to the consistency condition (6.26), but condition (6.27) is changed to

$$S = t + 4c. \tag{6.29}$$

Using (6.29), equation (6.28) can be rewritten in the following form

$$K = -(T - \frac{1}{4}tg), \tag{6.30}$$

where

$$K = -(\text{Ric} - \frac{1}{4}Sg) \tag{6.31}$$

is the trace-free part of the Ricci tensor of g. We call equation (6.30) **generalized field equations** of gravitation. We now show that these equations arise naturally in a geometric formulation of Einstein's equations. We begin by defining a tensor of **curvature type**.

Definition 6.2 *Let C be a tensor of type $(4,0)$ on M. We can regard C as a quadrilinear mapping (pointwise) so that for each $x \in M$, C_x can be identified with a multilinear map*

$$C_x : T_x^*(M) \times T_x^*(M) \times T_x^*(M) \times T_x^*(M) \to \mathbf{R}.$$

We say that the tensor C is of curvature type if C_x satisfies the following conditions for each $x \in M$ and for all $\alpha, \beta, \gamma, \delta \in T_x^*(M)$;

1. $C_x(\alpha, \beta, \gamma, \delta) = -C_x(\beta, \alpha, \gamma, \delta)$,
2. $C_x(\alpha, \beta, \gamma, \delta) = -C_x(\alpha, \beta, \delta, \gamma)$,
3. $C_x(\alpha, \beta, \gamma, \delta) + C_x(\alpha, \gamma, \delta, \beta) + C_x(\alpha, \delta, \gamma, \beta) = 0$.

From the above definition it follows that a tensor C of curvature type also satisfies the following condition:

$$C_x(\alpha, \beta, \gamma, \delta) = C_x(\gamma, \delta, \alpha, \beta), \qquad \forall x \in M.$$

Example 6.3 *The Riemann–Christoffel curvature tensor is of curvature type. Indeed, the definition of tensors of curvature type is modeled after this fundamental example. Another important example of a tensor of curvature type is the tensor G defined by*

$$G_x(\alpha, \beta, \gamma, \delta) = g_x(\alpha, \gamma)g_x(\beta, \delta) - g_x(\alpha, \delta)g_x(\beta, \gamma), \qquad \forall x \in M,$$

where g is the fundamental or metric tensor of M.

We now define the curvature product of two symmetric tensors of type $(2, 0)$ on M. The curvature product was introduced in [264] and used in [273] to obtain a geometric formulation of Einstein's equations.

Definition 6.3 *Let g and T be two symmetric tensors of type $(2, 0)$ on M. The **curvature product** of g and T, denoted by $g \times_c T$, is a tensor of type $(4, 0)$ defined by*

$$(g \times_c T)_x(\alpha, \beta, \gamma, \delta) := \tfrac{1}{2}\big[g(\alpha, \gamma)T(\beta, \delta) + g(\beta, \delta)T(\alpha, \gamma)$$
$$-g(\alpha, \delta)T(\beta, \gamma) - g(\beta, \gamma)T(\alpha, \delta)\big],$$

for all $x \in M$ and $\alpha, \beta, \gamma, \delta \in T_x^*(M)$.

In the following proposition we collect some important properties of the curvature product and tensors of curvature type.

Proposition 6.9 *Let g and T be two symmetric tensors of type $(2, 0)$ on M and let C be a tensor of curvature type on M. Then we have the following:*

1. $g \times_c T = T \times_c g$;
2. $g \times_c T$ is a tensor of curvature type:
3. $g \times_c g = G$, where G is the tensor defined in Example 6.3;
4. G_x induces a pseudo-inner product on $\Lambda_x^2(M), \forall x \in M$;
5. C_x induces a symmetric, linear transformation of $\Lambda_x^2(M), \forall x \in M$.

We denote the Hodge star operator on $\Lambda_x^2(M)$ by J_x. The fact that M is a Lorentz 4-manifold implies that J_x defines a complex structure on $\Lambda_x^2(M), \forall x \in M$. Using this complex structure we can give a natural structure of a complex vector space to $\Lambda_x^2(M)$. Then we have the following proposition.

Proposition 6.10 *Let* $L : \Lambda_x^2(M) \to \Lambda_x^2(M)$ *be a real, linear transformation. Then the following are equivalent:*

1. L *commutes with* J_x;
2. L *is a complex linear transformation of the complex vector space* $\Lambda_x^2(M)$;
3. *The matrix of* L *with respect to a* G_x-*orthonormal basis of* $\Lambda_x^2(M)$ *is of the form*

$$
\begin{pmatrix} A & B \\ -B & A \end{pmatrix} \tag{6.32}
$$

where A, B *are real* 3×3 *matrices.*

We now define the gravitational tensor W_g, of curvature type, which includes the source term.

Definition 6.4 *Let* M *be a space-time manifold with fundamental tensor* g *and let* T *be a symmetric tensor of type* $(2,0)$ *on* M. *Then the* **gravitational tensor** W_g *is defined by*

$$
W_g := R + g \times_c T, \tag{6.33}
$$

where R *is the Riemann–Christoffel curvature tensor of type* $(4,0)$.

We are now in a position to give a geometric formulation of the generalized field equations of gravitation.

Theorem 6.11 *Let* W_g *denote the gravitational tensor defined by (6.33) with source tensor* T. *We also denote by* W_g *the linear transformation of* $\Lambda_x^2(M)$ *induced by* W_g. *Then the following are equivalent:*

1. g *satisfies the generalized field equations of gravitation (6.30);*
2. W_g *commutes with* J_x;
3. W_g *is a complex linear transformation of the complex vector space* $\Lambda_x^2(M)$.

We shall call the triple (M, g, T) a **generalized gravitational field** if any one of the conditions of Theorem 6.11 is satisfied. Generalized gravitational field equations were introduced by the author in [264]. Their mathematical properties have been studied in [275, 266, 287]. Solutions of Marathe's generalized gravitational field equations that are not solutions of Einstein's equations are discussed in [69]. We note that the above theorem and the last condition in Proposition 6.8 can be used to discuss the Petrov classification of gravitational fields (see Petrov [316]). The tensor W_g can be used in place of R in the usual definition of sectional curvature to define the gravitational sectional curvature on the Grassmann manifold of non-degenerate 2-planes over M and to give a further geometric characterization of gravitational field equations. We observe that the generalized field equations of gravitation contain Einstein's equations with or without the cosmological constant as special cases. Solutions of the source-free generalized field equations are called **gravitational instantons**. If the base manifold is Riemannian, then gravitational instantons correspond to Einstein spaces. A detailed discussion of the structure of Einstein spaces and their moduli spaces may be found in [39].

We note that, over a compact, 4-dimensional, Riemannian manifold (M, g), the gravitational instantons that are not solutions of the vacuum Einstein equations are critical points of the quadratic, Riemannian functional or action $\mathcal{A}_2(g)$ defined by

$$\mathcal{A}_2(g) = \int_M S^2 dv_g.$$

In fact, using any polynomial in S of degree > 1 in the above action leads, generically to the gravitational instanton equations. Furthermore, the standard **Hilbert–Einstein action**

$$\mathcal{A}_1(g) = \int_M S dv_g$$

also leads to the generalized field equations when the variation of the action is restricted to metrics of volume 1.

There are several differences between the Riemannian functionals used in theories of gravitation and the Yang–Mills functional used to study gauge field theories. The most important difference is that the Riemannian functionals are dependent on the bundle of frames of M or its reductions, while the Yang–Mills functional can be defined on any principal bundle over M. However, we have the following interesting theorem [20].

Theorem 6.12 *Let (M, g) be a compact, 4-dimensional, Riemannian manifold. Let $\Lambda_+^2(M)$ denote the bundle of self-dual 2-forms on M with induced metric G_+. Then the Levi-Civita connection λ_g on M satisfies the gravitational instanton equations if and only if the Levi-Civita connection λ_{G_+} on $\Lambda_+^2(M)$ satisfies the Yang–Mills instanton equations.*

6.8 Geometrization Conjecture and Gravity

The classification problem for low-dimensional manifolds is a natural question after the success of the case of surfaces by the uniformization theorem. In 1905, Poincaré formulated his famous conjecture, which states, in the smooth case: A closed, simply connected 3-manifold is diffeomorphic to S^3, the standard sphere. A great deal of work in 3-dimensional topology the century that followed was motivated by this. In the 1980s Thurston studied hyperbolic manifolds. This led him to his "geometrization conjecture" about the existence of homogeneous metrics on all 3-manifolds. It includes the Poincaré conjecture as a special case. We already discussed this in Chapter 2. In the case of 4-manifolds there is at present no analogue of the geometrization conjecture. We now discuss briefly the main idea behind Perelman's proof of the geometrization conjecture and the relation of the perturbed Ricci flow equations to Einstein's equations in Euclidean gravity.

The Ricci flow equations

$$\frac{\partial g_{ij}}{\partial t} = -2R_{ij}$$

for a Riemannian metric g were introduced by Hamilton in [183]. They form a system of nonlinear second order partial differential equations. Hamilton proved that this equation has a unique solution for a short time for any smooth metric on a closed manifold. The evolution equation for the metric leads to the evolution equations for the curvature and Ricci tensors and for the scalar curvature. By developing a maximum principle for tensors Hamilton proved that the Ricci flow preserves the positivity of the Ricci tensor in dimension 3 and that of the curvature operator in dimension 4 [184]. In each of these cases he proved that the evolving metrics converge to metrics of constant positive curvature (modulo scaling). These and a series of further papers led him to conjecture that the Ricci flow with surgeries could be used to prove the Thurston geometrization conjecture. In a series of e-prints Perelman developed the essential framework for implementing the Hamilton program. We would like to add that the full Einstein equations with dilaton field as source play a fundamental role in Perelman's work (see, arXiv.math.DG/0211159, 0303109, 0307245 for details) on the geometrization conjecture. A corollary of this work is the proof of the long standing Poincaré conjecture. Perelman was awarded the Fields medal at the ICM 2006 in Madrid for his proof of the Poincaré and the geometrization conjectures. His ideas and methods have already found many applications in analysis and geometry. On March 18, 2010 Perelman was awarded the Clay Mathematics Institute's first millenium prize of one million dollars for his resolution of the Poincaré conjecture. A complete proof of the geometrization conjecture through application of the Hamilton–Perelman theory of the Ricci flow has now appeared in [70] in a special issue dedicated to the memory of S.-S. Chern,[5] one of the greatest mathematicians of the twentieth century.

The Ricci flow is perturbed by a scalar field, which corresponds in string theory to the dilaton. It is supposed to determine the overall strength of all interactions. The low energy effective action of the dilaton field coupled to gravity is given by the action functional

$$\mathcal{F}(g, f) = \int_M (R + |\nabla f|^2)e^{-f} dv.$$

---

[5] I first met Prof. Chern and his then newly arrived student S.-T. Yau in 1973 at the AMS summer workshop on differential geometry held at Stanford University. Chern was a gourmet and his conference dinners were always memorable. I attended the first one in 1973 and the last one in 2002 on the occassion of the ICM satellite conference at his institute in Tianjin. In spite of his advanced age and poor health he participated in the entire program and then continued with his duties as President of the ICM in Beijing.

Note that when f is the constant function the action reduces to the classical Hilbert–Einstein action. The first variation can be written as

$$\delta \mathcal{F}(g, f) = \int_M [-\delta g^{ij}(R_{ij} + \nabla_i f \nabla_j f) + (\tfrac{1}{2}\delta g^{ij}(g_{ij} - \delta f)(2\Delta f - |\nabla f|^2 + R)]dm,$$

where $dm = e^{-f}dv$. If $m = \int_M e^{-f}dv$ is kept fixed, then the second term in the variation is zero and then the symmetric tensor $-(R_{ij} + \nabla_i f \nabla_j f)$ is the L^2 gradient flow of the action functional $\mathcal{F}^m = \int_M (R + |\nabla f|^2)dm$. The choice of m is similar to the choice of a gauge. All choices of m lead to the same flow, up to diffeomorphism, if the flow exists. We remark that in the quantum field theory of the 2-dimensional nonlinear σ-model, the Ricci flow can be considered as an approximation to the renormalization group flow. This suggests gradient-flow-like behavior for the Ricci flow, from the physical point of view. Perelman's calculations confirm this result. The functional $\mathcal{F}^m$ has also a geometric interpretation in terms of the classical Bochner–Lichnerowicz formulas with the metric measure replaced by the dilaton twisted measure dm.

The corresponding variational equations are

$$R_{ij} - \tfrac{1}{2}Rg_{ij} = -(\nabla_i \nabla_j f - \tfrac{1}{2}(\Delta f)g_{ij}).$$

These are the usual Einstein equations with the energy-momentum tensor of the dilaton field as source. They lead to the decoupled evolution equations

$$(g_{ij})_t = -2(R_{ij} + \nabla_i \nabla_j f), \quad f_t = -R - \Delta f.$$

After applying a suitable diffeomorphism these equations lead to the gradient flow equations. This modified Ricci flow can be pushed through the singularities by surgery and rescaling. A detailed case by case analysis is then used to prove Thurston's geometrization conjecture [70]. This includes as a special case the classical Poincaré conjecture. A complete proof of the Poincaré conjecture without appealing to the Thurston geometrization conjecture may be found in the book [289] by Morgan and Tian. In fact, they prove a more general result which implies a closely related stronger conjecture called the 3-dimensional spherical space-form conjecture. This conjecture states that a closed 3-manifold with finite fundamental group is diffeomorphic to a 3-dimensional spherical space-form, i.e., the quotient of S^3 by free, linear action of a finite subgroup of the orthogonal group $O(4)$.

Chapter 7
Theory of Fields, II: Quantum and Topological

7.1 Introduction

Quantization of classical fields is an area of fundamental importance in modern mathematical physics. Although there is no satisfactory mathematical theory of quantization of classical dynamical systems or fields, physicists have developed several methods of quantization that can be applied to specific problems. Most successful among these is **QED (quantum electrodynamics)**, the theory of quantization of electromagnetic fields. The physical significance of electromagnetic fields is thus well understood at both the classical and the quantum level. Electromagnetic theory is the prototype of classical gauge theories. It is therefore natural to try to extend the methods of QED to the quantization of other gauge field theories. The methods of quantization may be broadly classified as non-perturbative and perturbative. The literature pertaining to each of these areas is vast. See for example, the two volumes [95,96] edited by Deligne, et al. which contain the lectures given at the Institute for Advanced Study, Princeton, during a special year devoted to quantum fields and strings; the book by Nash [298], and [41,354,89]. For a collection of lectures covering various aspects of quantum field theory, see, for example, [134,133,376].

Our aim in this chapter is to illustrate each of these methods by discussing some specific examples where a reasonably clear mathematical formulation is possible. A brief account of the non-perturbative methods in the quantization of gauge fields is given in Section 7.2. A widely used perturbative method in gauge theories is that of semiclassical approximation. We devote Section 7.3 to a mathematical formulation of semiclassical approximation in Euclidean Yang–Mills theory. This requires a detailed knowledge of the geometry of the moduli spaces of instantons and methods of regularization for the infinite-dimensional quantities. Two such methods are indicated in this section. Both classical and quantum field theories lead to topological invariants of base manifolds on which they are defined. We call these **topological field** theories, or **TFT**. They are subdivided into **topological classical field theories**, or **TCFT** and **topological quantum field theories**, or **TQFT**. The earliest result in TCFT is the Gauss formula for the linking number. We discuss this in Section 7.4. Donaldson's instanton invariants are also discussed here. Topological quantum field theories are discussed in Section 7.5. Interpretation of the Donaldson's polynomial invariants and the Jones polynomial via TQFT is then given. The Atiyah–Segal axioms for TQFT are also considered in this section. These topics are of independent interest and we will return to them in later chapters. They are included here because of their recent connections with topological quantum field theories, which now form an important branch of quantum field theory (or QFT for short). Ideas from QFT have already led to new ways of looking at old topological invariants of low-dimensional manifolds as well as to surprising new invariants.

7.2 Non-perturbative Methods

From the mathematical point of view, gauge field theories are classical field theories formulated on the infinite-dimensional space of connections on a principle bundle over a 4-dimensional space-time manifold M [273]. In physical applications the manifold M is usually a non-compact, Lorentzian 4-manifold. In many cases, M is taken to be the flat Minkowski space of special relativity. The corresponding quantum field theory is constructed by considering the space of classical fields as a configuration space $\mathcal{C}$ and defining the quantum expectation values of gauge invariant functions on $\mathcal{C}$ by using path integrals. This is usually referred to as the Feynman path integral method of quantization. Application of this method together with perturbative calculations has yielded some interesting results in the quantization of gauge theories. The starting point of this method is the choice of a Lagrangian defined on the configuration space of classical gauge fields. This Lagrangian is used to define the action functional that enters in the integrand of the Feynman path integral. Two important examples of the action functional considered are the Yang–Mills action and the Yang–Mills–Higgs action (see Chapter 8). Dimensional reduction allows us to think of a Yang–Mills–Higgs field as a Yang–Mills field on a higher dimensional manifold which is invariant under a certain symmetry group. Thus we may restrict our attention to the Yang–Mills case. Furthermore, we wish to consider only the Euclidean quantum field theory, where the pseudo-Riemannian space-time manifold is replaced by a Riemannian manifold. In physical literature the passage from a Lorentzian manifold to a Riemannian manifold is often referred to as a **Wick rotation** of the time coordinate. The Euclidean quantum field theory plays an important role in the calculation of tunneling amplitudes in quantum field theories (see Coleman [80] for a discussion of this and related aspects of quantization).

We consider the following mathematical setting for Euclidean quantum field theory. Let (M, g) be a compact, connected, oriented, Riemannian 4-manifold and let G be a compact, semisimple Lie group with Lie algebra $\mathfrak{g}$. From the physical point of view, the group G is the gauge group of the classical gauge field that we wish to quantize. Let $P(M, G)$ be a principal G-bundle. If $G = SU(n)$, then the isomorphism class of P is determined by its topological quantum number or the instanton number $k = -c_2(P) \in \mathbf{Z}$. We introduce a free parameter μ in the theory by considering the one-parameter family of G-invariant inner products $\langle X, Y \rangle_\mu$ on the Lie algebra $\mathbf{g}$ defined by

$$\langle X, Y \rangle_\mu := -\frac{1}{\mu} K(X, Y), \quad 0 < \mu \in \mathbf{R}, \tag{7.1}$$

where K is the Killing form on G.

The inner product defined by (7.1) and the metric g induce inner products on each fiber of the space $\Lambda^p(M, \mathrm{ad}(P)) = \Lambda^p(M) \otimes \mathrm{ad}(P)$ of p-forms on M with values in the bundle $\mathrm{ad}(P) = P \times_{\mathrm{ad}} \mathfrak{g}$. We denote these inner products by $\langle \, , \, \rangle_\mu^g$ or simply by $\langle \, , \, \rangle_\mu$ when the metric g is fixed. The corresponding pointwise norm is denoted by $| \, |_\mu^g$, or simply by $| \, |_\mu$. By integration over M these define the L^2 inner product

$$\langle\langle \alpha, \beta \rangle\rangle_\mu := \int_M \langle \alpha, \beta \rangle_\mu dv_g, \qquad \forall \alpha, \beta \in \Lambda^p(M, \mathrm{ad}(P)), \tag{7.2}$$

where dv_g is the volume form on M determined by the metric g. The norm of α corresponding to this L^2 inner product $\langle\langle \, , \, \rangle\rangle_\mu$ is denoted by $\|\alpha\|_\mu$. Thus,

$$\|\alpha\|_\mu^2 = \int_M |\alpha|_\mu^2 dv_g, \qquad \alpha \in \Lambda^p(M, \mathrm{ad}(P)). \tag{7.3}$$

In classical gauge field theories, the norms corresponding to different coupling constants μ lead to the same field equations up to trivial rescaling. However, as we shall show below, on quantization the different coupling constants μ lead to a one-parameter family of quantum field theories. By taking μ to be a suitable function of the other parameters of the given theory it may be possible to construct a renormalized version of the theory.

Recall that a connection ω on P determines the exterior covariant differential d^ω, its formal adjoint δ^ω, and the covariant derivative ∇^ω on the full tensor algebra with values in $\mathrm{ad}(P)$. These in turn determine the Laplacians on the various spaces $\Lambda^p(M, \mathrm{ad}(P))$. The ± 1 eigenspaces $\Lambda_\pm^2(M, \mathrm{ad}(P))$ of the Hodge operator give a decomposition of the curvature F_ω into its self-dual and anti-dual parts, i.e.,

$$F_\omega = F_\omega^+ + F_\omega^-. \tag{7.4}$$

In terms of these parts we can write the instanton number k and the Yang–Mills action S_μ as follows:

$$k = \frac{\mu}{8\pi^2}\left[||F_\omega^+||_\mu^2 - ||F_\omega^-||_\mu^2\right] = \frac{\mu}{8\pi^2}\int_M \left[|F_\omega^+|_\mu^2 - |F_\omega^-|_\mu^2\right]dv_g, \qquad (7.5)$$

$$S_\mu(\omega) = \frac{1}{8\pi^2}[||F_\omega^+||_\mu^2 + ||F_\omega^-||_\mu^2] = \frac{1}{8\pi^2}\int_M [|F_\omega^+|_\mu^2 + |F_\omega^-|_\mu^2]dv_g. \qquad (7.6)$$

From equation (7.6) it follows that μ rescales the Yang–Mills action, i.e., $S_\mu(\omega) = \frac{1}{\mu}S_1(\omega)$. We denote by $\mathcal{A}_P$ the space of connections on P and by $\mathcal{G}_P$ the group of gauge transformations. The Yang–Mills action S_μ defined on $\mathcal{A}_P$ is gauge invariant (i.e., invariant under the natural action of $\mathcal{G}_P$ on $\mathcal{A}_P$). We now define $\mathcal{A}(M)$ to be the disjoint union of the $\mathcal{A}_P$ over all equivalence classes of principal G-bundles over M. The Euclidean quantum field theory may be considered an assignment of the **quantum expectation** $\langle \Phi \rangle_\mu$ to each gauge invariant function $\Phi : \mathcal{A}(M) \to \mathbf{R}$. A gauge invariant function $\Phi : \mathcal{A}(M) \to \mathbf{R}$ is called an **observable** in quantum field theory. In the **Feynman path integral** approach to quantization, the quantum expectation $\langle \Phi \rangle_\mu$ of an observable is given by the following expression

$$\langle \Phi \rangle_\mu = \frac{\int_{\mathcal{A}(M)} e^{-S_\mu(\omega)}\Phi(\omega)\mathcal{D}\mathcal{A}}{\int_{\mathcal{A}(M)} e^{-S_\mu(\omega)}\mathcal{D}\mathcal{A}}, \qquad (7.7)$$

where $\mathcal{D}\mathcal{A}$ denotes a measure on $\mathcal{A}(M)$ whose precise definition is not known, in general. It is customary to express the quantum expectation $< \Phi >_\mu$ in terms of the **partition function** Z_μ defined by

$$Z_\mu(\Phi) := \int_{\mathcal{A}(M)} e^{-S_\mu(\omega)}\Phi(\omega)\mathcal{D}\mathcal{A}. \qquad (7.8)$$

Thus, we can write

$$\langle \Phi \rangle_\mu = \frac{Z_\mu(\Phi)}{Z_\mu(1)}. \qquad (7.9)$$

In the above equations we have written the quantum expectation as $\langle \Phi \rangle_\mu$ to indicate explicitly that, in fact, we have a one-parameter family of quantum expectations indexed by the coupling constant μ in the Yang–Mills action. In what follows we drop this subscript with the understanding that we do have a one-parameter family of quantum field theories when dealing with non-perturbative aspects of the theory. A mathematically precise definition of the Feynman path integral is not available at this time. Feynman (1918–1988) developed his diagrams and a set of rules to extract physically relevant information from the path integral in specific applications. He successfully applied these methods to his study of quantum electrodynamics. Feynman had a wide range of interests in and out of science. He received the Nobel Prize for physics for 1965, jointly with Schwinger and Tomonaga for their work in quantum electrodynamics.

There are several examples of gauge-invariant functions. For example, primary characteristic classes evaluated on suitable homology cycles give an important family of gauge invariant functions. The instanton number k of $P(M, G)$ belongs to this family, as it corresponds to the second Chern class evaluated on the fundamental cycle of M representing the fundamental class $[M]$. The pointwise norm $|F_\omega|_x$ of the gauge field at $x \in M$, the absolute value $|k|$ of the instanton number k, and the Yang–Mills action S_μ are also gauge-invariant functions. We now give two other important examples of gauge-invariant functions.

Example 7.1 (Instanton scale size λ) *In Chapter 9 we will show that an instanton solution of Yang–Mills equations on $\mathbf{R}^4$ (or on its compactification S^4) is characterized by three parameters, namely the* **instanton number** k, *the* **center** $x \in \mathbf{R}^4$, *and the* **scale size** λ. *We can extend the definition given there to the entire configuration space $\mathcal{A}(\mathbf{R}^4)$ as follows:*

$$\lambda(\omega) = \inf\{\rho(x)|x \in \mathbf{R}^4\}, \tag{7.10}$$

where

$$\rho(x) = \sup\left\{r \ \Big| \ \int_{S^3(x,r)} |F_\omega|^2 \le \tfrac{1}{2}S_\mu\right\} .$$

Thus, $\lambda(\omega)$ is the radius of the smallest sphere that contains half the Yang–Mills action. It is easy to see that the function $\lambda : \mathcal{A}(\mathbf{R}^4) \to \mathbf{R}$ is gauge-invariant. An expression for the semiclassical expectation of λ is given in the next section.

In the next example we introduce two families of gauge invariant functions that generalize the Wilson loop functional well-known in the physics literature.

Example 7.2 (Wilson loop functional) *Let ρ denote a representation of G on V. Let $\alpha \in \Omega(M, x_0)$ denote a loop at $x_0 \in M$. Let $\pi : P(M, G) \to M$ be the canonical projection and let $p \in \pi^{-1}(x_0)$. If ω is a connection on P then the parallel translation along α maps the fiber $\pi^{-1}(x_0)$ into itself. Let $\hat{\alpha}_\omega : \pi^{-1}(x_0) \to \pi^{-1}(x_0)$ denote this map. Since G acts transitively on the fibers, $\exists g_\omega \in G$ such that $\hat{\alpha}_\omega(p) = pg_\omega$. Now define*

$$\mathcal{W}_{\rho,\alpha}(\omega) := \mathrm{Tr}[\rho(g_\omega)]. \tag{7.11}$$

We note that g_ω and hence $\rho(g_\omega)$ change by conjugation if instead of p we choose another point in the fiber $\pi^{-1}(x_0)$, but the trace remains unchanged.

Alternatively, we can consider the vector bundle $P \times_\rho V$ associated to the principal bundle P and parallel displacement of its fibers induced by α. Let $\pi : P \times_\rho V \to M$ be the canonical projection. We note that in this case $\pi^{-1}(x_0) \cong V$. Now the map $\hat{\alpha}_\omega : \pi^{-1}(x_0) \to \pi^{-1}(x_0)$ is a linear transformation and we can define

$$\mathcal{W}_{\rho,\alpha}(\omega) := \mathrm{Tr}[\hat{\alpha}_\omega]. \tag{7.12}$$

We call these $\mathcal{W}_{\rho,\alpha}$ the Wilson loop functionals associated to the representation ρ and the loop α. In the particular case when $\rho = \mathrm{Ad}$, the adjoint representation of G on $\mathbf{g}$, our constructions reduce to those considered in physics.

We note that the gauge invariance of Φ makes the integral defining Z divergent, due to the infinite contribution coming from gauge-equivalent fields. To avoid this difficulty observe that the integrand is gauge-invariant and hence Z descends to the orbit space $\mathcal{O} = \mathcal{A}(M)/\mathcal{G}$ and can be evaluated by integrating over this orbit space $\mathcal{O}$. However, the mathematical structure of this space is essentially unknown at this time. Physicists have attempted to get around this difficulty by choosing a section $s : \mathcal{O} \rightarrow \mathcal{A}$ and integrating over its image $s(\mathcal{O})$ with a suitable weight factor such as the Faddeev–Popov determinant, which may be thought of as the Jacobian of the change of variables effected by $p_{|s(\mathcal{O})} : s(\mathcal{O}) \rightarrow \mathcal{O}$. As we saw in Chapter 6, this gauge fixing procedure does not work in general, due to the presence of the Gribov ambiguity. Also the Faddeev–Popov determinant is infinite-dimensional and needs to be regularized. This is usually done by introducing the anticommuting Grassmann variables called the **ghost** and **anti-ghost fields**. The Lagrangian in the action term is then replaced by a new Lagrangian containing these ghost and anti-ghost fields. This new Lagrangian is called the **effective Lagrangian**. The effective Lagrangian is not gauge-invariant, but it is invariant under a special group of transformations involving the ghost and anti-ghost fields. These transformations are called the **BRST** (Becchi–Rouet–Stora–Tyutin) transformations. On the infinitesimal level the BRST transformations correspond to cohomology operators and define what may be called the BRST cohomology. The non-zero elements of the BRST cohomology are called anomalies in the physics literature. At present there are several interesting proposals for studying these questions, proposals that make use of equivariant cohomology in the infinite-dimensional setting and which are closely related to the various interpretations of BRST cohomology (see, for example, [21, 193, 236, 404]). A detailed discussion of the material of this section from a physical point of view may be found in the books on quantum field theory referred in the introduction. A geometrical interpretation of some of these concepts may be found in [27, 86].

The general program of computing the curvature of connections on infinite dimensional bundles and of defining appropriate generalizations of characteristic classes was initiated by Isadore Singer in his fundamental paper [351] on Gribov ambiguity. Today this is an active area of research with strong links to quantum field theory (see, for example, [23, 61, 174, 233, 234, 352, 404]). We now give a brief description of some aspects of this program.

We proceed by analogy with the finite-dimensional case. To simplify considerations let us suppose that the group of gauge transformations $\mathcal{G}$ acts freely on the space of gauge connections $\mathcal{A}$. Then we can consider $\mathcal{A}$ to be a principal $\mathcal{G}$-bundle over the space $\mathcal{B} = \mathcal{A}/\mathcal{G}$ of gauge equivalence classes of connections. Define the map

$$F^- : \mathcal{A} \to \Lambda^2_-(M, \mathrm{ad}(P)) \quad \text{by} \quad \omega \mapsto F^-_\omega. \tag{7.13}$$

Then F^- is equivariant with respect to the standard action of $\mathcal{G}$ on $\mathcal{A}$ and the linear action of $\mathcal{G}$ on the vector space $\Lambda^2_-(M, \mathrm{ad}(P))$. Hence, we may think of F^- as defining a section of the vector bundle E associated to $\mathcal{A}$ with fiber $\Lambda^2_-(M, \mathrm{ad}(P))$. Thus, F^- is the infinite-dimensional analogue of a vector field and we may hope to obtain topological information about the base space $\mathcal{B}$ by studying the set of zeros of this vector field. In the generic finite-dimensional case, one can define an index of a vector field by studying its zero set and thereby obtain a topological invariant, namely the Euler characteristic. In our case, the zeros of F^- in $\mathcal{A}$ are precisely the Yang–Mills instantons and the zero set of the associated section s_{F^-} (see Theorem 6.2) is the moduli space $\mathcal{M}^+ \subset \mathcal{B}$. Thus

$$\mathcal{M}^+ = (F^-)^{-1}(\{0\})/\mathcal{G}. \tag{7.14}$$

In this general setting, there is no obvious way to define the index of s_{F^-} and to consider its relation with the analytic definition of the Euler characteristic, which should correspond to an integral of some "curvature form" analogous to the Chern–Gauss–Bonnet integrand.

In the case of Abelian gauge fields we have the following argument (see [280] for details). Once again we consider the finite-dimensional case. Here the Thom class, which lies in the cohomology of the vector bundle over M with compact support, is used to obtain the desired relation between the index and the Euler characteristic as follows. We assume that there exists a section s of the vector bundle E. Using s, we pull back the Thom class α to a cohomology class on the base M. In fact, we have a family of homologous sections $\lambda s, \lambda \in \mathbf{R}$, and the corresponding family of pull-backs $(\lambda s)^* \alpha$, which, for $\lambda = 0$, gives the Euler class of M and, for large λ ($\lambda \to \infty$) gives, in view of the compact support of α, the index of s. Thus, one would like to obtain a suitable generalization of the Thom class in the infinite-dimensional case. We shall refer to this idea—of interpolating between two different definitions of the Euler characteristic—as the **Mathai–Quillen formalism**. We consider the case of quantum electrodynamics where the gauge group G is $U(1)$. In this case the moduli space of instantons is 0-dimensional and the computation of the index can be carried out as in the finite-dimensional case. The index turns out to be related to an important new invariant due to Donaldson (see [109]). Even in this case there does not seem to be any version of the Thom class with compact support. Instead, we have an equivariant version with Gaussian asymptotics, which may be regularized. Locally, the action of $U(1)$ on $\mathbf{R}^2$ is given by the vector field

$$X = -x_1 \frac{\partial}{\partial x_2} + x_2 \frac{\partial}{\partial x_1}.$$

If ω is a gauge connection form on $P(M, U(1))$, then a representative of an equivariant Thom class α is given by

$$\alpha = \tfrac{1}{\pi}e^{-|x|^2}(u + \tfrac{1}{2}d\omega + (dx_1 + \omega x_2) \wedge (dx_2 - \omega x_1)), \qquad (7.15)$$

where u is a degree-2 indeterminate, which enters in the formulation of equivariant cohomology. The precise relation of equation (7.15) to the physical computation of invariants in quantum electrodynamics or its role in topological quantum field theory are yet to be understood.

On the principal $\mathcal{G}$-bundle $\mathcal{A} = P(\mathcal{B}, \mathcal{G})$, we can define a natural connection as follows. For $\omega \in \mathcal{A}$, the tangent space $T_\omega \mathcal{A}$ is identified with $\Lambda^1(M, \mathrm{ad}(P))$ and hence carries the inner product as defined in Definition 6.1 with $p = 1$. With respect to this inner product we have the orthogonal splitting

$$T_\omega \mathcal{A} = V_\omega \oplus H_\omega, \qquad (7.16)$$

where the vertical space V_ω is identified as the tangent space to the fiber $\mathcal{G}$ or, alternatively, as the image of the covariant differential $d^\omega : \Lambda^0(M, \mathrm{ad}(P)) \to \Lambda^1(M, \mathrm{ad}(P))$. The horizontal space H_ω can be identified with $\ker \delta^\omega$, where δ^ω is the formal adjoint of d^ω. The horizontal distribution is equivariant with respect to the action of $\mathcal{G}$ on $\mathcal{A}$ and defines a connection on $\mathcal{A}$. If $\mathcal{LG}(= \Lambda^0(M, \mathrm{ad}(P)))$ denotes the Lie algebra of the infinite-dimensional Hilbert Lie group $\mathcal{G}$ (see Section 6.3), the connection form $\hat{\omega} : T\mathcal{A} \to \mathcal{LG}$ of this connection is given by

$$\hat{\omega}(X) = G^\alpha(\delta^\alpha X), \quad X \in T_\alpha \mathcal{A}, \qquad (7.17)$$

where G^α is the **Green operator**, which inverts the Laplacian

$$\Delta^0_\alpha = \delta^\alpha d^\alpha : \Lambda^0(M, \mathrm{ad}(P)) \to \Lambda^0(M, \mathrm{ad}(P)).$$

The curvature $\Omega_{\hat{\omega}}$ of this natural connection form $\hat{\omega}$ is the 2-form with values in $\Lambda^0(M, \mathrm{ad}(P))$ given by the usual formula

$$\Omega_{\hat{\omega}} = d^{\hat{\omega}}\hat{\omega}$$

and can be expressed locally, in terms of the Green operator by

$$\Omega_{\hat{\omega}}(X, Y) = -2G^\alpha(g^{ij}[X_i, Y_j]), \qquad (7.18)$$

where locally $X = X_i dx^i$, $Y = Y_j dx^j$, and $g^{ij} = g(dx^i, dx^j)$. We remark that $\Omega_{\hat{\omega}}$ should be considered as defined on a local slice at $[\alpha] \in \mathcal{B}$ transversal to the vertical gauge orbit through α. We now explain the relation of the natural connection $\hat{\omega}$ defined above to a connection defined in [23]. Regarding $\mathcal{G}$ as a subgroup of $\mathrm{Aut}(P)$, we have a natural action of $\mathcal{G}$ on $P(M, G)$ and hence on $P(M, G) \times \mathcal{A}$. The quotient of this action can be regarded as a principal bundle $\mathcal{P}$ over $M \times \mathcal{B}$ called the **Poincaré bundle**. Pulling back the natural connection $\hat{\omega}$ to the Poincaré bundle $\mathcal{P}$, we get a connection that coincides with the one defined in [23]. The Poincaré bundle with this connection can be regarded as a universal bundle with respect to deformations of the bundle $P(M, G)$.

In addition to the gauge connection on the bundle $P(M, G)$ and associated differential operators, there are other bundles and operators which enter naturally into many mathematical and physical applications. For example, on M there is the frame bundle $L(M)$ and the Levi-Civita connection on iti, which are fundamental in gravitation. Another important example is that of the Dirac operator defined on spinor bundles over M, when M is a spin manifold. More generally, we can consider an elliptic operator or a sequence of elliptic operators on sections of real or complex vector bundles over M. For definiteness, we consider a fixed elliptic operator $\mathcal{D} : \Gamma(V) \to \Gamma(V)$, where V is a vector bundle over M. If E is a vector bundle associated to $P(M, G)$, then each connection $\omega \in \mathcal{A}$ induces a connection and corresponding covariant derivative operator on sections of E. By coupling the elliptic operator $\mathcal{D}$ with this gauge connection ω, we obtain an elliptic operator $\mathcal{D}_\omega : \Gamma(V \otimes E) \to \Gamma(V \otimes E)$. Then $\omega \mapsto \mathcal{D}_\omega$ defines a family $\hat{\mathcal{D}}$ of elliptic operators indexed by $\omega \in \mathcal{A}$. From ellipticity of $\mathcal{D}_\omega$ it follows that $\ker \mathcal{D}_\omega$ and $\operatorname{coker} \mathcal{D}_\omega$ are finite dimensional. By the Atiyah–Singer index theorem, it follows that the numerical index

$$n(\omega) = \dim \ker \mathcal{D}_\omega - \dim \operatorname{coker} \mathcal{D}_\omega \tag{7.19}$$

is independent of ω. The family $\hat{\mathcal{D}}$ can be used to define two bundles over $\mathcal{A}$, whose fibers over $\omega \in \mathcal{A}$ are $\ker \mathcal{D}_\omega$ and $\operatorname{coker} \mathcal{D}_\omega$, and a **virtual bundle** (in the sense of K-theory), whose fiber over ω is $\ker \mathcal{D}_\omega - \operatorname{coker} \mathcal{D}_\omega$. This virtual bundle is called the **index bundle** $\operatorname{Ind} \hat{\mathcal{D}}$ of $\hat{\mathcal{D}}$. In fact, since $\mathcal{G}$ acts equivariantly on the operators $\mathcal{D}_\omega$, the index bundle descends to the quotient and can be regarded as an element of the Grothendieck group $K(\mathcal{B})$. In particular, its restriction to the finite-dimensional instanton moduli space $\mathcal{M} \subset \mathcal{B}$ defines an element of $K(\mathcal{M})$. In this general set up, we cannot apply the index theorem for families to $\mathcal{B}$. However, in the special case when $\mathcal{D}$ is the Dirac operator, we have the following formula

$$ch(\operatorname{Ind} \hat{\mathcal{D}}) = \int_M \hat{a}(M) \, ch(\mathcal{E}), \tag{7.20}$$

where $\hat{a}(M)$ is the characteristic class of the spinor bundle whose value on the fundamental cycle of M is the $\hat{A}$-genus and $\mathcal{E}$ is a certain vector bundle associated to the Poincaré bundle $\mathcal{P}$ (see [23] for further details). The natural connection $\hat{\omega}$ on the principal bundle $\mathcal{A} = P(\mathcal{B}, \mathcal{G})$ restricts to define a connection on the index bundle in the case that it corresponds to a real vector bundle. For example, this happens if $\operatorname{coker} \mathcal{D}_\omega = 0$, $\forall \omega \in \mathcal{A}$ (see [204]). In this case there is also a well-defined second fundamental form and one can apply the Gauss–Codazzi equation to this situation to derive information on the geometry of $\mathcal{B}$.

Several of the topics mentioned above are now considered to be parts of topological field theory, which may be viewed as the study of topological structures related to infinite-dimensional manifolds and infinite-dimensional bundles over them. The mathematical foundations of this theory are not yet precisely formulated. In the next section, we discuss one of the most widely used perturbative techniques, namely, the semiclassical approximation to the partition functions for Yang–Mills theories.

7.3 Semiclassical Approximation

It is well-known that an asymptotic description of physical field theories has become an indispensable tool in deciding the validity and usefulness of these theories, which is done by our comparing their predictions with experimental measurements. In fact, the actual calculations carried out in quantum electrodynamics as well as in other areas of physics have been verified to a high degree of accuracy, even though fundamental theoretical justification for them is not always evident. These calculations involve various methods of perturbation theory. For Hamiltonian systems a geometric description of perturbation theory suitable for computer simulation and symbol manipulation techniques is given in Omohundro [307]. The mathematical formulation suitable for quantum field theories is currently under development.

A widely used method of approximation in quantum field theories is the so-called **semiclassical**, or the **one loop, approximation to** the Feynman path integrals. We give a brief account of the semiclassical approximation for Yang–Mills theory based on [174]. It is well-known that the partition functions of quantum Yang–Mills theory have an expansion in powers of the coupling constant. The leading order term in this expansion is called the semiclassical approximation. As we observed in the previous section we have a one parameter family of quantum expectations indexed by the coupling constant μ in the Yang–Mills action. As $\mu \to 0$ the integrals defining the quantum expectation $< \Phi >$ of the observable Φ have an asymptotic expansion in powers of μ. The leading order term in the resulting expansion of $[< \Phi >]$ is called the **semiclassical expectation** of Φ and is denoted by $< \Phi >_{sc}$. The calculation of $[< \Phi >_{sc}]$ requires a detailed study of some elliptic operators. Two important techniques that are required for this study are (i) the zeta function regularization and (ii) study of the associated heat family. These techniques are also useful in other applications. We therefore, give a slightly more general discussion of these techniques than is necessary for the statement of results.

7.3.1 Zeta Function Regularization

Let Q be a symmetric, positive operator on a finite-dimensional real Hilbert space H. We identify H with $\mathbf{R}^n$ and define the positive definite quadratic form $\hat{Q}$ on $\mathbf{R}^n$ associated to Q by $\hat{Q}(x) = <Q(x), x>$, $\forall x \in \mathbf{R}^n$. Then the **Gaussian integral** associated to Q is given by

$$\int_{\mathbf{R}^n} \exp(-\pi\hat{Q}(x))\,dx = (\det Q)^{-1/2}, \qquad (7.21)$$

where $Q(x) = \sum q_{ij}x^i x^j$, $\det Q$ is the determinant of the symmetric matrix (q_{ij}), and dx is the standard volume form on $\mathbf{R}^n$. We are interested in generalizing this Gaussian integral to the case when Q is a symmetric, positive operator on an infinite-dimensional Hilbert space. We could define the integral by the same formula as formula (7.21) provided $\det Q$ is well-defined. We now consider the method of zeta function regularization, which allows us to define $\det Q$ for a large class of operators. This class includes the families of Laplace operators which enter in the computation of semiclassical approximations in Yang–Mills theory. Recall first that in the finite-dimensional case we can write

$$\det Q = \prod_{i=1}^n \lambda_i = \exp\sum_{i=1}^n \ln(\lambda_i), \qquad (7.22)$$

where λ_i, $1 \le i \le n$, is the complete set of eigenvalues of Q. Now let Q be an operator on an arbitrary separable Hilbert space such that Q has a discrete set of positive eigenvalues λ_i, $i \in \mathbf{N}$. We define the **generalized zeta function** ζ_Q by

$$\zeta_Q(z) := \sum_{i=1}^{\infty} (\lambda_i)^{-z}. \qquad (7.23)$$

It can be shown (see [345]) that under certain conditions on Q, the sum in formula (7.23) converges for $\mathrm{Re}(z) >> 0$ (i.e., for sufficiently large $\mathrm{Re}(z)$) and has a meromorphic continuation to the entire complex z-plane with only simple poles. Furthermore, zero is a regular point of $\zeta_Q(z)$ and hence $\zeta_Q'(0)$ is well-defined. From equation (7.23) it follows that

$$\zeta_Q'(z) := -\sum_{i=1}^{\infty} \ln(\lambda_i)(\lambda_i)^{-z}. \qquad (7.24)$$

Thus formally one can write

$$\zeta_Q'(0) := -\sum_{i=1}^{\infty} \ln(\lambda_i). \qquad (7.25)$$

In view of this equation we define $\det Q$ by

$$\det Q := \exp(-\zeta_Q'(0)). \qquad (7.26)$$

We note that equation (7.26) reduces to the usual definition of determinant in the finite-dimensional case.

Example 7.3 *Consider a real scalar field ϕ on a compact Riemannian manifold (M, g) with Lagrangian density*

$$\mathcal{L} = \tfrac{1}{2}(g^{ij}\partial_i\phi\partial_j\phi - m^2\phi^2) - V(\phi), \qquad (7.27)$$

where m denotes the mass and V the potential. Then the vacuum to vacuum amplitude Z in the absence of sources is given by

$$Z = \int_{\mathcal{H}} \exp\left(-\tfrac{1}{2}\int_M <\phi, A\phi> dv_g\right) d\mu(\phi), \qquad (7.28)$$

where μ is a measure on the space $\mathcal{H}$ of all scalar fields, dv_g is the volume form on M determined by the metric g, and A is an operator on $\mathcal{H}$ defined by

$$A := -g^{ij}\partial_i\partial_j + m^2 + \frac{\partial V}{\partial \phi}. \qquad (7.29)$$

Under certain conditions $\mathcal{H}$ is a separable Hilbert space, A is a positive, symmetric operator with a discrete set of eigenvalues λ_i, and $\mathcal{H}$ has an orthonormal basis ψ_i consisting of eigenvectors of A. Then the zeta function regularization of the $\det A$ allows us to evaluate the integrals defining Z to obtain

$$Z = \exp\left(\tfrac{1}{2}\zeta_A'(0)\right). \qquad (7.30)$$

The discussion given above is implicit in most evaluations of such amplitudes.

We note that the zeta function regularization can be applied also in the fermionic case. This requires an extension of usual calculus to Grassmann or anti-commuting variables. An introduction to this subject may be found in Berezin [38].

7.3.2 Heat Kernel Regularization

Let E be a vector bundle over a compact Riemannian manifold M of dimension $m = 2n$. Let $D : \Gamma(E) \to \Gamma(E)$ be a self-adjoint, non-negative elliptic operator on the space $\Gamma(E)$ of the sections of E over M. From the standard theory of elliptic operators we know that D extends to the Hilbert space completion $L^2(E)$ of $\Gamma(E)$ as a self-adjoint, non-negative elliptic operator and that $L^2(E)$ has a complete orthonormal basis of eigenvectors $\{\phi_i \mid D\phi_i = \lambda_i\phi_i\}$. Furthermore, the spectrum $\{\lambda_i\}$ is discrete and non-negative and each eigenspace is finite-dimensional. In physics literature ker D with basis $\{\phi_i \mid D\phi_i = 0\}$ is called the space of **zero modes**. One often factors out the zero modes to obtain an operator D^+ with positive spectrum. In what follows we shall consider that this has been done and denote D^+ just by D.

The **heat family** associated with D is the unique family of operators H_t on $L^2(E)$ satisfying the heat equation

$$\frac{\partial}{\partial t}(H_t) = -DH_t, \tag{7.31}$$

with $H_0 = id$. It is customary to write formally $H_t = e^{-tD}$. We note that for $z \in \mathbf{C}$, $\mathrm{Re}(z) >> 0$, the operator D^{-z} is well-defined and is of trace class. This complex power of the operator D, its zeta function ζ_D, and the heat family are related by

$$\zeta_D(z) = \mathrm{Tr}(D^{-z}) = \frac{1}{\Gamma(z)} \int_0^\infty t^{z-1} \mathrm{Tr}(e^{-tD})dt, \tag{7.32}$$

where $\Gamma(z)$ is the usual gamma function. The operators H_t are themselves of trace class and are given by convolution with the heat kernel K_D. Heat equation was used by Patodi in his study of the index theorem. It can also be used to study harmonic forms and the Hodge decomposition theorem, (see, for example, [214]). It is the detailed study of the heat kernel that enters into the computation of the semiclassical measure when applied to suitable Laplace operators on $\Lambda^k(M, \mathrm{ad}(P))$.

The calculations of the semiclassical expectation are quite involved. We state the result for the expectations of $|k|$ and λ in the case that $M = S^4$ and $G = SU(2)$ and refer to [174] for further details.

Theorem 7.1 *Let $M = S^4$ and $G = SU(2)$. Let $\mathcal{M}_1$ denote the moduli space of $k = 1$ instantons over M. Then for any smooth, bounded, gauge-invariant function Φ, the* **semiclassical partition function**

$$Z_{sc}(\Phi) = \int_{\mathcal{M}_1} \Phi(\omega)e^{-S_\mu(\omega)}\, d\nu \tag{7.33}$$

is finite. Moreover, one has an explicit formula for the semiclassical measure $d\nu$ on $\mathcal{M}_1(S^4)$. In particular, the semiclassical expectations of $|k|$ and λ, where k is the instanton number and λ the instanton scale size, are given by

$$\langle |k| \rangle_{sc} \approx 0.05958 C\mu^2 e^{-8\pi^2/\mu} \tag{7.34}$$

and

$$\langle \lambda \rangle_{sc} \approx 0.03693 C\mu^2 e^{-8\pi^2/\mu}, \tag{7.35}$$

where $C = 2^{18}3^{-5/2}\pi^{23/2}$.

We note that, even though both expectations decrease exponentially as $\mu \to 0$, their ratio is non-zero and we have

$$\frac{\langle|k|\rangle_{sc}}{\langle\lambda\rangle_{sc}} \approx 0.6198.$$

On the other hand the ratio of classical expectations is given by

$$\frac{\langle|k|\rangle_c}{\langle\lambda\rangle_c} \approx 0.3112.$$

These ratios can be interpreted as conditional expectations: the expected scale size of a gauge potential, given that it has instanton number ± 1. The difference in the semiclassical and the classical values is due to the difference between the semiclassical measure and the classical measure used in computing these expectations. It remains to be seen how these results can be extended to non-compact Lorentz manifolds to compute quantum field theoretic expectation values of physically significant quantum variables.

Several of the topics mentioned above are now considered to be parts of topological field theories. The earliest example of a TFT result occurs in the well-known formula for the linking number by Gauss. In the next section we discuss this as an example of TCFT.

7.4 Topological Classical Field Theories (TCFTs)

We discuss the invariants of knots and links in 3-manifolds in Chapter 11. Here we consider one of the earliest investigations in combinatorial knot theory, contained in several unpublished notes written by Gauss between 1825 and 1844 and published posthumously as part of his *Nachlass*[1]. They deal mostly with his attempts to classify "Tractfiguren," or closed curves in the plane with a finite number of transverse self-intersections. As we shall see in Chapter 11, such figures arise as regular plane projections of knots in $\mathbf{R}^3$. However, one fragment of Gauss's notes deals with a pair of linked knots:

Es seien die Koordinaten eines unbestimmten Punkts der ersten Linie x, y, z; der zweiten x', y', z' und[2]

$$\int \int [(x'-x)^2 + (y'-y)^2 + (z'-z)^2]^{-3/2} [(x'-x)(dydz' - dzdy')$$
$$+ (y'-y)(dzdx' - dxdz') + (z'-z)(dxdy' - dydx')] = V$$

---

[1] Estate.

[2] Let the coordinates of an arbitrary point on the first curve be x, y, z; of the second x', y', z' and let

dann ist dies Integral durch beide Linien ausgedehnt

$$= 4\pi m$$

und m die Anzahl der Umschlingungen.

Der Werth ist gegenseitig, d.h. er bleibt derselbe, wenn beide Linien gegen einander umgetauscht werden,[3]   *1833. Jan. 22.*

In this fragment, Gauss had given an analytic formula for the linking number of a pair of knots. This number is a combinatorial topological invariant. As is quite common in Gauss's work, there is no indication of how he obtained this formula. The title of the note "Zur Elektrodynamik" ("On Electrodynamics") and his continuing work with Weber on the properties of electric and magnetic fields lead us to guess that it originated in the study of the magnetic field generated by an electric current flowing in a curved wire. Recall that the magnetic field due to a unit current flowing along a wire C generates the magnetic field $B(r')$ at a point $r' \in C'$ given by

$$B(r') = \frac{1}{4\pi} \int_C \frac{(r' - r) \times dr}{|r' - r|^3} \ ,$$

where we have used the vector notation $r = (x, y, z)$ and $r' = (x', y', z')$ and $\times$ is the vector product. The work W done by this magnetic field in moving a unit magnetic pole around the wire C' is given by

$$W = \int_{C'} B(r') \cdot dr = V/4\pi.$$

Maxwell knew Gauss's formula for the linking number and its topological significance and its origin in electromagnetic theory. In fact, in commenting on this formula, he wrote:

It was the discovery by Gauss of this very integral expressing the work done on a magnetic pole while describing a closed curve in the presence of a closed electric current and indicating the geometric connection between the two closed curves, that led him to lament the small progress made in the Geometry of Position since the time of Leibnitz, Euler and Vandermonde. We now have some progress to report, chiefly due to Riemann, Helmholtz and Listing.

In obtaining a topological invariant by using a physical field theory, Gauss had anticipated topological field theory by almost 150 years. Even the term "topology" was not used then. It was introduced in 1847 by J. B. Listing, a

---

[3] then this integral taken along both curves is $= 4\pi m$ and m is the number of intertwinings (linking number in modern terminology). The value (of the integral) is common (to the two curves), i.e., it remains the same if the curves are interchanged,

student and protegé of Gauss, in his essay "Vorstudien zur Topologie" ("Preliminary Studies on Topology"). Gauss's linking number formula can also be interpreted as the equality of topological and analytic degree of a suitable function. Starting with this a far-reaching generalization of the Gauss integral to higher self-linking integrals can be obtained. We discuss this work in Chapter 11. This result forms a small part of the program initiated by Kontsevich [235] to relate topology of low-dimensional manifolds, homotopical algebras, and non-commutative geometry with topological field theories and Feynman diagrams in physics.

7.4.1 Donaldson Invariants

Electromagnetic theory is the prototype of gauge theories with Abelian gauge group $U(1)$. Its generalization to non-Abelian gauge group $SU(2)$ was obtained by Yang and Mills in 1954. A spectacular application of Yang–Mills theory as TCFT came thirty years later in Donaldson theory. We discuss this theory in detail in Chapter 9. Here we simply indicate its interpretation as a topological field theory based on the classical instanton solutions of Yang–Mills equations. Donaldson's theorem on the topology of smooth, closed, 1-connected 4-manifolds provides a new obstruction to smoothability of these topological manifolds. A surprising ingredient in his proof of this theorem was the moduli space $\mathcal{I}_1$ of $SU(2)$-instantons on a manifold M. This theorem has been applied to obtain a number of new results in topology and geometry and has been extended to other manifolds. The space $\mathcal{I}_1$ is a subspace of the moduli space $\mathcal{M}_1$ of Yang–Mills fields with instanton number 1. The space $\mathcal{M}_1$ in turn is a subspace of the moduli space of $\mathcal{A}/\mathcal{G}$ of all Yang–Mills fields on M. In fact, we have

$$\mathcal{A}/\mathcal{G} = \bigcup_k \mathcal{M}_k.$$

Donaldson shows that the space $\mathcal{I}_1$ (or a suitable perturbation of it) is a 5-dimensional manifold with singularities and with one boundary component homeomorphic to the original base space M. By careful study of the remaining boundary components Donaldson obtained the following theorem.

Theorem 7.2 (Donaldson) *Let M be a smooth, closed, 1-connected, oriented manifold of dimension 4 with positive definite intersection form ι_M. Then $\iota_M \cong b_2(1)$, the identity form of rank b_2, the second Betti number of M.*

This theorem is the genesis of what can be called **gauge-theoretic topology**. In his later work, Donaldson used the homology of spaces $\mathcal{M}_k$, for sufficiently large k, to obtain a family of new invariants of a smooth 4-manifold M, satisfying a certain condition on its intersection form. We now describe these invariants, which are known as **Donaldson's polynomial invariants**, or simply **Donaldson polynomials**. The Donaldson polynomials are defined by polarization of a family q_k of symmetric, multilinear maps

$$q_k : \underbrace{H_2(M) \times \cdots \times H_2(M)}_{d \text{ times}} \to \mathbf{Q},$$

where k is the instanton number of $P(M, SU(2))$ and d is a certain function of k. We shall also refer to the maps q_k as Donaldson polynomials. A basic tool for the construction of Donaldson polynomials is a map that transfers the homology of M to the cohomology of the orbit space $\mathcal{O}_{ir}$ of irreducible connections on the $SU(2)$-bundle P. The details of this construction are given in Chapter 9. Donaldson polynomials are examples of TCFT invariants. It is interesting to note that in [404] Witten has given a TQFT interpretation for them. We comment on this in the next section, where an introduction to TQFT is given. Before that we give two more examples of TCFT.

7.4.2 Topological Gravity

Einstein's theory of gravity is the most extensively studied and experimentally supported theory of gravity at this time. However, Einstein was not content with it. We have already considered his introduction of the cosmological constant and indicated a generalized theory of gravity, which replaces the cosmological constant by the cosmological function. Over the years many alternative theories of gravitation have been proposed. Several of these start with a variational principle with some geometric function as the Lagrangian. Lanczos [242] in his study of Lagrangians for generalized gravitational field equations observed that one of his Lagrangians led to the integral that was invariant under the action of the group of diffeomorphisms of the space-time manifold and so did not contain any dynamics. Lanczos had obtained the expression for the Euler class of a 4-manifold as an integral of a polynomial in the Riemann curvature, but he did not recognize its topological significance. The general formula for the Euler class of an arbitrary oriented manifold was obtained by Chern, generalizing the earlier work for hypersurfaces by Weil and Allendoerfer. It was in studying this generalization that Chern was led to his famous characteristic classes, now called the Chern classes. All Chern classes and not just the Euler class are topological invariants and can be expressed as integrals of polynomials in the Riemann curvature. We can, therefore, consider them topological gravity invariants if we think of these polynomials as Lagrangians of some generalized gravitational field in arbitrary dimension.

7.4.3 Chern–Simons (CS) Theory

Chern–Simons theory is a classical gauge field theory formulated on an odd-dimensional manifold. We discuss it in detail in Chapter 10 and apply it to consider Witten's QFT interpretation of the Jones polynomial. Chern–Simons theory became widely known after Witten's paper was published. It gives new invariants as TCFT also. We discuss this after reviewing the 3-dimensional CS theory. Let M be a compact, connected, oriented 3- manifold and let $P(M)$ be a principal bundle over M with structure group $SU(n)$. Then the Chern–Simons action $\mathcal{A}_{CS}$ is defined by

$$\mathcal{A}_{CS} = \frac{k}{4\pi} \int_M \text{tr}(A \wedge F + \frac{2}{3}A \wedge A \wedge A), \qquad (7.36)$$

where $k \in \mathbf{R}$ is a coupling constant, A denotes the pull-back to M of the gauge potential ω by a section of P, and $F = F_\omega = d^\omega A$ is the gauge field on M corresponding to the gauge potential A. We note that the bundle P always admits a section over a 3-manifold. The action is manifestly covariant since the integral involved in its definition is independent of the metric on M. It is in this sense that the Chern–Simons theory is a topological field theory. The field equations obtained by variation of the action turn out to be flat connections. A detailed calculation showing this is given in Chapter 10. Under a gauge transformation g the action transforms as follows:

$$\mathcal{A}_{CS}(A^g) = \mathcal{A}_{CS}(A) + 2\pi k \mathcal{A}_{WZ}, \qquad (7.37)$$

where

$$\mathcal{A}_{WZ} := \frac{1}{24\pi^2} \int_M \epsilon^{\alpha\beta\gamma} \text{tr}(\theta_\alpha \theta_\beta \theta_\gamma) \qquad (7.38)$$

is the Wess–Zumino action functional. It can be shown that the Wess–Zumino functional is integer-valued and hence if the Chern–Simons coupling constant k is taken to be an integer, then we have

$$e^{i\mathcal{A}_{CS}(A^g)} = e^{i\mathcal{A}_{CS}(A)}.$$

It follows that the path integral quantization of the Chern–Simons model is gauge-invariant. This conclusion holds more generally for any compact simple group. In the next subsection we discuss briefly the development of topological QFT. Chern–Simons theory played a fundamental role in this development. It also gives a new way of looking at the Casson invariant. This is an example of TCFT. It rests on the observation that the moduli space $\mathcal{M}_f(N, H)$ of flat H-bundles over a manifold N can be identified with the set $\text{hom}(\pi_1(N), H)/H$. The moduli space $\mathcal{M}_f(N, H)$ and the set $\text{hom}(\pi_1(N), H)$ have a rich mathematical structure, which has been extensively studied. These spaces appear in the definition of the Casson invariant for a homology 3-sphere and its gauge theoretic interpretation was given by

Taubes [366]. Recall that the Casson invariant of an oriented homology 3-sphere Y is defined in terms of the number of irreducible representations of $\pi_1(Y)$ into $SU(2)$. As indicated above, this space can be identified with the moduli space $\mathcal{M}_f(Y, SU(2))$ of flat connections in the trivial $SU(2)$-bundle over Y. The map $F : \omega \mapsto *F_\omega$ defines a natural 1-form and its dual vector field V_{CS} on $\mathcal{A}/\mathcal{G}$. Thus, the zeros of this vector field V_{CS} are just the flat connections. We note that since $\mathcal{A}/\mathcal{G}$ is infinite-dimensional, it is necessary to use suitable Fredholm perturbations to get simple zeros and to count the index of the vector field with appropriate signs. Taubes showed that this index equals the Casson invariant of Y. In classical geometric topology of a compact manifold M, the Poincaré–Hopf theorem tells us that the index of a vector field equals the Euler characteristic of M. This theorem does not apply to the infinite-dimensional case considered here. Is there some homology theory associated to the above situation whose Euler characteristic would be equal to the Casson invariant? The surprising affirmative answer to this question is provided by Floer homology. We discuss it briefly in the next section and in detail in Chapter 10.

7.5 Topological Quantum Field Theories (TQFTs)

In recent developments in low-dimensional topology of manifolds, geometric analysis of partial differential equations plays an important role. These equations have their origins in physical field theories. A striking example of this is provided by Donaldson's work on smoothability of 4-manifolds, where the moduli space of instantons is used to obtain a cobordism between the base manifold M and a certain manifold N whose topology is well understood, thereby enabling one to study the topology of M. In physics the moduli spaces of these classical, non-Abelian gauge fields are used to study the problem of quantization of fields. Even though a precise mathematical formulation of QFT is not yet available, the methods of QFT have been applied successfully by Witten for studying the topology and geometry of low-dimensional manifolds. It seems reasonable to say that his work has played a fundamental role in the creation of topological QFT (see [41] and references therein for a review of this fast developing field).

Let Σ_g be a compact Riemann surface of genus g. Then the moduli space $\mathcal{M}_f(\Sigma_g, H)$ of flat connections has a canonical symplectic structure ι. We now discuss an interesting physical interpretation of the symplectic manifold $(\mathcal{M}_f(\Sigma_g, H), \iota)$. Consider a Chern–Simons theory on the principal bundle $P(M, H)$ over the $(2 + 1)$-dimensional space-time manifold $(M = \Sigma_g \times \mathbf{R})$ with gauge group H and with time-independent gauge potentials and gauge transformations. Let $\mathcal{A}$ (resp., $\mathcal{H}$) denote the space (resp., group) of these gauge connections (resp., transformations). It can be shown that the curvature F_ω defines an $\mathcal{H}$-equivariant moment map

$$\mu : \mathcal{A} \to \mathcal{LH} \cong \Lambda^1(M, \mathrm{ad}\, P), \qquad \text{by } \omega \mapsto *F_\omega,$$

where $\mathcal{LH}$ is the Lie algebra of $\mathcal{H}$. The zero set $\mu^{-1}(0)$ of this map is precisely the set of flat connections and hence

$$\mathcal{M}_f \cong \mu^{-1}(0)/\mathcal{H}$$

is the reduced phase space of the theory, in the sense of the Marsden–Weinstein reduction. We denote this reduced phase space by $\mathcal{A}//\mathcal{H}$ and call it the **symplectic quotient** of $\mathcal{A}$ by $\mathcal{H}$. Marsden–Weinstein reduction and symplectic quotient are fundamental constructions in geometrical mechanics and geometric quantization. They also arise in many other mathematical and physical applications.

A situation similar to that described above also arises in the geometric formulation of canonical quantization of field theories. One proceeds by analogy with the geometric quantization of finite-dimensional systems. For example, $Q = \mathcal{A}/\mathcal{H}$ can be taken as the configuration space and T^*Q as the corresponding phase space. The associated Hilbert space is obtained as the space of L^2 sections of a complex line bundle over Q. For physical reasons this bundle is taken to be flat. Inequivalent flat $U(1)$-bundles are said to correspond to distinct sectors of the theory. Thus we see that at least formally these sectors are parametrized by the moduli space

$$\mathcal{M}_f(Q, U(1)) \cong \hom(\pi_1(Q), U(1))/U(1) \cong \hom(\pi_1(Q), U(1)),$$

since $U(1)$ acts trivially on $\hom(\pi_1(Q), U(1))$.

Now let Y be a homology 3-sphere. By considering a one-parameter family $\{\omega_t\}_{t\in I}$ of connections of Y defines a connection on $Y \times I$ and the corresponding Chern–Simons action $\mathcal{A}_{CS}$. This is invariant under the connected component of the identity in $\mathcal{G}$ but changes by the Wess–Zumino action under the full group $\mathcal{G}$. In [403], Witten showed how the standard Morse theory (see Morse and Cairns [292] and Milnor [284]) can be modified by consideration of the gradient flow of the Morse function f between pairs of critical points of f. One may think of this as a sort of relative Morse theory. Witten was motivated by the phenomenon of the quantum mechanical tunneling effect between the states represented by the critical points. The resulting **Witten complex** (C_*, δ) can be used to define the **Floer homology groups** $HF_n(Y)$, $n \in \mathbf{Z}_8$ (see [130, 131]). This is an example of TQFT. The Euler characteristic of Floer homology equals the Casson invariant. Thus TQFT methods give a resolution or **categorification** of the Casson invariant in terms of a new homology theory. Floer homology is discussed in detail in Chapter 11.

Consideration of the Heegaard splitting of Y leads to interesting connections with conformal field theory. On the other hand given a 4-manifold M we can always decompose it along a homology 3-sphere Y into a pair of 4-manifolds M^+ and M^-. Then one can study instantons on M by studying a pair of instantons on M^+ and M^- and matching them along the boundary Y. This procedure can be used to relate the Floer homology to the polynomial invariants of M defined by Donaldson.

A geometrical interpretation of the Jones polynomial invariant of a link was provided by Witten [405], who applied ideas from QFT to the Chern–Simons Lagrangian. In fact, Witten's model allows us to consider the knot and link invariants in any compact 3-manifold M. Let $P(M, G)$ be a principal bundle over M, with compact semisimple Lie group G. The state space is taken to be the space of gauge potentials $\mathcal{A}_P$. The partition function Z of the theory is defined by

$$Z(\Phi) := \int_{\mathcal{A}_P} e^{-i\mathcal{A}_{CS}(\omega)} \Phi(\omega) \mathcal{D}\mathcal{A},$$

where $\Phi : \mathcal{A}_P \to \mathbf{R}$ is a quantum observable of the theory and $\mathcal{A}_{CS}$ is the Chern–Simons action. The expectation value $\langle \Phi \rangle$ of the observable Φ is given by

$$\langle \Phi \rangle = \frac{\int_{\mathcal{A}_P} e^{-i\mathcal{A}_{CS}(\omega)} \Phi(\omega) \mathcal{D}\mathcal{A}}{\int_{\mathcal{A}_P} e^{-i\mathcal{A}(\omega)} \mathcal{D}\mathcal{A}}.$$

Taking for Φ the Wilson loop functional $\mathcal{W}_{\rho,\kappa}$, where ρ is a suitably chosen representation of G and κ is the link under consideration, one is led to the following interpretation of the Jones polynomial:

$$\langle \Phi \rangle = V_\kappa(q), \qquad \text{where } q = e^{2\pi i/(k+2)}.$$

Witten's ideas have been used by several authors to obtain a geometrical interpretation of various knot and link invariants and to discover new invariants of knots, links, and 3-manifolds (see, for example, [87, 178]).

In [404] it is shown that the Donaldson polynomial invariants of a 4-manifold M appear as expectation values of certain observables in a topological QFT. The Lagrangian of this topological QFT has also been obtained through consideration of an infinite-dimensional version of the classical Chern–Gauss–Bonnet theorem, in [21]. The general program of computing the curvature of connections on infinite-dimensional bundles and of defining appropriate generalizations of characteristic classes was initiated by I. Singer in his fundamental paper [351] on Gribov ambiguity. We proceed by analogy with the finite-dimensional case. To simplify considerations, let us suppose that the group of gauge transformations $\mathcal{G}$ acts freely on the space of gauge connections $\mathcal{A}$. Then we can consider $\mathcal{A}$ as a principle $\mathcal{G}$-bundle over the space $\mathcal{B} = \mathcal{A}/\mathcal{G}$ of gauge equivalence classes of connections. Define the map

$$F^- : \mathcal{A} \to \Lambda^2_-(M, \mathrm{ad}(P)) \qquad \text{by } \omega \mapsto F^-_\omega. \tag{7.39}$$

Then F^- is equivariant with respect to the standard action of $\mathcal{G}$ on $\mathcal{A}$ and the linear action of $\mathcal{G}$ on the vector space $\Lambda^2_-(M, \mathrm{ad}(P))$. Hence, we may think of F^- as defining a section of the vector bundle E associated to $\mathcal{A}$ with fiber

$\Lambda^2_-(M, \mathrm{ad}(P))$. Thus F^- is the infinite-dimensional analogue of a vector field and we may hope to obtain topological information about the base space $\mathcal{B}$ by studying the set of zeros of this vector field. In the generic finite-dimensional case, one can define an index of a vector field by studying its zero set and thus obtain a topological invariant, namely the Euler characteristic. In our case, the zeros of F^- in $\mathcal{A}$ are precisely the Yang–Mills instantons and the zero set of the associated section s_{F^-} is the moduli space $\mathcal{M}^+ \subset \mathcal{B}$. Thus,

$$\mathcal{M}^+ = (F^-)^{-1}(\{0\})/\mathcal{G}. \tag{7.40}$$

In this general setting there is no obvious way to define the index of s_{F^-} and consider its relation with the analytic definition of the Euler characteristic, which should correspond to an integral of some "curvature form" analogous to the Chern–Gauss–Bonnet integrand.

Once again we consider the finite-dimensional case. Here the Thom class, which lies in the cohomology of the vector bundle over M with compact support, is used to obtain the desired relation between the index and the Euler characteristic as follows. We assume that there exists a section s of the vector bundle E. Using s, we pull back the Thom class α to a cohomology class on the base M. In fact, we have a family of homologous sections $\lambda s, \lambda \in \mathbf{R}$, and the corresponding family of pull-backs $(\lambda s)^* \alpha$ which, for $\lambda = 0$, gives the Euler class of M and, for large λ ($\lambda \to \infty$) gives, in view of the compact support of α, the index of s. Thus, one would like to obtain a suitable generalization of the Thom class in the infinite-dimensional case. We shall refer to this idea, of interpolating between two different definitions of the Euler characteristic, as the **Mathai–Quillen formalism**. On the principal $\mathcal{G}$-bundle $\mathcal{A} = P(\mathcal{B}, \mathcal{G})$, we can define a natural connection as follows. For $\omega \in \mathcal{A}$, the tangent space $T_\omega \mathcal{A}$ is identified with $\Lambda^1(M, \mathrm{ad}(P))$ and hence carries the inner product as defined in Definition 6.1. With respect to this inner product we have the orthogonal splitting

$$T_\omega \mathcal{A} = V_\omega \oplus H_\omega, \tag{7.41}$$

where the vertical space V_ω is identified as the tangent space to the fiber $\mathcal{G}$ or, alternatively, as the image of the covariant differential $d^\omega : \Lambda^0(M, \mathrm{ad}(P)) \to \Lambda^1(M, \mathrm{ad}(P))$. The horizontal space H_ω can be identified with $\ker \delta^\omega$, where δ^ω is the formal adjoint of d^ω. The horizontal distribution is equivariant with respect to the action of $\mathcal{G}$ on $\mathcal{A}$ and defines a connection on $\mathcal{A}$. If $L\mathcal{G}(= \Lambda^0(M, \mathrm{ad}(P)))$ denotes the Lie algebra of the infinite-dimensional Hilbert Lie group $\mathcal{G}$ then the connection form $\hat{\omega} : T\mathcal{A} \to L\mathcal{G}$ of this connection is given by

$$\hat{\omega}(X) = G^\alpha(\delta^\alpha X), \qquad X \in T_\alpha \mathcal{A}, \tag{7.42}$$

where G^α is the Green operator, which inverts the Laplacian

$$\Delta^0_\alpha = \delta^\alpha d^\alpha : \Lambda^0(M, \mathrm{ad}(P)) \to \Lambda^0(M, \mathrm{ad}(P)).$$

The curvature $\Omega_{\hat{\omega}}$ of this natural connection form $\hat{\omega}$ is the 2-form with values in $\Lambda^0(M, \mathrm{ad}(P))$ given by the usual formula

$$\Omega_{\hat{\omega}} = d^{\hat{\omega}}\hat{\omega}$$

and can be expressed locally, in terms of the Green operator, by

$$\Omega_{\hat{\omega}}(X, Y) = -2G^{\alpha}(g^{ij}[X_i, Y_j]), \tag{7.43}$$

where locally $X = X_i dx^i$, $Y = Y_j dx^j$, and $g^{ij} = g(dx^i, dx^j)$. We remark that $\Omega_{\hat{\omega}}$ should be considered as defined on a local slice at $[\alpha] \in \mathcal{B}$ transversal to the vertical gauge orbit through α. Let $\mathcal{P}$ denote the Poincaré bundle over $M \times \mathcal{B}$. Pulling back the natural connection $\hat{\omega}$ to the Poincaré bundle $\mathcal{P}$, we get a connection which coincides with the one defined in [23]. The Poincaré bundle with this connection can be regarded as a universal bundle with respect to deformations of the bundle $P(M, G)$. If $\mathcal{F}$ denotes the curvature of the canonical connection on $\mathcal{P}$, then the Künneth formula applied to $\mathcal{F}$ gives its (i, j) component

$$\mathcal{F}_{(i,j)} \in \Lambda^i(P) \otimes \Lambda^j(\mathcal{A}), \qquad i + j = 4.$$

Let $(p, \omega) \in P \times \mathcal{A}$. Then for $X, Y \in T_p P$ and $\alpha, \beta \in T_\omega \mathcal{A} = \Lambda^1(M, \mathrm{ad}(P))$ we have

$$\mathcal{F}_{(2,0)}(X, Y) = \Omega_\omega(X, Y) \tag{7.44}$$
$$\mathcal{F}_{(1,1)}(X, \alpha) = \alpha(X) \tag{7.45}$$
$$\mathcal{F}_{(0,2)}(\alpha, \beta) = (\alpha, \beta)_{\mathfrak{g}} \tag{7.46}$$

where $(\alpha, \beta)_{\mathfrak{g}}$ is the inner product on forms induced by the metric and bracket on the Lie algebra $\mathfrak{g}$ (identified with the fiber of $\mathrm{ad}(P)$). Let $c_2 = \mathrm{tr}(\mathcal{F} \wedge \mathcal{F})$ denote the second Chern class of the canonical connection on $\mathcal{P}$; then the Künneth formula applied to c_2 gives its (i, j) component

$$c_i^j \in \Lambda^i(M) \otimes \Lambda^j(\mathcal{A}), \qquad i + j = 4.$$

Explicit expressions for c_i^j in terms of the components of the curvature are given by (see [33] for details)

$$c_0^4 = \mathrm{tr}\left(\mathcal{F}_{(0,2)} \wedge \mathcal{F}_{(0,2)}\right), \tag{7.47}$$
$$c_1^3 = 2\,\mathrm{tr}\left(\mathcal{F}_{(0,2)} \wedge \mathcal{F}_{(1,1)}\right), \tag{7.48}$$
$$c_2^2 = \mathrm{tr}\left(\mathcal{F}_{(1,1)} \wedge \mathcal{F}_{(1,1)}\right) + 2\,\mathrm{tr}\left(\mathcal{F}_{(2,0)} \wedge \mathcal{F}_{(0,2)}\right), \tag{7.49}$$
$$c_3^1 = 2\,\mathrm{tr}\left(\mathcal{F}_{(2,0)} \wedge \mathcal{F}_{(1,1)}\right), \tag{7.50}$$
$$c_4^0 = \mathrm{tr}\left(\mathcal{F}_{(2,0)} \wedge \mathcal{F}_{(2,0)}\right) = \mathrm{tr}(F_\omega \wedge F_\omega). \tag{7.51}$$

The TQFT interpretation of the Donaldson polynomials makes essential use of the Mathai–Quillen formalism to obtain the expectation value of a quantum observable η as

$$\langle \eta \rangle = \lim_{t \to \infty} \int_{\mathcal{M}_k} \eta \omega_{ts},$$

where $\mathcal{M}_k$ is the moduli space of $SU(2)$-instantons on M, s is the self-dual part of the curvature, and ω_s is a form that is, essentially, the exponential of Witten's Lagrangian. At least on a formal level, the Mathai–Quillen formalism can be used to obtain an expression for the equivariant Euler class e of the Poincaré bundle $\mathcal{P}$. In [404] Witten constructs a Lagrangian out of bosonic and fermionic variables and defines a set of forms $W_i, 0 \leq i \leq 4$. The Lagrangian L is a function of the bosonic fields $\phi, \lambda \in \Lambda^0(M, \mathrm{ad}(P))$ and $A \in \Lambda^1(M, \mathrm{ad}(P))$ and the fermionic fields $\psi \in \Lambda^1(M, \mathrm{ad}(P))$, $\eta \in \Lambda^0(M, \mathrm{ad}(P))$ and $\chi \in \Lambda^2(M, \mathrm{ad}(P))$ and their covariant derivatives ∇ with respect to the connection ω (corresponding to the local gauge potential A) on P as well as the curvature F_ω and the metric g. In local coordinates L has the expression

$$L = \mathrm{tr}\left(\tfrac{1}{4} F_{ab} F^{ab} + \tfrac{1}{2} \phi \nabla_a \nabla^a \lambda + i \nabla_a \psi_b \chi^{ab} - i\eta \nabla_a \psi^a \right.$$
$$\left. - \tfrac{i}{8} \phi[\chi_{ab}, \chi^{ab}] - \tfrac{i}{2} \lambda[\psi_b, \psi^b] + \tfrac{ic}{2} \phi[\eta, \eta] + \tfrac{c}{8}[\phi, \lambda]^2 \right),$$

where c is a real coupling constant. The TQFT action S is defined by

$$S(g, \omega, \phi\lambda, B, \psi, \eta, \chi) = \int_M L' \sqrt{\det(g)} dx^1 dx^2 dx^3 dx^4,$$

where

$$L' = L + \tfrac{1}{8} \mathrm{tr}(F \wedge F).$$

The additional term is a multiple of the instanton number and is a topological invariant. It is inserted so that the action becomes invariant under an odd supersymmetry operator Q. When restricted to gauge-invariant functionals, Q is a boundary operator. The Lagrangian and the corresponding energy-momentum tensor turn out to be Q-exact. In fact, this property plays a fundamental role in the TQFT interpretation of the Donaldson invariants. The forms W_i are defined by

$$W_0 = \tfrac{1}{2} \mathrm{tr}(\phi^2), \tag{7.52}$$
$$W_1 = \mathrm{tr}(\phi \wedge \psi), \tag{7.53}$$
$$W_2 = \mathrm{tr}(\tfrac{1}{2} \psi \wedge \psi + i\phi \wedge F), \tag{7.54}$$
$$W_3 = i \, \mathrm{tr}(\psi \wedge F), \tag{7.55}$$
$$W_4 = -\tfrac{1}{2} \mathrm{tr}(F \wedge F). \tag{7.56}$$

Then, the expectation value of Witten's observables

$$\int_{\gamma_i} W_i$$

in the path integral formalism in the corresponding TQFT coincides with the integral of the Euler class e against the j-form

$$\int_{\gamma_i} c_i^j \in \Lambda^j(\mathcal{A}),$$

where γ_i is an i-cycle. More generally, we can define observables $\mathcal{O}$ by

$$\mathcal{O} := \Pi_{i=1}^r \int_{\gamma_{k_i}} W_{k_i},$$

where γ_{k_i} is a k_i-cycle and

$$\sum_{i=1}^r (4 - k_i) = 2d(k) = \dim(\mathcal{M}_k).$$

Then one can show that the expectation value of $\mathcal{O}$ is a topological invariant. If we set $k_i = 2$, $\forall i$, then this is, essentially, Donaldson's definition of his polynomial invariants. A similar interpretation of the Casson invariant is also given in [21].

Another important topic that gave an early indication of the importance of topological methods in field theory is that of anomalies. In the physical literature the word **anomaly** is used in a generic sense as denoting an obstruction to the lifting of a given structure to another structure. For example, we already referred to anomalies in our discussion of the BRST cohomolgy in the previous section. The term **quantum anomaly** in field theory refers to the situation in which a certain symmetry or invariance of the classical action is not preserved at the quantum level. The requirement of anomaly cancellation imposes strong restrictions on the construction of field theoretic models of gauge and associated fields. A detailed discussion of anomalies from a physical point of view and their geometrical interpretation can be found in [187, 325, 349].

As we have indicated earlier, there are formidable mathematical difficulties in obtaining physically significant results from evaluating the full Feynman path integral. To overcome these difficulties, physicists have developed several approximation procedures, which allow them to extract some information from such integrals. We remark that the vacuum expectation values of Wilson loop observables in the Chern–Simons theory have been computed recently up to second order of the inverse of the coupling constant. These calculations have provided a quantum field-theoretic definition of certain invariants of knots and links in 3-manifolds [87, 178]. A geometric formulation of the quantization of Chern–Simons theory is given in [25]. In the next subsection we discuss the Atiyah–Segal axioms for TQFT, which arose out of attempts to obtain a mathematical formulation of the physical methods of QFT as applied to topological calculations.

7.5.1 Atiyah–Segal Axioms for TQFT

The Atiyah–Segal axioms for TQFT [16, 346] arose from an attempt to give a mathematical formulation of the non-perturbative aspects of quantum field theory in general and to develop, in particular, computational tools for the Feynman path integrals that are fundamental in the Hamiltonian approach to QFT. They generalize the formulation given earlier by Segal for conformal field theories. The most spectacular application of the non-perturbative methods has been in the definition and calculation of the invariants of 3-manifolds with or without links and knots. In most physical applications however, it is the perturbative calculations that are predominantly used. Recently, perturbative aspects of the Chern–Simons theory in the context of TQFT have been considered in [30]. For other approaches to the invariants of 3-manifolds see [145, 156, 221, 223, 294, 380, 383].

Let $\mathcal{C}_n$ denote the category of compact, oriented, smooth n-dimensional manifolds with morphism given by oriented cobordism. Let $\mathcal{V}_{\mathbf{C}}$ denote the category of finite-dimensional complex vector spaces. An $(n+1)$-dimensional TQFT is a functor $\mathcal{T}$ from the category $\mathcal{C}_n$ to the category $\mathcal{V}_{\mathbf{C}}$ that satisfies the following axioms.

A1. Let $-\Sigma$ denote the manifold Σ with the opposite orientation of Σ and let V^* be the dual vector space of $V \in \mathcal{V}_{\mathbf{C}}$. Then

$$\mathcal{T}(-\Sigma) = (\mathcal{T}(\Sigma))^*, \qquad \forall \Sigma \in \mathcal{C}_n.$$

A2. Let $\sqcup$ denote disjoint union. Then

$$\mathcal{T}(\Sigma_1 \sqcup \Sigma_2) = \mathcal{T}(\Sigma_1) \otimes \mathcal{T}(\Sigma_2), \qquad \forall \Sigma_1, \Sigma_2 \in \mathcal{C}_n.$$

A3. Let $Y_i : \Sigma_i \to \Sigma_{i+1}$, $i = 1, 2$, be morphisms. Then

$$\mathcal{T}(Y_1 Y_2) = \mathcal{T}(Y_2)\mathcal{T}(Y_1) \in \hom(\mathcal{T}(\Sigma_1), \mathcal{T}(\Sigma_3)),$$

where $Y_1 Y_2$ denotes the morphism given by composite cobordism $Y_1 \cup_{\Sigma_2} Y_2$.

A4. Let ϕ_n be the empty n-dimensional manifold. Then

$$\mathcal{T}(\phi_n) = \mathbf{C}.$$

A5. For every $\Sigma \in \mathcal{C}_n$

$$\mathcal{T}(\Sigma \times [0, 1]) : \mathcal{T}(\Sigma) \to \mathcal{T}(\Sigma)$$

is the identity endomorphism.

We note that if Y is a compact, oriented, smooth $(n+1)$-manifold with compact, oriented, smooth boundary Σ; then

$$T(Y) : T(\phi_n) \to T(\Sigma)$$

is uniquely determined by the image of the basis vector $1 \in \mathbf{C} \equiv T(\phi_n)$. In this case the vector $T(Y) \cdot 1 \in T(\Sigma)$ is often denoted also by $T(Y)$. In particular, if Y is closed, then

$$T(Y) : T(\phi_n) \to T(\phi_n) \quad \text{and} \quad T(Y) \cdot 1 \in T(\phi_n) \equiv \mathbf{C}$$

is a complex number, which turns out to be an invariant of Y. Axiom A3 suggests a way of obtaining this invariant by a cut and paste operation on Y as follows. Let $Y = Y_1 \cup_\Sigma Y_2$ so that Y_1 (resp., Y_2) has boundary Σ (resp., $-\Sigma$). Then we have

$$T(Y) \cdot 1 = < T(Y_1) \cdot 1, T(Y_2) \cdot 1 >, \tag{7.57}$$

where $< \,,\, >$ is the pairing between the dual vector spaces $T(\Sigma)$ and $T(-\Sigma) = (T(\Sigma))^*$. Equation (7.57) is often referred to as a **gluing formula**. Such gluing formulas are characteristic of TQFT. They arise in Fukaya–Floer homology theory of 3-manifolds (see Chapter 10), Floer–Donaldson theory of 4-manifold invariants (see Chapter 9), as well as in 2-dimensional conformal field theory. For specific applications the Atiyah axioms given above need to be refined, supplemented and modified. For example, one may replace the category $\mathcal{V}_{\mathbf{C}}$ of complex vector spaces by the category of finite-dimensional Hilbert spaces. This is in fact the situation of the $(2+1)$-dimensional Jones–Witten theory. In this case it is natural to require the following additional axiom.

A6. Let Y be a compact oriented 3-manifold with $\partial Y = \Sigma_1 \sqcup (-\Sigma_2)$. Then the linear transformations

$$T(Y) : T(\Sigma_1) \to T(\Sigma_2) \quad \text{and} \quad T(-Y) : T(\Sigma_2) \to T(\Sigma_1)$$

are mutually adjoint.

For a closed 3-manifold Y axiom A6 implies that

$$T(-Y) = \overline{T(Y)} \in \mathbf{C}.$$

It is this property that is at the heart of the result that in general the Jones polynomials of a knot and its mirror image are different, i.e.,

$$V_\kappa(t) \neq V_{\kappa_m}(t),$$

where κ_m is the mirror image of the knot κ.

An important example of a $(3+1)$-dimensional TQFT is provided by the Floer–Donaldson theory. The functor T goes from the category $\mathcal{C}$ of compact, oriented homology 3-spheres to the category of $\mathbf{Z}_8$-graded Abelian groups. It is defined by

$$\mathcal{T} : Y \to HF_*(Y), \qquad Y \in \mathcal{C}.$$

For a compact, oriented, 4-manifold M with $\partial M = Y$, $\mathcal{T}(M)$ is defined to be the vector $q(M, Y)$

$$q(M, Y) := (q_1(M, Y), q_2(M, Y), \ldots),$$

where the components $q_i(M, Y)$ are the relative polynomial invariants of Donaldson defined on the relative homology group $H_2(M, Y; \mathbf{Z})$.

The axioms also suggest algebraic approaches to TQFT. The most widely studied of these approaches are those based on quantum groups, operator algebras, and Jones's theory of subfactors. See, for example, books [227, 229, 209, 381, 119] and articles [380, 383, 384]. Turaev and Viro have given an algebraic construction of such a TQFT. Ocneanu's method [304] starts with a special type of subfactor to generate the data, which can be used with the Turaev and Viro construction.

The correspondence between geometric (topological) and algebraic structures has played a fundamental role in the development of modern mathematics. Its roots can be traced back to the classical work of Descartes. Recent developments in low-dimensional geometric topology have raised this correspondence to a new level bringing in ever more exotic algebraic structures such as quantum groups, vertex algebras, and monoidal and higher categories. This broad area is now often referred to as **quantum topology**. See, for example, [414, 259].

Chapter 8
Yang–Mills–Higgs Fields

8.1 Introduction

Yang–Mills equations, originally derived for the isospin gauge group $SU(2)$, provide the first example of gauge field equations for a non-Abelian gauge group. This gauge group appears as an **internal** or **local symmetry group** of the theory. In fact, the theory can be extended easily to include the other classical Lie groups as gauge groups. Historically, the classical Lie groups appeared in physical theories, mainly in the form of **global symmetry groups** of dynamical systems. Noether's theorem established an important relation between symmetry and conservation laws of classical dynamical systems. It turns out that this relationship also extends to quantum mechanical systems. **Weyl** made fundamental contributions to the theory of representations of the classical groups [401] and to their application to quantum mechanics. The Lorentz group also appears first as the **global symmetry group** of the Minkowski space in the special theory of relativity. It then reappears as the structure group of the principal bundle of orthonormal frames (or the inertial frames) on a space-time manifold M in Einstein's general theory of relativity. In general relativity a gravitational field is defined in terms of the Lorentz metric of M and the corresponding Levi-Civita connection on M. Thus, a gravitational field is essentially determined by geometrical quantities intrinsically associated with the space-time manifold subject to the gravitational field equations. This **geometrization of gravity** must be considered one of the greatest events in the history of mathematical physics.

Weyl sought a geometric setting for a unified treatment of electromagnetism and gravitation. He proposed local scale invariance as the origin of electromagnetism. While this proposal was rejected on physical grounds, it contained the fundamental idea of gauge invariance and local symmetry. Indeed, replacing the local scale invariance by a local phase invariance leads to the interpretation of Maxwell's equations as gauge field equations with gauge group $U(1)$. In Section 8.2 we discuss the theory of electromagnetic fields

K. Marathe, *Topics in Physical Mathematics*, DOI 10.1007/978-1-84882-939-8_8,   235
© Springer-Verlag London Limited 2010

from this point of view. This theory is the prototype of **Abelian gauge theories**, i.e., theories with Abelian gauge group. Here, a novel feature is the discussion of the geometrical implications of Maxwell's equations and the use of universal connections in obtaining their solutions. This last method also yields solutions of pure Yang–Mills field equations, which are described in Section 8.3. We also introduce here the instanton equations, which correspond to self-dual Yang–Mills fields; however a detailed discussion of these equations and their applications is postponed until Chapter 9. The removable nature of isolated singularities in the solutions of Yang–Mills equations is also discussed in this section. In Section 8.4 we give a brief discussion of the non-dual solutions of the Yang–Mills equations. Yang–Mills–Higgs equations with various couplings to potential are studied in Section 8.5. The Higgs mechanism of symmetry-breaking is introduced in Section 8.6, and its application to the electroweak theory is discussed in Section 8.7. The full standard model including QCD is also considered here. Invariant connections are important in physical theories where both local and global symmetries are present. We discuss them briefly in Section 8.8.

8.2 Electromagnetic Fields

A **source-free**, or simply **free, electromagnetic field** is the prototype of Yang–Mills fields. We will show that a source-free electromagnetic field is a gauge field with gauge group $U(1)$. Let $P(M^4, U(1))$ be a principal $U(1)$-bundle over the Minkowski space M^4. Any principal bundle over M^4 is trivializable. We choose a fixed trivialization of $P(M^4, U(1))$ and use it to write $P(M^4, U(1)) = M^4 \times U(1)$. The Lie algebra $u(1)$ of $U(1)$ can be identified with $i\mathbf{R}$. Thus a connection form on P may be written as $i\omega$, $\omega \in \Lambda^1(P)$, by choosing i as the basis of the Lie algebra $i\mathbf{R}$. The gauge field can be written as $i\Omega$, where $\Omega = d\omega \in \Lambda^2(P)$. The Bianchi identity $d\Omega = 0$ is an immediate consequence of this result. The bundle $\mathrm{ad}(P)$ is also trivial and we have $\mathrm{ad}(P) = M^4 \times u(1)$. Thus, the gauge field $F_\omega \in \Lambda^2(M^4, \mathrm{ad}(P))$ on the base M^4 can be written as iF, $F \in \Lambda^2(M^4)$. Using the global gauge $s : M^4 \to P$ defined by $s(x) = (x, 1)$, $\forall x \in M^4$, we can pull the connection form $i\omega$ on P to M^4 to obtain the gauge potential $iA = is^*\omega$. Thus in this case we have a global potential $A \in \Lambda^1(M^4)$ and the corresponding gauge field $F = dA$. The Bianchi identity $dF = 0$ for F follows from the exactness of the 2-form F. The field equations $\delta F = 0$ are obtained as the **Euler–Lagrange equations** minimizing the action $\int |F|^2$, where $|F|$ is the pseudo-norm induced by the Lorentz metric on M^4 and the trivial inner product on the Lie algebra $u(1)$. We note that the action represents the total energy of the electromagnetic field. The two equations

$$dF = 0, \quad \delta F = 0 \qquad (8.1)$$

are Maxwell's equations for a source-free electromagnetic field.

A gauge transformation f is a section of $\mathrm{Ad}(P) = M^4 \times U(1)$. It is completely determined by the function $\psi \in \mathcal{F}(M^4)$ such that

$$f(x) = (x,\ e^{i\psi(x)}) \in \mathrm{Ad}(P), \qquad \forall x \in M^4.$$

If iB denotes the potential obtained by the action of the gauge transformation f on iA, then we have

$$iB = e^{-i\psi}(iA)e^{i\psi} + e^{-i\psi}de^{i\psi} \quad \text{or} \quad B = A + d\psi,$$

which is the classical formulation of the gauge transformation f. We observe that the group $\mathcal{G}$ of gauge transformations acts transitively on the solution space of gauge connections $\mathcal{A}$ of equation (8.1) and hence the moduli space $\mathcal{A}/\mathcal{G}$ reduces to a single point. This observation plays a fundamental role in the path integral approach to QED (quantum electrodynamics).

The above considerations can be applied to any $U(1)$-bundle over an arbitrary pseudo-Riemannian manifold M. We now consider the Euclidean version of Maxwell's equations to bring out the relation of differential geometry, topology and analysis with electromagnetic theory as an example of gauge theory. Let (M, g) be a compact, simply connected, oriented, Riemannian 4-manifold with volume form v_g. A connection ω on $P(M, U(1))$ is called a **Maxwell connection** or **potential** if it minimizes the **Maxwell action** $\mathcal{A}_M(\omega)$ defined by

$$\mathcal{A}_M(\omega) = \frac{1}{8\pi^2} \int\limits_M |F_\omega|_x^2 dv_g . \tag{8.2}$$

The corresponding Euler–Lagrange equations are

$$dF_\omega = 0, \quad \delta F_\omega = 0. \tag{8.3}$$

A solution of equations (8.3) is called a **Maxwell field** or a source-free Euclidean electromagnetic field on M. We note that equations (8.3) are equivalent to the condition that F_ω be **harmonic**. Thus, a Maxwell connection is characterized by its curvature 2-form being harmonic.

Now from topology we know that

$$H^2(M, \mathbf{Z}) = [M, \mathbf{CP}^\infty],$$

where $[M, \mathbf{CP}^\infty]$ is the set of homotopy classes of maps from M to $\mathbf{CP}^\infty$. But, as discussed earlier in Chapter 5, $\mathbf{CP}^\infty$ is the classifying space for $U(1)$-bundles. Thus, each element of $[M, \mathbf{CP}^\infty]$ determines a principal $U(1)$-bundle over M by pulling back the universal $U(1)$-bundle S^∞ over $\mathbf{CP}^\infty$. Hence each element $\alpha \in H^2(M, \mathbf{Z})$ corresponds to a unique isomorphism class of $U(1)$-bundles P_α over M. Moreover, the first Chern class of P_α equals α. Note that the natural embedding of $\mathbf{Z}$ into $\mathbf{R}$ induces an embedding of $H^2(M, \mathbf{Z})$

into $H^2(M, \mathbf{R})$. Thus, we can regard $H^2(M, \mathbf{Z})$ as a subset of $H^2(M, \mathbf{R})$. Now $H^2(M, \mathbf{R})$ can be identified with the second cohomology group of the de Rham complex $H^2_{\text{deR}}(M, \mathbf{R})$ of the base manifold M. Thus, an element

$$\alpha \in H^2(M, \mathbf{Z}) \subset H^2(M, \mathbf{R}) = H^2_{\text{deR}}(M, \mathbf{R})$$

corresponds to the class of a closed 2-form on M, which we also denote simply by α. By applying Hodge theory we can identify $H^2_{\text{deR}}(M, \mathbf{R})$ with the space of harmonic 2-forms with respect to the **Hodge Laplacian** Δ_2 on 2-forms defined by $\Delta_2 = d\delta + \delta d$. Thus there exists a unique harmonic 2-form β on M such that $\alpha = [\beta]$. It can be shown that β is the curvature (gauge field) of a gauge connection on the $U(1)$-bundle P_α over M. We note that $\Delta\beta = 0$ is equivalent to the set of two equations $d\beta = 0$, $\delta\beta = 0$. Thus, the harmonic form β is a Maxwell field, i.e., a source-free electromagnetic field. The above discussion proves the following theorem.

Theorem 8.1 *Let $P(M, U(1))$ be a principal bundle over a compact, simply connected, oriented, Riemannian manifold M. Then the Maxwell field is the unique harmonic 2-form representing the Euler class or the first Chern class $c_1(P)$.*

The role played by the various areas of topology and geometry in the proof of the above theorem is as follows. Let $\mathcal{I}$ denote the set of isomorphism classes of $U(1)$-bundles over M, $\mathcal{H}_2$ the set of harmonic 2-forms on M, $\mathcal{EG}$ the set of gauge equivalence classes of gauge fields, and $\mathcal{CI}$ the set of connections on $U(1)$-bundles with integral curvature. Then we have the following diagram:

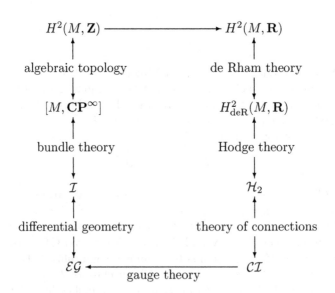

The fact that an integral Chern class $c_1(P)$ is represented by the curvature of a gauge connection in a complex line bundle associated to P also plays an important role in geometric quantization. If M is 4-dimensional then the Maxwell field can be decomposed into its self-dual and anti-dual parts under the Hodge star operator. The corresponding fields are related to certain topological invariants of the manifold M such as the Seiberg–Witten invariants.

Theorem 8.1 suggests where we should look for examples of source-free electromagnetic fields. Since every $U(1)$-bundle with a connection is a pull-back of a suitable $U(1)$-universal bundle with universal connection, it is natural to examine this bundle first. The **Stiefel bundle** $V_{\mathbf{R}}(n+k,k)$ over the **Grassmann manifold** $G_{\mathbf{R}}(n+k,k)$ is k-classifying for $SO(k)$. In particular, for $k = 2$, we get $V_{\mathbf{R}}(n+2,2) = SO(n+2)/SO(n)$ and $G_{\mathbf{R}}(n+2,2) = SO(n+2)/(SO(n) \times SO(2))$. Similarly, $V_{\mathbf{C}}(n+1,1) = U(n+1)/U(n) = S^{2n+1}$ and $G_{\mathbf{C}}(n+1,1) = U(n+1)/(U(n) \times U(1)) = \mathbf{CP}^n$, which is the well-known **Hopf fibration**. Recall that the first Chern class classifies these principal $U(1)$-bundles and is an integral class. When applied to the base manifold $\mathbf{CP}^1 \cong S^2$ this classification corresponds to the **Dirac quantization condition** for a monopole. Example 6.1 corresponds to the above Hopf fibration with $n = 1$. The natural (or universal) connections over these bundles satisfy source-free Maxwell's equations. We note that the pull-back of these universal connections do not, in general, satisfy Maxwell's equations. However, we do get new solutions in the following situation.

Example 8.1 *If M is an analytic submanifold of $\mathbf{CP}^n$, then the $U(1)$-bundle S^{2n+1}, pulled back by the embedding $i : M \hookrightarrow \mathbf{CP}^n$, gives a connection on M whose curvature satisfies Maxwell's equations. For example, if $M = \mathbf{CP}^1 = S^2$, then for each positive integer n, we have the following well-known embedding:*

$$f_n : \mathbf{CP}^1 \hookrightarrow \mathbf{CP}^n$$

given in homogeneous coordinates z_0, z_1 on $\mathbf{CP}^1$ by

$$f_n(z_0, z_1) = (z_0^n, c_1 z_0^{n-1} z_1, \ldots, c_m z_0^{n-m} z_1^m, \ldots, z_1^n),$$

where $c_i = \binom{n}{i}^{1/2}$. The electromagnetic field on $\mathbf{CP}^n$ is pulled back by f_n to give a field on $\mathbf{CP}^1 = S^2$, which corresponds to a magnetic monopole of strength $n/2$. Moreover the corresponding principal $U(1)$-bundle is isomorphic to the lens space $L(n,1)$.

8.2.1 Motion in an Electromagnetic Field

We discussed above the geometric setting that characterizes source-free electromagnetic fields. On the other hand, the existence of an electromagnetic field has consequences for the geometry of the base space and this in turn affects the motion of test particles. Indeed, these effects are well-known in classical physics only for the electromagnetic field. At this time there are no known physical effects associated with classical non-Abelian gauge fields such as the Yang–Mills field. We now give a brief discussion of these aspects of the electromagnetic field.

First recall that an electrostatic field E determines the difference in potential between two points by integration along a path joining them. It is therefore reasonable to think of E as a 1-form on $\mathbf{R}^3$. Maxwell's fundamental idea was to introduce a quantity, called **electric displacement** D, which has the property that its integral over a closed surface measures the charge enclosed by the surface. One should thus think of D as a 2-form. In a uniform medium characterized by dielectric constant ϵ, one usually writes the **constitutive equations** relating D and E as

$$D = \epsilon E.$$

However, our discussion indicates that this equation only makes sense if E is replaced by a 2-form. In fact, if we are given a metric on $\mathbf{R}^3$, there is a natural way to associate with E a 2-form, namely $*E$, where $*$ is the Hodge operator. Then the above equation can be written in a mathematically precise form

$$D = \epsilon(*E).$$

Conversely, requiring such a relation between D and E, i.e., specifying the operator $* : \Lambda^1(\mathbf{R}^3) \to \Lambda^2(\mathbf{R}^3)$, is equivalent to a Euclidean metric on $\mathbf{R}^3$. Similarly, one can interpret the **magnetic tension** or **induction** as a 2-form B and Faraday's law then implies that B is closed, i.e., $dB = 0$. The law of motion of a charged test particle of charge e, moving in the presence of B, is obtained by a modification of the canonical symplectic form ω on $T^*\mathbf{R}^3$ by the pull-back of B by the canonical projection $\pi : T^*\mathbf{R}^3 \to \mathbf{R}^3$, i.e., by use of the form

$$\omega_{e,B} = \omega + e\pi^* B.$$

The distinct theories of electricity and magnetism were unified by Maxwell in his electromagnetic theory. To describe this theory for arbitrary electromagnetic fields, it is convenient to consider their formulation on the 4-dimensional Minkowski space-time M^4. We define the two 2-forms F and G by

$$F := B + E \wedge dt, \quad G := D - H \wedge dt,$$

where B and H are the magnetic tension and magnetic field, respectively and E and D are the electric field and displacement, respectively. If J denotes the current 3-form, then Maxwell's equations with source J are given by

$$dF = 0$$
$$dG = 4\pi J.$$

The two fields F and G are related by the constitutive equationsm which depend on the medium in which they are defined. In a uniform medium (in particular in a vacuum) we can write the constitutive equations as

$$G = (\epsilon/\mu)^{1/2} * F.$$

We note that, specifying the operator $* : \Lambda^2(M^4) \rightarrow \Lambda^2(M^4)$ determines only the conformal class of the metric. If we choose the standard Minkowski metric, then we can write the source-free Maxwell's equations in a uniform medium as

$$dF = 0, \quad d * F = 0.$$

The second equation is equivalent to $\delta F = 0$; we thus obtain the gauge field equations discussed earlier.

It is well known that **Hamilton's equations** of motion of a particle in classical mechanics can be given a geometrical formulation by using the phase space P of the particle. The phase space P is, at least locally, the cotangent space T^*Q of the configuration space Q of the particle. For a geometrical formulation of classical mechanics see Abraham and Marsden [1]. We now show that this formalism can be extended to the motion of a charged particle in an electromagnetic field and leads to the usual equations of motion with the Lorentz force. We choose the configuration space Q of the particle as the usual Minkowski space. The law of motion of a charged test particle with charge e is obtained by considering the symplectic form

$$\omega_{e,F} = \omega + e\pi^* F,$$

where ω is the canonical symplectic form of T^*Q and $\pi : T^*Q \rightarrow Q$ is the canonical projection. The **Hamiltonian** function is given by

$$\mathcal{H}(q,p) = \frac{1}{2m} g^{ij} p_i p_j,$$

where g_{ij} are the components of the Lorentz metric and p_i are the components of 4-momentum. In the usual system of coordinates we can write the metric and the Hamiltonian as

$$ds^2 = dq_0^2 - dq_1^2 - dq_2^2 - dq_3^2,$$

$$\mathcal{H}(q,p) = \frac{1}{2m}(p_0^2 - p_1^2 - p_2^2 - p_3^2).$$

The matrix of the symplectic form $\omega_{e,F}$, in local coordinates $x^\alpha = (q^i, p_i)$ on T^*Q, can be written in block matrix form as

$$\omega_{e,F} = \begin{pmatrix} eF & -I \\ I & 0 \end{pmatrix}.$$

The Hamiltonian vector field is given by

$$X_{\mathcal{H}} = (d\mathcal{H})^{\sharp} = \frac{\partial \mathcal{H}}{\partial x^{\alpha}} \omega_{e,F}^{\alpha\beta} \partial_{\beta},$$

where $\omega_{e,F}^{\alpha\beta}$ are the elements of $(\omega_{e,F})^{-1}$. The corresponding Hamilton's equations are given by

$$\frac{dx^{\alpha}}{dt} = X_{\mathcal{H}}^{\alpha} = \omega_{e,F}^{\alpha\beta} \frac{\partial \mathcal{H}}{\partial x^{\beta}},$$

i.e.,

$$\frac{dq^{i}}{dt} = \frac{\partial \mathcal{H}}{\partial p_{i}}, \tag{8.4}$$

$$\frac{dp_{i}}{dt} = -\frac{\partial \mathcal{H}}{\partial q^{i}} + eF_{ij} \frac{\partial \mathcal{H}}{\partial p_{i}}. \tag{8.5}$$

Equation (8.5) implies that the 3-momentum $p = (p_1, p_2, p_3)$ of the particle satisfies the equation

$$\frac{dp}{dt} = \frac{e}{2m}(E + p \times B), \tag{8.6}$$

where E and B are respectively the electric and the magnetic fields. The equation (8.6) is the well-known equation of motion of a charged particle in an electromagnetic field subject to the Lorentz force. Now the electromagnetic field is a gauge field with Abelian gauge group $U(1)$. The orbits of contragredient action of $U(1)$ on the dual of its Lie algebra $u(1)$ are trivial. Identifying $u(1)$ and its dual with $\mathbf{R}$, we see that an orbit through $e \in \mathbf{R}$ is the point e itself. Thus, in this case, the choice of an orbit is the same thing as the choice of the unit of charge. This construction can be generalized to gauge fields with arbitrary structure group and, in particular, to the Yang–Mills fields to obtain the equations of motion of a particle moving in a Yang–Mills field. A detailed discussion of this is given in Guillemin and Sternberg [179].

8.2.2 The Bohm–Aharonov Effect

We discussed above the effect of the geometry of the base space on the properties of the electromagnetic fields defined on it. We now discuss a property of the electromagnetic field that depends on the topology of the base space. In Example 6.1 of the Dirac monopole we saw that the topology of the base space may require several local gauge potentials to describe a single global gauge field. In fact, this is the general situation. In classical theory only the electromagnetic field was supposed to have physical significance while the potential was regarded as a mathematical artifact. However, in topologically non-trivial spaces the potential also becomes physically significant. For example, in non-simply connected spaces the equation $dF = 0$, satisfied by a 2-form F, defines not only a local potential but a global topological property of belonging to a given cohomology class. Bohm and Aharonov ([8, 7]) suggested that in quantum theory the non-local character of electromagnetic potential $A = A_i dx^i$, in a multiply connected region of space-time, should have a further kind of significance that it does not have in the classical theory. They proposed detecting this topological effect by computing the **phase shift** $\oint A_i dx^i$ around a closed curve not homotopic to the identity and computing its effect in an electron interference experiment. We now discuss the special case of a non-relativistic charged particle moving through a vector potential corresponding to zero magnetic field to bring out the effect of potentials in the absence of fields.

Consider a long solenoid placed along the z-axis with its center at the origin. Then for motion near the origin the space may be considered to be $\mathbf{R}^3$ minus the z-axis. A loop around the solenoid is then homotopically non-trivial (i.e., not homotopic to the identity). Thus, two paths c_1, c_2 respectively joining two points p_1, p_2 on opposite sides of the solenoid are not homotopic. If we send particle beams along these paths from p_1 to p_2, we can observe the resulting interference pattern at p_2. Let ψ_i, $i = 1, 2$, denote the wave function corresponding to the beam passing along c_i when the vector potential is zero. When the field is switched on in the solenoid, the paths c_1, c_2 remain in the field-free region, but the vector potential in this region is not zero. The wave function ψ_j is now changed to $\psi_j e^{iS_j/\hbar}$, where

$$S_j = q \int_{c_j} A_k dx^k, \quad j = 1, 2.$$

It follows that the total wave function of the system is changed as follows:

$$\psi_1 + \psi_2 \rightarrow \psi_1 e^{iS_1/\hbar} + \psi_2 e^{iS_2\hbar}. \tag{8.7}$$

The interference effect of the recombined beams will thus depend on the **phase factor**

$$e^{i(S_1 - S_2)/\hbar} = e^{(iq/\hbar) \int_c A_k dx^k},$$

where c is the closed path $c_1 - c_2$. The predicted effect, called the **Bohm–Aharonov effect**, was confirmed by experimental observations. This result firmly established the physical significance of the gauge potential in quantum theory. Extensions of the Bohm–Aharonov effect to non-Abelian gauge fields have been proposed, but there is no physical evidence for this effect at this time. The prototype of non-Abelian gauge theories is the theory of Yang–Mills fields. Many of the mathematical considerations given above for the electromagnetic fields extend to Yang–Mills fields, which we discuss in the next section.

8.3 Yang–Mills Fields

Yang–Mills fields form a special class of gauge fields that have been extensively investigated. In addition to the references given for gauge fields we give the following references which deal with Yang–Mills fields. They are Douady and Verdier [114], [59, 60, 386], and [410, 411].

Let M be a connected manifold and let $P(M, G)$ be a principal bundle over M with Lie group G as the gauge group. In this section we restrict ourselves to the space $\mathcal{A}(P)$ of gauge connections as our configuration space. If ω is a gauge connection on P then, in a local gauge $t \in \Gamma(U, P)$, we have the corresponding local gauge potential

$$A_t = t^* \omega \in \Lambda^1(U, \mathrm{ad}(P)).$$

Locally, a gauge transformation $g \in \mathcal{G}$ reduces to a G-valued function g_t on U, and its action on A_t is given by

$$g_t \cdot A_t = (\mathrm{ad}\, g_t) \circ A_t + g_t^* \Theta,$$

where Θ is the canonical 1-form on G. It is customary to indicate this gauge transformation as follows

$$A^g := g \cdot A = g^{-1}Ag + g^{-1}dg$$

when the section t is understood.

The gauge field F_ω is the unique 2-form on M with values in the bundle $\mathrm{ad}(P)$ satisfying

$$F_\omega = s_\Omega,$$

where Ω is the curvature of the gauge connection ω. In the local gauge t we can write

$$F_\omega = d^\omega A_s = dA_s + \tfrac{1}{2}[A_s, A_s],$$

where the bracket is taken as the bracket of bundle-valued forms.

We note that it is always possible to introduce a Riemannian metric on a vector bundle over M. The manifold M itself admits a Riemannian metric. We assume that metrics are chosen on M and the bundles over M, and the norm is defined on sections of these bundles as an L^2-norm if M is not compact. However, to simplify the mathematical considerations, we assume in the rest of this chapter that M is a compact, connected, oriented, Riemannian manifold and that G is a compact, semisimple Lie group, unless otherwise indicated. For a given M and G, the principal G-bundles over M are classified by $[M; BG]$, the set of homotopy classes of maps of M into the classifying space BG for G. For $f \in [M; BG]$, let P_f be a representative of the isomorphism class of principal bundles corresponding to f. Let $\mathcal{A}(P_f)$ be the space of gauge potentials on P_f. Then, the Yang–Mills configuration space $\mathcal{A}_M$ is the disjoint union of the spaces $\mathcal{A}(P_f)$, i.e.,

$$\mathcal{A}_M = \cup\{\mathcal{A}(P_f) \mid f \in [M; BG]\}.$$

The isomorphism class $[P_f]$ is uniquely determined by the characteristic classes of the bundle P_f. For example, if $\dim M = 4$ and $G = SU(2)$, then $c_1(P_f) = 0$ and hence $[P_f]$ is determined by the second Chern class $c_2(P_f) = c_2(V_f)$, where V_f is the complex vector bundle of rank 2 associated to P_f by the defining representation of $SU(2)$. Now the class $c_2(P_f)$, evaluated on the fundamental cycle of M, is integral, i.e., $c_2(P_f)[M] \in \mathbf{Z}$. Thus, in this case P_f is classified by an integer $n(f)$ and we can write $\mathcal{A}(P_f)$ as $\mathcal{A}_{n(f)}$, or simply as $\mathcal{A}_n$. In the physics literature, this number n is referred to as the **instanton number** of P_f. From both the physical and mathematical point of view, this is the most important case.

We now restrict ourselves to a fixed member $\mathcal{A}(P_f)$ of the disjoint union $\mathcal{A}_M$ and write simply P for P_f. Then the **Yang–Mills action** or (**functional**) $\mathcal{A}_{YM}$ is defined by

$$\mathcal{A}_{YM}(\omega) = \frac{1}{8\pi^2} \int_M |F_\omega|_x^2 dv_g, \qquad \forall \omega \mathcal{A}(P). \tag{8.8}$$

To find the corresponding Euler–Lagrange equations, we take into account the fact that the space of gauge potentials is an affine space and hence it is enough to consider variations along the straight lines through ω of the form

$$\omega_t = \omega + tA, \text{ where } A \in \Lambda^1(M, \operatorname{ad} P).$$

Direct computation shows that the gauge field, corresponding to the gauge potential ω_t, is given by

$$F_{\omega_t} = F_\omega + td^\omega A + t^2(A \wedge A). \tag{8.9}$$

Using equations (8.8) and (8.9) we obtain

$$\frac{d}{dt} \mathcal{A}_{YM}(\omega_t)_{|t=0} = \frac{d}{dt} \Big(\int_M |F_\omega + td^\omega A + t^2(A \wedge A)|_x^2 dv_g \Big)_{|t=0}$$
$$= 2 \int_M < F_\omega, d^\omega A > dv_g$$
$$= 2 \int_M < \delta^\omega F_\omega, A > dv_g \ .$$

The above result is often expressed in the form of a variational equation

$$\delta \mathcal{A}_{YM}(\omega)(A) = 2\langle\langle \delta^\omega F_\omega, A \rangle\rangle. \tag{8.10}$$

A gauge potential ω is called a **critical point** of the Yang–Mills functional if

$$\delta \mathcal{A}_{YM}(\omega)(A) = 0, \quad \forall A \in \Lambda^1(M, \operatorname{ad} P). \tag{8.11}$$

The critical points of the Yang–Mills functional are solutions of the corresponding Euler–Lagrange equation

$$\delta^\omega F_\omega = 0. \tag{8.12}$$

The equations (8.12) are called the **pure** (or **sourceless**) **Yang–Mills equations**. A gauge potential ω satisfying the Yang–Mills equations is called the **Yang–Mills connection** and its gauge field F_ω is called the **Yang–Mills field**. Using local expressions for d^ω and δ^ω it is easy to show that $\delta^\omega = \pm * d^\omega *$. From this it follows that ω satisfies the Yang–Mills equations (8.12) if and only if it is a solution of the equation

$$d^\omega * F_\omega = 0. \tag{8.13}$$

In the following theorem we have collected various characterizations of the Yang–Mills equations.

Theorem 8.2 *Let ω be a connection on the bundle $P(M, G)$ with finite Yang–Mills action. Then the following statements are equivalent:*

1. *ω is a critical point of the Yang–Mills functional;*
2. *ω satisfies the equation $\delta^\omega F_\omega = 0$;*
3. *ω satisfies the equation $d^\omega * F_\omega = 0$;*
4. *ω satisfies the equation $\nabla^\omega F_\omega = 0$, where $\nabla^\omega = d^\omega \delta^\omega + \delta^\omega d^\omega$ is the Hodge Laplacian on forms.*

The equivalence of the first three statements follows from the discussion given above. The equivalence of the statements 2 and 4 follows from the identity

$$\langle\langle \nabla^\omega F_\omega, F_\omega \rangle\rangle = \|d^\omega F_\omega\|^2 + \|\delta^\omega F_\omega\|^2 = \|\delta^\omega F_\omega\|^2.$$

The second equality in the above identity follows from the Bianchi identity

$$d^\omega F_\omega = 0. \tag{8.14}$$

This identity is a consequence of the Cartan structure equations and expresses the fact that, locally, F_ω is derived from a potential. It is customary to consider the pair (8.12) and (8.14) or (8.13) and (8.14) as the Yang–Mills equations. This is consistent with the fact that they reduce to Maxwell's equations for the electromagnetic field F, when the gauge group G is $U(1)$ and M is the Minkowski space.

In local orthonormal coordinates, the Yang–Mills equations can be written as

$$\frac{\partial F_{ij}}{\partial x^i} + [A_i, F_{ij}] = 0, \tag{8.15}$$

where the components F_{ij} of the 2-form F_ω are given by

$$F_{ij} = \frac{\partial A_j}{\partial x^i} - \frac{\partial A_i}{\partial x^j} + [A_i, A_j]. \tag{8.16}$$

Thus, we see that the Yang–Mills equations are a system of non-linear, second order, partial differential equations for the components of the gauge potential A. The presence of quadratic and cubic terms in the potential represents the self-interaction of the Yang–Mills field.

In (8.8) we defined the Yang–Mills action $\mathcal{A}_{YM}$ on the domain $\mathcal{A}(P)$. This domain is an affine space and hence is contractible. But there is a large symmetry group acting on this space. This is the group $\mathcal{G}$ of gauge transformations of P. By definition, a gauge transformation $\phi \in \mathcal{G}$ is an automorphism of the bundle P, which covers the identity map of M and hence leaves each $x \in M$ fixed. Under this gauge transformation, the gauge field transforms as

$$F_\omega \to F_\omega^\phi := \phi^{-1} F_\omega \phi.$$

Therefore, the pointwise norm $|F_\omega|_x$ is gauge-invariant. It follows that the Yang–Mills action $\mathcal{A}_{YM}(\omega)$ is gauge-invariant and hence induces a functional on the moduli space $\mathcal{M} = \mathcal{A}/\mathcal{G}$ of gauge connections. This functional is also called the Yang–Mills functional (or action). In order to relate the topology of the moduli space $\mathcal{M}$ to the critical points of the Yang–Mills functional, it is necessary to compute the **second variation** or the **Hessian** of the Yang–Mills action

$$\delta^2 \mathcal{A}_{YM}(\omega) := \frac{d^2}{dt^2} \mathcal{A}_{YM}(\omega_t)_{|t=0}.$$

One can verify that the Hessian, viewed as a symmetric, bilinear form, is given by the following expression:

$$\delta^2 \mathcal{A}_{YM}(\omega)(A, B) = 2 \int_M (\langle d^\omega A, d^\omega B \rangle + \langle F_\omega, A \wedge B + B \wedge A \rangle).$$

Analysis of the Yang–Mills equations then proceeds by our obtaining various estimates using these variations.

We note that if $\phi \in \mathrm{Aut}(P)$ covers a conformal transformation of M (also denoted by ϕ), then this $\phi \in \mathrm{Diff}(M)$ induces a conformal change of metric. It is given by

$$\phi^* g = e^{2f} g, \quad f \in \mathcal{F}(M). \tag{8.17}$$

Neither the pointwise norm $|F_\omega|_x$ nor the Riemannian volume form v_g are invariant under this conformal change, and the integrand in the Yang–Mills action transforms as follows:

$$|F_\omega|_x^2 \, dv_g \to (e^{-4f} |F_\omega|_x^2)(e^{mf} \, dv_g). \tag{8.18}$$

It follows that, in the particular case of $m = 4$, the Yang–Mills action is invariant under the generalized gauge transformations that cover conformal transformations of M.

A large number of solutions of the Yang–Mills equations are known for special manifolds. For example, the universal connections on Stiefel bundles provide solutions of pure (sourceless) Yang–Mills equations. As we observed in the previous section, the natural connections associated to the complex Hopf fibration satisfy source-free Maxwell's equations. As is well known, Hopf fibrations are particular cases of Stiefel bundles. In Chapter 4, we gave a unified treatment of real, complex, and quaternionic Stiefel bundles. Those results can be used to arrive at solutions of pure Yang–Mills equations as follows (for further details see [302, 377]).

Let A be the local gauge potential corresponding to the universal connection ω on the Stiefel bundle $V_k(F^n)(G_k(F^n), U_F(k))$ with gauge group $U_F(k)$, and let $F_\omega = d^\omega A$ be the corresponding gauge field. The basic computation involves an explicit local expression for $*F_\omega$, the Hodge dual of the gauge field F_ω. In the complex case it can be shown that $*F_\omega$ is represented by an invariant polynomial of degree $kl - 1$, $l = n - k$ in F_ω, i.e.,

$$*F = a \underbrace{F \wedge F \wedge \cdots \wedge F}_{kl-1 \text{ times}},$$

where a is a constant. This expression, together with the Bianchi identities, imply

$$d^\omega * F_\omega = 0. \tag{8.19}$$

In the quaternionic case, a similar expression for $*F_\omega$ leads to equation (8.19). In the real case the expression for $*F_\omega$ also involves the gauge potential explicitly. These terms can be written as a wedge product of certain l-forms ψ, which satisfy the condition $d^A \psi = 0$. As a consequence of the Maurer–Cartan equation this, together with the Bianchi identities, implies (8.19) as in the complex case.

Let M be a compact, oriented Riemannian manifold of dimension 4. Let $P(M, G)$ be a principal bundle over M with compact semisimple Lie group G as structure group. Recall that the Hodge star operator on $\Lambda(M)$ has a natural extension to bundle valued forms. A form $\alpha \in \Lambda^2(M, \mathrm{ad}(P))$ is said to be **self-dual** (resp., **anti-dual**) if

$$*\alpha = \alpha \quad (\text{resp.}, \ *\alpha = -\alpha).$$

We define the **self-dual part** α_+ of a form $\alpha \in \Lambda^2(M, \mathrm{ad}(P))$ by

$$\alpha_+ := \frac{1}{2}(\alpha + *\alpha).$$

Similarly, the **anti-dual part** α_- of a form $\alpha \in \Lambda^2(M, \mathrm{ad}(P))$ is defined by

$$\alpha_- := \frac{1}{2}(\alpha - *\alpha).$$

Over a 4-dimensional base manifold, the second Chern class and the Euler class of P are equal and we define the instanton number k of the bundle P by

$$k := -c_2(P)[M] = -\chi(P)[M]. \tag{8.20}$$

Recall that $F = F_\omega$ is the curvature form of the gauge connection ω, and hence, by the theory of characteristic classes, we have

$$k := -c_2(P)[M] = -\frac{1}{8\pi^2} \int_M \mathrm{Tr}(F \wedge F). \tag{8.21}$$

Decomposing F into its self-dual part F_+ and anti-dual part F_-, we get

$$k = \frac{1}{8\pi^2} \int_M (|F_+|^2 - |F_-|^2). \tag{8.22}$$

Using F_+ and F_- we can rewrite the Yang–Mills action (8.8) as follows:

$$\mathcal{A}_{YM}(\omega) = \frac{1}{8\pi^2} \int_M (|F_+|^2 + |F_-|^2). \tag{8.23}$$

Comparing equations (8.22) and (8.23) above we see that the Yang–Mills action is bounded below by the absolute value of the instanton number k, i.e.,

$$\mathcal{A}_{YM}(\omega) \geq |k|, \qquad \forall \omega \in \mathcal{A}(P). \tag{8.24}$$

When M is 4-dimensional, we can associate to the pure Yang–Mills equations the **first order instanton** (resp., **anti-instanton**) **equations**

$$F_\omega = *F_\omega \quad (\text{resp.}, F_\omega = -*F_\omega). \tag{8.25}$$

The Bianchi identities imply that any solution of the instanton equations is also a solution of the Yang–Mills equations. The fields satisfying $F_\omega = *F_\omega$ (resp., $F_\omega = -*F_\omega$) are also called self-dual (resp., anti-dual) Yang–Mills fields. We note that a gauge connection satisfies the instanton or the anti-instanton equation if and only if it is the absolute minimum of the Yang–Mills action. It can be shown that the instanton equations are also invariant under the generalized gauge transformations, which cover conformal transformations of M. This result plays a crucial role in the construction of the moduli space of instantons. The solutions of the instanton equations in the case of $M = S^4$ were originally called "instantons," but this term is now used to denote any solution of the instanton equations over a Riemannian manifold. We study these solutions and consider their topological and geometrical applications in Chapter 9.

We now give two examples of Yang–Mills fields on S^4 obtained by gluing two potentials along the equator.

Example 8.2 *We consider the two standard charts U_1, U_2 on S^4 (see Chapter 1) and define the $SU(2)$ principal bundle $P_1(S^4, SU(2))$ by the transition function*

$$g : U_1 \cap U_2 \to SU(2), \quad g(x) = x/|x|,$$

where $x \in \mathbf{R}^4$ is identified with a quaternion. Essentially, the transition function g defines a map of the equator S^3 into $SU(2)$, i.e., an element of the homotopy group $[g] \in \pi_3(SU(2)) \cong \mathbf{Z}$. In fact, $[g]$ is non-trivial, corresponding to the non-triviality of the bundle P_1. A connection $A \in \mathcal{A}(P_1(S^4, SU(2))$ is specified by data consisting of a pair of $su(2)$-valued 1-forms A^i on U_i, $i = 1, 2$, which are restricted to $U_1 \cap U_2$ to obey the cocycle condition

$$A^1(x) = g(x) A^2 g^{-1}(x) + g(x) dg^{-1}(x).$$

For each $\lambda \in (0, 1)$ define the connection $A^\lambda = (A_1^\lambda, A_2^\lambda)$ by

$$A_1^\lambda = \mathrm{Im}\left(\frac{x d\bar{x}}{\lambda^2 + |x|^2}\right), \quad A_2^\lambda = \mathrm{Im}\left(\frac{\lambda^2 \bar{x} dx}{|x|^2 (\lambda^2 + |x|^2)}\right).$$

The connection A^λ is self-dual with instanton number 1. The curvature of this connection A^λ is $F^{\lambda+} = (F_1^{\lambda+}, F_2^{\lambda+})$ given by

$$F_1^{\lambda+} = \mathrm{Im}\left(\frac{\lambda^2 dx \wedge d\bar{x}}{(\lambda^2 + |x|^2)^2}\right), \quad F_2^{\lambda+} = \mathrm{Im}\left(\frac{\lambda^2 \bar{x} dx \wedge d\bar{x} x}{|x|(\lambda^2 + |x|^2)^2 |x|}\right).$$

The basic anti-instanton over $\mathbf{R}^4$ is described as follows. The principal bundle $P_{-1}(S^4, SU(2))$ is defined by the transition function

$$\bar{g} : U_1 \cap U_2 \to SU(2), \quad \bar{g}(x) = \bar{x}/|x|.$$

For each $\lambda \in (0, 1)$ define the connection $B^\lambda = (B_1^\lambda, B_2^\lambda)$ by

$$B_1^\lambda = \mathrm{Im}\left(\frac{\bar{x} dx}{\lambda^2 + |x|^2}\right), \quad B_2^\lambda = \mathrm{Im}\left(\frac{\lambda^2 x d\bar{x}}{|x|^2 (\lambda^2 + |x|^2)}\right).$$

The connection B^λ is anti-dual with instanton number -1. We note that $[\bar{g}] = -[g] \in \pi_3(SU(2)) \cong \mathbf{Z}$ and thus $[\bar{g}]$ is also non-trivial. The curvature of this connection A^λ is $F^{\lambda-} = (F_1^{\lambda-}, F_2^{\lambda-})$ given by

$$F_1^{\lambda-} = \mathrm{Im}\left(\frac{\lambda^2 d\bar{x} \wedge dx}{(\lambda^2 + |x|^2)^2}\right), \quad F_2^{\lambda-} = \mathrm{Im}\left(\frac{\lambda^2 x d\bar{x} \wedge dx \bar{x}}{|x|(\lambda^2 + |x|^2)^2 |x|}\right).$$

In the above example we have given an explicit construction of two basic Yang–Mills fields. In fact, these fields can be obtained by applying Uhlenbeck's removable singularities theorem to a suitable connection on $S^4 \setminus \{(0, 0, 0, 0, 1)\} \subset \mathbf{R}^5$ to extend it to all of S^4. We use this method in our construction of the BPST instanton in Chapter 9.

An explicit construction of the full $(8k - 3)$-parameter family of solutions was given by Atiyah, Drinfeld, Hitchin, and Manin ([19]). An alternative construction was given by Atiyah and Ward using the Penrose correspondence. Several solutions to the Yang–Mills equations on special manifolds of various dimensions have been obtained in [26, 175, 186, 243, 302, 377]. Although the solutions' physical significance is not clear, they may prove to be useful in understanding the mathematical aspects of the Yang–Mills equations.

Gauge fields and their associated fields arise naturally in the study of physical fields and their interactions. The solutions of these equations are often obtained locally. The question of whether finite energy solutions of the coupled field equations can be obtained globally is of great significance for both the physical and mathematical considerations. The early solutions of $SU(2)$ Yang–Mills field equations in the Euclidean setting had a finite number of point singularities when expressed as solutions on the base manifold $\mathbf{R}^4$. The fundamental work of Uhlenbeck ([385, 386, 387]) showed that these point singularities in gauge fields are removable by suitable gauge transformations and that these solutions can be extended from $\mathbf{R}^4$ to its compactification S^4 as singularity-free solutions of finite energy. The original proof of the removable singularities theorem is greatly simplified by using the blown-up manifold technique. This proof also applies to arbitrary Yang–Mills fields and, in particular, to non-dual solutions. Note that we call a connection **non-dual** if it is neither self-dual ($*F = F$) nor anti-dual ($*F = -F$), i.e., if it is not an absolute minimum of the Yang–Mills action. Specifically, one obtains in this way a local solution of the source-free Yang–Mills equations on the open ball B^4 by removing the singularity at the origin.

Theorem 8.3 *Let B_g^4 be the open ball $B^4 \subset \mathbf{R}^5$ with some metric g (not necessarily the standard metric induced from $\mathbf{R}^5$). Let ω be a solution of the Yang–Mills equations in $B^4 \setminus \{0\}$ with finite action, i.e.,*

$$\int_{B^4} |F_\omega|^2 < \infty.$$

Assume that the local potential $A \in H_1(B^4 \setminus \{0\})$ of ω has the property that, for every smooth, compactly supported function $\phi \in C_0^\infty(B^4 \setminus \{0\})$, $\phi A \in H_1(B^4 \setminus \{0\})$. Then ω is gauge-equivalent to a connection $\tilde{\omega}$, which extends smoothly across the singularity to a smooth connection on B^4.

By a grafting procedure the result of this theorem can be extended to manifolds with a finite number of singularities as follows.

Theorem 8.4 *Let M be a compact, oriented, Riemannian 4-manifold. Let $\{p_1, p_2, \ldots, p_k\}$ be any finite set of points in M. Let P_k be an $SU(2)$-bundle defined over $M \setminus \{p_1, p_2, \ldots, p_k\}$ and let ω_k be a Yang–Mills connection on P. Then the bundle P_k extends to an $SU(2)$-bundle defined over M and the connection ω_k extends to a Yang–Mills connection ω on P.*

These theorems can be extended to finite energy solutions of coupled field equations on Riemannian base manifolds. It was shown in [386] that the removable singularities theorems for pure gauge fields fail in dimensions greater than 4. In dimension 3 the removable nature of isolated singularities is discussed for the Yang–Mills and Yang–Mills–Higgs equations in Jaffe and Taubes [207].

8.4 Non-dual Solutions

In general, the standard steepest descent method for finding the global minima of the Yang–Mills action is not successful in 4 dimensions, even though this technique does work in dimensions 2 and 3. In fact, the problem of finding critical points of the Yang–Mills action seems to be similar to the 2-dimensional harmonic map problem as well as to conformally invariant variational problems. These similarities had led some researchers to believe in the non-existence of non-dual solutions. It can be shown that local Sobolev connections, on $SU(2)$-bundles over $\mathbf{R}^4 \setminus M$ (M a smoothly embedded compact 2-manifold in $\mathbf{R}^4$) with finite Yang–Mills action, satisfy a certain holonomy condition. If the singular set has codimension greater than 2, then the techniques used for removing the point singularities can be applied to remove these singularities. Thus, a codimension 2 singular set, such as an S^2 embedded in $\mathbf{R}^4$, provides an appropriate setting for new techniques and results. For example, the holonomy condition implies that there exist flat connections in a principal bundle over $\mathbf{R}^4 \setminus S^2$ that cannot be extended to a neighborhood of the singular set S^2 even though the bundle itself may be topologically trivial. It was also known that connections satisfying certain stability or symmetry properties are either self-dual or anti-dual. In view of these results, it was widely believed that no non-dual solutions with gauge group $SU(2)$ existed on the standard four sphere S^4. Analogies with harmonic maps of S^2 to S^2 also appeared to support the non-existence of such solutions. In fact, it turns out that a family of such connections may be used to obtain a non-trivial connection that can be extended to a neighborhood of the singular set. These ideas, together with the work on monopoles in hyperbolic 3-space, have been used in [350] to prove the existence of non-dual (and hence non-minimal) solutions of Yang–Mills equations. The existence of other non-dual solutions has been discussed in [332, 388, 389].

In [332] it is shown that non-dual Yang–Mills connections exist on all $SU(2)$-bundles over S^4 with second Chern class different from ± 1. The results are obtained by studying connections that are equivariant with respect to the symmetry group $G = SU(2)$ that acts on $S^4 \subset \mathbf{R}^5$ via the unique irreducible representation. The principal orbits are 3-dimensional, reducing the Yang–Mills equations and the instanton equations to a system of ordinary differential equations following [389]. The inequivalent lifts of this G-action to

$SU(2)$-bundles over S^4 are called **quadrupole bundles**. These bundles are naturally classified by a pair of odd positive integers (n_+, n_-). The principal bundle corresponding to the pair (n_+, n_-) is denoted by $P_{(n_+,n_-)}$. Then we have

$$c_2(P_{(n_+,n_-)}) = (n_+^2 - n_-^2)/8.$$

Thus the second Chern class of the quadrupole bundle $P_{(n_+,n_-)}$ assumes all integral values. The main result is:

Theorem 8.5 *On every quadrupole bundle $P_{(n_+,n_-)}$ with $n_+ \neq 1$ and $n_- \neq 1$ there exists a non-dual $SU(2)$ Yang–Mills connection.*

It is not known whether any non-dual connections exist for $c_2 = \pm 1$. The non-dual Yang–Mills connections obtained in this work on the trivial bundle over S^4 (i.e., with $c_2 = 0$) are different from those of [350]. The proof and further discussion may be found in [332].

Recently, non-dual solutions of the Yang–Mills equations over $S^2 \times S^2$ and $S^1 \times S^3$ with gauge group $SU(2)$ have been obtained in [395]. We give a brief discussion of these results below.

Theorem 8.6 *Let (m, n) be a pair of integers with $|m| \neq |n|$, which satisfy the following conditions:*

1. *If $|m| > |n|$, then $|m| \neq |n|(2l + 1) + l(l + 1)$, for $l = 0, 1, 2, \ldots$.*
2. *If $|n| > |m|$, then $|n| \neq |m|(2l + 1) + l(l + 1)$, for $l = 0, 1, 2, \ldots$.*

Then there exists a positive integer K_0 such that, for any positive odd number $k > K_0$, there exists an irreducible $SU(2)$-connection $A(m, n, k)$ over $S^2 \times S^2$ with instanton number $2mn$, which is a non-minimal solution of the Yang–Mills equations. In fact, the Yang–Mills action defined by equation (8.9) satisfies the inequality

$$2(m^2 + n^2 + 2k) - \epsilon < \mathcal{A}_{YM}(A(m, n, k)) < 2(m^2 + n^2 + 2k) + \epsilon$$

for some $\epsilon \leq 1$.

Proof: It is well known that S^2 is diffeomorphic to the complex projective space $\mathbf{CP}^1$ viewed as the set of 1-dimensional linear subspaces in $\mathbf{C}^2$. There exists a tautological line bundle L over S^2 whose first Chern number is

$$\int_{\mathbf{CP}^1} c_1(L) = -1,$$

where $c_1(L)$ is the first Chern class of L. Consider the standard metric on $S^2 \cong \mathbf{CP}^1$; then the first Chern class is written as

$$c_1(L) = -\frac{1}{4\pi}\omega,$$

where ω is the volume form on $\mathbf{CP}^1$. Suppose that A_0 is the canonical connection on L; then the curvature of A_0 is

$$F_{A_0} = i\omega/2.$$

Set $L(m,n) = \pi_1^* L^m \otimes \pi_2^* L^m \to \mathbf{CP}^1 \times \mathbf{CP}^1$, which is a complex line bundle over the product manifold $\mathbf{CP}^1 \times \mathbf{CP}^1$, where $m, n \in \mathbf{Z}$, and π_1, π_2 are the natural projections from $\mathbf{CP}^1 \times \mathbf{CP}^1$ to the first and second factor, respectively. The first Chern class of $L(m,n)$ is

$$c_1(L(m,n)) = -\frac{1}{4}(m\omega_1 + n\omega_2),$$

where $\omega_i = \pi_i^* \omega$, $i = 1, 2$. Let $A(m,n) = \pi_1^*(\otimes^m A_0) \otimes \pi_2^*(\otimes^n A_0)$. Then the corresponding curvature is

$$F(m,n) = F_{A(m,n)} = \frac{i}{2}(m\omega_1 + n\omega_2).$$

Hence $L(m,n) \oplus L(m,n)^{-1} \to \mathbf{CP}^1 \times \mathbf{CP}^1$ is a reducible $SU(2)$-bundle over $\mathbf{CP}^1 \times \mathbf{CP}^1$, and there exists a reducible $SU(2)$-connection whose curvature is

$$F(m,n) = \frac{1}{2}(m\omega_1 + n\omega_2)\begin{pmatrix} i & 0 \\ 0 & -i \end{pmatrix}.$$

We note that $c_1(L(m,n) \oplus L(m,n)^{-1}) = 0$ and

$$-c_2(L(m,n) \oplus L(m,n)^{-1}) = \frac{1}{4\pi^2}\det(F(m,n)) = \frac{mn}{8\pi^2}\omega_1 \wedge \omega_2.$$

It follows that if $m = n$ (resp., $m = -n$), then $A(m,n)$ is a reducible self-dual (resp., anti-dual) $SU(2)$-connection on $L(m,n) \oplus L(m,n)^{-1}$ with instanton number $2m^2$ (resp. $-2m^2$). When $|m| \neq |n|$, then $A(m,n)$ is a reducible non-minimal Yang–Mills connection on $L(m,n) \oplus L(m,n)^{-1}$ with degree $2mn$ since $F(m,n)$ is neither self-dual nor anti-dual. Grafting the standard instanton and anti-instanton solutions on these non-dual solutions gives the required family of solutions. $\square$

For Yang–Mills fields on $S^1 \times S^3$ we have the following result.

Theorem 8.7 *There exists a positive integer K_0 such that, for any positive odd number $k > K_0$, there exists an irreducible $SU(2)$-connection $A(k)$ over $S^1 \times S^3$ with instanton number 0 that is a non-minimal solution of the Yang–Mills equations and its action satisfies the inequality*

$$\mathcal{A}_{YM}(A(k)) > 4k.$$

A special non-dual Yang–Mills field on $S^1 \times S^3$ was used in [269] as the source field in the construction of generalized gravitational instantons. From a physical point of view, one can consider the family of non-dual solutions to be obtained by studying the interaction of instantons, anti-instantons, and the field of a fixed background connection, which is an isolated non-minimal solution of the Yang–Mills equations.

8.5 Yang–Mills–Higgs Fields

In this section we give a brief discussion of the most extensively studied coupled system, namely, the system of Yang–Mills–Higgs fields, and touch upon some related areas of active current research. From the physical point of view the Yang–Mills–Higgs system is the starting point for the construction of realistic models of field theories, which describe interactions with massive carrier particles. From the mathematical point of view the Yang–Mills–Higgs system provides a set of non-linear partial differential equations, whose solutions have many interesting properties. In fact, these solutions include solitons, vortices, and monopoles as particular cases. Establishing the existence of solutions of the Yang–Mills–Higgs system of equations has required the introduction of new techniques, since the standard methods in the theory of non-linear partial differential equations are, in general, not applicable to this system. We now give a brief general discussion of various couplings of Yang–Mills field to Higgs fields, with or without self-interaction.

Let (M, g) be a compact Riemannian manifold and $P(M, G)$ a principal bundle with compact semisimple gauge group G. Let h denote a fixed bi-invariant metric on G. The metrics g and h induce inner products and norms on various bundles associated to P and their sections. We denote all these different norms by the same symbol, since the particular norm used is clear from the context. The **Yang–Mills–Higgs configuration space** is defined by

$$\mathcal{C}_H := \mathcal{A}(P) \times \Gamma(\mathrm{ad}\, P),$$

i.e.,

$$\mathcal{C}_H \subset \{(\omega, \phi) \in \Lambda^1(P, \mathfrak{g}) \times \Lambda^0(M, \mathrm{ad}\, P)\},$$

where ω is a gauge connection and ϕ is a section of the **Higgs bundle** $\mathrm{ad}\, P$. In the physics literature ϕ is called the **Higgs field** in the adjoint representation. In the rest of this section we consider various coupled systems with such Higgs field.

A **Yang–Mills–Higgs action** $\mathcal{A}_S$, with self-interaction potential $V :$ $\mathbf{R}^+ \to \mathbf{R}^+$, is defined on the configuration space $\mathcal{C}$ by

$$\mathcal{A}_S(\omega, \phi) = c(M) \int_M \left[|F_\omega|^2 + c_1 |d^\omega \phi|^2 + c_2 V(|\phi|^2) \right] , \qquad (8.26)$$

where $c(M)$ is a **normalizing constant** that depends on the dimension of M ($c(M) = \frac{1}{8\pi^2}$ for a 4-dimensional manifold) and c_1, c_2 are the **coupling constants.** The constant c_1 measures the relative strengths of the gauge field and its interaction with the Higgs field. When $c_1 \neq 0$, the constant c_2/c_1 measures the relative strengths of the Higgs field self-interaction and the gauge field–Higgs field interaction. A **Yang–Mills–Higgs system with self-interaction potential** V is defined as a critical point $(\omega, \phi) \in C$ of the action $\mathcal{A}_S$. The corresponding Euler–Lagrange equations are

$$\delta^\omega F_\omega + c_1[\phi, d^\omega \phi] = 0, \tag{8.27}$$

$$\delta^\omega d^\omega \phi + c_2 V'(|\phi|^2)\phi = 0, \tag{8.28}$$

where $V'(x) = dV/dx$. Equations (8.27), (8.28) are called the Yang–Mills–Higgs field equations with self-interaction potential V. In the physics literature, it is customary to define the current J by

$$J = -c_1[\phi, d^\omega \phi]. \tag{8.29}$$

Using this definition of the current we can rewrite equation (8.27) as follows:

$$\delta^\omega F_\omega = J. \tag{8.30}$$

Note that in these equations δ^ω is the formal L^2-adjoint of the corresponding map d^ω. Thus, in (8.27) δ^ω is a map

$$\delta^\omega : \Lambda^2(M, \operatorname{ad} P) \to \Lambda^1(M, \operatorname{ad} P)$$

and in (8.28) δ^ω is a map

$$\delta^\omega : \Lambda^1(M, \operatorname{ad} P) \to \Lambda^0(M, \operatorname{ad} P) = \Gamma(\operatorname{ad} P).$$

The pair (ω, ϕ) also satisfies the following Bianchi identities

$$d^\omega F_\omega = 0, \tag{8.31}$$
$$d^\omega \cdot d^\omega \phi = [F_\omega, \phi]. \tag{8.32}$$

Note that these identities are always satisfied whether or not equations (8.27), (8.28) are satisfied.

We note that no solution in closed form of the Yang–Mills–Higgs equations with self-interaction potential is known for $c_2 > 0$, but existence of spherically symmetric solutions is known. Also, existence of solutions for the system is known in dimensions 2 and 3. It can be shown that the solutions of the system of equations (8.27), (8.28) on $\mathbf{R}^n$ satisfy the following relation (Jaffe and Taubes [207]):

$$(n-4)|F_\omega|^2 + (n-2)c_1|d^\omega \phi|^2 + nc_2|V(|\phi|^2)| = 0.$$

From this relation it follows that for $c_1 \geq 0$, $c_2 \geq 0$ there are no non-trivial solutions for $n > 4$, and for $n = 4$ every solution decouples (i.e., is equivalent to a pure Yang–Mills solution.) For the Yang–Mills–Higgs system on $\mathbf{R}^3$ with $c_1 = 1$ and $c_2 \geq 0$ we have the following result [169].

Theorem 8.8 *Let* $V(t) = (1 - t)^2(1 + at)$, $a \geq 0$. *Then for* c_2 *sufficiently small, there exists a positive action solution to equations (8.27), (8.28) which is not gauge equivalent to a spherically symmetric solution. Furthermore, for* c_2 *smaller still, there exists a solution that has the above properties and is, in addition, not a local minimum of the action.*

Writing $c_1 = 1$ and $c_2 = 0$ in (8.26) we get the usual Yang–Mills–Higgs action

$$\mathcal{A}_H(\omega, \phi) = c(M) \int_M [|F_\omega|^2 + |d^\omega \phi|^2]. \tag{8.33}$$

The corresponding Yang–Mills–Higgs equations are

$$\delta^\omega F_\omega + [\phi, d^\omega \phi] = 0, \tag{8.34}$$

$$\delta^\omega d^\omega \phi = 0. \tag{8.35}$$

In the above discussion the Higgs field was defined to be a section of the Lie algebra bundle $\operatorname{ad} P$. We now remove this restriction by defining the **generalized Higgs bundle**

$$H_\rho := P \times_\rho H$$

to be the vector bundle associated to P by the representation ρ of the gauge group G on the vector space H (which may be real or complex). We assume that H admits a G-invariant metric. A section $\psi \in \Gamma(H_\rho)$ is called the **generalized Higgs field**. In what follows we shall drop the adjective "generalized" when the representation ρ is understood.

8.5.1 Monopoles

Finite action solutions of equations (8.27), (8.28) are called **solitons**. Locally, these solutions correspond to time-independent, finite-energy solutions on $\mathbf{R} \times M$. The Yang–Mills instantons discussed earlier are soliton solutions of equations (8.27), (8.28) corresponding to $\phi = 0$, $c_2 = 0$. The soliton solutions on a 2-dimensional base are known as **vortices** and those on a 3-dimensional base are known as **monopoles**. Vortices and monopoles have many properties that are qualitatively similar to those of the Yang–Mills instantons.

When M is 3-dimensional, we can associate to the Yang–Mills–Higgs equations the first order **Bogomolnyi equations** [46]

$$F_\omega = \pm * d^\omega \phi. \tag{8.36}$$

Equation (8.36) is also referred to as the **monopole equation**. The Bianchi identities imply that each solution (ω, ϕ) of the Bogomolnyi equations (8.36) is a solution of the second order Yang–Mills–Higgs equations. In fact, one can show that such solutions of the Yang–Mills–Higgs equations are global minima on each connected component of the Yang–Mills–Higgs **monopole configuration space** $\mathcal{C}_m$ defined by

$$\mathcal{C}_m = \{(\omega, \phi) \in \mathcal{C} \mid \mathcal{A}_H(\omega, \phi) < \infty, \ \lim_{y \to \infty} \sup\{\hat{\phi}(x)\| \mid |x| \ge y\} = 0\},$$

where $\hat{\phi}(x) = |1 - |\phi(x)||$. Locally, the Bogomolnyi equations are obtained by applying the reduction procedure to the instanton or anti-instanton equations on $\mathbf{R}^4$. No such first order equations corresponding to the Yang–Mills–Higgs equations with self-interaction potential are known. In particular, a class of solutions of the Bogomolnyi equations on $\mathbf{R}^3$ has been studied extensively. They are the most extensively studied special class of the Yang–Mills monopole solutions. If the gauge group is a compact, simple Lie group, then every solution (ω, ϕ) of the Bogomolnyi equations, satisfying certain asymptotic conditions, defines a gauge-invariant set of integers. These integers are topological invariants corresponding to elements of the second homotopy group $\pi_2(G/J)$, where J is a certain subgroup of G obtained by fixing the boundary conditions. In the simplest example, when the gauge group $G = SU(2)$ and $J = U(1)$, we have $\pi_2(G/J) \cong \pi_2(S^2) \cong \mathbf{Z}$ and there is only one integer $N(\omega, \phi) \in \mathbf{Z}$ that classifies the monopole solutions. It is called the **monopole number** or the **topological charge** and is defined by

$$N(\omega, \phi) = \frac{1}{4\pi} \int\limits_{R^3} d^\omega \phi \wedge F , \qquad (8.37)$$

where we have written F for F_ω. It can be shown that, with suitable decay of $|\phi|$ in $\mathbf{R}^3$, $N(\omega, \phi)$ is an integer and we have

$$N(\omega, \phi) = \frac{1}{4\pi} \int\limits_{R^3} d^\omega \phi \wedge F$$

$$= \lim_{r \to \infty} \frac{1}{4\pi} \int\limits_{S_r^2} \langle \phi, F \rangle$$

$$= \deg\{\phi/|\phi| : S_r^2 \to SU(2)\},$$

where $\langle \ , \ \rangle$ is the inner product in the Lie algebra. In fact, Groisser [170, 169] has proved that in classical $SU(2)$ Yang–Mills–Higgs theories on $\mathbf{R}^3$ with a Higgs field in the adjoint representation, an integer-valued monopole number is canonically defined for any finite action smooth configuration and that the monopole configuration space essentially has the homotopy type of $\text{Maps}(S^2, S^2)$ (regarded as maps from the sphere at infinity to the sphere in the Lie algebra $su(2)$) with infinitely many path components, labeled by the monopole number.

The Yang–Mills equations and the Yang–Mills–Higgs equations share several common features. As we noted above, locally, the Yang–Mills–Higgs equations are obtained by a dimensional reduction of the pure Yang–Mills equations. Both equations have solutions, which are classified by topological invariants. For example, the G-instanton solutions over S^4 are classified by $\pi_3(G)$. For simple Lie groups G this classification goes by the integer defined by the Pontryagin index or the instanton number, whereas the monopole solutions over $\mathbf{R}^3$ are classified by $\pi_2(G/J)$. The first order instanton equations correspond to the first order Bogomolnyi equations and both have solution spaces that are parametrized by manifolds with singularities or moduli spaces. However, there are important global differences in the solutions of the two systems that arise due to different boundary conditions. For example, no translation-invariant non-trivial connection over $\mathbf{R}^4$ can extend to S^4. Extending the analytical foundations laid in [344, 386, 387], Taubes proved the following theorem:

Theorem 8.9 *There exists a solution to the $SU(2)$ Yang–Mills–Higgs equations that is not a solution to the Bogomolnyi equations.*

The corresponding problem regarding the relation of the solutions of the full Yang–Mills equations and those of the instanton equations has been solved in [350], where the existence of non-dual solutions to pure Yang–Mills equations over S^4 is also established. Other non-dual solutions have been obtained in [332]. The basic references for material in this section are Atiyah and Hitchin [18] and Jaffe and Taubes [207]. For further developments see [169, 194].

8.6 Spontaneous Symmetry Breaking

We shall consider the spontaneous symmetry breaking in the context of a Lagrangian formulation of field theories. We therefore begin by discussing some examples of standard Lagrangian for matter fields, gauge fields, and their interactions. An example of a Lagrangian for spin $1/2$ fermions interacting with a scalar field ϕ_1 and a pseudo-scalar field ϕ_2 is given by

$$L = -\bar{\psi}(\mathcal{D}\psi + m\psi) + y_1\phi_1\bar{\psi}\psi + iy_2\phi_2\bar{\psi}\gamma_5\psi, \qquad (8.38)$$

where ψ is the 4-component spinor field of the fermion and $\bar{\psi}$ is its conjugate spinor field and $\mathcal{D} = \gamma^\mu\partial_\mu$ is the usual Dirac operator. Thus, the first term in the Lagrangian corresponds to the Dirac Lagrangian. The second and third terms, representing couplings of scalar and pseudo-scalar fields to fermions are called **Yukawa couplings**. The constants y_1, y_2 are called the **Yukawa coupling constants**. They are introduced to give mass to the fermions. The field equations for the Lagrangian (8.38) are given by

$$(\mathcal{D} + m)\psi = y_1\phi_1\psi + iy_2\phi_2\gamma_5\psi \tag{8.39}$$

and the conjugate of equation (8.39). The Lagrangian for a spin 1/2 fermion interacting with an Abelian gauge field (e.g., electromagnetic field) is given by

$$L = -\bar{\psi}(\mathcal{D}^\omega\psi + m\psi) - \frac{1}{4}F_{\mu\nu}F^{\mu\nu}, \tag{8.40}$$

where $\mathcal{D}^\omega$ is the Dirac operator coupled to the gauge potential ω and F_ω is the corresponding gauge field. A local expression for $\mathcal{D}^\omega$ is obtained by the minimal coupling to the local gauge potential A by

$$\mathcal{D}^\omega = \gamma^\mu\nabla_\mu^\omega \equiv \gamma^\mu(\partial_\mu - iqA_\mu). \tag{8.41}$$

We recall that the principle of minimal coupling to a gauge potential signifies the replacement of the ordinary derivative ∂_μ in the Lagrangian by the gauge-covariant derivative ∇_μ^ω. In the physics literature it is customary to denote $\gamma^\mu\partial_\mu$ by $\not{\partial}$ for any vector ∂_μ (∂_μ is regarded as a vector operator). The field equations corresponding to (8.40) are given by

$$\nabla_\lambda^\omega F_{\mu\nu} + \nabla_\mu^\omega F_{\nu\lambda} + \nabla_\nu^\omega F_{\lambda\mu} = 0, \tag{8.42}$$

$$\partial^\nu F_{\mu\nu} = J_\mu := iq\bar{\psi}\gamma_\mu\psi, \tag{8.43}$$

$$(\gamma^\mu\partial_\mu + m)\psi = iq\gamma^\mu A_\mu\psi \tag{8.44}$$

and the conjugate of equation (8.44). We note that (8.42) is often expressed in the equivalent forms

$$\delta^\omega F = 0 \text{ or } \nabla^\omega * F = 0,$$

where δ^ω is the formal adjoint of ∇^ω. Equation (8.43) implies the conservation of current $\partial_\mu J^\mu = 0$. The Lagrangian for spin 1/2 fermions interacting with a non-Abelian gauge field (e.g., an $SU(N)$-gauge field) has the same form as in equation (8.40) but with the operator D^ω defined by

$$\mathcal{D}^\omega = \gamma^\mu\nabla_\mu^\omega \equiv \gamma^\mu(\partial_\mu - gT_aA_\mu^a), \tag{8.45}$$

where $\{T_a, 1 \leq a \leq N^2 - 1\}$ is a basis for the Lie algebra $su(N)$. The expression for the current is given by

$$(J_a)_\mu := g\bar{\psi}\gamma_\mu T_a\psi.$$

The law of conservation of current takes the form

$$\nabla_\mu^\omega J_a^\mu = 0 \text{ or } \partial_\mu J_a^\mu - f_{ab}^c A_\mu^b J_c^\mu = 0,$$

where f_{ab}^c are the structure constants of the Lie algebra of $su(N)$.

We now discuss the idea of spontaneous symmetry breaking in classical field theories by considering the Higgs field $\psi \in \Gamma(H_\rho)$ with self-interaction and then in interaction with a gauge field. We note that most of the observed symmetries in nature are only approximate. From classical spectral theories as well as their quantum counterparts dealing with particle spectra, we know that broken symmetries remove spectral degeneracies. Thus, for example, one consequence of spontaneous symmetry breaking of isospin is the observed mass difference between the proton and neutron masses (proton and neutron are postulated to form an isospin doublet with isospin symmetric Lagrangian). In general, if we have a field theory with a Lagrangian that admits an exact symmetry group K but gives rise to a ground state that is not invariant under K, then we say that we have **spontaneous symmetry breaking**. The ground state of a classical field theory represents the vacuum state (i.e., the state with no particles) in the corresponding quantum field theory. We now illustrate these ideas by considering an example of a classical Higgs field with self-interaction potential.

Let H_ρ be a complex line bundle. Then $\psi \in \Gamma(H_\rho)$ is called a **complex scalar field**. We shall be concerned only with a local section in some coordinate neighborhood with local coordinates (x^j). Let us suppose that the dynamics of this field is determined by the Lagrangian L defined by

$$L := \frac{1}{2}(\partial_j \bar{\psi})(\partial^j \psi) - V(\psi), \tag{8.46}$$

where $\bar{\psi} = \psi_1 - i\psi_2$ is the complex conjugate of ψ. The potential V is usually taken to be independent of the derivatives of the field. Let us consider the following form for the potential:

$$V(\psi) = m^2|\psi|^2 + \frac{m^2}{2a^2}|\psi|^4. \tag{8.47}$$

The Lagrangian L admits $U(1)$ as the global symmetry group with action given by

$$\psi \mapsto e^{i\alpha}\psi, \quad e^{i\alpha} \in U(1).$$

We note that if we regard ψ as a real doublet, this action is just a rotation of the vector ψ by the angle α and hence this Lagrangian is often called **rotationally symmetric**. A classical state in which the energy attains its absolute minimum value is called the **ground state** of the system. The corresponding quantum state of lowest energy is called the **vacuum state**. For the potential given by equation (8.47) the lowest energy state corresponds to the field configuration defined by $\psi = 0$. By considering small oscillations around the corresponding vacuum state and examining the quadratic terms in the Lagrangian we conclude that the field ψ represents a pair of massive bosons each with mass m. Let us now modify the potential V defined by equation (8.47) so that the ground state will be non-zero by changing the sign of the first term; i.e., we assume that

$$V(\psi) = -m^2|\psi|^2 + \frac{m^2}{2a^2}|\psi|^4. \tag{8.48}$$

As in the previous case the new Lagrangian admits $U(1)$ as the global symmetry group. By rewriting the potential V of equation (8.48) in the form

$$V(\psi) = \frac{m^2}{2a^2}(|\psi|^2 - a^2)^2 - \frac{m^2a^2}{2}, \tag{8.49}$$

it is easy to see that the absolute minimum of the energy is $(-m^2a^2/2)$. The manifold of states of lowest energy is the circle $\psi_1^2 + \psi_2^2 = a^2$. A vacuum state $\hat{\psi}$ corresponding to ψ has expectation value given by

$$|\langle\hat{\psi}\rangle|^2 = a^2.$$

Thus, we have a degenerate vacuum. Choosing a vacuum state breaks the $U(1)$ symmetry. The particle masses corresponding to the given system are determined by considering the spectra of small oscillations about the vacuum state. We shall carry out this analysis for the classical field by choosing the ground state $\psi_0 := a$ and expressing the Lagrangian in terms of the **shifted field** $\phi := \psi - \psi_0$. A simple calculation shows that the only term quadratic in the field variable ϕ is

$$L_{quad} = 2m^2\phi_1^2. \tag{8.50}$$

The absence of a quadratic term in ϕ_2 in (8.50) and hence in the Lagrangian of the shifted field is interpreted in physics as corresponding to a massless boson field ϕ_2. The field ϕ_1 corresponds to a massive boson field of mass $\sqrt{2}m$. The massless boson arising in this calculation is called the **Nambu–Goldstone boson**. The appearance of one or more such bosons is in fact a feature of a large class of field theories, with Lagrangians that admit a global continuous symmetry group K, but which have ground states invariant under a proper subgroup $J \subset K$. The assertion of such a statement is called **Goldstone's theorem** in the physics literature. The example that we have given above is a special case of a version of Goldstone's theorem that we now discuss.

Theorem 8.10 *Let L denote a Lagrangian that is a function of the field ϕ and its derivatives. Suppose that L admits a k-dimensional Lie group K as a global symmetry group. Let J (dim $J = j$) be the isotropy subgroup of K at a fixed ground state ϕ_0. Then by considering small oscillations about the fixed ground state one obtains a particle spectrum containing $k - j$ massless Nambu–Goldstone bosons.*

However, there is no experimental evidence for the existence of such massless bosons with the exception of the photon. Therefore, Lagrangian field theories that predict the existence of these massless bosons were considered unsatisfactory. We note that Goldstone type theorems do not apply to gauge field theories. In fact, as we observed earlier, gauge field theories predict the existence of massless gauge mesons, which are also not found in nature. We now consider the phenomenon of spontaneous symmetry breaking for the coupled system of gauge and matter fields in interaction. It is quite remarkable that in such a theory the massless Goldstone bosons do not appear and the massless gauge mesons become massive. This phenomenon is generally referred to as the **Higgs mechanism**, or **Higgs phenomenon**, although several scientists arrived at the same conclusion in considering such coupled systems (see, for example, Coleman [80]). A geometric formulation of spontaneous symmetry breaking and the Higgs mechanism is given in [148].

8.7 Electroweak Theory

We now discuss spontaneous symmetry breaking in the case of a Higgs field with self-interaction minimally coupled to a gauge field and explain the corresponding Higgs phenomena. For the Higgs field we choose an isodoublet field ψ (i.e., $\psi \in \Gamma(H_\rho)$, where ρ is the fundamental 2-dimensional complex representation of the isospin group $SU(2)$) in interaction with an $(SU(2) \times U(1))$-gauge field. This choice illustrates all the important features of the general case and also serves as a realistic model of the electroweak theory of Glashow, Salam, and Weinberg. Experimental verification of the gauge bosons predicted by their theory as carrier particles of the weak force earned them the Nobel Prize for physics in 1979.

We take the base M to be a 4-dimensional pseudo-Riemannian manifold. Since we work in a local coordinate chart we may in fact consider the theory to be based on $\mathbf{R}^4$ in the Euclidean case and on the Minkowski space M^4 in the case of Lorentz signature. The principal bundle P over M with the gauge group $SU(2) \times U(1)$ and all the associated bundles are trivial in this case. We choose a fixed trivialization of these bundles and omit its explicit mention in the rest of this section. Let us suppose that the dynamics is governed by the free field Lagrangian

$$L := \frac{1}{2}(\partial_j \psi^\dagger)(\partial^j \psi) - V(\psi), \qquad (8.51)$$

where $\psi^\dagger$ is the Hermitian conjugate of ψ and the potential V is given by

$$V(\psi) = \frac{m^2}{2a^2}(|\psi|^2 - a^2)^2 - \frac{m^2 a^2}{2}. \qquad (8.52)$$

Let us write the field ψ as a pair of complex fields

$$\psi = \begin{pmatrix} \pi_1 + i\pi_2 \\ \nu_1 + i\nu_2 \end{pmatrix}. \qquad (8.53)$$

Then we have
$$|\psi|^2 = \pi_1^2 + \pi_2^2 + \nu_1^2 + \nu_2^2.$$

We want to minimally couple the field ψ to an $(SU(2) \times U(1))$-gauge field. The gauge potential ω takes values in the Lie algebra $su(2) \oplus u(1)$ of the gauge group and the pull-back of ω to the base splits naturally into a pair (A, B) of gauge potentials; A with values in $su(2)$ and B with values in $u(1)$. Let F (resp., G) denote the gauge field corresponding to the gauge potential A (resp., B). Then the Lagrangian of the coupled fields is given by

$$L_S(\omega, \psi) = c(M)[|F|^2 + |G|^2 + c_1(|\nabla^\omega \psi|^2 - V(\psi))] , \qquad (8.54)$$

where $c(M)$ is a normalizing constant that depends on the dimension of M ($c(M) = 1/(8\pi^2)$ for a 4-dimensional manifold) and the constant c_1 measures the relative strengths of the gauge field and its interaction with the Higgs field. In what follows we put $c_1 = 1$. The covariant derivative operator ∇^ω is defined by

$$\nabla_j^\omega \psi := (\partial_j + ig_1 A_j^a \sigma_a + ig_2 B_j \iota)\psi , \qquad (8.55)$$

where the subscript j corresponds to the coordinate x^j, constants g_1, g_2 are the **gauge coupling constants**, while the σ_a, $a = 1, 2, 3$, are the Pauli spin matrices, which form a basis for the Lie algebra $su(2)$ and ι is the 2×2 identity matrix. It is easy to see that $E_{\min}$, the absolute minimum of the energy of the coupled system, is given by

$$E_{\min} = -\frac{m^2 a^2}{2}.$$

The manifold of states of lowest energy is the 3-sphere

$$A^a = 0, \quad B = 0, \quad \pi_1^2 + \pi_2^2 + \nu_1^2 + \nu_2^2 = a^2.$$

A vacuum state $\hat{\psi}_0$ corresponding to the lowest energy state ψ_0 has expectation value given by
$$|\langle \hat{\psi}_0 \rangle|^2 = a^2.$$

Thus we have a degenerate vacuum. Choosing a vacuum state breaks the $SU(2) \times U(1)$ symmetry. Let us choose

$$\psi_0 = \begin{pmatrix} 0 \\ a \end{pmatrix} . \qquad (8.56)$$

It is easy to verify that $Q\psi_0 = 0$, where

$$Q := \frac{1}{2}(\iota + \sigma_3). \qquad (8.57)$$

Thus we see that the symmetry is not completely broken. The stability group of ψ_0, generated by the **charge operator** Q defined in (8.57), is isomorphic to

$U(1)$. Thus, in the free field situation, there would be three Goldstone bosons. However, in coupling to the gauge field, these Goldstone bosons disappear as we now show. As before, we consider small oscillations about the vacuum to determine the particle spectrum. The shifted fields are A, B, and $\phi = \psi - \psi_0$. We look for quadratic terms in these shifted fields in the Lagrangian. The potential contributes a quadratic term in ν_1, which corresponds to a massive scalar boson of mass $\sqrt{2}m$. The only other quadratic terms arise from the term involving the covariant derivative. Since the charge operator annihilates the ground state ψ_0, we rewrite the covariant derivative as follows:

$$\nabla_j^\omega := \partial_j + ig_1 A_j^1 \sigma_1 + ig_1 A_j^2 \sigma_2 + \tfrac{i}{2}(-g_1 A_j^3 + g_2 B_j)(\iota - \sigma_3)$$
$$+ i(g_1 A_j^3 \sigma_a + ig_2 B_j)Q. \tag{8.58}$$

Using this expression, it is easy to see that quadratic terms in the gauge fields are

$$g_1^2 a^2 (|A_j^1|^2 + |A_j^2|^2) + a^2 |(-g_1 A_j^3 + g_2 B_j)|^2. \tag{8.59}$$

To remove the mixed terms in expression (8.59) we introduce two new fields Z^0, N by rotating in the A^3, B-plane by an angle θ as follows:

$$Z_j^0 = A_j^3 \cos\theta - B_j \sin\theta , \tag{8.60}$$
$$N_j = A_j^3 \sin\theta + B_j \cos\theta , \tag{8.61}$$

where the angle θ is defined in terms of the gauge coupling constants by

$$g_1 = g\cos\theta, \qquad g_2 = g\sin\theta, \tag{8.62}$$

or, equivalently,

$$\tan\theta = g_2/g_1, \qquad g^2 = g_1^2 + g_2^2. \tag{8.63}$$

The angle θ (also denoted by θ_{ew} or θ_w) is called the **mixing angle** of the electroweak theory. Using the fields Z^0, N the quadratic terms in the gauge fields can be rewritten as

$$g_1^2 a^2 (|A^1|^2 + |A^2|^2) + g^2 a^2 |Z^0|^2. \tag{8.64}$$

It is customary to define two complex fields $W^\pm$ by

$$W^+ := \frac{1}{\sqrt{2}}(A^1 + iA^2), \qquad W^- := \frac{1}{\sqrt{2}}(A^1 - iA^2) \tag{8.65}$$

and to write the quadratic terms in the Lagrangian as follows:

$$L_{\text{quad}} = 2m^2 \nu_1^2 + g_1^2 a^2 (|W^+|^2 + |W^-|^2) + g^2 a^2 |Z^0|^2. \tag{8.66}$$

The terms in equation (8.66) allow us to identify the complete particle spectrum of the theory. The absence of a term in N indicates that it corresponds to a massless field, which can be interpreted as the electromagnetic field. In the covariant derivative N is coupled to the charge operator Q by the term $g \sin 2\theta Q N_j$. In the usual electromagnetic theory, this term is eQN_j, where e is the electric charge. Comparing these, we get the following relation for the mixing angle θ:

$$g \sin 2\theta = e. \tag{8.67}$$

The electric charge e is related to the fine structure constant $\alpha \approx 1/137$ by the relation $e = \sqrt{4\pi\alpha}$. There are three massive vector bosons, two W^+, W^- of equal mass $m_W = g_1 a$ and the third Z^0 of mass $m_Z = ga$. In the electroweak theory the carrier particles W^+, W^- are associated with the charged intermediate bosons, which mediate the β-decay processes, while the Z^0 is associated with the neutral intermediate boson, which is supposed to mediate the neutral current processes. The masses of the W and Z bosons are related to the mixing angle by the relation

$$m_W = m_Z \cos\theta. \tag{8.68}$$

The term containing ν_1 in equation (10.39) represents the massive Higgs particle. The mass of this particle is called the **Higgs mass** and is denoted by m_H. It can be expressed, in terms of the parameters in the Higgs potential (10.24), as

$$m_H = \sqrt{2}m.$$

We note that, while there is strong experimental evidence for the existence of the W and the Z particles, there is no evidence for the existence of the Higgs particle.

Up to this point our calculations have been based on the Lagrangian of equation (8.54). This Lagrangian is part of the **standard model Lagrangian**, which forms the basis of the **standard model** of electroweak interactions. We conclude this section with a brief discussion of this model.

The phenomenological basis for the formulation of the standard model of the electroweak theory evolved over a period of time. The four most important features of the theory, based on experimental observations, are the following:

1. The family structure of the fermions under the **electroweak gauge group** $SU_w(2) \times U_Y(1)$ consists of left-handed doublets and right-handed singlets characterized by the **quantum numbers** I, I_3 of the **weak isospin group** $SU_w(2)$ and the **weak hypercharge** Y. For the leptons we have

$$\begin{pmatrix} e \\ \nu_e \end{pmatrix}_L, \begin{pmatrix} \mu \\ \nu_\mu \end{pmatrix}_L, \begin{pmatrix} \tau \\ \nu_\tau \end{pmatrix}_L, \ e_R, \ \mu_R, \ \tau_R$$

and for the quarks we have

$$\begin{pmatrix} u \\ d \end{pmatrix}_L, \begin{pmatrix} c \\ s \end{pmatrix}_L, \begin{pmatrix} t \\ b \end{pmatrix}_L, \ u_R, \ d_R, \ c_R, \ \dots .$$

The subscript L (resp., R) in these terms is used to denote the left-handed fermions (resp., right-handed fermions).

2. The physical fields are also subject to finite symmetry groups. Three fundamental symmetries of order 2 are the **charge conjugation** C, **parity** P, and **time reversal** T. These symmetries may be preserved or violated individually or in combination. The electroweak theory is left-right asymmetric. This result is called the **parity violation**. Thus, contrary to what many physicists believed, nature does distinguish between left and right. Lee and Yang had put forth an argument predicting that parity could be violated in weak interactions. They also suggested a number of experiments to test this violation. This parity violation was verified experimentally by Chien-Shiung Wu in 1957. Lee and Yang were awarded the Nobel Prize for physics in the year 1957 for their fundamental work on the topic. However, it is Yang's work on the matrix-valued generalization of the Maxwell equations to the Yang–Mills equations and the relation of these equations with the theory of connections in a principal bundle over the space-time manifold that has had the most profound effect on recent developments in geometric topology.

If Φ_L (resp., Φ_R) denotes the set of all left-handed (resp., right-handed) fermions, then the assignment of the hypercharge quantum number must satisfy the following conditions:

$$\sum_{\Phi_{L2}} Y = 0, \qquad \sum_{\Phi_L} Y^3 = \sum_{\Phi_R} Y^3,$$

where Φ_{L2} denotes the set of left-handed fermion doublets.

3. The quantum numbers of the fermions with respect to the electroweak gauge group and their electric charge Q satisfy the **Gell-Mann–Nishijima relation**

$$Q = I_3 + \frac{1}{2}Y. \tag{8.69}$$

4. There are four vector bosons;

$$\gamma, \ W^+, \ W^-, \ Z^0.$$

These act as carriers of the electroweak force. The first of these is identified as the photon, which is massless, while the other three, collectively called the **weak intermediate bosons**, are massive.

Taking into account the above features and applying the general principles of constructing gauge-invariant field theory with spontaneous symmetry breaking, one arrives at the following standard model Lagrangian for electroweak theory:

$$L_{\text{ew}} = L_G + L_H + L_F + L_Y.$$

L_Y is the Yukawa Lagrangian, and L_G, the electroweak gauge field Lagrangian, is given by

$$L_G = c(M)[|F|^2 + |G|^2].$$

The Lagrangian L_H of the Higgs field in interaction with the electroweak gauge field is given by

$$L_H = c(M)[|\nabla^\omega \psi|^2 - V(\psi)],$$

and the Lagrangian L_F of the fermion fields in interaction with the electroweak gauge field is given by

$$L_F = \sum \bar{\psi}^L i\gamma_j \nabla_j^\omega \psi^L + \sum \bar{\psi}^R i\gamma_j \nabla_j^\omega \psi^R.$$

In the definition of L_F the first sum extends over all left-handed doublets and the second sum extends over all right-handed singlets and the electroweak gauge covariant derivative ∇^ω is given by equation (8.55). In addition to the above terms it is necessary to introduce Yukawa couplings to fermions, to give masses to charged fermions. We denote by g_f the Yukawa coupling constant to fermion f. Thus, there is one Yukawa coupling constant for each family of leptons and quarks. These are contained in the Yukawa Lagrangian L_Y, whose detailed expression is not relevant to our discussion. Using the standard model Lagrangian, one can construct models for the weak and electromagnetic interactions of various types of known particles, such as leptons, and predict the outcome of processes mediated by the heavy intermediate bosons. A striking experimental confirmation of the predictions of electroweak theory in the late 1970s must be viewed as a big boost to the gauge theory point of view in elementary particle physics. Several physicists had been working on the electroweak theory. Glashow, Salam, and Weinberg working independently arrived at essentially similar results including the mass ratios of the electroweak bosons. The three shared the 1979 Nobel Prize for physics for this work, paving the way for the current version of the standard model.

8.7.1 The Standard Model

The above considerations on electroweak theory can be extended to construct the standard model of a unified theory of the strong, weak, and electromagnetic interactions as a gauge theory with the gauge group G_{SM} defined by

$$G_{\mathrm{SM}} := SU_c(3) \times SU_L(2) \times U_Y(1).$$

The gauge potential of the theory corresponds to spin 1 vector bosons. These split naturally into the gauge bosons of the electroweak theory and the gluons $G^a, 1 \leq a \leq 8$, which correspond to the color gauge group $SU_c(3)$ of the strong interaction with coupling constant g_3. Quantized Yang–Mills theory with the full gauge group G_{SM} is the foundation of the theory of elementary particles. However, the mechanism of mass generation for gluons, the carrier particles of the strong interaction, is not via the Higgs mechanism. It uses a quantum mechanical property called the "mass gap" in the physics literature. This property was discovered by experimental physicists but its mathematical foundation is unclear. Putting the quantum Yang–Mills theory on a mathematical foundation and explaining the mass gap is one of the seven millennium problems announced by the Clay Mathematics Institute. We refer the reader to their web page at www.claymath.org for more details. The full Lagrangian L_{SM} of the theory can be written along the lines of the electroweak Lagrangian given above. The Lagrangian L_{SM} contains the following, rather large, number of free parameters:

1. three gauge coupling constants g_1, g_2, g_3
2. Higgs coupling constant
3. Yukawa coupling constants
4. Higgs particle mass m_H
5. quark-mixing matrix elements
6. number of matter families or generations (there is strong evidence that this number is 3).

In Tables 6.1 and 6.2 we have already given the masses and electric charges of the fermions and bosons that enter into the description of the standard model. Initially the isospin group $SU(2)$ of the Yang–Mills theory was expected to provide an explanation of the spectrum of strongly interacting particles. However, it turned out that a larger group, $SU(3)$, was needed to explain the observed particles. The representations of $SU(3)$ have been well known in mathematics since the work of Weyl in the 1930s. In the 1950s, however, Weyl's work was not included in physicists' training. In fact, Pauli had dubbed group theory "Gruppenpest," and most physicists found the mathematical developments quite hard to understand and felt that they would be of little relevance to their work. The mathematicians also did not understand physicists work and did not feel that it would contribute to a better understanding of group theory. When **Murray Gell-Mann** started to study the representations of $SU(3)$, he never discussed his problem with his lunchtime companion J.-P. Serre, one of the leading experts in representation theory. Upon returning to Caltech from Paris, where he had discussions with his mathematician colleagues, Gell-Mann was able to show that the observed particles did fit into representations of $SU(3)$. The simplest representation he could use was 8-dimensional and he started to call this symmetry the **eightfold way** (perhaps an allusion to Buddha's eight steps to Nirvana). He could fit 9 of the particles into a 10-dimensional representation and was able to predict the properties of the missing 10th particle. When this particle was found at Brookhaven, it made the theory of group representations an essential tool for particle theory. Gell-Mann's work had not used the fundamental or defining representation of $SU(3)$ as it predicted particles with charges equal to a fraction of the electron charge. Such particles have never been observed.

Gell-Mann was nevertheless convinced of the existence of these unobserved or confined particles as the ultimate building blocks of matter. He called them **quarks**. Gell-Mann expects the quarks to be unobservable individually. They would be confined in groups making the fundamental particles.

The single mixing angle of the electroweak theory must be replaced by a 3×3 matrix in the study of strong interactions. The **Cabibbo–Kobayashi–Maskawa (CKM) matrix**, also known as the **quark mixing matrix**, is a unitary matrix, which contains information on the **flavor changing** of quarks under weak interaction. Kobayashi and Maskawa shared the 2008 Nobel Prize in physics with Nambu for their work on the origin of broken symmetry, which predicts the existence of at least three families of quarks. Their work was based on the earlier work of Cabibbo, who was not honored with the Prize. An **up type quark**, labeled (u, c, t), can decay into a **down type quark**, labeled (d, s, b). The entries of the CKM matrix are denoted by $V_{\alpha\beta}$, where α is an up type quark and β is a down type quark. The **transition probability** of going from quark α to quark β is proportional to $|V_{\alpha\beta}|^2$. Using this notation, the CKM matrix V can be written as follows:

$$V := \begin{pmatrix} V_{ud} & V_{us} & V_{ub} \\ V_{cd} & V_{cs} & V_{cb} \\ V_{td} & V_{ts} & V_{tb} \end{pmatrix}. \tag{8.70}$$

The CKM matrix is a generalization of the **Cabibbo angle**, which was the only parameter needed when only two generations of quarks were known. The Cabibbo angle is a mixing angle between two generations of quarks. The current model with three generations has three mixing angles. Unitarity of the CKM matrix implies three relations involving only the absolute values of the entries of the matrix V. These relations are

$$|V_{\alpha d}|^2 + |V_{\alpha s}|^2 + |V_{\alpha b}|^2 = 1, \qquad \alpha = u, c, t. \tag{8.71}$$

The relations of (8.71) imply that the sum of all couplings of any of the up type quarks to all the down type quarks is the same for all generations. This was first pointed out by Cabibbo. These relations are called the **weak universality**. Unitarity of the CKM matrix also implies the following three phase-dependent relations.

$$V_{\alpha d}V_{\beta d} + V_{\alpha s}V_{\beta s} + V_{\alpha b}V_{\beta b} = 0, \tag{8.72}$$

where α, β are any two distinct up type quarks.

These phase-dependent relations are crucial for an understanding of the origin of **CP violation**. For fixed α, β each of the relations says that the sum of the three terms is zero. This implies that they form a triangle in the complex plane. Each of these triangles is called a **unitary triangle**, and their areas are the same. The area is zero for the specific choice of parameters in the standard model for which there is no CP violation. Non-zero area indicates

CP violation. The CPM matrix elements are not determined by theory at this time; various experiments for their determination are being carried out at laboratories around the world. It was the observation of CP violation in weak interactions that led to the existence of a third generation of quarks. Experimental evidence for the existence of a third generation b-quark came in 1977 through the work of the Fermi Lab, Chicago group led by Leon Lederman. For this work he was awarded the Noble Prize in 1988.

Predictions of the standard model have been verified with great accuracy and present there are no experimental results that contradict this model. However, there are a number of unsatisfactory features of this theory, e.g., the problem of the origin of mass generation, prediction of the Higgs particle mass, and so on.

We conclude this section with a brief look at some consistency checks of grand unified theories from recent experimental results. Since 1989, the Z-factory at LEP (large electron–positron collider at CERN, Geneva) has produced a large number of electron–positron collisions for the resonance production of Z^0 gauge bosons to perform a number of high-precision experiments for determinatinig of the mass of **asthenons** (weak intermediate vector bosons $W^{\pm}$ and Z^0) and to find lower bounds on the mass of **higgsons**[1] (Higgs bosons) and the top quark. These same LEP data have been used to perform consistency checks of grand unified theories and to study the problem of unification of the three coupling constants of the standard model (see, for example, [10]). It has been shown that the standard model is in excellent agreement with all LEP data. However, an extrapolation of the three independent running coupling constants to high energies indicates that their unification within the standard model is highly unlikely. On the other hand, the minimal supersymmetric extension of the standard model leads to a unification of couplings at a scale compatible with the present data. As Richard Feynman once observed "In physics, the best one can hope for is to be temporarily not wrong."

8.8 Invariant Connections

In many physically interesting situations, in addition to the local internal symmetry provided by the gauge group, an external global action of another Lie group exists on the space of phase factors. Mathematically, we can describe the situation as follows. We consider, as before, a principal bundle $P(M, G)$ with gauge group G, which represents the space of phase factors. For the external symmetry group we take a Lie group K acting on P by bundle automorphisms. Then every element of K induces a diffeomorphism of M. Thus, K may be regarded as a group of generalized gauge transfor-

---

[1] I thank Prof. Ugo Amaldi of the University Milan for suggesting these particle names and for useful discussions on the analysis of LEP data.

mations as defined in Chapter 6. We fix a point $u_0 \in P$ as a reference point and let $\pi(u_0) = x_0 \in M$. Let J be the set of all elements $f \in K$ such that $f \cdot u_0 \in \pi^{-1}(x_0)$. Thus the induced map $f_0 \in \mathrm{Diff}(M)$ fixes the point x_0. We note that the set J is a closed subgroup of K. We call J the **isotropy subgroup** of K at x_0. Define a map $\psi : J \to G$ as follows. For each $f \in J$, $\psi(f)$ is the unique element $a \in G$ such that $f u_0 = u_0 a$. It is easy to see that ψ is a Lie group homomorphism and hence induces a Lie algebra homomorphism $\bar{\psi}$ of the Lie algebra $\mathbf{j}$ of J to the Lie algebra $\mathbf{g}$ of G.

Definition 8.1 *Using the notation of the above paragraph, let Γ be a connection in P with connection 1-form ω. We say that Γ is a K-**invariant connection** if each $f \in K$ preserves the horizontal distribution of Γ, or, equivalently, leaves the form ω invariant, i.e., $f^*\omega = \omega$.*

In physical literature the group K is referred to as the **external symmetry group** of the gauge potential ω.

Theorem 8.11 *Let K be a group of bundle automorphisms of $P(M, G)$ and Γ a K-invariant connection in P with connection 1-form ω. Let Ψ be the linear map from the Lie algebra $\mathbf{k}$ of K to $\mathbf{g}$ defined by*

$$\Psi(A) = \omega_{u_0}(\hat{A}), \quad A \in \mathbf{k}, \tag{8.73}$$

where $\hat{A}$ is the vector field on P induced by A. Then we have

1. $\Psi(A) = \bar{\psi}(A)$, $A \in \mathbf{j}$;
2. *if ad_J is the restriction of the adjoint representation of K in $\mathbf{k}$ to the subgroup J and ad is the adjoint representation of G in $\mathbf{g}$ (see Section 1.6), then*

$$\Psi(\mathrm{ad}_J f(A)) = \mathrm{ad}(\psi(f))(\Psi(A)), \quad f \in J, \ A \in \mathbf{k};$$

3. $2\Omega_{u_0}(\hat{A}, \hat{B}) = [\Psi(A), \Psi(B)] - \Psi([A, B])$, $A, B \in \mathbf{k}$.

We say that K acts **fiber-transitively** on P if, for any two fibers of P, there is an element of K that maps one fiber into the other or, equivalently, if the induced action on the base M is transitive. In this case M is isomorphic to the homogeneous space K/J, where J is the isotropy subgroup of K at x_0.

Theorem 8.12 *Let K be a **fiber-transitive** group of bundle automorphisms of P. Then there is a one-to-one correspondence between the set of K-invariant connections in P and the set of linear maps $\Psi : \mathfrak{k} \to \mathfrak{g}$ satisfying the first two conditions of Theorem 8.11, the correspondence being given by equation (8.73).*

If K is fiber-transitive on P, the curvature form Ω of the invariant connection ω is a tensorial form of type (ad, G) that is K-invariant. It is completely determined by the values $\Omega_{u_0}(\hat{A}, \hat{B})$, where $A, B \in \mathbf{k}$. The following corollary is an immediate consequence of condition 3 of Theorem 8.11.

Corollary 8.13 *The K-invariant connection in P defined by Ψ is flat if and only if $\Psi : \mathbf{k} \to \mathbf{g}$ is a Lie algebra homomorphism.*

The following particular situation is often useful for the construction of invariant connections. Suppose that $\mathbf{k}$ admits an ad_J-invariant subspace $\mathbf{m}$ that is complementary to $\mathbf{j}$, i.e., $\mathbf{k} = \mathbf{j} \oplus \mathbf{m}$. Then we have the following.

Theorem 8.14 *There is a one-to-one correspondence between the set of K-invariant connections in P and the set of linear maps $\Psi_{\mathbf{m}} : \mathbf{m} \to \mathbf{g}$ such that*

$$\Psi_{\mathbf{m}}(\mathrm{ad}_J f(A)) = \mathrm{ad}(\psi(f))(\Psi_{\mathbf{m}}(A)), \quad f \in J, A \in \mathbf{m}.$$

Furthermore, the curvature form Ω of this K-invariant connection satisfies, $\forall X, Y \in \mathbf{m}$, the condition

$$2\Omega_{u_0}(\hat{A}, \hat{B}) = [\Psi_{\mathbf{m}}(A), \Psi_{\mathbf{m}}(B)] - \Psi_{\mathbf{m}}([A, B]_{\mathbf{m}}) - \Psi([A, B]_{\mathbf{j}}),$$

where $[A, B]_{\mathbf{m}}$ (resp., $[A, B]_{\mathbf{j}}$) is the $\mathbf{m}$-component (resp., $\mathbf{j}$-component) of $[A, B]$ in $\mathbf{k}$.

The K-invariant connection in P defined by $\Psi_{\mathbf{m}} = 0$ is called the **canonical connection** in P with respect to the decomposition $\mathbf{k} = \mathbf{j} \oplus \mathbf{m}$. An important special case that arises in many applications is the following. Let H be a closed subgroup of a connected Lie group K and let $M = K/H$ and $P = K$ (i.e., we consider the principal bundle $K(K/H, H)$). If there exists an $\mathrm{ad}\,H$-invariant subspace $\mathbf{m}$ complementary to $\mathbf{h}$ in $\mathbf{k}$, then the $\mathbf{h}$-component ω of the canonical 1-form θ of K with respect to the decomposition $\mathbf{k} = \mathbf{h} \oplus \mathbf{m}$ defines a K-invariant connection in the bundle K (the action of K being that by left multiplication). Conversely, any such K-invariant connection (if it exists) determines a decomposition $\mathbf{k} = \mathbf{h} \oplus \mathbf{m}$. Furthermore, the curvature form Ω of the invariant connection defined by ω is given by

$$\Omega(A, B) = -\frac{1}{2}[A, B]_{\mathbf{h}}.$$

($[A, B]_{\mathbf{h}}$ is the $\mathbf{h}$-component of $[A, B]$ in $\mathbf{k}$), where A, B are left-invariant vector fields on K belonging to $\mathbf{m}$. The invariant connection ω defined above coincides with the canonical connection defined at the beginning of this paragraph.

At least locally, the Yang–Mills–Higgs equations (8.27), (8.28) may be regarded as obtained by dimensional reduction from the pure Yang–Mills equations. To see this, consider a Yang–Mills connection α on the trivial principal bundle $\mathbf{R}^{m+1} \times G$ over $\mathbf{R}^{m+1}$ and let

$$A_i \, dx^i + A_{m+1} \, dx^{m+1}, \quad 1 \le i \le m,$$

be the gauge potential on $\mathbf{R}^{m+1}$ with values in $\mathbf{g}$. Suppose that this potential does not depend on x^{m+1}. Define $\phi := A_{m+1}$; then ϕ can be regarded as a Higgs potential on $\mathbf{R}^m$ with values in $\mathbf{g}$ and

$$A = A_i \, dx^i, \quad 1 \le i \le m,$$

can be regarded as the gauge potential on $\mathbf{R}^m$ with values in $\mathbf{g}$. Let ω denote the gauge connection on $\mathbf{R}^m \times G$ corresponding to A. Then we have

$$(F_\omega)_{ij} = (F_\alpha)_{ij}, \quad (F_\alpha)_{i\,m+1} = (d^\omega \phi)_i \qquad 1 \le i, j \le m$$

and the pure Yang–Mills action of F_α on $\mathbf{R}^{m+1}$ reduces to the Yang–Mills–Higgs action (8.33). It is easy to see that this reduction is a consequence of the translation invariance of the pure Yang–Mills system in the x^{m+1}-direction. In general, suppose that T is a Lie group that acts on $P(M, G)$ as a subgroup of $\mathrm{Diff}_M(P)$ with induced action on M. We say that T is a **local symmetry group of the connection** ω on P or that ω is a **locally T-invariant connection** on P if the condition

$$L_{\hat{A}} \omega = 0, \qquad \forall A \in \mathbf{t},$$

is satisfied, where $\hat{A}$ is the fundamental vector field corresponding to the element A in the Lie algebra $\mathbf{t}$ of T. Under certain conditions P/T is a principal bundle over M/T and a system of equations on the original bundle can be reduced to a coupled system on the reduced base. Invariant connections and their applications to gauge theories over Riemannian manifolds are discussed in [129, 388, 389]. For a particular class of self-interaction potentials, such reduction and the consequent symmetry breaking are responsible for the Higgs mechanism [192]. For a geometrical description of dimensional reduction and its relation to the Kaluza–Klein theories, see [206] and the book by Coquereaux and Jadczyk [85].

Chapter 9
4-Manifold Invariants

9.1 Introduction

The concept of moduli space was introduced by Riemann in his study of the conformal (or equivalently, complex) structures on a Riemann surface. Let us consider the simplest non-trivial case, namely, that of a Riemann surface of genus 1 or the torus T^2. The set of all complex structures $\mathcal{C}(T^2)$ on the torus is an infinite-dimensional space acted on by the infinite-dimensional group $\mathrm{Diff}(T^2)$. The quotient space

$$\mathcal{M}(T^2) := \mathcal{C}(T^2)/\mathrm{Diff}(T^2)$$

is the moduli space of complex structures on T^2. Since T^2 with a given complex structure defines an elliptic curve, $\mathcal{M}(T^2)$ is, in fact, the moduli space of elliptic curves. It is well known that a point $\omega = \omega_1 + i\omega_2$ in the upper half-plane H ($\omega_2 > 0$) determines a complex structure and is called the modulus of the corresponding elliptic curve. The modular group $SL(2, \mathbf{Z})$ acts on H by modular transformations and we can identify $\mathcal{M}(T^2)$ with $H/SL(2, \mathbf{Z})$. This is the reason for calling $\mathcal{M}(T^2)$ the space of moduli of elliptic curves or simply the moduli space. The topology and geometry of the moduli space has rich structure. The natural boundary $\omega_2 = 0$ of the upper half plane corresponds to singular structures. Several important aspects of this classical example are also found in the moduli spaces of other geometric structures. Typically, there is an infinite-dimensional group acting on an infinite-dimensional space of geometric structures with quotient a "nice space" (for example, a finite-dimensional manifold with singularities). For a general discussion of moduli spaces arising in various applications see, for example, [193].

We shall be concerned with the construction of the moduli spaces of instantons and in studying their topological and geometric properties in this chapter. The simplest of these moduli spaces is the space $\mathcal{M}_1(M)$ of instantons with instanton number 1 on a compact, simply connected Riemannian manifold M with positive definite intersection form. Donaldson showed that

K. Marathe, *Topics in Physical Mathematics*, DOI 10.1007/978-1-84882-939-8_9,

this moduli space $\mathcal{M}_1(M)$ is a compact, oriented, 5-dimensional manifold with singularities and that it provides a cobordism between M and a sum of a certain number of copies of $\mathbf{CP}^2$. This result has led to the existence of new obstructions to smoothability of 4-manifolds and to some surprising and unexpected results about exotic differential structures on the standard Euclidean space $\mathbf{R}^4$. In Section 9.2 we give an explicit construction of the moduli space $\mathcal{M}_1$ of the BPST-instantons of instanton number 1 and indicate the construction of the moduli space $\mathcal{M}_k$ of the complete $(8k - 3)$-parameter family of instanton solutions over S^4 with gauge group $SU(2)$ and instanton number k. The moduli spaces of instantons on an arbitrary Riemannian 4-manifold with semisimple Lie group as gauge group are also discussed here.

A brief account of Donaldson's theorem on the topology of moduli spaces of instantons and its implications for smoothability of 4-manifolds can be found in Section 9.3. The investigation of the Riemannian geometry of these moduli spaces was begun after Donaldson's work. The results obtained for the metric and curvature of $\mathcal{M}_1$ are given and a geometrical interpretation of Donaldson's results is also indicated in this section. Donaldson's polynomial invariants are discussed in Section 9.4. The generating function of these invariants and its relation to basic classes is also given there. In the summer of 1994 Witten announced in his lecture at the International Congress on Mathematical Physics in Paris that the Seiberg–Witten theory can be used to obtain all the information contained in the Donaldson invariants. His paper [407] gave an explicit formula relating the SW invariants and the Donaldson invariants. We discuss this work in Section 9.5. Witten's derivation of this fantastic formula was based on physical reasoning. There is now a mathematical proof for a large class of manifolds. We indicate this result and some of its applications in the same section.

9.2 Moduli Spaces of Instantons

A complete set of solutions of instanton equations on S^4 was obtained by the methods of algebraic geometry and the theory of complex manifolds. These methods have also been used in the study of the Yang–Mills equations over Riemann surfaces. Since this approach is not discussed here, we simply indicate below some references where this and related topics are developed. They are [17,19,51,67,256]. See also the book [143] by Friedman and Morgan.

Let M be a compact, connected, oriented Riemannian manifold of dimension 4. Let $P(M, G)$ be a principal bundle over M with compact, semisimple Lie group G as structure group. Recall that the Hodge star operator on $\Lambda(M)$ has a natural extension to bundle valued forms and this is used to define the self-dual and anti-dual forms in $\Lambda^2(M, \operatorname{ad} P)$ and to decompose a form $\alpha \in \Lambda^2(M, \operatorname{ad} P)$ into its self-dual and anti-dual parts. A **G-instanton** (resp., **G-anti-instanton**) over M is a self-dual (resp., anti-dual) Yang–Mills field on the principal bundle P. If $F = F_\omega$ is the gauge field corresponding to the gauge connection ω, then

$$*F = F \tag{9.1}$$

is called the **instanton equation** and

$$*F = -F \tag{9.2}$$

is called the **anti-instanton equation**. Over a 4-dimensional base manifold M, the second Chern class and the Euler class of P are equal and we define the **instanton number** k of a G-instanton by

$$k := -c_2(P)[M] = -\chi(P)[M]. \tag{9.3}$$

Recall that using the decomposition of $F = F_\omega$ into its self-dual part F_+ and anti-dual part F_- (see Chapter 8) we get

$$k = \frac{1}{8\pi^2} \int_M (|F_+|^2 - |F_-|^2) \tag{9.4}$$

and

$$\mathcal{A}_{YM}(\omega) = \frac{1}{8\pi^2} \int_M (|F_+|^2 + |F_-|^2). \tag{9.5}$$

Comparing equations (9.4) and (9.5) above we see that the Yang–Mills action is bounded below by the absolute value of the instanton number k, i.e.,

$$\mathcal{A}_{YM}(\omega) \geq |k|, \qquad \forall \omega \in \mathcal{A}(P) \tag{9.6}$$

and the connections satisfying the instanton or the anti-instanton equations are the absolute minima of the action. In particular, for a self-dual ($*F = F$) Yang–Mills field, i.e., a G-instanton, we have

$$\mathcal{A}_{YM} = \frac{1}{8\pi^2} \int_M (\|F_+\|^2) = |k|.$$

Similarly, for an anti-dual ($*F = -F$) Yang–Mills field, i.e., a G-anti-instanton, we have

$$\mathcal{A}_{YM} = \frac{1}{8\pi^2} \int_M (\|F_-\|^2) = -|k|.$$

We now discuss the most well known family of instanton solutions, namely the **BPST instantons (Belavin–Polyakov–Schwartz–Tyupkin instantons)** [36]. It can be shown that the BPST instanton solution over the base manifold $\mathbf{R}^4$ can be extended to the conformal compactification of $\mathbf{R}^4$, i.e., S^4. This extension is characterized by a self-dual connection on a non-trivial

$SU(2)$-bundle over S^4. This bundle is the quaternionic Hopf fibration of $\mathbf{H}^2$ over $\mathbf{HP}^1$, where $\mathbf{H}$ is the space of quaternions and $\mathbf{HP}^1$ is the quaternionic projective space, of quaternionic lines through the origin in the quaternionic plane $\mathbf{H}^2$. We identify $\mathbf{H}$ with $\mathbf{R}^4$ by the map $\mathbf{H} \to \mathbf{R}^4$ given by

$$x = x_0 + ix_1 + jx_2 + kx_3 \longmapsto (x_0, x_1, x_2, x_3).$$

Then $\mathbf{H}^2$ is isomorphic to $\mathbf{R}^8$ and each quaternionic line intersects the 7-sphere $S^7 \subset \mathbf{H}^2$ in S^3. On the other hand, the base $\mathbf{HP}^1$ can be identified with S^4. Thus, the quaternionic Hopf fibration leads to the bundle

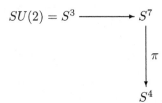

As indicated in the diagram, the fiber S^3 can be identified with the group $SU(2)$ of unit quaternions and its action on $\mathbf{H}^2$ by right multiplication restricts to S^7. This makes it into a principal $SU(2)$-bundle over S^4. This follows from the observation that $\alpha \in SU(2)$ and $(x, y) \in S^7 \subset \mathbf{H}^2$ imply that $(\alpha x, \alpha y) \in S^7$ and that the $SU(2)$-action is free. This principal bundle is clearly non-trivial and admits a canonical connection ω_1 (also called the universal connection), whose curvature Ω_1 is self-dual and hence satisfies the instanton equation (and hence also the full Yang–Mills equations). It corresponds to a BPST instanton of instanton number 1. The entire BPST family of instantons can be generated from this solution as follows. The group $SO(5,1)$ acts on S^4 by conformal transformations and this action induces an action on ω_1. If $g \in SO(5,1)$, then we denote the induced action on ω_1 also by g. Then $g\omega_1$ is also an $SU(2)$-connection over S^4 and it has self-dual curvature. Since the Yang–Mills action is conformally invariant, the solution generated by $g\omega_1$ also has instanton number 1. The connection $g\omega_1$ is gauge-equivalent to ω_1 (and hence determines the same point in the moduli space) if and only if g is an isometry of S^4, i.e., if and only if $g \in SO(5) \subset SO(5,1)$. Thus, the space of gauge-inequivalent, self-dual, $k = 1$, $SU(2)$-connections on S^4, or the moduli space $\mathcal{M}_1(S^4, SU(2))$, is given by

$$\mathcal{M}_1(S^4, SU(2)) = SO(5,1)/(SO(5)).$$

We note that the quotient space $SO(5,1)/(SO(5))$ can be identified with the hyperbolic 5-space H_5.

We now give an explicit local formulation of the BPST family. Consider the chart defined by the map ψ_{e_5}

$$\psi_{e_5} : S^4 - \{e_5\} \to \mathbf{R}^4,$$

where ψ_{e_5} is obtained by projecting from $e_5 = (0,0,0,0,1) \in \mathbf{R}^5$ onto the tangent hyperplane to S^4 at $-e_5$. This chart gives conformal coordinates on S^4. Identifying $\mathbf{R}^4$ with $\mathbf{H}$, the metric can be written as

$$ds^2 = 4|dx|^2/(1+|x|^2),$$

where $|x|^2 = x\bar{x}$, $|dx|^2 = dxd\bar{x}$. Identifying the Lie algebra $su(2)$ with the set of pure imaginary quaternions, we can write the gauge potential $A^{(1)}$ corresponding to the universal connection ω_1, as follows:

$$A^{(1)} = \mathrm{Im}\left(\frac{xd\bar{x}}{1+|x|^2}\right)$$

It is possible to give a similar expression for the potential in the chart obtained by projecting from $-e_5$ and to show that these expressions are compatible, under a change of charts and hence define a global connection with corresponding Yang–Mills field. However, we apply the removable singularities theorem of Uhlenbeck (see [386]) to guarantee the extension of the local connection to all of S^4. This procedure is also useful in the general construction of multi-instantons, where there is a finite number of removable singularities. The Yang–Mills field $F^{(1)}$, corresponding to the gauge potential $A^{(1)}$, is given by

$$F^{(1)} = \frac{dx \wedge d\bar{x}}{(1+|x|^2)^2}.$$

Using the formula

$$\mathrm{vol}(S^{2m}) = \frac{2^{2m+1}\pi^m m!}{(2m)!}$$

and calculating the Euclidean norm $|F^{(1)}|^2_{Eu} = 3$, we can evaluate the Yang–Mills action

$$\mathcal{A}_{YM} = \frac{1}{8\pi^2}\int_{S^4} |F^{(1)}|^2_{Eu} = 3 \cdot \mathrm{vol}(S^4)/(8\pi^2) = 1.$$

On the other hand, $\mathcal{A}_{YM} = k$ for a self-dual instanton with instanton number k. Thus we see that the above solution corresponds to $k = 1$. We now apply the conformal dilatation

$$f_\lambda : \mathbf{R}^4 \to \mathbf{R}^4 \text{ defined by } x \mapsto x/\lambda, \quad 0 < \lambda < 1,$$

to obtain induced connections $A^{(\lambda)} = f_\lambda^* A^{(1)}$ and the corresponding Yang–Mills fields $F^{(\lambda)} = f_\lambda^* F^{(1)}$. Because of conformal invariance of the action $\mathcal{A}_{YM}$, the fields $F^{(\lambda)}$ have the same instanton number $k = 1$. The local expressions for $A^{(\lambda)}$ and $F^{(\lambda)}$ are given by

$$A^{(\lambda)} = \text{Im}\left(\frac{xd\bar{x}}{\lambda^2 + |x|^2}\right),$$

and

$$F^{(\lambda)} = \frac{\lambda^2 dx \wedge d\bar{x}}{(\lambda^2 + |x|^2)^2}.$$

If we write the above expressions in terms of the quaternionic components we recover the formulas for the BPST instanton. The connections corresponding to different λ are gauge-inequivalent, since $|F|^2$ is a gauge-invariant function. Moreover, choosing an arbitrary point $q \in S^4$ and considering the chart obtained by projecting from this point, we obtain gauge inequivalent connections for different choices of q. In fact, we have a 5-parameter family of instantons parametrized by the pairs (q, λ), where $q \in S^4$ and $\lambda \in (0, 1)$. Two pairs (q_1, λ_1) and (q_2, λ_2) give gauge-equivalent connections if and only if $q_1 = q_2$ and $\lambda_1 = \lambda_2$. It is customary to call q the **center** of the instanton and λ its **size**. The map defined by

$$(q, \lambda) \mapsto (1 - \lambda)q \quad \text{from } S^4 \times (0, 1) \text{ to } \mathbf{R}^5$$

is an isomorphism onto the punctured open ball $B^5 - \{0\}$. The universal connection $A^{(1)}$ corresponds to the origin. Thus, the moduli space of gauge inequivalent-instantons is identified with the open unit ball B^5, which is the Poincaré model of the hyperbolic 5-space H_5. The connection corresponding to (q, λ) as $\lambda \to 0$ can be identified with a boundary point of the open ball B^5. This realizes S^4 as the boundary of the ball B^5. Thus our base space appears as the boundary of the moduli space. This is one of the key ideas of Donaldson in his work on the topology of the moduli space of instantons [106]. The moduli space of the fundamental BPST instantons or self-dual $SU(2)$ Yang–Mills fields with instanton number 1 over the Euclidean 4-sphere S^4 is denoted by $\mathcal{M}_1^+$. It can be shown [20] that the action of the group $SO(5, 1)$ of conformal diffeomorphisms of S^4 induces a transitive action of $SO(5, 1)$ on the moduli space $\mathcal{M}_1^+$ with isotropy group $SO(5)$. Thus $\mathcal{M}_1^+$ is diffeomorphic to the homogeneous hyperbolic 5-space $SO(5, 1)/SO(5)$. In particular, the topology of $\mathcal{M}_1^+$ is the same as that of $\mathbf{R}^5$. A more general result of Donaldson, discussed in the next section, shows that any 1-connected 4-manifold M with positive definite intersection form, can be realized as a boundary of a suitable moduli space.

Most of the solutions of the pure Yang–Mills equations that have been constructed are in fact solutions of the self-dual or anti-dual instanton equations, i.e., instantons or anti-instantons. The instanton and anti-instanton solutions are also called collectively **pseudo-particle solutions**. The first such solutions, consisting of a 5-parameter family of self-dual Yang–Mills fields on $\mathbf{R}^4$, was constructed by Belavin, Polyakov, Schwartz, and Tyupkin ([36]) in 1975 and it is these solutions or their extension to S^4 that are commonly referred to as the **BPST instantons**. We will show that the BPST instanton solutions correspond to the gauge group $SU(2)$ over the base manifold S^4 and have instanton number $k = 1$. In this case the instanton number k is defined by

$$k := -c_2(P(S^4, SU(2)))[S^4],$$

where c_2 denotes the second Chern class of the bundle P and $[S^4]$ denotes the fundamental cycle of the manifold S^4. The BPST solution was generalized to the so-called **multi-instanton solutions**, which correspond to the self-dual Yang–Mills fields with instanton number k. A $5k$-parameter family of solutions was obtained by 't Hooft (unpublished) and a $(5k+4)$-parameter family was obtained by Jackiw, Nohl, and Rebbi in 1977. For a given instanton number k the maximum number of parameters in the corresponding instanton solution can be identified with the dimension of the space of gauge inequivalent solutions (the moduli space). For $SU(2)$-instantons over S^4 this dimension of the moduli space was computed by Atiyah, Hitchin, and Singer ([20]) by using the Atiyah–Singer index theorem and turns out to be $8k - 3$. Thus, for $k = 1$ the moduli space has dimension 5 and the 5-parameter BPST solutions correspond to this space. For $k > 1$, the 't Hooft and Jackiw, Nohl, Rebbi solutions do not give all the possible instantons with instanton number k. Instead of describing these solutions we will describe briefly the construction of the most general $(8k - 3)$-parameter family of solutions, which include the above solutions as special cases.

The construction of the $(8k-3)$-parameter family of instantons was carried out by Atiyah, Drinfeld, Hitchin, and Manin [19] by using the methods of analytic and algebraic geometry. It is called the **ADHM construction**. It starts by generalizing the **twistor space** construction of Penrose (see Wells [399] for a discussion of twistor spaces and their applications in field theory) as follows.

Atiyah, Hitchin and Singer [20] studied the properties of self-dual connections over self-dual base manifolds M, which have positive scalar curvature. In this case the bundle of projective anti-dual spinors PV_- has a complex structure and the **The Ward correspondence:** holds:

Let E be a Hermitian vector bundle with self-dual connection over a compact, self-dual 4-manifold M. Let PV_- denote the bundle of projective anti-dual spinors on M. Then we have the following commutative diagram

$$
\begin{array}{ccc}
F & \xrightarrow{\hat{h}} & E \\
\Big\downarrow{\scriptstyle h^*p} & & \Big\downarrow{\scriptstyle p} \\
PV_- & \xrightarrow{h} & M
\end{array}
$$

where $F := h^*E$ is the pull-back of the bundle E to PV_-. Then

1. F is holomorphic on PV_- with holomorphically trivial fibers.

2. Let $\rho : PV_- \to PV_-$ denote the real structure on PV_-. Then there is a holomorphic bundle isomorphism $\sigma : \rho^* \bar{F} \to F$, which induces a positive definite Hermitian structure on the space of holomorphic sections of F on each fiber.

3. Every bundle F on PV_- satisfying the above two conditions is the pullback of a Hermitian vector bundle E over M, with self-dual connection.

When $M = S^4$, the bundle PV_- can be identified with $\mathbf{CP}^3$. This is the original twistor space of Penrose. In this case standard techniques from algebraic geometry can be used to study the bundle F and to obtain all self-dual connections on G-bundles over S^4. It turns out that for $G = SU(2)$, the general solution can be described explicitly by starting with a vector $\lambda = (\lambda_1, \ldots, \lambda_k) \in \mathbf{H}^k$ and a $k \times k$ symmetric matrix B over $\mathbf{H}$ satisfying the following conditions:

1. $(B^\dagger B + \lambda^\dagger \lambda)$ is a real matrix ($\dagger$ is the quaternionic conjugate transpose).
2. $\operatorname{rank} \begin{pmatrix} \lambda \\ B - xI \end{pmatrix} = k, \quad \forall x \in \mathbf{H}$.

The local expression for the gauge potential $A^{(\lambda, B)}$ determined by the pair (λ, B), $\lambda \in \mathbf{H}^k$, $B \in \mathbf{H}_{k \times k}$, is given by

$$A^{(\lambda, B)} = f^*(A(u)),$$

where $u = (u_1, \ldots, u_k) \in \mathbf{H}^k$, $f : \mathbf{H} \to \mathbf{H}^k$ is defined by $f(x) = [\lambda(B - xI)^{-1}]^\dagger, \forall x \in \mathbf{H}$, and $A(u)$ is the $SU(2)$-gauge potential on the space $\mathbf{H}^k$ given by

$$A(u) = \operatorname{Im} \left(\frac{u \, du^\dagger}{1 + |u|^2} \right).$$

The gauge field $F^{(\lambda, B)}$ corresponding to the gauge potential $A^{(\lambda, B)}$ contains a finite number of removable singularities and hence, by applying Uhlenbeck's theorem on removable singularities, the solution can be extended smoothly to S^4. The potentials determined by (λ, B) and (λ', B') are gauge-equivalent if and only if there exists $\alpha \in SU(2)$ and $T \in O(k)$ such that

$$\lambda' = \alpha \lambda T \qquad \text{and} \qquad B' = T^{-1} B T.$$

We may carry out a naïve counting of free real parameters as follows. The pair (λ, B) gives

$$4k + 4 \left[\tfrac{1}{2} k(k+1) \right] = 2k^2 + 6k \tag{9.7}$$

real parameters. The reality condition (i) above involves $3k(k-1)/2$ parameters. The condition (ii) on the rank does not restrict the number of parameters. The gauge equivalence further reduces the number of parameters in (9.7) by 3 (by the $SU(2)$-action) and by $k(k-1)/2$ (by the $O(k)$-action). Thus, we have the following count for free real parameters.

$$2k^2 + 6k - \tfrac{3}{2} k(k-1) - 3 - \tfrac{k}{2}(k-1) = 8k - 3.$$

It can be shown that this family of $(8k - 3)$-parameter solutions exhausts all the possible solutions up to gauge equivalence. Thus, the space of these solutions may be identified as the moduli space $\mathcal{M}_k(S^4, SU(2))$ of k-instantons (i.e., instantons with instanton number k) over S^4. In the theorem below we give an alternative characterization of this moduli space.

Theorem 9.1 *Let k be a positive integer. Write*

$$B(q) = A_1 q_1 + A_2 q_2,$$

where, $q = (q_1, q_2) \in \mathbf{H}^2$, and $A_i \in \mathrm{Hom}_{\mathbf{H}}(\mathbf{H}^k, \mathbf{H}^{k+1})$, $i = 1, 2$. Define the manifold M and Lie group G by

$$M := \{(A_1, A_2) \mid B(q)^\dagger B(q) \in GL(k, \mathbf{R}), \ \forall q \neq 0\},$$

$$G := (Sp(k+1) \times GL(k, \mathbf{R}))/\mathbf{Z}_2.$$

Then $M \neq \emptyset$, G acts freely and properly on M by the action

$$(a, b) \cdot (A_1, A_2) := (aA_1 b^{-1}, aA_2 b^{-1}),$$

and the quotient space M/G is a manifold of dimension $8k - 3$, which is isomorphic to the moduli space $\mathcal{M}_k(S^4, SU(2))$ of instantons of instanton number k.

9.2.1 Atiyah–Jones Conjecture

Soon after the differential geometric setting of the Yang–Mills theory was established, Atiyah and Jones [22] took up the study of the topological aspects of the theory over the 4-sphere S^4. They proved the following theorem.

Theorem 9.2 *Let $\mathcal{B}_k$ (resp., $\mathcal{M}_k$) denote the moduli space of based gauge-equivalence classes of connections (resp., instantons) on the principal bundle $P_k(S^4, SU(2))$ with instanton number k. Then $\mathcal{B}_k$ is homotopy equivalent to the third loop space $\Omega_k^3(SU(2))$, and the natural forgetful map*

$$\theta_k : \mathcal{M}_k \to \mathcal{B}_k \cong \Omega_k^3(SU(2))$$

induces a surjection $\hat{\theta}_k$ in homology through the range $q(k)$; i.e., there exists a function $q(k)$ such that

$$\hat{\theta}_k^i : H_i(\mathcal{M}_k) \to H_i\left(\Omega_k^3(SU(2))\right) \quad for \ i \leq q(k) << k$$

is a surjective homomorphism.

In its strong form, the Atiyah–Jones conjecture states that θ_k is a homotopy equivalence through the range $q(k)$ and that the function $q(k)$ can be explicitly determined as a function of k. The stable version of the conjecture was proved by Taubes in [365]. In fact, he proved the following theorem for the general case of the gauge group $SU(n)$.

Theorem 9.3 *Let $\mathcal{M}_\infty$ (resp., θ) be the direct limit of the moduli spaces $\mathcal{M}_k$ (resp., maps θ_k). Then*

$$\theta : \mathcal{M}_\infty \to \Omega_0^3(SU(n)), \qquad n \geq 2$$

is a homotopy equivalence.

Boyer et al. [63] proved the following theorem, settling the Atiyah–Jones conjecture in the affirmative.

Theorem 9.4 *For all $k > 0$ the natural inclusion map*

$$\theta_k : \mathcal{M}_k \to \Omega_k^3(SU(2))$$

is a homotopy equivalence through dimension $q(k) = [k/2] - 2$, i.e., θ_k induces an isomorphism $\hat{\theta}_k$ in homotopy

$$\hat{\theta}_k^i : \pi_i(\mathcal{M}_k) \to \pi_i(\Omega_k^3(SU(2))) = \pi_{i+3}((SU(2)) \qquad \forall i \leq q(k).$$

In particular, using the known homotopy groups of $SU(2)$ we can conclude that the low-dimensional homotopy groups of $\mathcal{M}_k$ are finite.

The result of this theorem has been extended to the gauge group $SU(n)$ in [375] (see also [61,62]).

The long period between the formulation of the Atiyah–Jones conjecture and its proof is an indication of the difficulty involved in the topology and geometry of the moduli spaces of gauge potentials, even for the case when the topology and the geometry of the base manifold is well known. In fact, an explicit description of moduli spaces of instantons on S^4 available through the ADHM formalism [19] did not ease the situation. A surprising advance did come through Donaldson's study of the moduli space $\mathcal{M}_1(M, SU(2))$, where M is a positive definite 4-manifold. He showed that $\mathcal{M}_1$ is a 5-manifold with singularities and used the topology of $\mathcal{M}_1$ to obtain new obstructions to smoothability of M.

We begin by considering Taubes's fundamental contributions to the analysis of Yang–Mills equations. His work on the existence of self-dual and anti-dual solutions of Yang–Mills equations on various classes of 4-manifolds paved the way for many later contributions including Donaldson's. A direct application of the steepest descent method to find the global minima of the Yang–Mills functional $\mathcal{A}_{YM}$ does not work in dimension 4, although this technique does work in dimensions 2 and 3. In fact, the problem of finding critical points of $\mathcal{A}_{YM}$ in 4 dimensions is a conformally invariant variational problem. It has some features in common with sigma models (the harmonic map problem in 2 dimensions). Taubes used analytic techniques to study the problem of existence of self-dual connections on a manifold. This allowed him to remove some of the restrictions on the base manifolds that were required in earlier work. He does impose a topological condition on the base manifold M, namely that there be no anti-dual harmonic 2-forms on M. This condition can be written using the second cohomology group of the de Rham complex:

$$p_-(H^2_{deR}(M; \mathbf{R})) = 0.$$

This condition is equivalent to the statement that the intersection form ι_M of M be positive definite. It is precisely this condition that appears in Donaldson's fundamental theorem, which gives a new obstruction to smoothability of a 4-manifold. We discuss this theorem in the next section.

In the rest of this section we take M to be a compact, connected, oriented, 4-dimensional, Riemannian manifold with a compact, connected, semisimple Lie group G as gauge group. We begin by recalling the classification of principal G-bundles over M. The isomorphism classes of principal bundles $P(M, G)$ are in one-to-one correspondence with the elements of the set $[M; BG]$ of homotopy classes of maps of M into the classifying space BG for G. Using this isomorphism, we can consider the isomorphism class $[P]$ as an element of the set $[M; BG]$. Since G is semisimple, its Lie algebra $\mathfrak{g}$ is a direct sum of a finite number of non-trivial simple ideals, i.e., there exists a positive integer k such that

$$\mathfrak{g} = \mathfrak{g}_1 \oplus \mathfrak{g}_2 \oplus \cdots \oplus \mathfrak{g}_k.$$

In this case, we can write the first Pontryagin class of the vector bundle $\mathrm{ad}(P)$ as a vector

$$\mathbf{p} = (p^1, p^2, \ldots, p^k) \in \mathbf{Z}^k.$$

Each isomorphism class of P uniquely determines a set of multipliers $\{r_{\mathfrak{g}_i}\}$, one for each simple ideal $\mathfrak{g}_i$ of $\mathfrak{g}$, which we write as a vector

$$\mathbf{r} = (r_1, r_2, \ldots, r_k) \in \mathbf{Q}^k.$$

where we have written r_i for $r_{\mathfrak{g}_i}$. The values of $r_\mathfrak{h}$ when $\mathfrak{h}$ is the Lie algebra of the simple group H are given in Table 9.1

Table 9.1 Multipliers for the Pontryagin classes of P

H	$SU(n)$	$Spin(n)$	$Sp(n)$	G_2	F_4	E_6	E_7	E_8
$r_\mathfrak{g}$	$4n$	$4n-2$	$4n+4$	16	36	48	72	120

The classification of principal bundles by Pontryagin classes can be described as follows.

Proposition 9.5 *Define the map*

$$\phi : [M; BG] \to \mathbf{Z}^k \quad by \ [P] \mapsto (r_1^{-1}p^1, r_2^{-1}p^2, \ldots, r_k^{-1}p^k).$$

Then we have the following result. The map ϕ is a surjection and there exists a map η

$$\eta : [M; BG] \to H^2(M; \pi_1(G))$$

such that ϕ restricted to the kernel of η is an isomorphism. In particular, if G is simply connected, then ϕ is an isomorphism.

We now state two of Taubes' theorems on the existence of irreducible, self-dual, Yang–Mills connections on 4-manifolds.

Theorem 9.6 *Let M be a compact, connected, oriented, Riemannian, 4-manifold with $p_- H^2_{deR}(M) = 0$ and let G be a compact, semisimple Lie group. Then there exists a principal bundle $P(M, G)$ which admits irreducible self-dual connections.*

Moreover, we have the following theorem.

Theorem 9.7 *Let M, G be as in the previous theorem and let $P(M, G)$ be a principal bundle with non-negative Pontryagin classes such that $[P] \in \ker \eta$. Then we have the following.*

1. *The space $\mathcal{A}(P)$ contains a self-dual connection.*
2. *Let $Q(S^4, G)$ be a principal bundle with the same Pontryagin classes as P. If Q admits an irreducible, self-dual connection B, then there exists an irreducible connection A on P. Furthermore, such an A has a neighborhood in $\mathcal{A}(P)/\mathcal{G}(P)$ in which the moduli space of irreducible self-dual connections is a manifold of dimension*

$$p_1(\mathrm{ad}\, P) - \tfrac{1}{2}(\dim G)(\chi + \sigma),$$

where $p_1(\mathrm{ad}\, P) = \sum_{i=1}^k p^i$ is the sum of the Pontryagin classes of $\mathrm{ad}\, P$, χ is the Euler characteristic of M, and σ is the signature of M.

The analytical tools developed in these and related theorems by Taubes play a fundamental role in the construction and analysis of Yang–Mills fields and their moduli spaces on several classes of manifolds. We now discuss a special class of these moduli spaces, namely the moduli spaces of instantons.

We restrict ourselves to considering self-dual Yang–Mills fields on M with gauge group G, i.e., with G-instantons. Similar considerations apply to anti-dual fields or anti-instantons. Since both the instanton and anti-instanton solutions are used in the literature, we shall state important results for both. The configuration space of G-instantons is denoted by $\mathcal{C}^+(P)$ and is defined by

$$\mathcal{C}^+(P) := \{\omega \in \mathcal{A} \mid F_\omega = *F_\omega\},$$

where $\mathcal{A}$ is the space of gauge connections. The group $\mathcal{G}$ of gauge transformations acts on $\mathcal{C}^+(P)$ and the quotient space under this action is called the

moduli space of instantons with instanton number k. It is denoted by $\mathcal{M}_k^+(M, G)$, or simply by $\mathcal{M}_k(M, G)$. Thus, we have

$$\mathcal{M}_k(M, G) := \mathcal{C}^+(P(M, G))/\mathcal{G}.$$

We now briefly indicate how the dimension of the moduli space $\mathcal{M}_k(M, G)$ can be computed by applying the Atiyah–Singer index theorem. The fundamental elliptic complex comes from a modification of the generalized de Rham sequence. For a vector bundle E over M associated to the principal bundle $P(M, G)$, the generalized de Rham sequence can be written as

$$0 \longrightarrow \Lambda^0(M, E) \xrightarrow{d^\omega} \Lambda^1(M, E) \xrightarrow{d^\omega} \Lambda^2(M, E) \xrightarrow{d^\omega} \cdots,$$

where $\Lambda^p(M, E) = \Gamma(\Lambda^p(T^*M) \otimes E)$. On the oriented Riemannian 4-manifold M the space $\Lambda^2(T^*M)$ splits under the action of the Hodge $*$ operator into a direct sum of self-dual and anti-dual 2-forms. This splitting extends to $\Lambda^2(M, E)$ so that

$$\Lambda^2(M, E) = \Lambda^2_+(M, E) \oplus \Lambda^2_-(M, E),$$

where $\Lambda^2_+(M, E)$ (resp., $\Lambda^2_-(M, E)$) is the space of self-dual (resp., anti-dual) 2-forms with values in the vector bundle E. Let

$$p_\pm : \Lambda^2(M, E) \to \Lambda^2_\pm(M, E)$$

be the canonical projections defined by

$$p_\pm := \tfrac{1}{2}(1 \pm *).$$

Then we have the following diagram

$$
\begin{array}{c}
\hspace{6cm} \Lambda^2_+(M, E) \\
\hspace{6cm} \uparrow{\scriptstyle p_+} \\
0 \longrightarrow \Lambda^0(M, E) \xrightarrow{d^\omega} \Lambda^1(M, E) \xrightarrow{d^\omega} \Lambda^2(M, E) \xrightarrow{d^\omega} \cdots \\
\hspace{6cm} \downarrow{\scriptstyle p_-} \\
\hspace{6cm} \Lambda^2_-(M, E)
\end{array}
$$

We now consider the special case of $E = \mathrm{ad}(P) = P \times_{\mathrm{ad}} \mathfrak{g}$, the Lie algebra bundle associated to P with fiber $\mathfrak{g}$, and define

$$d_{\pm}^{\omega} : \Lambda^1(M, E) \to \Lambda^2_{\pm}(M, E) \quad \text{by } d_{\pm}^{\omega} := p_{\pm} \circ d^{\omega}.$$

Now the curvature 2-form F_{ω} splits into its self-dual part $(F_{\omega})_+ \in \Lambda^2_+(M, E)$ and its anti-dual part $(F_{\omega})_- \in \Lambda^2_-(M, E)$. Thus, for a self-dual (resp., anti-dual) gauge connection ω

$$d_-^{\omega} \circ d^{\omega} = (F_{\omega})_- = 0 \quad (\text{resp., } d_+^{\omega} \circ d^{\omega} = (F_{\omega})_+ = 0). \tag{9.8}$$

Using equation (9.8) and a suitable part of the above diagram we obtain the fundamental **instanton deformation complex** (or simply the **instanton complex**)

$$0 \longrightarrow \Lambda^0(M, E) \xrightarrow{d^{\omega}} \Lambda^1(M, E) \xrightarrow{d_-^{\omega}} \Lambda^2_-(M, E) \longrightarrow 0$$

Similarly, one has the **anti-instanton deformation complex** (or simply the **anti-instanton complex**). It can be shown that the instanton (resp., anti-instanton) complex is an elliptic complex. The formal dimension of the instanton moduli space is determined by the cohomology of this complex. We define the **twisted Dirac operator**

$$\mathcal{D}_t : \Lambda^1(M, E) \to \Lambda^0(M, E) \oplus \Lambda^2_-(M, E)$$

by

$$\mathcal{D}_t(\alpha) := \delta^{\omega}\alpha \oplus d_-^{\omega}\alpha, \quad \alpha \in \Lambda^1(M, E),$$

where δ^{ω} is the formal adjoint of d^{ω}. It can be shown that the twisted Dirac operator $\mathcal{D}_t$ is elliptic. We now compute its analytic index by studying the cohomology of the original instanton complex. There are three Hodge–de Rham Laplacians

1. $\Delta_0^{\omega} := \delta^{\omega} d^{\omega}$ on $\Lambda^0(M, E)$,
2. $\Delta_1^{\omega} := d^{\omega} \delta^{\omega} + \delta^{\omega} d_-^{\omega}$ on $\Lambda^1(M, E)$,
3. $\Delta_2^{\omega} := d_-^{\omega} \delta^{\omega}$,

and their associated harmonic spaces, which determine the cohomology spaces H^i, $0 \le i \le 2$, of the instanton complex. We denote by h_i the dimension of H_i for all i. Let $\mathcal{M}_k \subset \mathcal{A}/\mathcal{G}$ denote the moduli space of irreducible instantons of instanton number k. For a self-dual connection ω the tangent space $T_{\omega}\mathcal{M}_k$ can be identified with $\ker d_-^{\omega} / \mathrm{Im}\, d^{\omega}$. Thus, at least formally $\dim \mathcal{M}_k = h_1$. Under suitable conditions on M and the gauge group, one can show that $h_0 = 0 = h_2$ and that $\mathcal{M}_k$ is a manifold. Thus, the index $\mathrm{Ind}(\mathcal{D}_t) = -h_0 + h_1 - h_2 = h_1$ equals the formal dimension of the moduli space $\mathcal{M}_k(M, G)$. This index can be computed by using the Atiyah–Singer index theorem and leads to the following result:

$$\dim \mathcal{M}_k(M, G) = 2\, ch(P \times_{\mathrm{ad}} \mathbf{g_C})[M] - \tfrac{1}{2} \dim G(\chi(M) + \sigma(M)), \qquad (9.9)$$

where ch is the Chern character, $\mathbf{g_C} = \mathbf{g} \otimes \mathbf{C}$ is the complexification of the Lie algebra $\mathbf{g}$, $\chi(M)$ is the Euler characteristic of M and $\sigma(M)$ is the Hirzebruch signature of M. If M is 1-connected then $b_0 = 1 = b_4$ and $b_1 = 0 = b_3$. Thus, $\chi(M) = 2 + b_2 = 2 + b_2^+ + b_2^-$ and $\sigma(M) = b_2^+ - b_2^-$. This gives

$$\dim \mathcal{M}_k = c(P) - \tfrac{1}{2} \dim G(2 + 2b_2^+) = c(P) - \dim G(1 + b_2^+),$$

where $c(P)$ is a constant depending on P. For example, if P is an $SU(N)$-bundle over M with instanton number k, then

$$\dim \mathcal{M}_k = 4kN - (N^2 - 1)(1 + b_2^+).$$

In the particular case of $N = 2$, we get

$$\dim \mathcal{M}_k(M, SU(2)) = 8k - 3(1 + b_2^+).$$

We shall use this result in our discussion of the Donaldson polynomial invariants of M later in this chapter.

If $M = S^4$ then $\chi(M) = 2$ and $\sigma(M) = 0$; thus formula (9.9) reduces to

$$\dim \mathcal{M}_k(S^4, G) = 2\, ch(P \times_{\mathrm{ad}} \mathbf{g_C})[S^4] - \dim G. \qquad (9.10)$$

We now use this formula to obtain dimensions of moduli spaces of instantons over S^4 for some standard non-Abelian gauge groups.

Proposition 9.8 *Let k denote the instanton number of the principal bundle $P(S^4, G)$. Then we have the following results.*

1. Let $G = SU(n), \; n \geq 2$. Then

$$\dim G = n^2 - 1 \; \text{ and } \; ch(P \times_{\mathrm{ad}} \mathbf{g_C})[S^4] = 2nk.$$

Thus, equation (9.10) becomes

$$\dim \mathcal{M}_k(S^4, SU(n)) = 4nk - n^2 + 1. \qquad (9.11)$$

Applying formula (9.11) to the particular case of $G = SU(2)$ we obtain

$$\dim \mathcal{M}_k(S^4, SU(2)) = 8k - 3.$$

For instanton number $k = 1$, we get

$$\dim \mathcal{M}_1(S^4, SU(2)) = 5,$$

corresponding to the BPST family of solutions. In the last section we have given an explicit geometric construction of the $(8k - 3)$-parameter family of instantons with instanton number k.

2. Let $G = Spin(n), n > 3$, where $Spin(n)$ is the universal covering group of $SO(n)$. Then

$$\dim G = \tfrac{n}{2}(n-1) \quad and \quad ch(P \times_{\mathrm{ad}} \mathbf{g}_{\mathbf{C}})[S^4] = 2(n-2)k.$$

Thus, equation (9.10) becomes

$$\dim \mathcal{M}_k(S^4, Spin(n)) = 4(n-2)k - \tfrac{n}{2}(n-1), \quad k \geq n/2. \tag{9.12}$$

For small values of n some of the gauge groups are locally isomorphic. For example, $Spin(6)$ is locally isomorphic to $SU(4)$. Thus, (9.11) and (9.12) lead to the same dimension for the corresponding moduli spaces in this case.

3. *Let $G = Sp(n)$. Then*

$$\dim G = n(2n+1), \quad and \quad ch(P \times_{\mathrm{ad}} \mathbf{g}_{\mathbf{C}})[S^4] = 2(n+1)k,$$

and we obtain

$$\dim \mathcal{M}_k(S^4, Sp(n)) = 4(n+1)k - n(2n+1), \quad k \geq n/4. \tag{9.13}$$

We note that the groups $SU(2)$, $Spin(3)$, and $Sp(1)$ are isomorphic and hence the corresponding moduli spaces are homeomorphic. Each of these groups have been used to study the topology and geometry of the fundamental moduli space of BPST instantons from different perspectives (see, for example, [103, 172, 173, 181]).

9.3 Topology and Geometry of Moduli Spaces

Study of the differential geometric and topological aspects of the moduli space of Yang–Mills instantons on a 4-dimensional manifold was initiated by Donaldson [104]. Building on the analytical work of Taubes [364, 363] and Uhlenbeck [385, 386, 387], Donaldson studied the space $\mathcal{M}_1^+(M)$, where M is a compact, simply connected, differential 4-manifold with positive definite intersection form. He showed that for such M the intersection form is equivalent to the unit matrix. Freedman had proved a classification theorem for topological 4-manifolds [136], which shows that every positive definite form occurs as the intersection form of a topological manifold. A spectacular application of this classification theorem is his proof of the 4-dimensional Poincaré conjecture. Donaldson's result showed the profound difference in the differentiable and topological cases in dimension 4 to the known results in dimensions greater than 4. In particular, these results imply the existence of exotic 4-spaces, which are homeomorphic but not diffeomorphic to the standard Euclidean 4-space R^4. Soon many examples of such exotic $\mathbf{R}^4$

were found ([157]). A result of Taubes gives an uncountable family of exotic $\mathbf{R}^4$ and yet this list of examples is not exhaustive. The question of the existence of exotic differentiable structures on compact 4-manifolds has also been answered in the affirmative (see, for example [343]). Using instantons as a powerful new tool Donaldson has opened up a new area of what may be called **gauge-theoretic mathematics** (see the book by Donaldson and Kronheimer [112], Donaldson's papers [105, 107, 109] and [136]).

We now give a brief account of Donaldson's theorem and its implications for the classification of smooth 4-manifolds. In Chapter 2 we discussed the classification of a class of topological 4-manifolds. In addition to these there are two other categories of manifolds that topologists are interested in studying. These are the category PL of piecewise linear manifolds and the category $DIFF$ of smooth manifolds. We have the inclusions

$$DIFF \subset PL \subset TOP.$$

In general, these are strict inclusions, however, it is well known that every piecewise linear 4-manifold carries a unique smooth structure compatible with its piecewise linear structure. Freedman's classification of closed, 1-connected, oriented 4-manifolds does not extend to smooth manifolds. In the smooth category the situation is much more complicated. In fact, we have the following theorem:

Theorem 9.9 (Rochlin) *Let M be a smooth, closed, 1-connected, oriented, spin manifold of dimension 4. Then 16 divides $\sigma(M)$, the signature of M.*

Now as we observed earlier, 8 always divides the signature of an even form but 16 need divide it. Thus, we can define the **Rochlin invariant** $\rho(\mu)$ of an even form μ by

$$\rho(\mu) := \tfrac{1}{8}\sigma(\mu) \pmod 2.$$

We note that the Rochlin invariant and the Kirby–Siebenmann invariant are equal in this case, but for non-spin manifolds the Kirby–Siebenmann invariant is not related to the intersection form and thus provides a further obstruction to smoothability. From Freedman's classification and Rochlin's theorem it follows that a topological manifold with non-zero Rochlin invariant is not smoothable. For example, the topological manifold $|E_8| := \iota^{-1}(E_8)$ corresponds to the equivalence class of the form E_8 and has signature 8 (Rochlin invariant 1) and hence is not smoothable.

For nearly two decades very little progress was made beyond the result of the above theorem in the smooth category. Then, in 1982, through his study of the topology and geometry of the moduli space of instantons on 4-manifolds Donaldson discovered the following unexpected result, which has led to a number of important results, including the existence of uncountably many exotic differentiable structures on the standard Euclidean topological space $\mathbf{R}^4$.

Theorem 9.10 (Donaldson) *Let M be a smooth, closed, 1-connected, oriented manifold of dimension 4 with positive definite intersection form ι_M. Then $\iota_M \cong b_2(1)$, the diagonal form of rank b_2, the second Betti number of M.*

Proof: We given only a sketch of the main ideas involved in the proof. The theorem is proved by consideration of the solution space of the instanton field equations

$$*F_\omega = F_\omega,$$

where ω is the gauge potential on a principal $SU(2)$-bundle P over M with instanton number (Euler class) 1 and F_ω is the corresponding $SU(2)$-gauge field on M with values in the bundle $\mathrm{ad}(P)$. The instanton equations are a semi-elliptic system of nonlinear partial differential equations, which are a nonlinear generalization of the well-known Dirac equation on M. The group $\mathcal{G}$ of gauge transformations acts as a symmetry group on the solution space of the instanton equations. The quotient space under this group action is the moduli space $\mathcal{M}_1^+(M)$ (or simply $\mathcal{M}$) of instantons with instanton number 1. Recall that the action of $\mathcal{G}$ is not, in general, free due to the existence of reducible connections. The space $\mathcal{M}$ depends on the choice of the metric on the base M. By perturbing the metric or the instanton equations we obtain a nearby moduli space (also denoted by $\mathcal{M}$), which can be shown to be an oriented 5-dimensional manifold with boundary and with a finite number k of singular points. In fact, moduli space $\mathcal{M}$ has the following properties:

1. Let k be half the number of solutions to the equation

$$\iota_M(\alpha, \alpha) = 1, \quad \alpha \in H^2(M, \mathbf{Z}).$$

   Then for almost all metrics on M, there exists a set of k points $p_1, p_2, \ldots, p_k$ in $\mathcal{M}$ such that $\mathcal{M} \setminus \{p_1, p_2, \ldots, p_k\}$ is a smooth 5-manifold. The points p_j are in one-to-one correspondence with equivalence classes of reducible connections.
2. Each point p_j has a neighborhood $U_j \subset \mathcal{M}$, which is homeomorphic to a cone on $\mathbf{CP}^2$.
3. The moduli space $\mathcal{M}$ is orientable.
4. There exists a collar $(0, 1] \times M \subset \mathcal{M}$. Attaching M to the open end of the collar at 0 we obtain the space $\bar{\mathcal{M}} := \mathcal{M} \cup M$. This space is a compact manifold with the manifold M appearing as part of the boundary.

The proofs of the properties listed are quite involved and we refer the reader to the books by Freed and Uhlenbeck [135], Lawson [247], and Donaldson and Kronheimer [112] for details. Using the above properties, one can show that the moduli space $\bar{\mathcal{M}}$ provides a cobordism between the manifold M and a disjoint union X of k copies of $\mathbf{CP}^2$. Since the intersection form is a cobordism invariant, M and X have isomorphic intersection forms. Thus, the intersection form of M is equivalent to the identity form. $\square$

The above theorem, combined with Freedman's classification of topological 4-manifolds, provides many examples of non-smoothable 4-manifolds with zero Kirby–Siebenmann invariant. In fact, we have the following classification of smooth 4-manifolds up to homeomorphism.

Theorem 9.11 *Let M be a smooth, closed, 1-connected, oriented 4-manifold. Let $\cong_h$ denote the relation of homeomorphism and let ι_M denote the intersection form of M. If $\iota_M = \emptyset$ then $M \cong_h S^4$ and if $\iota_M \neq \emptyset$ then we have the following cases:*

1. *M is non-spin with odd intersection form*

$$\iota_M \cong j(1) \oplus k(-1), \quad j, k \geq 0,$$

and

$$M \cong_h j(\mathbf{CP}^2) \# k(\overline{\mathbf{CP}}^2);$$

i.e., M is homeomorphic to the connected sum of j copies of $\mathbf{CP}^2$ and k copies of $\overline{\mathbf{CP}}^2$, that is, $\mathbf{CP}^2$ with the opposite complex structure and orientation.

2. *M is spin with even intersection form*

$$\iota \cong m\sigma_1 \oplus pE_8, \qquad m, p \geq 0,$$

where

$$\sigma_1 = \begin{pmatrix} 0 & 1 \\ 1 & 0 \end{pmatrix}$$

is a Pauli spin matrix and E_8 is the matrix associated to the exceptional Lie group E_8 in Cartan's classification of simple Lie groups. i.e.,

$$E_8 = \begin{pmatrix} 2 & -1 & 0 & 0 & 0 & 0 & 0 & 0 \\ -1 & 2 & -1 & 0 & 0 & 0 & 0 & 0 \\ 0 & -1 & 2 & -1 & 0 & 0 & 0 & 0 \\ 0 & 0 & -1 & 2 & -1 & 0 & 0 & 0 \\ 0 & 0 & 0 & -1 & 2 & -1 & 0 & -1 \\ 0 & 0 & 0 & 0 & -1 & 2 & -1 & 0 \\ 0 & 0 & 0 & 0 & 0 & -1 & 2 & 0 \\ 0 & 0 & 0 & 0 & -1 & 0 & 0 & 2 \end{pmatrix}$$

and

$$M \cong_h m(S^2 \times S^2) \# p(|E_8|).$$

That is, M is homeomorphic to the connected sum of m copies of $S^2 \times S^2$ and p copies of $|E_8|$, the unique topological manifold corresponding to the intersection form E_8.

Gauge-theoretic methods have become increasingly important in the study of the geometry and topology of low-dimensional manifolds. New invariants are being discovered and new links with various physical theories are being established. In this chapter we are studying their application to 4-manifolds. We shall consider applications to invariants of 3-manifolds and links in them in later chapters.

9.3.1 Geometry of Moduli Spaces

In the Feynman path integral approach to quantum field theory, one is interested in integrating a suitable function of the classical action over the space of all gauge-inequivalent fields. In addition, one assumes that an "analytic continuation" can be made from the Lorentz manifold to a Riemannian manifold, the integration carried out and then the results transferred back to the physically relevant space-time manifold. Although the mathematical aspects of this program are far from clear, it has served as a motivation for the study of Euclidean Yang–Mills fields, i.e., fields over a Riemannian base manifold. Thus, for the quantization of Yang–Mills field, the space over which the Feynman integral is to be evaluated turns out to be the Yang–Mills moduli space. Evaluation of such integrals requires a detailed knowledge of the geometry of the moduli space. We have very little information on the geometry of the general Yang–Mills moduli space. However, we know that the dominant contribution to the Feynman integral comes from solutions that absolutely minimize the Yang–Mills action, i.e., from the instanton solutions. If $\mathcal{Y}$ denotes the Yang–Mills moduli space, then $\mathcal{Y} = \cup \mathcal{Y}_k$, where $\mathcal{Y}_k$ is the moduli space of fields with instanton number k. The moduli space $\mathcal{M}_k^+$ of self-dual Yang–Mills fields or instantons of instanton number k is a subspace of $\mathcal{Y}_k$. Thus, one hopes to obtain some information by integrating over the space $\mathcal{M}_k^+$. Several mathematicians [103, 172, 173, 255, 281] have studied the geometry of the space $\mathcal{M}_k^+$ and we now have detailed results about the Riemannian metric, volume, form and curvature of the most basic moduli space $\mathcal{M}_1^+$. We give below a brief discussion of these results.

Let (M, g) be a compact, oriented, Riemannian 4-manifold. Let $P(M, G)$ be a principal G-bundle, where G has a bi-invariant metric h. The metrics g and h induce the inner products $\langle \ , \ \rangle_{(g,h)}$ on the spaces $\Lambda^k(M, \mathrm{ad}\, P)$ of k-forms with values in the vector bundle $\mathrm{ad}\, P$. We can use these inner products to define a Riemannian metric on the space of gauge connections $\mathcal{A}(P)$ as follows. Recall that the space $\mathcal{A}(P)$ is an affine space, so that for each $\omega \in \mathcal{A}(P)$ we have the canonical identification between the tangent space $T_\omega \mathcal{A}(P)$ and $\Lambda^1(M, \mathrm{ad}\, P)$. Using this identification we can transfer the inner product $\langle \ , \ \rangle_{(g,h)}$ on $\Lambda^1(M, \mathrm{ad}\, P)$ to $T_\omega \mathcal{A}(P)$. Integrating this pointwise inner product against the Riemannian volume form we obtain an L^2 inner product on the space $\mathcal{A}(P)$. This inner product is invariant under the action of the group $\mathcal{G}$ on $\mathcal{A}$ and hence we get an inner product on the moduli space $\mathcal{A}/\mathcal{G}$. Recall that $\mathcal{G}$ does not act freely on $\mathcal{A}$, but $\mathcal{G}/Z(\mathcal{G})$ acts freely on the open dense subset

$\mathcal{A}_{ir}$ of irreducible connections. After suitable Sobolev completions of all the relevant spaces, the moduli space $\mathcal{O}_{ir} = \mathcal{A}_{ir}/\mathcal{G}$ of irreducible connections can be given the structure of a Hilbert manifold. The L^2 metric on $\mathcal{A}$ restricted to $\mathcal{A}_{ir}$ induces a weak Riemannian metric on $\mathcal{O}_{ir}$ by requiring the canonical projection $\mathcal{A}_{ir} \to \mathcal{O}_{ir}$ to be a Riemannian submersion. The space $\mathcal{I}_k$ of irreducible instantons of instanton number $k = -c_2(P)$, is defined by

$$\mathcal{I}_k = \mathcal{M}_k^+ \cap \mathcal{O}_{ir},$$

where $k = -c_2(P)$. The space $\mathcal{I}_k$ is a finite dimensional manifold with singularities and the weak Riemannian metric on $\mathcal{O}_{ir}$ restricts to a Riemannian metric on $\mathcal{I}_k$.

The space $\mathcal{I}_1$ was studied by Donaldson ([106]) in the case where M is a simply connected manifold with positive definite intersection form and where the gauge group $G = SU(2)$. He proved that there exists a compact set $K \in \mathcal{I}_1$ such that $\mathcal{I}_1 - K$ is a union of a finite number of components, one of which is diffeomorphic to $M \times (0,1)$ and all the others are diffeomorphic to $\mathbf{CP}^2 \times (0,1)$. Using the Riemannian geometry of $\mathcal{I}_1$ discussed above we can put this result in its geometric perspective as follows:

Theorem 9.12 *Let M be a simply connected manifold with positive definite intersection form and let the gauge group $G = SU(2)$. Then*

1. *$\mathcal{I}_1$ is an incomplete manifold with finite diameter and volume.*
2. *Let $\bar{\mathcal{I}}_1$ be the metric completion of $\mathcal{I}_1$. Then $\bar{\mathcal{I}}_1 - \mathcal{I}_1$ is the disjoint union of a finite set of points $\{p_i\}$ and a set X diffeomorphic to M. There exists an $\epsilon > 0$ such that $M \times [0, \epsilon)$ is diffeomorphic to a neighborhood of X in $\bar{\mathcal{I}}_1$ with the pull-back metric asymptotic to a product metric and for each p_i there is a neighborhood diffeomorphic to $\mathbf{CP}^2 \times (0, \epsilon)$ with the pull-back metric asymptotic to a cone metric with base $\mathbf{CP}^2$.*

See [173, 172] for a proof and further details.

In general, the L^2 metric on a moduli space cannot be calculated explicitly, since it depends on global analytic data about M. However, for the fundamental BPST instanton moduli space $\mathcal{I}_1$, the metric and the curvature have been computed by several people (see, for example, [103, 172, 181, 203]). We have the following theorem.

Theorem 9.13 *There exists a diffeomorphism $\phi : \mathbf{R}^5 \to \mathcal{I}_1$ for which the pull-back metric has the form*

$$\phi^* g_1 = \psi^2(r)g,$$

where g_1 is the L^2 metric on $\mathcal{I}_1$ and g is the standard Euclidean metric on $\mathbf{R}^5$, and ψ is a smooth function of the distance r.

The explicit formula for the function ψ is quite complicated and is given in [172]. The geometry of the moduli space of instanton solutions in the particular case of $\mathbf{CP}^2$ has been studied in detail in [67, 171]. The moduli spaces

of Yang–Mills connections on various base manifolds have also been studied
in [201, 202, 203] and in [171]. We now state the result on the curvature of
the moduli space $\mathcal{M}_{ir}$ of irreducible instantons on a compact, Riemannian
manifold with compact, semisimple gauge group. The calculation of this curvature is based on a generalization of O'Neill's formula for the curvature of
a Riemannian submersion.

Theorem 9.14 *Let* $X, Y \in T_\alpha \mathcal{A}$ *be the horizontal lifts of tangent vectors*
$X_1, Y_1 \in T_{[\alpha]} \mathcal{M}_{ir}$, *where* $[\alpha]$ *is the equivalence class of gauge connections*
that are gauge-equivalent to α. *Then the sectional curvature* R *of* $\mathcal{M}_{ir}$ *at* $[\alpha]$
is given by

$$\langle R(X_1, Y_1)Y_1, X_1 \rangle = 3 \langle b_X^*(Y), G_\alpha^0(b_X^*(Y)) \rangle$$
$$+ \langle b_X^-(X), G_\alpha^2(b_Y^-(Y)) \rangle$$
$$- \langle b_X^-(Y), G_\alpha^2(b_X^-(Y)) \rangle,$$

where b *is bracketing on bundle-valued forms,* b^* *its adjoint, and* b^- *is* b
followed by orthogonal projection onto $\Lambda_-^2(M, \operatorname{ad} P)$ *and* G^i *are the Green*
operators of the corresponding Laplacians on bundle-valued forms.

9.4 Donaldson Polynomials

Donaldson's theorem on the topology of smooth, closed, 1-connected 4-
manifolds provides a new obstruction to smoothability of these topological
manifolds. A surprising ingredient in his proof of this theorem was the moduli space $\mathcal{I}_1$ of $SU(2)$-instantons on a manifold M. This theorem has been
applied to obtain a number of new results in topology and geometry and has
been extended to other manifolds. The space $\mathcal{I}_1$ is a subspace of the moduli
space $\mathcal{M}_1$ of Yang–Mills instantons with instanton number 1. The space $\mathcal{M}_1$
in turn is a subspace of the moduli space $\mathcal{Y}_1$ of all Yang–Mills fields on M.
In fact, we have

$$\bigcup_k \mathcal{M}_k \subset \bigcup_k \mathcal{Y}_k \subset \mathcal{A}/\mathcal{G}.$$

Donaldson has used the homology of these spaces $\mathcal{M}_k$, for sufficiently large
k, to obtain a family of new invariants of a smooth 4-manifold M, satisfying
a certain condition on its intersection form. We now describe these invariants
known as **Donaldson's polynomial invariantsi**, or simply as the **Donaldson polynomials**.

The Donaldson polynomials are defined by polarization of a family q_k of
symmetric, multilinear maps

$$q_k : \underbrace{H_2(M) \times \cdots \times H_2(M)}_{d(k) \text{ times}} \to \mathbf{Q},$$

where k is the instanton number of $P(M, SU(2))$ and the function $d(k)$ is given by

$$d(k) := 4k - \tfrac{3}{4}(\chi(M) + \sigma(M)) = 4k - \tfrac{3}{2}(1 + b_2^+), \qquad (9.14)$$

where $\chi(M)$ is the Euler characteristic of M, $\sigma(M)$ is the signature of M, and b_2^+ is the dimension of the space of self-dual harmonic 2-forms on M. We shall also refer to the maps q_k as Donaldson polynomials.

A basic tool for the construction of Donaldson polynomials is a map that transfers the homology of M to the cohomology of the orbit space $\mathcal{O}_{ir}$ of irreducible connections on the $SU(2)$-bundle P. To describe this map we begin by recalling the Künneth formula for cohomology of a product of two manifolds

$$H^k(M_1 \times M_2) \cong \bigoplus_{k=i+j} H^i(M_1) \times H^j(M_2).$$

If $\alpha \in H^k(M_1 \times M_2)$ we denote by $\alpha_{(i,j)} = (\beta, \gamma)$, its component in $H^i(M_1) \times H^j(M_2)$. For a fixed $i \leq k$ we define a map

$$f : H_i(M_1) \times H^k(M_1 \times M_2) \to H^j(M_2)$$

by

$$f(a, \alpha) = \int_a \alpha_{(i,j)} := \left(\int_a \beta \right) \gamma.$$

Given a fixed element $\alpha \in H^k(M_1 \times M_2)$, the above map induces the map μ_α

$$(\mu_\alpha)_i : H_i(M_1) \to H^{k-i}(M_2),$$

defined by $(\mu_\alpha)_i(a) = f(a, \alpha)$.

We now apply this formula to the case when M_1 is a smooth, closed, 1-connected 4-manifold M, M_2 is the moduli space $\mathcal{O}_{ir}$ of irreducible instantons on M, $k = 4$, $i = 2$, and $\alpha = c_2(\mathcal{P})$, the second Chern class of the Poincaré bundle $\mathcal{P}$ over $M \times \mathcal{O}_{ir}$ with structure group $SU(2)$, to obtain Donaldson's map

$$\mu := (\mu_{c_2(\mathcal{P})})_2 : H_2(M) \to H^2(\mathcal{O}_{ir}). \qquad (9.15)$$

We note that the rational cohomology ring of $\mathcal{O}_{ir}$ is generated by cohomology classes in dimensions 2 and 4 and that the map μ and the map $(\mu_{c_2(\mathcal{P})})_0$ can be used to express these generators in terms of the homology of M. Using the map μ we define maps μ^d, $d \geq 1$,

$$\mu^d : \underbrace{H_2(M) \times \cdots \times H_2(M)}_{d \text{ times}} \to H^{2d}(\mathcal{O}_{ir}), \qquad d \geq 1,$$

by

$$\mu^d(a_1, a_2, \ldots, a_d) = \mu(a_1) \wedge \mu(a_2) \wedge \cdots \wedge \mu(a_d). \qquad (9.16)$$

This is the first step in the definition of q_k. The second step is to pair the cohomology class on the right hand side of (9.16) with a suitable homology cycle. We now describe this homology cycle. Observe first that the formal (or virtual) dimension, of the moduli space of anti-instantons of instanton number k is given by

$$\dim \mathcal{M}_k(M, SU(2)) = 8k - 3(1 + b_2^+). \tag{9.17}$$

We note that the instanton equations depend on the choice of a metric g on M through the Hodge $*$-operator. It follows that the instanton moduli space also depends on this choice. We indicate this dependence of $\mathcal{M}_k$ on the metric g by writing $\mathcal{M}_k((M, g), SU(2))$ or simply $\mathcal{M}_k(g)$ for the moduli space. If $b_2^+ = 2a + 1$, $a > 0$, then for a generic metric g the moduli space $\mathcal{M}_k(g)$ has even dimension $2d$, where $d = 4k - 3(1 + a)$. This requirement is incorporated in the following definition.

Definition 9.1 *A C-manifold is a pair (M, β) where*

1. *M is a smooth, compact, 1-connected, oriented 4-manifold with $b_2^+(M) = 2a + 1$, $a > 0$;*
2. *β is an orientation of a maximal positive subspace $H_2^+ \subset H^2(M, \mathbf{R})$ for the intersection form of M.*

In the following theorem we collect some properties of the moduli spaces over a C-manifold that play a fundamental role in the definition of Donaldson polynomials.

Theorem 9.15 *Let M be a C-manifold. Let $\mathcal{RM}(M)$ denote the set of Riemannian metrics on M. Then there exists a second category subset $\mathcal{GM}(M) \subset \mathcal{RM}(M)$ (called the set of **generic metrics**) such that*

1. *For $g \in \mathcal{GM}(M)$ and $k > 0$ the moduli space $\mathcal{M}_k(g)$ is a manifold with boundary. The part $\mathcal{M}_k^*(g)$ corresponding to irreducible instantons is a smooth submanifold of the space $\mathcal{B}_k^*$ of all irreducible gauge potentials, cut out transversely by the instanton equations and hence is of dimension equal to the virtual dimension $8k - 6(1 + a)$.*
2. *For $g_0, g_1 \in \mathcal{GM}(M)$ there exists a smooth path of generic metrics g_t, $t \in [0, 1]$ joining g_0 and g_1 such that the **parametrized moduli space** $\mathcal{N}$ defined by*

$$\mathcal{N} = \{(\omega, t) \in \mathcal{B}_k^* \times [0, 1] \mid \omega \in \mathcal{M}_k^*(g_t)\}$$

   is a smooth manifold with boundary, the boundary consisting of the disjoint union of $\mathcal{M}_k^*(g_0)$ and $\mathcal{M}_k^*(g_1)$. The manifold $\mathcal{N}$ has dimension $8k - 6a - 5$ and is a cobordism between $\mathcal{M}_k^*(g_0)$ and $\mathcal{M}_k^*(g_1)$.
3. *The orientation β induces an orientation on all the moduli spaces $\mathcal{M}_k(g)$ so that changing β to $-\beta$ reverses the orientation of all the moduli spaces. Given an orientation on all the moduli spaces $\mathcal{M}_k(g)$, the parametrized moduli space $\mathcal{N}$ can be given an orientation so that it provides an oriented cobordism between $\mathcal{M}_k^*(g_0)$ and $\mathcal{M}_k^*(g_1)$.*

In the rest of this section we take M to be a C-manifold. There are a number of technical points to be addressed before we give the definition of Donaldson polynomials. Briefly, we know that $\mathcal{M}_k(g)$ is a manifold of dimension $2d$. If we can associate a $2d$-dimensional homology class h_{2d} to this manifold then we can evaluate the $2d$-form defined in (9.16) on the class h_{2d} to define Donaldson polynomials. Theorem 9.15 can then be used to show that the definition is independent of the metric and the orientation of $\mathcal{M}_k(g)$. In view of this we drop the reference to the metric and orientation of the moduli spaces in what follows. For a compact manifold, the fundamental homology class is well-defined. However, the moduli spaces $\mathcal{M}_k$ are, in general, not compact. They do have a natural compactification $\overline{\mathcal{M}}_k$ defined as follows. Let $s^i(M)$ denote the ith symmetric power of M. Define the topology $\mathcal{T}_D$ on

$$T_k := \mathcal{M}_k \bigcup_{i=1}^{k} \left(\mathcal{M}_{k-i} \times s^i(M) \right) \tag{9.18}$$

by the following notion of convergence. We say that a sequence $[\omega_n]$ in $\mathcal{M}_k$ converges to a point $([\omega], (x_1, \ldots, x_i)) \in \mathcal{M}_{k-i} \times s^i(M)$ if

$$\lim_{n \to \infty} [\omega_n] = [\omega] \quad \text{on } M \setminus \{x_1, \ldots, x_i\}$$

and the corresponding sequence of Yang–Mills integrands

$$|F_{\omega_n}|^2 \text{ converges to } |F_\omega|^2 + 8\pi^2 \sum_{j=1}^{i} \delta_{x_j}.$$

By extending the analytical results of Uhlenbeck on the compactification of Yang–Mills moduli spaces, Donaldson has proved the following theorem:

Theorem 9.16 *Let M be a C-manifold. Then the closure $\overline{\mathcal{M}}_k$ of $\mathcal{M}_k$ in the topological space $(\mathcal{M}_k \cup_{i=1}^{k} \mathcal{M}_{k-i} \times s^i(M), \mathcal{T}_D)$ is a compact stratified space with principal stratum $\mathcal{M}_k$. Moreover, if*

$$4k > 3(b_2^+(M) + 1) \tag{9.19}$$

then the secondary strata

$$\overline{\mathcal{M}}_k \cap (\mathcal{M}_{k-i} \times s^i(M)), \qquad i > 1,$$

have codimension at least 2 in $\overline{\mathcal{M}}_k$. In particular, this implies that the compactified moduli space $\overline{\mathcal{M}}_k$ carries the fundamental homology class.

The compactification $\overline{\mathcal{M}}_k$ is called the **Donaldson compactification** of $\mathcal{M}_k$. The values of k satisfying the inequality (9.19) are said to be in the **stable range**. The fundamental class of the space $\mathcal{M}_k$ defines a $2d$-dimensional homology class $[\mathcal{M}_k]$ in the homology $H_*(\mathcal{O}_{ir})$. The Donaldson polynomial q_k

$$q_k : \underbrace{H_2(M) \times \cdots \times H_2(M)}_{d \text{ times}} \to \mathbf{Q}$$

is then defined by

$$q_k(a_1, \ldots, a_d) = (\mu^d(a_1, \ldots, a_d))[\mathcal{M}_k]. \tag{9.20}$$

An important alternative definition of the map μ can be given by considering a representative of the homology class $a \in H_2(M)$ by an embedded surface S as follows. The embedding map $\iota : S \to M$ defines the map

$$r : \mathcal{A}_{ir}(M) \to \mathcal{A}(S)$$

obtained by restricting an irreducible connection over M to a connection over S. Coupling the Dirac operator $\mathcal{D}$ on S, to the restriction of gauge fields on M to S gives a family of coupled Dirac operators. This family defines a determinant line bundle L over $\mathcal{A}(S)$. We denote by L_r the pull-back r^*L of this bundle to $\mathcal{A}_{ir}(M)$. We then define

$$\mu(a) := c_1(L_r^{-1}) \in H^2(\mathcal{A}_{ir}(M)).$$

One can show that this definition of μ leads to the same cohomology class as the previous definition.

In [237] the definition of Donaldson polynomial invariants has been extended beyond the stable range as well as to manifolds M with $b_2^+(M) = 1$ (see also [113]).

In [404] it is shown that the Donaldson polynomial invariants of a 4-manifold M appear as expectation values of certain observables in a topological QFT. As we have discussed in Chapter 7, this is a physical interpretation of Donaldson's polynomial invariants. We note that this TQFT formulation or its later variants have not led to any new insight into the structure of these invariants. As we have indicated above, Donaldson polynomials are independent of the metric on M and depend only on its differential structure. Thus we have the following theorem.

Theorem 9.17 *Let M be a smooth, oriented, compact, simply connected 4-manifold with $b_2^+(M) = 2a + 1$, $a \geq 1$. For each k in the stable range $(4k > 3(b_2^+(M)+1))$, the polynomial q_k defined by equation (9.20) is a differential topological invariant of M, which is natural with respect to orientation-preserving diffeomorphisms.*

When M is a connected sum of manifolds M_1 and M_2, one can construct instantons on M by gluing instantons on M_1 and M_2. By carefully analyzing the corresponding moduli spaces Donaldson proved the following important vanishing theorem.

Theorem 9.18 *Let M be a smooth, oriented, compact, simply connected 4-manifold with $b_2^+(M) = 2a + 1$, $a \geq 1$. If M can be written as a smooth,* *oriented connected sum $M = M_1 \# M_2$ and $b_2^+ (M_i) > 0$, for $i = 1, 2$, then the polynomial invariant q_k is identically zero for all k.*

The case when the base manifold carries a compatible complex structure is usually referred as the **integrable case**. In particular, when M is a compact Kähler manifold with Kähler form ω, there exists a one-to-one correspondence between the following two spaces of equaivalence classes:

1. space of irreducible $SU(2)$ anti-instantons,
2. space of holomorphic $SL(2, \mathbf{C})$ bundles that satisfy a certain stability condition with respect to the polarization induced by the Kähler form.

In the integrable case one has the following positivity property of the invariants, which has strong influence on the differential topology of complex surfaces even in cases where the invariants cannot be explicitly calculated.

Theorem 9.19 *Let M be a compact Kähler manifold with Kähler form ω. Let $\Sigma \in H_2(M; \mathbf{Z})$ be the Poincaré dual of the Kähler class $[\omega] \in H^2(M; \mathbf{Z})$. Then $q_k(\Sigma, \ldots, \Sigma) > 0$ for all sufficiently large k.*

The integrable case has been studied extensively (see, for example, [108, 113], [141] and [150]). For algebro-geometric analogues of Donaldson polynomials see J. Morgan and K. O'Grady [291]. For applications to complex manifolds and algebraic varieties see R. Friedman and J. Morgan [142] and [290].

9.4.1 Structure of Polynomial Invariants

Donaldson's polynomial invariants are defined in terms of the 2-dimensional cohomology classes in the image of the μ map. The moduli space $\mathcal{O}_{ir}$ also carries a distinguished 4-dimensional cohomology class ν defined by the image of the generator $1 \in H_0(M)$ under the map

$$(\mu_{c_2(\mathcal{P})})_0 : H_0(M) \to H^4(\mathcal{O}_{ir}). \tag{9.21}$$

We note that the cohomology class ν is essentially the first Pontryagin class of the principal $SO(3)$-bundle $\hat{\mathcal{O}}_{ir}$ over $\mathcal{O}_{ir}$ generated by the base point fibration. Using the class ν we can define a larger class $q_{(i,j)}$

$$q_{(i,j)} : \underbrace{H_2(M) \times \cdots \times H_2(M)}_{i \text{ times}} \to \mathbf{Q}$$

of invariants of M by

$$q_{(i,j)}(a_1, \ldots, a_i) = (\mu^i(a_1, \ldots, a_i) \wedge \nu^j)[\mathcal{M}_k], \tag{9.22}$$

where i, j are non-negative integers such that $i + 2j = d(k)$. We say that M is of **KM simple type**, or just simple type, if

$$q_{(i,j)} = 4q_{(i,j+2)} \qquad (9.23)$$

for all i, j for which the invariants are defined. It is known that condition (9.23) holds for several classes of 4-manifolds including simply connected elliptic surfaces, complete intersections, and some branched covers. It is not known whether the condition holds for all simply connected 4-manifolds.

In the rest of this section we restrict ourselves to manifolds of simple type. Then the evaluation of the invariants $q_{(i,j)}$ is reduced to two cases. Case one has $j = 0$ and corresponds to the original polynomial invariants. In case two $j = 1$ and in this case, following [238, 239] we define new invariants $q_{d(k)-2}$ by

$$q_{d(k)-2} = \tfrac{1}{2} q_{(d(k)-2,1)} \qquad (9.24)$$

For non-negative n we define q_n by equation (9.20) if $n = d(k)$, by equation (9.24) if $n = d(k) - 2$, and set $q_n = 0$ for all other n. Then we define a function q by a series involving q_n

$$q : H_2(M; \mathbf{Z}) \to \mathbf{R} \text{ by } q(a) := \sum_{n=0}^{\infty} \frac{q_n(a)}{n!}. \qquad (9.25)$$

The following structure theorem for the invariant q was proved by Kronheimer and Mrówka [239].

Theorem 9.20 *Let M be a simply connected 4-manifold of simple type. Then there exist finitely many cohomology classes $K_1, \ldots, K_p \in H^2(M; \mathbf{Z})$ and non-zero rational numbers $r_1, \ldots, r_p$ such that*

$$q = e^{(\iota_M)/2} \sum_{i=1}^{p} r_i e^{K_i}, \qquad (9.26)$$

where the equality is to be understood as equality of analytic functions on $H_2(M; \mathbf{R})$. The classes K_i, $1 \le i \le p$, are called the **KM basic classes** *and each K_i is an integral lift of the second Stiefel–Whitney class $w_2(M) \in H^2(M; \mathbf{Z})$.*

The proof of the above theorem depends in an essential way on studying the moduli spaces of singular gauge fields on M with codimension two singular surface Σ. One starts by specifying the following data:

1. a principal $SU(2)$ bundle over $M \setminus \Sigma$ with instanton number k;
2. a complex line bundle L over Σ with monopole number $l = -c_1(L)[\Sigma]$;
3. a non-trivial holonomy α around Σ such that $\alpha \in (0, 1/2)$.

Fix a background connection A^α and define the configuration space $\mathcal{A}^\alpha$ by

$$\mathcal{A}^a = \{A^\alpha + a \mid \exists p > 2 \text{ such that } \nabla_{A^\alpha} a \in L^p\}.$$

Then define the moduli space:

$$M_{k,l}^\alpha = \{A \in \mathcal{A}^\alpha \mid *F_A = -F_A\}/\mathcal{G},$$

where $\mathcal{G}$ is the group of gauge transformations. We call $M_{k,l}^\alpha$ a moduli space of connections on M with fixed singularity along Σ. In general, $M_{k,l}^\alpha$ is a manifold with singularities, and

$$\dim M_{k,l}^\alpha = 8k - 3(b_2^+(M) - b^1(M) + 1) + 4l - (2g - 2),$$

where g is the genus of Σ. One can define a family of invariants of Donaldson type using these moduli spaces. These invariants interpolate between the usual Donaldson invariants to provide the recurrence relations and also furnish information on their structure. In particular, we have the following theorem:

Theorem 9.21 *Let M be a simply connected 4-manifold of simple type and let $K_1, \ldots, K_p \in H^2(M; \mathbf{Z})$ be a complete set of KM basic classes given by Theorem 9.20. Let Σ be any smoothly embedded, connected, essential surface in M of genus g with normal bundle of non-negative degree. Then g satisfies the inequality*

$$\iota_M(\Sigma) + \max_i \langle K_i, \Sigma \rangle \leq 2g - 2. \tag{9.27}$$

This result is closely related to the classical Thom conjecture about embedded surfaces in $\mathbf{CP}^2$. While the singular gauge theory methods do not apply to $\mathbf{CP}^2$, they do lead to interesting new results. As an example of such a result we state the following theorem due to Kronheimer and Mrówka.

Theorem 9.22 *Let M be a manifold with $b_2^+ \geq 3$ and non-zero Donaldson invariants. Let $\Sigma \subset M$ be an essential surface with $g > 0$. Then*

$$2g - 2 \geq \iota_M(\Sigma).$$

The condition that M be of simple type is very mysterious. It is not known whether every simply connected 4-manifold is of simple type. Simple type and basic classes also arise in the Seiberg–Witten theory, as we will see later in this chapter.

9.4.2 Relative Invariants and Gluing

Let Y be an oriented compact manifold for which the Fukaya–Floer homology $FF_*(Y)$ is defined. Let M be an oriented, simply connected compact manifold with boundary $\partial M = Y$. Then one can define a relative version of Donaldson's polynomial invariants of M with values in the Fukaya–Floer

homology $FF_*(Y)$. This definition can be used to obtain gluing formulas
for Donaldson polynomials for 4-manifolds, which can be written as generalized connected sums split along a common boundary Y. If $\bar{Y}$ is Y with the
opposite orientation, then there is a dual pairing between the graded homology groups $FF_*(Y)$ and $FF_*(\bar{Y})$. Let N be an oriented simply connected
closed 4-manifold, which can be written as the generalized connected sum
$N = N_1 \#_Y N_2$. If q_1 (resp., q_2) is a relative polynomial invariant with values
in $FF_*(Y)$ (resp., $FF_*(\bar{Y})$) then q_1, q_2 can be glued to obtain a polynomial
invariant q of N by the "formula"

$$q = \langle q_1, q_2 \rangle.$$

The definition of relative invariants and detailed structure of the gluing formula in this generality is not yet known. Some specific examples have been
considered in [110, 369].

We give a definition of relative invariants in the special case when Y is
an integral homology sphere and the representation space $\mathcal{R}^*(Y)$ is regular.
Let M be an oriented simply connected 4-manifold with boundary Y. Let
P be a trivial $SU(2)$ bundle over M and fix a trivialization of P to write
$P = M \times SU(2)$ and let θ_M be the corresponding trivial connection. Let θ_Y
denote the induced trivial connection on $P_{|Y} = Y \times SU(2)$. Let $\alpha \in \mathcal{R}^*(Y)$
and let $\mathcal{M}(M; \alpha)$ denote the moduli space of self-dual connections on P which
equal α on the boundary Y of M. The index of the corresponding instanton
deformation complex gives the formal (or virtual) dimension of the moduli
space $\mathcal{M}(M; \alpha)$. Let n be the dimension of a non-empty component $\mathcal{M}^n(\alpha)$
of $\mathcal{M}_g(M; \alpha)$ for a suitable generic metric g on M. Then

$$n \equiv -3(1 + b_2^+(M)) - sf(\alpha, \theta) \quad (\text{mod } 8). \tag{9.28}$$

If $n = 2m$ we can define the relative polynomial invariants q_m

$$q_m : \underbrace{H_2(M, Y; \mathbf{Z}) \times \cdots H_2(M, Y; \mathbf{Z})}_{m \text{ times}} \to \mathcal{R}^*(Y)$$

by

$$q_m(a_1, \ldots, a_m) = (\mu^m(a_1, \ldots, a_m)[\mathcal{M}^{2m}(\alpha)])\alpha. \tag{9.29}$$

Let $n_\alpha \in \mathbf{Z}_8$ be the congruence class of the spectral flow $sf(\alpha, \theta)$. Then one
can show that the right hand side of equation (9.29) depends only on the
Floer homology class of $\alpha \in HF_{n_\alpha}$. Thus the relative polynomial invariants
of M take values in the Floer homology of the boundary $\partial M = Y$.

9.5 Seiberg–Witten Theory

I first learned about the Seiberg–Witten theory from Witten's lecture at the 1994 International Congress on Mathematical Physics (ICMP 1994) in Paris. His earlier formulation of the Jones polynomial using quantum field theory had given a new and geometrical way of looking at the Jones polynomial and it led to the WRT invariants of 3-manifolds. It was thus natural to consider a similar interpretation of the Donaldson polynomials. Witten's results were given a more mathematical reformulation in [21]. However, these results provided no new insight or method of computation beyond the well known-methods used in physics and mathematics. Another idea Witten used successfully was a one-parameter family of supersymmetric Hamiltonians $(H_t,\ t \geq 0)$ to relate Morse theory and de Rham cohomology. Large values of t lead to Morse theory while small t give the de Rham cohomology. In physics these limiting theories are considered dual theories and are referred to as the **infrared limit** and **ultraviolate limit**, respectively. , Witten used the ultraviolet limit of $N = 2$ supersymmetric Yang–Mills theory to write the Donaldson invariants as QFT correlation functions. As these invariants do not depend on the choice of a generic metric, they could also be calculated in the infrared limit. The infrared behavior of the $N = 2$ supersymmetric Yang–Mills theory was determined by Seiberg and Witten in 1994. The equations of the theory dual to the $SU(2)$ gauge theory are the Seiberg–Witten or the monopole equations. They involve $U(1)$ gauge fields coupled to monopoles. Thus the new theory should be expected to give information on the Donaldson invariants of 4-manifolds. Witten told me that his paper [407] giving further details should appear soon.

After returning to America, Witten gave a lecture at MIT discussing his new work. He remarked that the **monopole invariants** (now known as the **Seiberg–Witten invariants** or the **SW-invariants**) could be used to compute the **Donaldson invariants** and vice versa. Gauge theory underlying the SW-invariants has gauge group $U(1)$, the Abelian group of Maxwell's electromagnetic field theory. The moduli space of the solutions of the SW-equations has a much simpler structure than the instanton moduli spaces used in the Donaldson theory. A great deal of hard analysis is required to deal with the problem of compactification of moduli spaces and the existence of reducible connections to extract topological information about the base manifold in Donaldson's theory. It was (and still is) hard to believe that one can bypass the hard analysis if one uses the SW-moduli space and that doing so gives a much simpler approach to the known results. It also leads to proofs of several results that have seemed intractable via the Donaldson theory.[1]

---

[1] Taubes described what happened after the lecture: Several people working in geometric topology gathered at Bott's house. Most of us were thinking of obtaining a counterexample to Witten's assertion of equivalence of the monopole and instanton invariants. No such example was found by the time we broke up late that night. It was agreed that anyone who finds a counterexample would communicate it to the others. (Personal communication)

Witten's assertion of equivalence of the monopole and instanton invariants seems incredible. However, as of this writing no counterexample has been found to Witten's assertion that monopole and instanton invariants are equivalent. In fact, the equivalence of the monopole and instanton invariants has been established for a large class of 4-manifolds. SW-equations arose as a byproduct of the solution of $N = 2$ supersymmetric Yang–Mills equations in theoretical physics. It was the physical concept of S-duality that led Witten to his formula relating the SW and Donaldson invariants. Thus, SW theory provides the most exciting recent example of physical mathematics. SW theory is an active area of current research with vast literature including monographs and texts. The book [277] by Marcolli gives a very nice introduction to various aspects of SW theory. A comprehensive introduction can be found in Nicolaescu [299]. The relation between the Seiberg–Witten theory and the Gromov–Witten theory is discussed in detail by Taubes in [362].

Spin structures and Dirac operators on spinor bundles are now reviewed. These are used to define the SW-equations, SW-moduli space, and SW-invariants. We then indicate some applications of the SW-invariants and state their relation to Donaldson's polynomial invariants.

9.5.1 Spin Structures and Dirac Operators

Dirac's 1928 discovery of his relativistic equation for the electron is considered one of the great achievements in theoretical physics in the first half of the twentieth century. This equation introduces a first order differential operator acting on spinors defined over the Minkowski space. This is the original Dirac operator, which is a square root of the D'Alembertian (or the Minkowski space Laplacian). The equation also had a solution corresponding to a positively charged electron. No such particle was observed at that time. The Dirac gamma matrices in the definition of the Dirac operator have a natural interpretation in the setting of the Clifford algebra of the Minkowski space. In fact, given a finite-dimensional real vector space V with a pseudo-inner product g, we can construct a unique (up to isomorphism) Clifford algebra associated to the pair (V, g). The definitions of the gamma matrices, spinors and the Dirac operator all have natural extensions to this general setting. For mathematical applications we want to consider Riemannian manifolds which admit **spinor bundles**. The Dirac operator is then defined on sections of these bundles. As we discussed in Chapter 2 and Appendix D, there are topological obstructions to a manifold admitting a spin structure. However, a compact oriented Riemannian manifold of even dimension admits a $Spin^c$-structure which can be used to construct spinor bundles on it. We now discuss this structure on a fixed compact oriented Riemannian manifold M of dimension m. The group $Spin^c(m)$ is defined by

$$Spin^c(m) := (Spin(m) \times U(1))/\mathbf{Z}_2 . \tag{9.30}$$

Definition 9.2 *A Spin^c-structure on a compact oriented Riemannian manifold M is a lift of the principal $SO(m)$-bundle of oriented orthonormal frames of M to a principal Spin$^c(m)$-bundle over M.*

As with the existence of *Spin*-structure, that of the *Spin^c*-structure has topological obstruction that can be expressed in terms of the second Stiefel–Whitney class $w_2(M)$. Using standard homology theory argument, it can be shown that a manifold M admits a *Spin^c*-structure if $w_2(M)$ is the reduction of an integral homology class (mod 2). Moreover, a compact oriented even dimensional Riemannian manifold admits a *Spin^c*-structure. Recall that the Clifford algebra of the Euclidean space $\mathbf{R}^{2n}$ has a unique irreducible complex representation on a Hermitian vector space S of rank 2^n. It is called the (complex) **spinor representation** . The elements of S are called (complex) **spinors** . The space S has a natural $\mathbf{Z}_2$-graded or superspace structure under which it can be written as a vector space direct sum $S = S^+ \oplus S^-$. The elements of S^+ (resp. S^-) are called **positive spinors** (resp. **negative spinors**) . Clifford multiplication by $v \in \mathbf{R}^{2n}$ acts as a skew-Hermitian operator on the vector space S interchanging the positive and negative spinors.

9.5.2 The Seiberg–Witten (SW) Invariants

In what follows we restrict ourselves to a compact oriented Riemannian 4-manifold M. The manifold M admits a *Spin^c*-structure. We choose a fixed *Spin^c*-structure and the corresponding S as discussed above. A **spin connection** ∇ on S is a connection compatible with the Levi-Civita connection ∇^g on the orthonormal frame bundle of M; i.e.,

$$\nabla_X(Y \cdot \psi) = \nabla^g_X(Y) + \nabla_X(\psi),$$

where X, Y are vector fields on M and ψ is a section of the spinor bundle S. The corresponding Dirac operator D acting on a spinor ψ is the first order operator whose local expression is given by

$$D(\psi) = \sum_{k=1}^{4} e_k \cdot \nabla_{e_k}(\psi),$$

where $e_k, 1 \leq k \leq 4$ is a local orthonormal basis for the tangent bundle TM. The Dirac operator maps positive spinors to negative spinors and vice versa. The determinant line bundles of S^+ and S^- are canonically isomorphic. We denote this complex line bundle by L. The spin connection ∇ on S leaves the bundles S^+ and S^- invariant and hence induces a connection on L. We denote by A the gauge potential and by F_A the corresponding gauge field of the induced connection on L. An important property of the Dirac operator that is used in the study of the SW equations is given by the following theorem.

Theorem 9.23 (Lichnerowicz–Weitzenböck formula) *Let ∇^* be the formal adjoint of the spin connection (covariant derivative operator) ∇. Let R be the scalar curvature of the manifold (M, g). Then the Dirac operator satisfies the following Lichnerowicz–Weitzenböck formula:*

$$D^2(\psi) = \left(\nabla^*\nabla + \tfrac{1}{4}(R - iF_A)\right)\psi, \qquad \forall \psi \in \Gamma(S^+).$$

In the case we are considering, the manifold M admits *Spin^c* structures and they are classified by $H^2(M, \mathbf{Z})$. Given $\alpha \in H^2(M, \mathbf{Z})$ there is a principal $U(1)$-bundle P_α with connection A and curvature F_A whose first Chern class equals α. The associated complex line bundle L is the determinant bundle of the positive and negative spinors. Given a positive spinor ψ, the Seiberg–Witten or SW equations are a system of differential equations for the pair (A, ψ) given by

$$D\psi = 0 , \quad F_A^+ = \tfrac{1}{4}\sigma(\psi), \tag{9.31}$$

where D is the Dirac operator of the spin connection, F_A^+ is the self-dual part of the curvature F_A, and $\sigma(\psi)$ is a pure imaginary valued self-dual 2-form with local coordinate expression

$$\sigma(\psi) = \langle e_i e_j \psi, \psi \rangle e^i \wedge e^j.$$

The SW equations are the absolute minima of the variational problem for the SW functional given by

$$\mathcal{A}_{SW}(A, \psi) = \int_M \left(||D\psi||^2 + ||F_A^+ - \frac{1}{4}\sigma(\psi)||^2\right) dv_g . \tag{9.32}$$

Using the Lichnerowicz–Weitzenböck formula the SW functional can be written as

$$\mathcal{A}_{SW}(A, \psi) = \int_M \left(||\nabla\psi||^2 + ||F_A^+||^2 + \frac{R}{4}||\psi||^2 + \frac{1}{8}||\psi||^4\right) dv_g . \tag{9.33}$$

This second form of the SW functional is useful in the study of the topology of the moduli space of the solutions of the SW equations. The group of gauge transformations given by $\mathcal{G} = \mathcal{F}(M, U(1))$ acts on the pair (A, ψ) as follows:

$$f(A, \psi) = (A - f^{-1}df, \ f\psi), \qquad \forall f \in \mathcal{G}. \tag{9.34}$$

This action induces an action on the solutions (A, ψ) of the SW equations. The moduli space $\mathcal{M}_{SW}$ of the solutions of the SW equations is defined by

$$\mathcal{M}_{SW} := \{(A, \ \psi) \mid (A, \ \psi) \in C\}/\mathcal{G},$$

where C is the space of solutions of the SW equation. The moduli space $\mathcal{M}_{SW}$ depends on the choice of $\alpha \in H^2(M, \mathbf{Z})$ and we indicate this by writing $\mathcal{M}_{SW}(\alpha)$ for the moduli space. To simplify our considerations we now take $b_2^+(M) > 1$. Then for a generic metric on M at a regular point (i.e., where $\psi \neq 0$) the dimension of the tangent space to $\mathcal{M}_{SW}(\alpha)$ can be computed by applying index theory to the linearization of the SW equations. This dimension d_α is given by

$$d_\alpha = \tfrac{1}{4}(c_1^2(L) - (2\chi + 3\sigma)).$$

It can be shown that the space $\mathcal{M}_{SW}(\alpha)$ (or the space obtained by perturbation) when non-empty is a smooth compact orientable manifold. We choose an orientation and define the SW-invariants as follows.

Definition 9.3 (SW Invariants) *Suppose that $\alpha \in H^2(M, \mathbf{Z})$ be such that $d_\alpha = 0$. Then the space $\mathcal{M}_{SW}(\alpha)$ consists of a finite number p_i, $1 \leq i \leq k$, of signed points. We define the SW-invariant $\iota_{SW}(\alpha)$ of α by*

$$\iota_{SW}(\alpha) := \sum_{i=1}^{k} s(p_i) , \tag{9.35}$$

where $s(p_i) = \pm 1$ is the sign of p_i.

Theorem 9.24 *The SW invariants $\iota_{SW}(\alpha)$ are independent of the metric and hence are differential topological invariants of M.*

It is possible to define SW-invariants when $d_\alpha > 0$ by a construction similar to that used in Donaldson theory. However, their significance is not clear yet. In most of the examples the SW moduli space turns out to be 0-dimensional. We now restrict the class of manifolds to those of SW simple type defined below.

Definition 9.4 *A class α is called an* **SW basic class** *if the invariant $\iota_{SW}(\alpha) \neq 0$. It can be shown that there are only finitely many basic classes. We say that M is of* **SW simple type** *if $d_\alpha = 0$ for every basic class α.*

Simple Type Conjecture: Let M be a closed simply connected oriented Riemannian 4-manifold with $b_2^+ > 1$. Then the simple type conjecture says that every such manifold is of KM and SW simple type. It is known that the conjecture holds if M is also a symplectic manifold. There are no known examples of manifolds that are not simple type. If M has non-negative scalar curvature then all the SW invariants are zero. The following theorem is similar to the one on invariants of connected sums in Donaldson theory.

Theorem 9.25 (Connected sum theorem) *Let M_i, $i = 1, 2$, be compact oriented 4-manifolds with $b_2^+(M_i) \geq 1$; then $SW(M_1 \# M_2, s) = 0$, where s is an arbitrary $Spin^c$ structure on $(M_1 \# M_2)$.*

A symplectic manifold M has a canonical $Spin^c$ structure for which the SW invariant is non-zero. It follows from the above theorem that such M cannot be a connected sum of two manifolds each with $b_2^+ \geq 1$ if $b_2^+(M) \geq 1$. In his theory of pseudo-holomorphic curves in symplectic manifolds, Gromov defined a set of invariants, which were given a physical interpretation in terms of sigma models by Witten. They are called the Gromov–Witten or GW-invariants. The following theorem due to Taubes gives a surprising relation between the SW- and the GW-invariants.

Theorem 9.26 (Taubes) *Let M be a compact symplectic manifold with $b_2^+ > 1$. Orient M and its positive determinant line bundle by using the symplectic structure. Define Seiberg–Witten invariants SW and Gromov–Witten invariants Gr as maps from $H^2(M; \mathbf{Z})$ to $\Lambda^* H^1(M; \mathbf{Z})$. Then $SW = Gr$.*

For a complete treatment of this theorem and relevant definitions we refer the reader to Taubes' book [362].

One of the earliest applications of SW-invariants was the proof by Kronheimer and Mrówka of the classical Thom conjecture about embedded surfaces in $\mathbf{CP}^2$. This was beyond reach of the regular or singular gauge theory methods. There is a version of SW equations for 3-manifolds and the corresponding Seiberg–Witten Floer homology (see, for example, Marcolli [277]). For other applications of the SW-invariants see [249, 293, 329, 367, 379].

9.6 Relation between SW and Donaldson Invariants

As we saw earlier, Donaldson used the moduli spaces of instantons to define a new set of invariants of M, which can be regarded as polynomials on the second homology $H_2(M)$. In [239] Kronheimer and Mrówka obtained a structure theorem for the Donaldson invariants in terms of their basic classes and introduced a technical property called "KM-simple type" for a closed simply connected 4-manifold M. Then, in 1994, the Seiberg–Witten (SW) equations were obtained as a byproduct of the study of super Yang–Mills equations. As we discussed earlier these equations are defined using a $U(1)$ monopole gauge theory, and the Dirac operator obtained by coupling to a $Spin^c$ structure. The moduli space of solutions of SW equations is used to define the SW-invariants. The structure of this SW moduli space is much simpler than the instanton moduli spaces used to define Donaldson's polynomial invariants. This led to a number of new results that met with insurmountable difficulties in the Donaldson theory (see, for example, [240, 368]). We also discussed the simple type condition and basic classes in SW theory. Witten used the idea of taking ultraviolet and infrared limits of $N = 2$ supersymmetric quantum Yang–Mills theory and metric independence of correlation functions to relate Donaldson and SW-invariants. The precise form of Witten's conjecture can be expressed as follows.

Witten's conjecture for relation between Donaldson and SW-invariants: A closed, simply connected 4-manifold M has KM-simple type if and only if it has SW-simple type. If M has simple type and if $D(\alpha)$ (resp., $SW(\alpha)$) denotes the generating function series for the Donaldson (resp., Seiberg–Witten) invariants with $\alpha \in H_2(M; \mathbf{R})$, then we have

$$D(\alpha) = 2^c e^{\iota_M(\alpha)/2} SW(\alpha), \qquad \forall \alpha \in H_2(M; \mathbf{R}). \tag{9.36}$$

In the above formula ι_M is the intersection form of M and c is a constant given by

$$c = 2 + \frac{7\chi(M) + 11\sigma(M)}{4}. \tag{9.37}$$

A mathematical approach to a proof of Witten's conjecture was proposed by Pidstrigatch and Tyurin (see dg-ga/9507004). In a series of papers, Feehan and Leness (see [122,123] and references therein) used similar ideas employing an $SO(3)$ monopole gauge theory that generalizes both the instanton and $U(1)$ monopole theories to prove the Witten conjecture for a large class of 4-manifolds. This result is an important ingredient in the proof of the property P conjecture that we next discuss. The problem of relating Feehan and Leness' proof to Witten's TQFT argument remains open.

9.6.1 Property P Conjecture

In early 1960s Bing tried to find a counterexample to the Poincaré conjecture by constructing 3-manifolds by surgery on knots. Bing and Martin later formalized this search by defining property P of a knot as follows:

Definition 9.5 *A non-trivial knot K has* **property P** *if every 3-manifold Y obtained by a non-trivial Dehn surgery on K has a non-trivial fundamental group.*

Using the above definition we can state the **property P conjecture** as follows: *Every non-trivial knot K has property P. In particular, Y is not a homotopy 3-sphere.*

Thus, to verify the property P conjecture, it is enough to show that $\pi_1(Y)$ is non-trivial if the knot K is non-trivial. It can be shown by topological arguments that $\pi_1(Y)$ is non-trivial if it is obtained by Dehn surgery with coefficient other than ± 1. The proof of the remaining cases was recently given by Kronheimer and Mrówka [241], thereby proving the property P conjecture. In fact, they showed that in these cases $\pi_1(Y)$ admits a non-trivial homomorphism to the group $SO(3)$. The proof uses several results from gauge theory, symplectic and contact geometry, and the proof of Witten's conjecture relating the Seiberg–Witten and Donaldson invariants for a special class of manifolds. The proof is quite complicated as it requires that methods and results from many different areas be fit together in a precise way. According to Gromov[2] a simplified proof of the theorem should be possible. We note that it is known that surgery on a non-trivial knot can never produce the manifold S^3. It follows that the Poincaré conjecture implies the property P conjecture. Thanks to Perelman's proof of the Poincaré conjecture, we now have a different proof of the property P conjecture closely related to ideas from gravitational physics.

---

[2] Gromov made this remark after the seminar by Mrówka at Columbia University explaining this work.

Chapter 10
3-Manifold Invariants

10.1 Introduction

In Chapter 9 we discussed the geometry and topology of moduli spaces of
gauge fields on a manifold. In recent years these moduli spaces have been
extensively studied for manifolds of dimensions 2, 3, and 4 (collectively referred to as low-dimensional manifolds). This study was initiated for the
2-dimensional case in [17]. Even in this classical case, the gauge theory perspective provided fresh insights as well as new results and links with physical
theories. We make only a passing reference to this case in the context of
Chern–Simons theory. In this chapter, we mainly study various instanton
invariants of 3-manifolds. The material of this chapter is based in part on
[263]. The basic ideas come from Witten's work on supersymmetric Morse
theory. We discuss this work in Section 10.2. In Section 10.3 we consider gauge
fields on a 3-dimensional manifold. The field equations are obtained from the
Chern–Simons action functional and correspond to flat connections. Casson
invariant is discussed in Section 10.4. In Section 10.5 we discuss the $\mathbf{Z}_8$-graded
instanton homology theory due to Floer and its relation to the Casson invariant. Floer's theory was extended to arbitrary closed oriented 3-manifolds
by Fukaya. When the first homology of such a manifold is torsion-free, but
not necessarily zero, Fukaya also defines a class of invariants indexed by the
integer s, $0 \leq s < 3$, where s is the rank of the first integral homology group
of the manifold. These invariants include, in particular, the Floer homology
groups in the case $s = 0$. The construction of these invariants is closely related to that of Donaldson polynomials of 4-manifolds, which we considered in
Chapter 9. As with the definition of Donaldson polynomials a careful analysis
of the singular locus (the set of reducible connections) is required in defining
the Fukaya invariants. Section 10.6 is a brief introduction to an extension of
Floer homology to a $\mathbf{Z}$-graded homology theory, due to Fintushel and Stern,
for homology 3-spheres. Floer also defined a homology theory for symplectic manifolds using Lagrangian submanifolds and used it in his proof of the

K. Marathe, *Topics in Physical Mathematics*, DOI 10.1007/978-1-84882-939-8_10,   313
© Springer-Verlag London Limited 2010

Arnold conjecture. We do not discuss this theory. For general information on various Morse homologies, see, for example, [29]. The WRT invariants, which arise as a byproduct of Witten's TQFT interpretation of the Jones polynomial are discussed in Section 10.7. Section 10.8 is devoted to a special case of the question of relating gauge theory and string theory where exact results are available. Geometric transition that is used to interpolate between these theories is also considered here. Some of the material of this chapter is taken from [263].

10.2 Witten Complex and Morse Theory

Classical Morse theory on a finite-dimensional, compact, differentiable manifold M relates the behavior of critical points of a suitable function on M with topological information about M. The relation is generally stated as an equality of certain polynomials as follows. Recall first that a smooth function $f : M \to \mathbf{R}$ is called a **Morse function** if its critical points are isolated and non-degenerate. If $x \in M$ is a **critical point** (i.e., $df(x) = 0$) then by Taylor expansion of f around x we obtain the **Hessian** of f at x defined by

$$\left\{ \frac{\partial^2 f}{\partial x^i \partial x^j}(x) \right\}.$$

Then the non-degeneracy of the critical point x is equivalent to the non-degeneracy of the quadratic form determined by the Hessian. The dimension of the negative eigenspace of this form is called the **Morse index**, or simply **index**, of f at x and is denoted by $\mu_f(x)$, or simply $\mu(x)$ when f is understood. It can be verified that these definitions are independent of the choice of the local coordinates. Let m_k be the number of critical points with index k. Then the **Morse series** of f is the formal power series

$$\sum_k m_k t^k.$$

Recall that the Poincaré series of M is given by $\sum_k b_k t^k$, where $b_k \equiv b_k(M)$ is the kth Betti number of M. The relation between the two series is given by

$$\sum_k m_k t^k = \sum_k b_k t^k + (1+t) \sum_k q_k t^k, \tag{10.1}$$

where q_k are non-negative integers. Comparing the coefficients of the powers of t in this relation leads to the well-known **Morse inequalities**

$$\sum_{k=0}^{i} m_{i-k}(-1)^k \geq \sum_{k=0}^{i} b_{i-k}(-1)^k , \quad 0 \leq i \leq n-1,$$

and to the equality

$$\sum_{k=0}^{n} m_{n-k}(-1)^k = \sum_{k=0}^{n} b_{n-k}(-1)^k \ .$$

The Morse inequalities can also be obtained from the following observation: Let C^* be the graded vector space over the set of critical points of f. Then the Morse inequalities are equivalent to the existence of a coboundary operator $\partial : C^* \to C^*$ so that $\partial^2 = 0$ and the cohomology of the complex (C^*, ∂) coincides with the de Rham cohomology of M.

In his fundamental paper [403] Witten arrives at precisely such a complex by considering a suitable supersymmetric quantum mechanical Hamiltonian. Witten showed how the standard Morse theory (see Morse and Cairns [292] and Milnor [284]) can be modified by consideration of the gradient flow of the Morse function f between pairs of critical points of f. One may think of this as a sort of relative Morse theory. Witten was motivated by the phenomenon of quantum mechanical tunneling. We now discuss this approach. From a mathematical point of view, supersymmetry may be regarded as a theory of operators on a Z_2-graded Hilbert space. In recent years this theory has attracted a great deal of interest even though as yet there is no physical evidence for its existence. In our general formulation of a supersymmetric theory we let E denote the Hilbert space of a supersymmetric theory, i.e., $E = E_0 \oplus E_1$, where the even (resp., odd) space E_0 (resp., E_1) is called the space of **bosonic states** (resp. **fermionic states**). These spaces are distinguished by the **parity operator** $S : E \to E$ defined by

$$Su = u, \quad \forall u \in E_0,$$

$$Sv = -v, \quad \forall v \in E_1.$$

The operator S is interpreted as counting the number of fermions modulo 2. A supersymmetric theory begins with a collection $\{Q_i \mid i = 1, \ldots, n\}$ of **supercharge operators** (or **supersymmetry operators**) on E that are of odd degree, that is, they anti-commute with S,

$$SQ_i + Q_i S = 0, \quad \forall i, \tag{10.2}$$

and satisfy the following anti-commutation relations

$$Q_i Q_j + Q_j Q_i = 0, \quad \forall i, j \text{ with } i \neq j. \tag{10.3}$$

The mechanics is introduced by the Hamiltonian operator H, which commutes with the supercharge operators and is usually required to satisfy additional conditions. For example, in the simplest non-relativistic theory one requires that

$$H = Q_i^2, \quad \forall i. \tag{10.4}$$

In fact, this simplest supersymmetric theory has surprising connections with Morse theory, as was shown by Witten in his fundamental paper [403]. It gave a new interpretation of this classical theory and paved the way for Floer's new homology theory. We consider Floer homology later in this chapter. We now discuss Witten's work.

Let M be a compact differentiable manifold and define E by

$$E := \Lambda(M) \otimes \mathbf{C}.$$

The natural grading on $\Lambda(M)$ induces a grading on E. We define

$$E_0 := \bigoplus_j \Lambda^{2j}(M) \otimes \mathbf{C} \quad (\text{resp., } E_1 := \bigoplus_j \Lambda^{2j+1}(M) \otimes \mathbf{C}),$$

the space of complex-valued even (resp., odd) forms on M. The exterior differential d and its formal adjoint δ have natural extension to odd operators on E and thus satisfy (10.2). We define supercharge operators Q_j, $j = 1, 2$, by

$$Q_1 = d + \delta, \tag{10.5}$$

$$Q_2 = i(d - \delta). \tag{10.6}$$

The Hamiltonian is taken to be the Hodge–de Rham operator extended to E, i.e.,

$$H = d\delta + \delta d. \tag{10.7}$$

The relations $d^2 = \delta^2 = 0$ imply the supersymmetry relations (10.3) and (10.4). We note that in this case bosonic (resp., fermionic) states correspond to even (resp., odd) forms. The relation to Morse theory arises in the following way. If f is a Morse function on M, a one-parameter family of operators is defined:

$$d_t = e^{-ft} d e^{ft}, \quad \delta_t = e^{ft} \delta e^{-ft}, \quad t \in \mathbf{R}, \tag{10.8}$$

and the corresponding supersymmetry operators

$$Q_{1,t}, \quad Q_{2,t}, \quad H_t$$

are defined as in equations (10.5), (10.6), and (10.7). It is easy to verify that $d_t^2 = \delta_t^2 = 0$ and hence $Q_{1,t}$, $Q_{2,t}$, H_t satisfy the supersymmetry relations (10.3) and (10.4). The parameter t interpolates between the de Rham cohomology and the Morse indices as t goes from 0 to $+\infty$. At $t = 0$ the number of linearly independent eigenvectors with zero eigenvalue is just the kth Betti number b_k when $H_0 = H$ is restricted to act on k-forms. In fact, these ground states of the Hamiltonian are just the harmonic forms. On the other hand, for large t the spectrum of H_t simplifies greatly with the eigenfunctions concentrating near the critical points of the Morse function. It is in this way that the Morse indices enter into this picture. We can write H_t

as a perturbation of H near the critical points. In fact, we have

$$H_t = H + t \sum_{j,k} f_{,jk}[\alpha^j, i_{X^k}] + t^2 \|df\|^2,$$

where $\alpha^j = dx^j$ acts by exterior multiplication, $X^k = \partial/\partial x^k$, and i_{X^k} is the usual action of inner multiplication by X^k on forms and the norm $\|df\|$ is the norm on $\Lambda^1(M)$ induced by the Riemannian metric on M. In a suitable neighborhood of a fixed critical point taken as origin, we can approximate H_t up to quadratic terms in x^j by

$$\overline{H}_t = \sum_j \left(-\frac{\partial^2}{\partial x_j^2} + t^2 \lambda_j^2 x_j^2 + t\lambda_j[\alpha^j, i_{X^j}] \right),$$

where λ_j are the eigenvalues of the Hessian of f. The first two terms correspond to the quantized Hamiltonian of a harmonic oscillator (see Appendix B for a discussion of the classical and quantum harmonic oscillator) with eigenvalues

$$t \sum_j |\lambda_j| (1 + 2N_j),$$

whereas the last term defines an operator with eigenvalues $\pm\lambda_j$. It commutes with the first and thus the spectrum of $\overline{H}_t$ is given by

$$t \sum_j [|\lambda_j| (1 + 2N_j) + \lambda_j n_j],$$

where N_j are non-negative integers and $n_j = \pm 1$. Restricting H to act on k-forms we can find the ground states by requiring all the N_j to be 0 and by choosing n_j to be 1 whenever λ_j is negative. Thus the ground states (zero eigenvalues) of H correspond to the critical points of Morse index k. All other eigenvalues are proportional to t with positive coefficients. Starting from this observation and using standard perturbation theory, one finds that the number of k-form ground states equals the number of critical points of Morse index k. Comparing this with the ground state for $t = 0$, we obtain the weak Morse inequalities $m_k \geq b_k$. As we observed in the introduction, the strong Morse inequalities are equivalent to the existence of a certain cochain complex that has cohomology isomorphic to $H^*(M)$, the cohomology of the base manifold M. Witten defines C_p, the set of p-chains of this complex, to be the free group generated by the critical points of Morse index p. He then argues that the operator d_t defined in (10.8) defines in the limit as $t \to \infty$ a coboundary operator

$$d_\infty : C_p \to C_{p+1}$$

and that the cohomology of this complex is isomorphic to the de Rham cohomology of Y.

Thus we see that in establishing both the weak and strong form of Morse inequalities a fundamental role is played by the ground states of the supersymmetric quantum mechanical system (10.5), (10.6), (10.7). In a classical system the transition from one ground state to another is forbidden, but in a quantum mechanical system it is possible to have tunneling paths between two ground states. In gauge theory the role of such tunneling paths is played by instantons. Indeed, Witten uses the prescient words "instanton analysis" to describe the tunneling effects obtained by considering the gradient flow of the Morse function f between two ground states (critical points). If β (resp., α) is a critical point of f of Morse index $p+1$ (resp., p) and Γ is a gradient flow of f from β to α, then by comparing the orientation of negative eigenspaces of the Hessian of f at β and α, Witten defines the signature n_Γ of this flow. By considering the set S of all such flows from β to α, he defines

$$n(\alpha, \beta) := \sum_{\Gamma \in S} n_\Gamma.$$

Now defining δ_∞

$$\delta_\infty : C_p \to C_{p+1} \qquad \text{by } \alpha \mapsto \sum_{\beta \in C_{p+1}} n(\alpha, \beta)\beta, \qquad (10.9)$$

he shows that (C_*, δ_∞) is a cochain complex with integer coefficients. Witten conjectures that the integer-valued coboundary operator δ_∞ actually gives the integral cohomology of the manifold M. The complex (C_*, δ_∞), with the coboundary operator defined by (10.9), is referred to as the **Witten complex**. As we will see later, Floer homology is the result of such "instanton analysis" applied to the gradient flow of a suitable Morse function on the moduli space of gauge potentials on an integral homology 3-sphere. Floer also used these ideas to study a "symplectic homology" associated to a manifold. A corollary of this theory proves the Witten conjecture for finite-dimensional manifolds (see [334] for further details), namely,

$$H^*(C_*, \delta_\infty) = H^*(M, \mathbf{Z}).$$

A direct proof of the conjecture may be found in the appendix to K. C. Chang [72]. A detailed study of the homological concepts of finite-dimensional Morse theory in anology with Floer homology may be found in M. Schwarz [342]. While many basic concepts of "Morse homology" can be found in the classical investigations of Milnor, Smale, and Thom, its presentation as an axiomatic homology theory in the sense of Eilenberg and Steeenrod [118] is given for the first time in [342]. One consequence of this axiomatic approach is the uniqueness result for so-called Morse homology and its natural equivalence with other axiomatic homology theories defined on a suitable category of topological spaces. Witten conjecture is then a corollary of this result.

A discussion of the relation of equivariant cohomolgy and supersymmetry may be found in Guillemin and Sternberg's book [180].

10.3 Chern–Simons Theory

Let M be a compact manifold of dimension $m = 2r + 1$, $r > 0$, and let $P(M, G)$ be a principal bundle over M with a compact, semisimple Lie group G as its structure group. Let $\alpha_m(\omega)$ denote the Chern–Simons m-form on M corresponding to the gauge potential ω on P; then the Chern–Simons action $\mathcal{A}_{CS}$ is defined by

$$\mathcal{A}_{CS} := c(G) \int_M \alpha_m(\omega), \qquad (10.10)$$

where $c(G)$ is a coupling constant whose normalization depends on the group G. In the rest of this paragraph we restrict ourselves to the case $r = 1$ and $G = SU(n)$. The most interesting applications of the Chern–Simons theory to low-dimensional topologies are related to this case. It has been extensively studied by both physicists and mathematicians in recent years. In this case the action (10.10) takes the form

$$\mathcal{A}_{CS} = \frac{k}{4\pi} \int_M \operatorname{tr}(A \wedge F - \tfrac{1}{3} A \wedge A \wedge A) \qquad (10.11)$$

$$= \frac{k}{4\pi} \int_M \operatorname{tr}(A \wedge dA + \tfrac{2}{3} A \wedge A \wedge A), \qquad (10.12)$$

where $k \in \mathbf{R}$ is a coupling constant, A denotes the pull-back to M of the gauge potential ω by a local section of P, and $F = F_\omega = d^\omega A$ is the gauge field on M corresponding to the gauge potential A. A local expression for (10.11) is given by

$$\mathcal{A}_{CS} = \frac{k}{4\pi} \int_M \epsilon^{\alpha\beta\gamma} \operatorname{tr}(A_\alpha \partial_\beta A_\gamma + \tfrac{2}{3} A_\alpha A_\beta A_\gamma), \qquad (10.13)$$

where $A_\alpha = A_\alpha^a T_a$ are the components of the gauge potential with respect to the local coordinates $\{x_\alpha\}$, $\{T_a\}$ is a basis of the Lie algebra $su(n)$ in the fundamental representation, and $\epsilon^{\alpha\beta\gamma}$ is the totally skew-symmetric Levi-Civita symbol with $\epsilon^{123} = 1$. We take the basis $\{T_a\}$ with the normalization

$$\operatorname{tr}(T_a T_b) = \tfrac{1}{2}\delta_{ab}, \qquad (10.14)$$

where δ_{ab} is the Kronecker δ function. Let $g \in \mathcal{G}$ be a gauge transformation regarded (locally) as a function from M to $SU(n)$ and define the 1-form θ by

$$\theta = g^{-1}dg = g^{-1}\partial_\mu g \, dx^\mu.$$

Then the gauge transformation A^g of A by g has the local expression

$$A_\mu^g = g^{-1}A_\mu g + g^{-1}\partial_\mu g. \tag{10.15}$$

In the physics literature the connected component of the identity $\mathcal{G}_{id} \subset \mathcal{G}$ is called the group of **small gauge transformations**. A gauge transformation not belonging to $\mathcal{G}_{id}$ is called a **large gauge transformation**. By a direct calculation one can show that the Chern–Simons action is invariant under small gauge transformations, i.e.,

$$\mathcal{A}_{CS}(A^g) = \mathcal{A}_{CS}(A), \qquad \forall g \in \mathcal{G}_{id}.$$

Under a large gauge transformation g, the action (10.13) transforms as follows:

$$\mathcal{A}_{CS}(A^g) = \mathcal{A}_{CS}(A) + 2\pi k \mathcal{A}_{WZ}, \tag{10.16}$$

where

$$\mathcal{A}_{WZ} := \frac{1}{24\pi^2} \int_M \epsilon^{\alpha\beta\gamma} \operatorname{tr}(\theta_\alpha \theta_\beta \theta_\gamma) \tag{10.17}$$

is the **Wess–Zumino action functional**. It can be shown that the Wess–Zumino functional is integer-valued and hence, if the Chern–Simons coupling constant k is taken to be an integer, then we have

$$e^{i\mathcal{A}_{CS}(A^g)} = e^{i\mathcal{A}_{CS}(A)}.$$

The integer k is called the **level** of the corresponding Chern–Simons theory. It follows that the path integral quantization of the Chern–Simons model is gauge-invariant. This conclusion holds more generally for any compact simple group if the coupling constant $c(G)$ is chosen appropriately. The action is manifestly covariant since the integral involved in its definition is independent of the metric on M. It is in this sense that the Chern–Simons theory is a topological field theory. We considered this aspect of the Chern–Simons theory in Chapter 7.

In general, the Chern–Simons action is defined on the space $\mathcal{A}_{P(M,G)}$ of all gauge potentials on the principal bundle $P(M,G)$. But when M is 3-dimensional P is trivial (in a non-canonical way). We fix a trivialization to write $P(M,G) = M \times G$ and write $\mathcal{A}_M$ for $\mathcal{A}_{P(M,G)}$. Then the group of gauge transformations $\mathcal{G}_P$ can be identified with the group of smooth functions from M to G and we denote it simply by $\mathcal{G}_M$. For $k \in \mathbf{N}$ the transformation law (10.16) implies that the Chern–Simons action descends to the quotient $\mathcal{B}_M = \mathcal{A}_M/\mathcal{G}_M$ as a function with values in $\mathbf{R}/\mathbf{Z}$. We denote this function by f_{CS}, i.e.,

$$f_{CS} : \mathcal{B}_M \to \mathbf{R}/\mathbf{Z}$$

is defined by

$$[\omega] \mapsto \mathcal{A}_{CS}(\omega), \qquad \forall [\omega] = \omega\mathcal{G} \in \mathcal{B}_M. \tag{10.18}$$

The field equations of the Chern–Simons theory are obtained by setting the first variation of the action to zero as

$$\delta \mathcal{A}_{CS} = 0.$$

We discuss two approaches to this calculation. Consider first a one-parameter family $c(t)$ of connections on P with $c(0) = \omega$ and $\dot{c}(0) = \alpha$. Differentiating the action $\mathcal{A}_{CS}(c(t))$ with respect to t and noting that differentiation commutes with integration and the tr operator, we get

$$\frac{d}{dt} \mathcal{A}_{CS}(c(t)) = \frac{1}{4\pi} \int_M \mathrm{tr}\left(2\dot{c}(t) \wedge dc(t) + 2(\dot{c}(t) \wedge c(t) \wedge c(t))\right)$$

$$= \frac{1}{2\pi} \int_M \mathrm{tr}\left(\dot{c}(t) \wedge (dc(t) + c(t) \wedge c(t))\right)$$

$$= \frac{1}{2\pi} \int_M \langle \dot{c}(t), *F_{c(t)} \rangle,$$

where the inner product on the right is as in Definition 2.1. It follows that

$$\delta \mathcal{A}_{CS} = \frac{d}{dt} \mathcal{A}_{CS}(c(t))_{|t=0} = \frac{1}{2\pi} \int_M \langle \alpha, *F_\omega \rangle. \tag{10.19}$$

Since α can be chosen arbitrarily, the field equations are given by

$$* F_\omega = 0 \quad \text{or equivalently,} \quad F_\omega = 0. \tag{10.20}$$

Alternatively, one can start with the local coordinate expression of equation (10.13) as follows:

$$\mathcal{A}_{CS} = \frac{k}{4\pi} \int_M \epsilon^{\alpha\beta\gamma} \, \mathrm{tr}(A_\alpha \partial_\beta A_\gamma + \tfrac{2}{3} A_\alpha A_\beta A_\gamma)$$

$$= \frac{k}{4\pi} \int_M \epsilon^{\alpha\beta\gamma} \, \mathrm{tr}(A_\alpha^a \partial_\beta A_\gamma^c T_a T_b + \tfrac{2}{3} A_\alpha^a A_\beta^b A_\gamma^c T_a T_b T_c)$$

and find the field equations by using the variational equation

$$\frac{\delta \mathcal{A}_{CS}}{\delta A_\rho^a} = 0. \tag{10.21}$$

This method brings out the role of commutation relations and the structure constants of the Lie algebra $su(n)$ as well as the boundary conditions used in the integration by parts in the course of calculating the variation of the action. The result of this calculation gives

$$\frac{\delta \mathcal{A}_{CS}}{\delta A_\rho^a} = \frac{k}{2\pi} \int_M \epsilon^{\rho\beta\gamma} \left(\partial_\beta A_\gamma^a + A_\beta^b A_\gamma^c f_{abc}\right), \tag{10.22}$$

where f_{abc} are the structure constants of $su(n)$ with respect to the basis T_a. The integrand on the right hand side of the equation (10.22) is just the local coordinate expression of $*F_A$, the dual of the curvature, and hence leads to the same field equations.

The calculations leading to the field equations (10.20) also show that the gradient vector field of the function f_{CS} is given by

$$\text{grad } f_{CS} = \frac{1}{2\pi} *F. \tag{10.23}$$

The gradient flow of f_{CS} plays a fundamental role in the definition of Floer homology. The solutions of the field equations (10.20) are called the **Chern–Simons connections**. They are precisely the flat connections. Next we discuss flat connections on a manifold N and their relation to the homomorphisms of the fundamental group $\pi_1(N)$ into the gauge group.

Flat Connections

Let H be a compact Lie group and $Q(N, H)$ be a principal bundle with structure group H over a compact Riemannian manifold N. A connection ω on Q is said to be **flat** if its curvature is zero, i.e., $\Omega_\omega = 0$. The pair (Q, ω) is called a **flat bundle**. Let $\Omega(N, x)$ be the loop space at $x \in N$. Recall that the horizontal lift h_u of $c \in \Omega(N, x)$ to $u \in \pi^{-1}(x)$ determines a unique element of H. Thus we have the map

$$h_u : \Omega(N, x) \to H.$$

It is easy to see that ω flat implies that this map h_u depends only on the homotopy class of the loop c and hence induces a map (also denoted by h_u)

$$h_u : \pi_1(N, x) \to H.$$

It is this map that is responsible for the Bohm–Aharonov effect. It can be shown that the map h_u is a homomorphism of groups. The group H acts on the set $\text{Hom}(\pi_1(N), H)$ by conjugation sending h_u to $g^{-1}h_u g = h_{ug}$. Thus, a flat bundle (Q, ω) determines an element of the quotient $\text{Hom}(\pi_1(N), H)/H$. If $a \in \mathcal{G}(Q)$, the group of gauge transformations of Q, then $a \cdot \omega$ is also a flat connection on Q and determines the same element of $\text{Hom}(\pi_1(N), H)/H$. Conversely, let $f \in \text{Hom}(\pi_1(N), H)$ and let (U, q) be the universal covering of N. Then U is a principal bundle over N with structure group $\pi_1(N)$. Define $Q := U \times_f H$ to be the bundle associated to U by the action f with standard fiber H. It can be shown that Q admits a natural flat connection and that f and $g^{-1}fg$, $g \in H$, determine isomorphic flat bundles. Thus, the moduli space $\mathcal{M}_f(N, H)$ of flat H-bundles over N can be identified with the set $\text{Hom}(\pi_1(N), H)/H$. The moduli space $\mathcal{M}_f(N, H)$ and the set $\text{Hom}(\pi_1(N), H)$ have a rich mathematical structure, which has been exten-

sively studied in the particular case when N is a compact Riemann surface
[17].

The **flat connection deformation complex** is the generalized de Rham
sequence

$$0 \longrightarrow \Lambda^0 \xrightarrow{d^\omega} \Lambda^1 \xrightarrow{d^\omega} \cdots \xrightarrow{d^\omega} \Lambda^n \longrightarrow 0.$$

The fact that in this case it is a complex follows from the observation that
ω flat implies $d^\omega \circ d^\omega = 0$. By rolling up this complex, we can consider the
rolled up deformation operator $d^\omega + \delta^\omega : \Lambda^{ev} \to \Lambda^{odd}$. By the index theorem
we have

$$\mathrm{Ind}(d^\omega + \delta^\omega) = \chi(N)\dim H$$

and hence

$$\sum_{i=0}^{n}(-1)^i b_i = \chi(N)\dim H, \tag{10.24}$$

where b_i is the dimension of the ith cohomology of the deformation complex.
Both sides are identically zero for odd n. For even n the formula can be used to
obtain some information on the virtual dimension of $\mathcal{M}_f$ $(= b_1)$. For example,
if $N = \Sigma_g$ is a Riemann surface of genus $g > 1$, then $\chi(\Sigma_g) = -2g + 2$,
while, by Hodge duality, $b_0 = b_2 = 0$ at an irreducible connection. Thus,
equation (10.24) gives

$$-b_1 = -(2g - 2)\dim H.$$

From this it follows that

$$\dim \mathcal{M}_f(\Sigma_g, H) = \dim \mathcal{M}_f = (2g - 2)\dim H. \tag{10.25}$$

In even dimensions greater than 2, the higher cohomology groups provide
additional obstructions to smoothability of $\mathcal{M}_f$. For example, for $n = 4$,
Hodge duality implies that $b_0 = b_4$ and $b_1 = b_3$ and (10.24) gives

$$b_1 = b_0 + \tfrac{1}{2}(b_2 - \chi(N)\dim H).$$

Equation (10.25) shows that $\dim \mathcal{M}_f$ is even. Identifying the first cohomology $H^1(\Lambda(M, \mathrm{ad}\,h), d^\omega)$ of the deformation complex with the tangent space
$T_\omega \mathcal{M}_f$ to $\mathcal{M}_f$, the intersection form defines a map $\iota_\omega : T_\omega \mathcal{M}_f \times T_\omega \mathcal{M}_f \to \mathbf{R}$
by

$$\iota(X, Y) = \int_{\Sigma_g} X \wedge Y, \qquad X, Y \in T_\omega \mathcal{M}_f. \tag{10.26}$$

The map ι_ω is skew-symmetric and bilinear. The map

$$\iota : \omega \mapsto \iota_\omega, \qquad \forall \omega \in \mathcal{M}_f, \tag{10.27}$$

defines a 2-form ι on $\mathcal{M}_f$. If $\mathbf{h}$ admits an H-invariant inner product, then this
2-form ι is closed and non-degenerate and hence defines a symplectic structure

on $\mathcal{M}_f$. It can be shown that, for a Riemann surface with $H = PSL(2, \mathbf{R})$, the form ι, restricted to the Teichmüller space, agrees with the well-known Weil–Petersson form.

We now discuss an interesting physical interpretation of the symplectic manifold $(\mathcal{M}_f(\Sigma_g, H), \iota)$. Consider a Chern–Simons theory on the principal bundle $P(M, H)$ over the $2+1$-dimensional space-time manifold $M = \Sigma_g \times \mathbf{R}$ with gauge group H and with time-independent gauge potentials and gauge transformations. Let $\mathcal{A}$ (resp., $\mathcal{H}$) denote the space (resp., group) of these gauge connections (resp. transformations). It can be shown that the curvature F_ω defines an $\mathcal{H}$-equivariant moment map

$$\mu : \mathcal{A} \to \mathcal{LH} \cong \Lambda^1(M, \operatorname{ad} P), \text{ by } \omega \mapsto *F_\omega,$$

where $\mathcal{LH}$ is the Lie algebra of $\mathcal{H}$. The zero set $\mu^{-1}(0)$ of this map is precisely the set of flat connections and hence

$$\mathcal{M}_f \cong \mu^{-1}(0)/\mathcal{H} := \mathcal{A}//\mathcal{H} \tag{10.28}$$

is the reduced phase space of the theory, in the sense of the Marsden–Weinstein reduction. We call $\mathcal{A}//\mathcal{H}$ the **symplectic quotient** of $\mathcal{A}$ by $\mathcal{H}$. Marsden–Weinstein reduction and symplectic quotient are fundamental constructions in geometrical mechanics and geometric quantization. They also arise in many other mathematical applications.

A situation similar to that described above also arises in the geometric formulation of canonical quantization of field theories. One proceeds by analogy with the geometric quantization of finite-dimensional systems. For example, $Q = \mathcal{A}/\mathcal{H}$ can be taken as the configuration space and T^*Q as the corresponding phase space. The associated Hilbert space is obtained as the space of L^2 sections of a complex line bundle over Q. For physical reasons this bundle is taken to be flat. Inequivalent flat $U(1)$-bundles are said to correspond to distinct sectors of the theory. Thus we see that at least formally these sectors are parametrized by the moduli space

$$\mathcal{M}_f(Q, U(1)) \cong \operatorname{Hom}(\pi_1(Q), U(1))/U(1) \cong \operatorname{Hom}(\pi_1(Q), U(1))$$

since $U(1)$ acts trivially on $\operatorname{Hom}(\pi_1(Q), U(1))$.

We note that the Chern–Simons theory has been extended by Witten to the cases when the gauge group is finite and when it is the complexification of compact real gauge groups [101, 406]. While there are some similarities between these theories and the standard CS theory, there are major differences in the corresponding TQFTs. New invariants of some hyperbolic 3-manifolds have recently been obtained by considering the complex gauge groups leading to the concept of arithmetic TQFT by Zagier and collaborators (see arXiv:0903.24272). See also Dijkgraaf and Fuji arXiv:0903.2084 [hep-th] and Gukov and Witten arXiv:0809.0305 [hep-th].

10.4 Casson Invariant

Let Y be a homology 3-sphere. Let D_1, D_2 be two unitary, unimodular representations of $\pi_1(Y)$ in $\mathbf{C}^2$. We say that they are equivalent if they are conjugate under the natural $SU(2)$-action on $\mathbf{C}^2$, i.e.,

$$D_2(g) = S^{-1} D_1(g) S, \qquad \forall g \in \pi_1(Y), \ S \in SU(2).$$

Let us denote by $\mathcal{R}(Y)$ the set equivalence classes of such representations. It is customary to write

$$\mathcal{R}(Y) := \mathrm{Hom}\{\pi_1(Y) \to SU(2)\}/\mathrm{conj.} \tag{10.29}$$

Let θ be the trivial representation and define

$$\mathcal{R}^*(Y) := \mathcal{R}(Y) \setminus \{\theta\}. \tag{10.30}$$

Fixing an orientation of Y, Casson showed how to assign a sign $s(\alpha)$ to each element $\alpha \in \mathcal{R}^*(Y)$. He showed that the set $\mathcal{R}^*(Y)$ is finite and defined a numerical invariant of Y by counting the signed number of elements of $\mathcal{R}^*(Y)$ via

$$c(Y) := \sum_{\alpha \in \mathcal{R}^*(Y)} s(\alpha).1, \tag{10.31}$$

where $s(\alpha).1$ is ± 1 depending on the sign of α. The integer $c(Y)$ is called the **Casson invariant** of Y.

A similar idea was used by Taubes [366] to give a new interpretation of the Casson invariant $c(Y)$ of an oriented homology 3-sphere Y, which is defined in terms of the number of irreducible representations of $\pi_1(Y)$ into $SU(2)$. As indicated above, this space can be identified with the moduli space $\mathcal{M}_f(Y, SU(2))$ of flat connections in the trivial $SU(2)$-bundle over Y. The map $F : \omega \mapsto F_\omega$ defines a natural 1-form on $\mathcal{A}/\mathcal{G}$ and its dual vector field v_F. The zeros of this vector field are just the flat connections. We note that since $\mathcal{A}/\mathcal{G}$ is infinite-dimensional, it is necessary to use suitable Fredholm perturbations to get simple zeros and to count them with appropriate signs. Let Z denote the set of zeros of the perturbed vector field (also denoted by v_F) and let $s(a)$ be the sign of $a \in Z$. Taubes showed that Z is contained in a compact set and hence the index i_F of the vector field v_F is well-defined by the classical formula

$$i_F := \sum_{a \in Z} s(a).1.$$

He then proved that this index is an invariant of Y and equals the Casson invariant, i.e.,

$$c(Y) = i_F = \sum_{a \in Z} s(a).1.$$

Now for a compact Riemannian manifold M the Poincaré–Hopf theorem states that the index of a vector field equals the Euler characteristic $\chi(M)$ defined in terms of the homology of M. It is thus natural to ask if the Casson invariant can be realized by some homology theory associated to the moduli space $\mathcal{M}_f(Y, SU(2))$ of flat connections on Y. This question was answered in the affirmative by Floer by constructing his instanton homology. We will discuss this homology in the next section. The main idea is to consider a one-parameter family $\{\omega_t\}_{t \in I}$ of connections of Y that defines a connection on $Y \times I$ and the corresponding Chern–Simons action $\mathcal{A}_{CS}$. This is invariant under the connected component of the identity in $\mathcal{G}$ but changes by the Wess–Zumino action under the full group $\mathcal{G}$. The Chern–Simons action $\mathcal{A}_{CS}$ can be modified (if necessary) to define a Morse function f on the moduli space of flat connections on Y. By considering the gradient flow of this Morse function f between pairs of critical points of f, the resulting **Witten complex** (C_*, δ) was used by Floer to define the **Floer homology groups** $FH_n(Y)$, $n \in \mathbf{Z}_8$ (see [130, 131]). Using Taubes' interpretation of the Casson invariant $c(Y)$, one can show that $c(Y)$ equals the Euler number of the Floer homology. This result has an interesting geometric interpretation in terms of topological quantum field theory. In fact, ideas from TQFT have provided geometric interpretation of invariants of knots and links in 3-manifolds and have also led to new invariants of 3-manifolds. We considered some of these results in Chapter 7.

Another approach to Casson's invariant involves symplectic geometry and topology. We conclude this section with a brief explanation of this approach. Let $Y_+ \cup_{\Sigma_g} Y_-$ be a Heegaard splitting of Y along the Riemann surface Σ_g of genus g. The space $\mathcal{R}(\Sigma_g)$ of conjugacy classes of representations of $\pi_1(\Sigma_g)$ into $SU(2)$ can be identified with the moduli space $\mathcal{M}_f(\Sigma_g, SU(2))$ of flat connections. This identification endows it with a natural symplectic structure that makes it into a $(6g-6)$-dimensional symplectic manifold. The representations which extend to Y_+ (resp., Y_-) form a $(3g-3)$-dimensional Lagrangian submanifold of $\mathcal{R}(\Sigma_g)$, which we denote by $\mathcal{R}(Y_+)$ (resp., $\mathcal{R}(Y_-)$). Casson's invariant is then obtained from the intersection number of the Lagrangian submanifolds $\mathcal{R}(Y_+)$ and $\mathcal{R}(Y_-)$ in the symplectic manifold $\mathcal{R}(\Sigma_g)$. How the Floer homology of Y fits into this scheme seems to be unknown at this time.

An $SU(3)$ Casson invariant of integral homology 3-spheres is defined in [45] and some of its properties are studied in [44].

10.5 Floer Homology

The idea of instanton tunneling and the corresponding Witten complex was extended by Floer to do Morse theory on the infinite-dimensional moduli space of gauge potentials on a homology 3-sphere Y and to define new topological invariants of Y. Fukaya generalized this work to apply to arbitrary oriented 3-manifolds. We shall refer to the invariants of Floer and Fukaya collectively as **Fukaya–Floer homology** (see [66, 147, 303]). A detailed discussion of the Floer homology groups in Yang–Mills theory is given in Donaldson's book [111]. Fukaya–Floer homology associates to an oriented connected closed smooth 3-dimensional manifold Y, a family of $\mathbf{Z}_8$-graded homology. We begin by introducing Floer's original definition, which requires Y to be a homology 3-sphere.

Let $\alpha \in \mathcal{R}^*(Y)$. We say that α is a **regular representation** if

$$H^1(Y, \mathrm{ad}(\alpha)) = 0. \tag{10.32}$$

The Chern–Simons functional has non-degenerate Hessian at α if α is regular. Fix a trivialization P of the given $SU(2)$-bundle over Y. Using the trivial connection θ on $P = Y \times SU(2)$ as a background connection on Y, we can identify the space of connections $\mathcal{A}_Y$ with the space of sections of $\Lambda^1(Y) \otimes su(2)$. In what follows we shall consider a suitable Sobolev completion of this space and continue to denote it by $\mathcal{A}_Y$.

Let $c : I \to \mathcal{A}_Y$ be a path from α to θ. The family of connections $c(t)$ on Y can be identified as a connection A on $Y \times I$. Using this connection we can rewrite the Chern–Simons action (10.11) as follows

$$\mathcal{A}_{CS} = \frac{1}{8\pi^2} \int_{Y \times I} \mathrm{tr}(F_A \wedge F_A). \tag{10.33}$$

We note that the integrand corresponds to the second Chern class of the pull-back of the trivial $SU(2)$-bundle over Y to $Y \times I$. Recall that the critical points of the Chern–Simons action are the flat connections. The gauge group $\mathcal{G}_Y$ acts on $\mathcal{A}_{CS} : \mathcal{A} \to \mathbf{R}$ by

$$\mathcal{A}_{CS}(\alpha^g) = \mathcal{A}_{CS}(\alpha) + \deg(g), \qquad g \in \mathcal{G}_Y.$$

It follows that $\mathcal{A}_{CS}$ descends to $\mathcal{B}_Y := \mathcal{A}_Y / \mathcal{G}_Y$ as a map $f_{CS} : \mathcal{B}_Y \to \mathbf{R}/\mathbf{Z}$ and we can take $\mathcal{R}(Y) \subset \mathcal{B}_Y$ as the critical set of f_{CS}. The gradient flow of this function is given by the equation

$$\frac{\partial c(t)}{\partial t} = *_Y F_{c(t)}. \tag{10.34}$$

Since Y is a homology 3-sphere, the critical points of the flow of $\mathrm{grad}\, f_{CS}$ and the set of reducible connections intersect at a single point, the trivial connection θ. If all the critical points of the flow are regular then it is a Morse–Smale flow. If not, one can perturb the function f_{CS} to get a Morse function.

In general the representation space $\mathcal{R}^*(Y) \subset \mathcal{B}_Y$ contains degenerate critical points of the Chern–Simons function f_{CS}. In this case Floer defines a set of perturbations of f_{CS} as follows. Let $m \in \mathbf{N}$ and let $\vee_{i=1}^m S_i^1$ be a bouquet of m copies of the circle S^1. Let Γ_m be the set of maps

$$\gamma : \bigvee_{i=1}^{m} S_i^1 \times D^2 \to Y$$

such that the restrictions

$$\gamma_x : \bigvee_{i=1}^{m} S_i^1 \times \{x\} \to Y \quad \text{and} \quad \gamma_i : S_i^1 \times D^2 \to Y$$

are smooth embeddings for each $x \in D^2$ and for each i, $1 \le i \le m$. Let $\hat{\gamma}_x$ denote the family of holonomy maps

$$\hat{\gamma}_x : A_Y \to \underbrace{SU(2) \times \cdots \times SU(2)}_{m \text{ times}}, \qquad x \in D^2.$$

The holonomy is conjugated under the action of the group of gauge transformations and we continue to denote by $\hat{\gamma}_x$ the induced map on the quotient $\mathcal{B}_Y = A_Y / \mathcal{G}$. Let $\mathcal{F}_m$ denote the set of smooth functions

$$h : \underbrace{SU(2) \times \cdots \times SU(2)}_{m \text{ times}} \to \mathbf{R}$$

which are invariant under the adjoint action of $SU(2)$. Floer's set of perturbations Π is defined as

$$\Pi := \bigcup_{m \in \mathbf{N}} \Gamma_m \times \mathcal{F}_m.$$

Floer proves that for each $(\gamma, h) \in \Pi$ the function

$$h_\gamma : \mathcal{B}_Y \to \mathbf{R} \quad \text{defined by} \quad h_\gamma(\alpha) - \int_{D^2} h(\hat{\gamma}_x(\alpha))$$

is a smooth function and that for a dense subset $\mathcal{P} \subset \mathcal{RM}(Y) \times \Pi$ the critical points of the perturbed function

$$f_{(\gamma,h)} := f_{CS} + h_\gamma$$

are non-degenerate. The corresponding moduli space decomposes into smooth oriented manifolds of regular trajectories of the gradient flow of the function $f_{(\gamma,h)}$ with respect to a generic metric $\sigma \in \mathcal{RM}(Y)$. Furthermore, the homology groups of the perturbed chain complex are independent of the choice of perturbation in $\mathcal{P}$. We shall assume that a suitable perturbation has been chosen. Let α, β be two critical points of the function f_{CS}. Considering the spectral flow (denoted by sf) from α to β we obtain the moduli space $\mathcal{M}(\alpha, \beta)$ as the moduli space of self-dual connections on $Y \times \mathbf{R}$ which are asymptotic to α and β (as $t \to \pm\infty$). Let $\mathcal{M}^j(\alpha, \beta)$ denote the component of dimension j in $\mathcal{M}(\alpha, \beta)$. There is a natural action of $\mathbf{R}$ on $\mathcal{M}(\alpha, \beta)$. Let $\hat{\mathcal{M}}^j(\alpha, \beta)$ denote the component of dimension $j-1$ in $\mathcal{M}(\alpha, \beta)/\mathbf{R}$. Let $\#\hat{\mathcal{M}}^1(\alpha, \beta)$ denote the signed sum of the number of points in $\hat{\mathcal{M}}^1(\alpha, \beta)$. Floer defines the Morse index of α by considering the spectral flow from α to the trivial connection θ. It can be shown that the spectral flow and hence the Morse index are defined modulo 8.

Now define the chain groups by

$$\mathcal{R}_n(Y) = \mathbf{Z}\{\alpha \in \mathcal{R}^*(Y) \mid sf(\alpha) = n\}, \ n \in \mathbf{Z}_8$$

and define the boundary operator ∂

$$\partial : \mathcal{R}_n(Y) \to \mathcal{R}_{n-1}(Y)$$

by

$$\partial \alpha = \sum_{\beta \in \mathcal{R}_{n-1}(Y)} \#\hat{\mathcal{M}}^1(\alpha, \beta)\beta. \tag{10.35}$$

It can be shown that $\partial^2 = 0$ and hence $(\mathcal{R}(Y), \partial)$ is a complex. This complex can be thought of as an infinite-dimensional generalization [131] of Witten's instanton tunneling and we will call it the **Floer–Witten Complex** of the pair $(Y, SU(2))$. Since the spectral flow and hence the dimensions of the components of $\mathcal{M}(\alpha, \beta)$ are congruent modulo 8, this complex defines the Floer homology groups $FH_j(Y)$, $j \in \mathbf{Z}_8$, where j is the spectral flow of α to θ modulo 8. If r_j denotes the rank of the Floer homology group $FH_j(Y)$, $j \in \mathbf{Z}_8$, then we can define the corresponding Euler characteristic $\chi_F(Y)$ by

$$\chi_F(Y) := \sum_{j \in \mathbf{Z}_8} (-1)^j r_j.$$

Combining this with Taubes' interpretation of the Casson invariant $c(Y)$ we get

$$c(Y) = \chi_F(Y) = \sum_{j \in \mathbf{Z}_8} (-1)^j r_j. \tag{10.36}$$

An important feature of Floer's instanton homology is that it can be regarded as a functor from the category of homology 3-spheres to the category of graded Abelian groups, with morphisms given by oriented cobordism. Let M be a smooth oriented cobordism from Y_1 to Y_2 so that $\partial M = Y_2 - Y_1$. By a careful analysis of instantons on M, Floer showed [130] that M induces a graded homomorphism

$$M_j : FH_j(Y_1) \to FH_{j+b(M)}(Y_2), \qquad j \in \mathbf{Z}_8, \tag{10.37}$$

where

$$b(M) = 3(b_1(M) - b_2(M)). \tag{10.38}$$

Then the homomorphisms induced by cobordism have the following functorial properties:

$$(Y \times \mathbf{R})_j = id, \qquad\qquad\qquad (10.39)$$

$$(MN)_j = M_{j+b(N)} N_j. \qquad\qquad (10.40)$$

An algorithm for computing the Floer homology groups for Seifert-fibered homology 3-spheres with three exceptional fibers (or orbits) is discussed in [127]. In the following example we present some of the results given in that article.

Example 10.1 *Let Y denote the Seifert-fibered integral homology 3-sphere with exceptional orbits of order a_1, a_2, a_3. Then $Y = \Sigma(a_1, a_2, a_3)$ can be represented as the Brieskorn homology 3-sphere*

$$\Sigma(a_1, a_2, a_3) := \{(z_1, z_2, z_3) \mid z_1^{a_1} + z_2^{a_2} + z_3^{a_3} = 0\} \cap S^5.$$

Floer homology is computed by using the moduli space of $SO(3)$-connections over Y. In this case the representation space $\mathcal{R}(Y)$ contains only regular representations and hence is a finite set. The Chern–Simons function

$$f_{CS} : \mathcal{R}(Y) \to \mathbf{R}/4\mathbf{Z}$$

is defined by

$$\alpha \mapsto (2e_\alpha^2/a_1 a_2 a_3) \pmod 4, \qquad\qquad (10.41)$$

where $e_\alpha \in \mathbf{Z}$ is the Euler class of the representation α. The algorithm to determine this number is one of the highlights of [127]. Moreover, it can be shown that α is a critical point of index n where

$$n \equiv \frac{2e_\alpha^2}{a_1 a_2 a_3} + \sum_{i=1}^{3} \frac{2}{a_i} \sum_{k=1}^{a_i - 1} \cot\left(\frac{\pi a k}{a_i^2}\right) \cot\left(\frac{\pi k}{a_i}\right) \sin^2\left(\frac{\pi e_\alpha k}{a_i}\right) \pmod 8.$$

$$(10.42)$$

It can be shown that the index n is odd for all critical points and hence the boundary operator ∂ defined in (10.35) is zero. It follows that the Floer homology groups $FH_i(\Sigma(a_1, a_2, a_3))$ are zero for odd i and are free for even i. If we denote by $r_i(Y)$ the rank of the ith Floer homology group of the manifold Y then the set of ranks $\{r_0(Y), r_2(Y), r_4(Y), r_6(Y)\}$ determines the Floer homology of Y in this case. In Table 10.1 we give some computations of these ranks (see [127] for further details).

An examination of Table 10.1 shows that Floer homology of Brieskorn spheres is periodic of period 4, i.e.,

$$FH_i(\Sigma(a_1, a_2, a_3)) = FH_{i+4}(\Sigma(a_1, a_2, a_3)), \qquad \forall i \in \mathbf{Z}_8.$$

It has been conjectured that this property of periodicity holds for Floer homology groups of any homology 3-sphere.

The result discussed in the above example can be extended to compute the Floer homology groups of any Seifert-fibered integral homology 3-sphere.

Table 10.1 Floer homology of Brieskorn spheres

$Y = \Sigma(a_1, a_2, a_3)$				
$(a_1, a_2, a_3), \quad (-1)^k$	$r_0(Y)$	$r_2(Y)$	$r_4(Y)$	$r_6(Y)$
$(2, 3, 6k \pm 1), \quad -1$	$(k \mp 1)/2$	$(k \pm 1)/2$	$(k \mp 1)/2$	$(k \pm 1)/2$
$(2, 3, 6k \pm 1), \quad 1$	$k/2$	$k/2$	$k/2$	$k/2$
$(2, 5, 10k \pm 1), \quad -1$	$(3k \mp 1)/2$	$(3k \pm 1)/2$	$(3k \mp 1)/2$	$(3k \pm 1)/2$
$(2, 5, 10k \pm 1), \quad 1$	$3k/2$	$3k/2$	$3k/2$	$3k/2$
$(2, 5, 10k \pm 3), \quad -1$	$(3k \pm 1)/2$	$(3k \pm 1)/2$	$(3k \pm 1)/2$	$(3k \pm 1)/2$
$(2, 5, 10k \pm 3), \quad 1$	$(3k \pm 2)/2$	$3k/2$	$(3k \pm 2)/2$	$3k/2$
$(2, 7, 14k \pm 1), \quad -1$	$3k \mp 1$	$3k \pm 1$	$3k \mp 1$	$3k \pm 1$
$(2, 7, 14k \pm 1), \quad 1$	$3k$	$3k$	$3k$	$3k$

Further examples of the computation of Floer homology groups can be found in [335].

Instanton Homology for Oriented 3-Manifolds

We now discuss briefly some of the results of Fukaya (see [147]) on the instanton homology for oriented 3-manifolds. Let Y be a closed connected oriented 3-manifold. We restrict ourselves to the special case of $H_1(Y; \mathbf{Z})$ torsion-free. Let $\gamma_1, \ldots, \gamma_k$ be a basis for $H_1(Y, \mathbf{Z})$, where $k = b_1(Y)$ is the first Betti number of Y. Fukaya constructs a function using this basis to modify the Chern–Simons function of the Floer theory. To define this function we begin by observing that there is a natural gauge-invariant function associated to a loop on Y which generalizes the Wilson loop functional well known in the physics literature. We recall a definition of this functional from Chapter 7. It is given in a more general form than we need. It is of independent interest in the TQFT interpretation of knot and link invariants, as we will show in Chapter 11.

We begin with a general definition of the Wilson loop functional that was introduced as an example of a quantum observable in Chapter 7. This functional is also used in the TQFT interpretation of the Jones polynomial in Chapter 11.

Definition 10.1 (Wilson loop functional) *Let ρ denote a representation of G on V. Let $\alpha \in \Omega(M, x_0)$ denote a loop at $x_0 \in M$. Let $\pi : P(M, G) \to M$ be the canonical projection and let $p \in \pi^{-1}(x_0)$. If ω is a connection on P, then the parallel translation along α maps the fiber $\pi^{-1}(x_0)$ into itself. Let $\hat{\alpha}_\omega : \pi^{-1}(x_0) \to \pi^{-1}(x_0)$ denote this map. Since G acts transitively on the fibers, $\exists g_\omega \in G$ such that $\hat{\alpha}_\omega(p) = p g_\omega$. Now define*

$$\mathcal{W}_{\rho,\alpha}(\omega) := \text{Tr}[\rho(g_\omega)], \qquad \forall \omega \in \mathcal{A}. \tag{10.43}$$

We note that g_ω and hence $\rho(g_\omega)$ change by conjugation if, instead of p, we choose another point in the fiber $\pi^{-1}(x_0)$, but the trace remains unchanged.

Alternatively, we can consider the vector bundle $P \times_\rho V$ associated to the principal bundle P and parallel displacement of its fibers induced by α. Let $\pi : P \times_\rho V \to M$ be the canonical projection. We note that in this case $\pi^{-1}(x_0) \cong V$. Now the map $\hat{\alpha}_\omega : \pi^{-1}(x_0) \to \pi^{-1}(x_0)$ is a linear transformation and we can define

$$\mathcal{W}_{\rho,\alpha}(\omega) := \text{Tr}[\hat{\alpha}_\omega], \qquad \forall \omega \in \mathcal{A}. \tag{10.44}$$

We call these $\mathcal{W}_{\rho,\alpha}$ the Wilson loop functionals associated to the representation ρ and the loop α. In the particular case when $\rho = \text{Ad}$, the adjoint representation of G on $\mathbf{g}$, our constructions reduce to those considered in physical applications.

A gauge transformation $f \in \mathcal{G}$ acts on $\omega \in \mathcal{A}$ by a vertical automorphism of P and therefore changes the holonomy by conjugation by an element of the gauge group G. This leaves the trace invariant and hence we have

$$\mathcal{W}_{\rho,f\cdot\alpha}(\omega) = \mathcal{W}_{\rho,\alpha}(\omega), \qquad \forall \omega \in \mathcal{A} \text{ and } f \in \mathcal{G}. \tag{10.45}$$

Equation (10.45) implies that the Wilson loop functional is gauge-invariant and hence descends to the moduli space $\mathcal{B} = \mathcal{A}/\mathcal{G}$ of gauge potentials. We shall continue to denote it by $\mathcal{W}$.

In the application that we want to consider, we are interested in the special case of gauge group $G = SU(2)$ and ρ the defining representation of $SU(2)$. It can be shown that if α, β are loops representing the same homology class then

$$\mathcal{W}_{\rho,\alpha} = \mathcal{W}_{\rho,\beta}. \tag{10.46}$$

Thus, $\mathcal{W}$ can be regarded as defined on the first homology $H_1(Y,\mathbf{Z})$. Let α_i be a loop representing the basis element γ_i of the homology $H_1(Y,\mathbf{Z})$. The loop $\alpha_i : S^1 \to Y$ can be extended to an embedding $\alpha_i^0 : S^1 \times D^2 \to Y$. Define the loop

$$\alpha_{i,x}^0 : S^1 \to Y$$

by

$$\alpha_{i,x}^0(\theta) = \alpha_i^0(\theta, x), \qquad \theta \in S^1.$$

Choose a non-negative function $u : D^2 \to \mathbf{R}$ with compact support such that

$$\int_{D^2} u(x)dx = 1.$$

We define $\mathcal{W}_i : D^2 \times \mathcal{B}_Y \to \mathbf{R}$ by

$$\mathcal{W}_i(x,\omega) := \mathcal{W}_{\rho,\alpha_{i,x}^0}(\omega), \qquad 1 \le i \le k. \tag{10.47}$$

Now fix an $\epsilon > 0$ and define $f_\epsilon : \mathcal{B}_Y \to \mathbf{R}$ by

$$f_\epsilon(\omega) := \epsilon \sum_{i=1}^k \int_{D^2} \mathcal{W}_i(x, \omega) u(x) dx. \tag{10.48}$$

It can be shown that for sufficiently small ϵ the function $f_{CS} - f_\epsilon$ on $\mathcal{B}_Y$ has finitely many non-degenerate critical points and that the Morse index of each critical point is well-defined modulo 8. We now proceed as in Floer theory by considering the gradient flow of the function $f_{CS} - f_\epsilon$ on $\mathcal{B}_Y$ defined by

$$\frac{\partial c(t)}{\partial t} = *_Y F_{c(t)} - \operatorname{grad}_{c(t)} f_\epsilon. \tag{10.49}$$

The corresponding chain complex is denoted by (C^0, ∂) and the Fukaya homology groups are denoted by I_n^0, $n \in \mathbf{Z}_8$. When Y is a homology 3-sphere the Fukaya theory goes over into the Floer theory and hence we refer to these collectively as the Fukaya–Floer homology groups. By tensoring the chains of the complex with symmetric powers of $H_1(Y, \mathbf{Z})$ Fukaya defines new chain complexes (C^s, ∂^s) indexed by non-negative integer $s < 3$. The main result of [147] is the following theorem.

Theorem 10.1 *Let I_n^s, $n \in \mathbf{Z}_8$, denote the homology groups of the complex (C^s, ∂^s) indexed by non-negative integer $s < 3$. Then the groups I_n^s do not depend on the choice of metrics, the bases γ_i for the homology $H_1(Y, \mathbf{Z})$, and the other choices made in perturbing the function $f_{CS} - f_\epsilon$ on $\mathcal{B}_Y$, and they are topological invariants of the manifold Y.*

The significance of these new invariants and their relation to other known invariants is not yet clear.

10.6 Integer-Graded Instanton Homology

In Section 10.4 we discussed the $\mathbf{Z}_8$-graded homology theory due to Floer and its extension to arbitrary closed oriented 3-manifolds by Fukaya. Fintushel and Stern [128] have extended Floer homology theory in another direction by defining a $\mathbf{Z}$-graded homology theory for homology 3-spheres. We now discuss this extension.

To extend FH to a $\mathbf{Z}$-graded homology we use the universal cover (infinite cyclic) $\tilde{\mathcal{B}}_Y$ of $\mathcal{B}_Y$ defined by

$$\tilde{\mathcal{B}}_Y = \mathcal{A}/\mathcal{G}_0. \tag{10.50}$$

The Chern–Simons action functional and the spectral flow (denoted by sf) on $\mathcal{A}$ descend to $\tilde{\mathcal{B}}_Y \to \mathbf{R}$. Let $\tilde{\mathcal{R}}^*(Y)$ be the cover of $\mathcal{R}^*(Y)$ and let

$$\mathbf{R}_Y = \mathbf{R} \setminus c(\tilde{\mathcal{R}}^*(Y)) \tag{10.51}$$

be the set of regular values of the Chern–Simons functional. We note that the set $c(\tilde{\mathcal{R}}^*(Y))$ is finite modulo $\mathbf{Z}$. Fix $\mu \in \mathbf{R}_Y$ and let $\alpha^{(\mu)} \in \tilde{\mathcal{R}}(Y) \subset \tilde{\mathcal{B}}_Y$ be the unique lift of $\alpha \in \mathcal{R}(Y)$ (see section 10.4) such that $c(\alpha^{(\mu)}) \in (\mu, \mu+1)$. Define

$$\mathcal{R}_n^{(\mu)}(Y) := \mathbf{Z}\{\alpha \in \mathcal{R}^*(Y) \mid sf(\alpha^{(\mu)}) = n\} \tag{10.52}$$

and define the boundary operator $\partial^{(\mu)}$

$$\partial^{(\mu)} : \mathcal{R}_n^{(\mu)}(Y) \to \mathcal{R}_{n-1}^{(\mu)}(Y)$$

by

$$\partial^{(\mu)}\alpha = \sum_{\beta \in \mathcal{R}_{n-1}^{(\mu)}(Y)} \#(\hat{\mathcal{M}}^1(\alpha, \beta))\beta, \tag{10.53}$$

where $\hat{\mathcal{M}}^1(\alpha, \beta)$ is defined by a construction similar to that in the Floer theory. It can be shown that $\partial^{(\mu)}\partial^{(\mu)} = 0$ and hence $(\mathcal{R}^{(\mu)}(Y), \partial^{(\mu)})$ is a complex. The resulting homology groups are denoted by $\mathcal{I}_n^{(\mu)}$, $n \in \mathbf{Z}$, and are called the **integer-graded instanton homology groups**. In general, not all the representations in $\mathcal{R}(Y)$ are regular. However, as in the Floer theory, one can define the integer-graded instanton homology groups by perturbing the Chern–Simons function. The resulting homology groups are independent of the perturbations and satisfy the following properties:

1. For $\mu \in [\mu_0, \mu_1] \subset \mathbf{R}_Y$,

$$\mathcal{I}_n^{(\mu_0)} = \mathcal{I}_n^{(\mu)} = \mathcal{I}_n^{(\mu_1)}, \qquad \forall n \in \mathbf{Z},$$

2. For $\mu \in \mathbf{R}_Y$,

$$\mathcal{I}_n^{(\mu)} = \mathcal{I}_n^{(\mu+1)}, \qquad \forall n \in \mathbf{Z}.$$

The homology groups $\mathcal{I}_n^{(\mu)}$, $n \in \mathbf{Z}$, determine the Floer homology groups by filtering the Floer chain complex. If $\varPhi_{s,n}^{(\mu)}(Y)$ denotes the filtration induced on the Floer homology then we have the following theorem.

Theorem 10.2 *Let $n \in \mathbf{Z}_8$, $s \in \mathbf{Z}$, and $s \equiv n$ (mod 8). Then there exists a spectral sequence $(E_{s,n}^r(Y), d^r)$ such that*

$$E_{s,n}^1(Y) \cong \mathcal{I}_s^{(\mu)}, \tag{10.54}$$

$$E_{s,n}^\infty(Y) \cong \varPhi_{s,n}^{(\mu)}(Y)/\varPhi_{s+8,n}^{(\mu)}(Y). \tag{10.55}$$

The groups $E_{s,n}^r(Y)$ of the spectral sequence are topological invariants of Y.

The groups $E_{s,n}^r(Y)$ as well as the instanton homology groups $\mathcal{I}_n^{(\mu)}$, $n \in \mathbf{Z}$, can be regarded as functors from the category of homology 3-spheres to the category of graded Abelian groups with morphisms given by oriented cobordism.

Theorem 10.3 *If M is an oriented cobordism from Y_1 to Y_2 so that $\partial M = Y_2 - Y_1$ then for $\mu \in \mathbf{R}_{Y_1} \cap \mathbf{R}_{Y_2}$, M induces graded homomorphisms*

$$M_s^{(\mu)} : E_{s,n}^r(Y_1) \to E_{s+b(M),n+b(M)}^r(Y_2),$$

where $b(M) = 3(b_1(M) - b_2(M))$. Moreover, we have the following relations:

$$(Y \times \mathbf{R})_s = id, \tag{10.56}$$

$$(MN)_s = M_{s+b(N)}N_s. \tag{10.57}$$

Proofs of the above two theorems are given in [128].

In Example 10.1, we presented some results on the Floer homology groups $FH_i(\Sigma(a_1, a_2, a_3))$, $i \in \mathbf{Z}_8$, where $\Sigma(a_1, a_2, a_3)$ is the Seifert-fibered homology 3-sphere with exceptional orbits of order a_1, a_2, a_3. An algorithm for computing these Floer homology groups discussed in [127] is extended in [128] to compute the integer-graded instanton homology groups $\mathcal{I}_i^{(\mu)}(\Sigma(a_1, a_2, a_3))$, $i \in \mathbf{Z}$. In the following example we present some of the results given in that article.

Example 10.2 *As in Example 10.1, we let Y denote the Seifert-fibered integral homology 3-sphere with exceptional orbits of order a_1, a_2, a_3. Integer-graded instanton homology is computed via the moduli space of $SO(3)$-connections over Y. The representation space $\mathcal{R}(Y)$ contains only regular representations. Let e_α denote the Euler class of the representation α. Let $e_\alpha^{(\mu)} \in (\mu, 4 + \mu)$ be defined by the congruence*

$$e_\alpha^{(\mu)} \equiv (2e_\alpha^2/a_1a_2a_3) \pmod 4.$$

Then α is a critical point of the lift of the Chern–Simons functional

$$f_{CS}^{(\mu)} : \mathcal{R}^{(\mu)}(Y) \to \mathbf{R}/4\mathbf{Z} \quad \text{defined by } \alpha \mapsto e_\alpha^{(\mu)}. \tag{10.58}$$

The index $l^{(\mu)}(\alpha)$ of α is given by

$$l^{(\mu)}(\alpha) = \frac{2(e_\alpha^{(\mu)})^2}{a_1a_2a_3} + \sum_{i=1}^{3} \frac{2}{a_i} \sum_{k=1}^{a_i-1} \cot\left(\frac{\pi ak}{a_i^2}\right) \cot\left(\frac{\pi k}{a_i}\right) \sin^2\left(\frac{\pi e_\alpha^{(\mu)} k}{a_i}\right).$$

It can be shown that the index $l^{(\mu)}(\alpha)$ is always an odd integer. It follows that the boundary operator $\partial^{(\mu)}$ is identically zero in this case and hence the integer-graded instanton homology is given by

$$\mathcal{I}_n^{(\mu)}(\Sigma(a_1, a_2, a_3)) \cong \begin{cases} \mathcal{R}_n^{(\mu)}(\Sigma(a_1, a_2, a_3)), & \text{for odd } n \\ 0, & \text{for even } n. \end{cases}$$

It can be shown that $\mathcal{I}_n^{(\mu)}(\Sigma(a_1, a_2, a_3))$ is free of finite rank for all n for which it is non-zero. For example, we can obtain the ranks of the homology groups

$\mathcal{I}_n^{(0)}(\Sigma(2,3,6k\pm1))$ *after some explicit computations. Using this information we can write down the Poincaré–Laurent polynomials* $p(\Sigma(2,3,6k\pm1))(t)$ *of* $\Sigma(2,3,6k\pm1)$. *A list of these polynomials for the values of* k *from 1 to 9 is given in Table 10.2. We have corrected some typographical errors in the list given in [128].*[1]

Table 10.2 Poincaré–Laurent polynomials of $\Sigma(2,3,6k\pm1)$

k	$p(\Sigma(2,3,6k-1))(t)$	$p(\Sigma(2,3,6k+1))(t)$
1	$t+t^5$	$t^{-1}+t^3$
2	$t+t^3+t^5+t^7$	$t^{-1}+t+t^3+t^5$
3	$t+t^3+2t^5+t^7+t^9$	$2t^{-1}+t+2t^3+t^5$
4	$t+2t^3+2t^5+2t^7+t^9$	$2t^{-1}+2t+2t^3+2t^5$
5	$t+t^3+3t^5+2t^7+2t^9+t^{11}$	$2t^{-1}+2t+3t^3+2t^5+t^7$
6	$t+2t^3+3t^5+3t^7+2t^9+t^{11}$	$t^{-1}+3t+3t^3+3t^5+2t^7$
7	$t+t^3+4t^5+3t^7+3t^9+2t^{11}$	$2t^{-1}+3t+4t^3+3t^5+2t^7$
8	$t+2t^3+3t^5+4t^7+3t^9+2t^{11}+t^{13}$	$t^{-1}+4t+4t^3+4t^5+3t^7$
9	$t+t^3+4t^5+4t^7+4t^9+3t^{11}+t^{13}$	$2t^{-1}+3t+5t^3+4t^5+3t^7+t^9$

At the end of Example 10.1 we saw that the conjecture that the Floer homology is periodic with period 4 is verified for the Brieskorn spheres. An examination of Table 10.2 shows that the integer-graded instanton homology groups of the Brieskorn spheres do not have this periodicity property. The integer-graded instanton homology groups can be thought of as a refinement of the $\mathbf{Z}_8$-graded Floer homology groups just as the Floer homology groups can be thought of as a refinement of the Casson invariant of an integral homology 3-sphere. An extension of the Casson invariant to a rational homology 3-sphere is defined in Walker [394] (see [64] for another approach).

A method for calculating the spectral flow in the Fukaya–Floer theory for a split manifold is given in [416]. Recall that a connected closed oriented 3-manifold is said to be **split** if

$$M = M_1\bigcup M_2, \quad M_1\bigcap M_2 = \partial M_1 = \partial M_2 = \Sigma_g, \quad g>1,$$

where M_1, M_2 are 0-codimension submanifolds of M and Σ_g is a connected, closed Riemann surface of genus g oriented as the boundary of M_1. Then we have the following theorem

Theorem 10.4 (Yoshida) *Let* M *be split manifold. Let* A_i, $i=1,2$, *be smooth irreducible flat connections on the trivial bundle* $M\times SU(2)$ *with* $\mathrm{Ker}(d^{A_i})=0$. *Assume that the connections* A_i *restrict to smooth irreducible*

---

[1] We would like to thank Ron Stern for confirming these corrections.

flat connections B_i on the trivial bundle $\Sigma_g \times SU(2)$. Then there exists a metric on M and a smooth generic path of connections A_t, $0 \leq t \leq 1$, on $M \times SU(2)$ such that A_t restricts to a path $B_t \times 1$ on a tubular neighborhood of Σ_g and defines a path $\hat{A}_t$ in the space of all Lagrangian pairs in the $6(g-1)$-dimensional symplectic vector space of the equivalence classes of representations of $\pi_1(\Sigma_g)$ into $SU(2)$. Let $\gamma(\hat{A}_t)$ be the corresponding Maslov index. Then

$$sf(A_1, A_0) = \gamma(\hat{A}_t) \tag{10.59}$$

where sf denotes the spectral flow.

This theorem can be viewed as a link between the Fukaya–Floer homology and the symplectic homology. In fact, similar ideas have been used by Taubes in his interpretation of the Casson invariant. An application of the above theorem leads to a calculation of the Floer homology groups of a class of manifolds as indicated in the following theorem (see [416] for a proof).

Theorem 10.5 *Let N_k, $0 > k \in \mathbf{Z}$, denote the integral homology 3-sphere obtained by the $(1/k)$-Dehn surgery along the figure eight knot in S^3. Then it can be shown that the Floer homology groups $FH_i(N_k)$ are zero for odd i and are free for even i. If we denote by $r_i(N_k)$ the rank of the ith Floer homology group of the manifold N_k then the set of ranks $\{r_0(N_k), r_2(N_k), r_4(N_k), r_6(N_k)\}$ determines the Floer homology of N_k. Computations of these ranks for all $k \in \mathbf{Z}$, $k < 0$, is given in the Table 10.3.*

Table 10.3 Floer homology of homology 3-spheres N_k, $k < 0$

k	$r_0(N_k)$	$r_2(N_k)$	$r_4(N_k)$	$r_6(N_k)$
$-4m$	$2m$	$2m$	$2m$	$2m$
$-4m+1$	$2m-1$	$2m$	$2m-1$	$2m$
$-4m+2$	$2m-1$	$2m-1$	$2m-1$	$2m-1$
$-4m+3$	$2m-2$	$2m-1$	$2m-2$	$2m-1$

This table can also be used to determine the Floer homology groups of the homology 3-spheres N_{-k}, $0 > k \in \mathbf{Z}$, by noting that N_{-k} is orientation-preserving diffeomorphic to $-N_k$ and that

$$FH_i(-N_k) = FH_{3-i}(N_k), \qquad \forall i \in \mathbf{Z}_8.$$

An examination of Table 10.3 shows that Floer homology of the homology 3-spheres N_k is periodic of period 4, i.e.,

$$FH_i(N_k) = FH_{i+4}(N_k), \qquad \forall i \in \mathbf{Z}_8.$$

In the above theorem and in Example 10.1 Floer's boundary operator is trivial. An example of a non-trivial boundary operator appears in the computation of the Floer homology for connected sums of homology 3-spheres. A Künneth formula also holds for $SO(3)$ Floer homology. Another important development in the topology of 3-manifolds is the use of Morse–Bott theory, which generalizes the classical Morse theory to functions whose critical points are not necessarily isolated. This is closely related to the definition of equivariant Floer (co)homology. Another approach to Donaldson–Floer theory that uses a simplicial model of the theory was proposed by Taubes [196].

We would like to add that there is a vast body of work on the topology and geometry of 3-manifolds which was initiated by Thurston [373, 374]. At present the relation of this work to the methods and results of the gauge theory approach to the study of 3-manifolds remains mysterious.

10.7 WRT Invariants

Witten's TQFT invariants of 3-manifolds were given a mathematical definition by Reshetikhin and Turaev in [326]. In view of this and with a suggestion from Prof. Zagier, I called them Witten– Reshetikhin–Turaev or WRT invariants in [260]. Several alternative approaches to WRT invariants are now available. We will discuss some of them later in this section. In the course of our discussion of Witten's interpretation of the Jones polynomial in Chapter 11 we will indicate an evaluation of a specific partition function (see equation (11.22)). This partition function provides a new family of invariants of S^3. Such a partition function can be defined for a more general class of 3-manifolds and gauge groups. More precisely, let G be a compact simply connected simple Lie group and let $k \in \mathbf{Z}$. Let M be a 2-framed closed oriented 3-manifold. We define the **Witten invariant** $\mathcal{T}_{G,k}(M)$ of the triple (M, G, k) by

$$\mathcal{T}_{G,k}(M) := \int_{\mathcal{B}_M} e^{-ifcs([\omega])} \mathcal{D}[\omega], \qquad (10.60)$$

where $\mathcal{D}[\omega]$ is a suitable measure on $\mathcal{B}_M$. We note that no precise definition of such a measure is available at this time and the definition is to be regarded as a formal expression. Indeed, one of the aims of TQFT is to make sense of such formal expressions. We define the **normalized Witten invariant** $\mathcal{W}_{G,k}(M)$ of a 2-framed closed oriented 3-manifold M by

$$\mathcal{W}_{G,k}(M) := \frac{\mathcal{T}_{G,k}(M)}{\mathcal{T}_{G,k}(S^3)}. \qquad (10.61)$$

Then we have the following "theorem":

Theorem 10.6 (Witten) *Let G be a compact simply connected simple Lie group. Let M, N be two 2-framed closed oriented 3-manifolds. Then we have the following results:*

$$\mathcal{T}_{G,k}(S^2 \times S^1) = 1, \tag{10.62}$$

$$\mathcal{T}_{SU(2),k}(S^3) = \sqrt{\frac{2}{k+2}} \sin\left(\frac{\pi}{k+2}\right), \tag{10.63}$$

$$\mathcal{W}_{G,k}(M\#N) = \mathcal{W}_{G,k}(M) \cdot \mathcal{W}_{G,k}(N). \tag{10.64}$$

If G is a compact semi-simple group then the **WRT invariant** $\mathcal{T}_{G,k}(S^3)$ can be given in a closed form in terms of the root and weight lattices associated to G. In particular, for $G = SU(n)$ we get

$$\mathcal{T} = \frac{1}{\sqrt{n(k+n)^{(n-1)}}} \prod_{j=1}^{n-1} \left[2\sin\left(\frac{j\pi}{k+n}\right) \right]^{n-j} .$$

We will show later that this invariant can be expressed in terms of the generating function of topological string amplitudes in a closed string theory compactified on a suitable Calabi–Yau manifold. More generally, if a manifold M can be cut into pieces over which the CS path integral can be computed, then the gluing rules of TQFT can be applied to these pieces to find $\mathcal{T}$. Different ways of using such a cut and paste operation can lead to different ways of computing this invariant. Another method that is used in both the theoretical and experimental applications is the perturbative quantum field theory. The rules for perturbative expansion around classical solutions of field equations are well understood in physics. It is called the stationary phase approximation to the partition function. It leads to the asymptotic expansion in terms of a parameter depending on the coupling constants and the group. If $\check{c}(G)$ is the dual Coxeter number of G then the asymptotic expansion is in terms of $\hbar = 2\pi i/(k + \check{c}(G))$. This notation in TQFT is a reminder of the Planck's constant used in physical field theories. The asymptotic expansion of $\log \mathcal{T}$ is then given by

$$\log \mathcal{T} = -b\log(\hbar) + \frac{a_0}{\hbar} + \sum_{n=1}^{\infty} a_{n+1}(\hbar)^n,$$

where a_i are evaluated on Feynman diagrams with i loops. The expansion may be around any flat connection and the dependence of a_i on the choice of a connection may be explicitly indicated if necessary. For Chern–Simons theory the above perturbative expansion is also valid for non-compact groups.

In addition to the results described above, there are several other applications of TQFT in the study of the geometry and topology of low-dimensional manifolds. In 2 and 3 dimensions, conformal field theory (CFT) methods have proved to be useful. An attempt to put the CFT on a firm mathemat-

ical foundation was begun by Segal in [347] (see also, [288]) by proposing a
set of axioms for CFT. CFT is a two-dimensional theory and it was necessary
to modify and generalize these axioms to apply to topological field theory in
any dimension. We discussed these TQFT axioms in Chapter 7.

10.7.1 CFT Approach to WRT Invariants

In [230] Kohno defines a family of invariants $\Phi_k(M)$ of a 3-manifold M by
using its Heegaard decomposition along a Riemann surface Σ_g and representations of the mapping class group of Σ_g. Kohno's work makes essential use of
ideas and results from conformal field theory. We now give a brief discussion
of Kohno's definition.

We begin by reviewing some information about the geometric topology
of 3-manifolds and their Heegaard splittings. Recall that two compact 3-
manifolds X_1, X_2 with homeomorphic boundaries can be glued together
along a homeomorphism $f : \partial X_1 \to \partial X_2$ to obtain a closed 3-manifold
$X = X_1 \cup_f X_2$. If X_1, X_2 are oriented with compatible orientations on the
boundaries, then f can be taken to be either orientation-preserving or reversing. Conversely, any closed orientable 3-manifold can be obtained by such a
gluing procedure where each of the pieces is a special 3-manifold called a **handlebody**. Recall that a handlebody of genus g is an orientable 3-manifold
obtained from gluing g copies of **1-handles** $D^2 \times [-1, 1]$ to the 3-ball D^3.
The gluing homeomorphisms join the $2g$ disks $D^2 \times \{\pm 1\}$ to the $2g$ pairwise
disjoint 2-disks in $\partial D^3 = S^2$ in such a way that the resulting manifold is
orientable. The handlebodies H_1, H_2 have the same genus and a common
boundary $H_1 \cap H_2 = \partial H_1 = \partial H_2$. Such a decomposition of a 3-manifold X is
called a **Heegaard splitting** of X of genus g. We say that X has **Heegaard
genus** g if it has some Heegaard splitting of genus g but no Heegaard splitting of smaller genus. Given a Heegaard splitting of genus g of X, there exists
an operation called **stabilization**, which gives another Heegaard splitting of
X of genus $g + 1$. Two Heegaard splittings of X are called **equivalent** if
there exists a homeomorphism of X onto itself taking one splitting into the
other. Two Heegaard splittings of X are called **stably equivalent** if they
are equivalent after a finite number of stabilizations. A proof of the following
theorem is given in [336].

Theorem 10.7 *Any two Heegaard splittings of a closed orientable 3-
manifold X are stably equivalent.*

The **mapping class group** $\mathcal{M}(M)$ of a connected compact smooth surface M is the quotient group of the group of diffeomorphisms $\mathrm{Diff}(M)$ of M
modulo the group $\mathrm{Diff}_0(M)$ of diffeomorphisms isotopic to the identity. i.e.,

$$\mathcal{M}(M) := \mathrm{Diff}(M) / \mathrm{Diff}_0(M)$$

If M is oriented, then $\mathcal{M}(M)$ has a normal subgroup $\mathcal{M}^+(M)$ of index 2 consisting of orientation-preserving diffeomorphisms of M modulo isotopies. The group $\mathcal{M}(M)$ can also be defined as $\pi_0(\mathrm{Diff}(M))$. Smooth closed orientable surfaces Σ_g are classified by their genus g, and in this case it is customary to denote $\mathcal{M}(\Sigma_g)$ by $\mathcal{M}_g$. In the applications that we have in mind it is this group $\mathcal{M}_g$ that we shall use. The group $\mathcal{M}_g$ is generated by **Dehn twists** along simple closed curves in Σ_g. Let c be a simple closed curve in Σ_g that forms one of the boundaries of an annulus. In local complex coordinate z we can identify the annulus with $\{z \mid 1 \le |z| \le 2\}$ and the curve c with $\{z \mid |z| = 1\}$. Then, the Dehn twist τ_c along c is an automorphism of Σ_g, which is the identity outside the annulus, while in the annulus it is given by the formula

$$\tau_c(re^{i\theta}) = re^{i(\theta + 2\pi(r-1))},$$

where

$$z = re^{i\theta}, \quad 1 \le r \le 2, \quad 0 \le \theta \le 2\pi.$$

Changing the curve c by an isotopic curve or changing the annulus gives isotopic twists. However, twists in opposite directions define elements of $\mathcal{M}_g$ that are the inverses of each other. Note that any two homotopic simple closed curves on Σ_g are isotopic. A useful description of $\mathcal{M}_g$ is given by the following theorem.

Theorem 10.8 *Let Σ_g be a smooth closed orientable surface of genus g. Then the group $\mathcal{M}_g$ is generated by the $3g - 1$ Dehn twists along the curves (called Lickororish generators) α_i, β_j, γ_k, $1 \le i \le g, 1 \le j \le g, 1 \le k < g$, where α_i , β_j are Poincaré dual to a basis of the first integral homology of Σ_g. See the book by Birman [40] for a construction of these generators.*

In [230] Kohno obtains a representation of the mapping class group $\mathcal{M}_g$ in the space of conformal blocks which arise in conformal field theory (see the book [232] by Kohno for more applications of conformal field theory to topology). Kohno then uses a special function for this representation and the stabilization to define a family of invariants $\Phi_k(M)$ of the 3-manifold M; these invariants are independent of its stable Heegaard decomposition. Kohno obtains the following formulas:

$$\Phi_k(S^2 \times S^1) = \left(\sqrt{\frac{2}{k+2}} \sin\left(\frac{\pi}{k+2}\right) \right)^{-1}, \tag{10.65}$$

$$\Phi_k(S^3) = 1, \tag{10.66}$$

$$\Phi_k(M \# N) = \Phi_k(M) \cdot \Phi_k(N). \tag{10.67}$$

Kohno's invariant coincides with the normalized Witten invariant with the gauge group $SU(2)$. Similar results were also obtained by Crane [88]. The agreement of these results with those of Witten may be regarded as strong evidence for the correctness of the TQFT calculations. In [230] Kohno also obtains the Jones–Witten polynomial invariants for a framed colored link in a 3-manifold M by using representations of mapping class groups via conformal field theory. In [231] the Jones–Witten polynomials are used to estimate the tunnel number of knots and the Heegaard genus of a 3-manifold. The monodromy of the Knizhnik–Zamolodchikov equation [224] plays a crucial role in these calculations.

10.7.2 WRT Invariants via Quantum Groups

Shortly after the publication of Witten's paper [405], Reshetikhin and Turaev [326] gave a precise combinatorial definition of a new invariant by using the representation theory of quantum group $U_q sl_2$ at the root of unity $q = e^{2\pi i/(k+2)}$. The parameter q coincides with Witten's $SU(n)$ Chern–Simons theory parameter t when $n = 2$ and in this case the invariant of Reshetikhin and Turaev is the same as the normalized Witten invariant. As we have observed at the begining of section 10.7, it is now customary to call the normalized Witten invariants the **Witten–Reshetikhin–Turaev invariants**, or **WRT invariants**. We now discuss their construction in the form given by Kirby and Melvin in [222].

Let $\mathbf{U}$ denote the universal enveloping algebra of $sl(2, \mathbf{C})$ and let $\mathbf{U}_h$ denote the quantized universal enveloping algebra of formal power series in h. Recall that $\mathbf{U}$ is generated by X, Y, H subject to relations as in the algebra $sl(2, \mathbf{C})$, i.e.,

$$[H, X] = 2X, \quad [H, Y] = -2Y, \quad [X, Y] = H.$$

In $\mathbf{U}_h$ the last relation is replaced by

$$[X, Y] = [H]_s := \frac{s^H - s^{-H}}{s - s^{-1}}, \qquad s = e^{h/2}.$$

It can be shown that $\mathbf{U}_h$ admits a Hopf algebra structure as a module over the ring of formal power series. However, the presence of divergent series makes this algebra unsuitable for representation theory. We construct a finite-dimensional algebra by using $\mathbf{U}_h$. Define

$$K := e^{hH/4} \quad \text{and} \quad \bar{K} := e^{-hH/4} = K^{-1}.$$

Fix an integer $r > 1$ ($r = k+2$ of the Witten formula) and set $q = e^h = e^{2\pi i/r}$. We restrict this to a subalgebra over the ring of convergent power series in h generated by $X, Y, K, \bar{K}$. This infinite-dimensional algebra occurs in the work of Jimbo (see the books [254] by Lusztig, [306] by Ohtsuki, and [381] by Turae). We take its quotient by setting

$$X^r = 0, \quad Y^r = 0, \quad K^{4r} = 1.$$

The representations of this quotient algebra $\mathcal{A}$ are used to define colored Jones polynomials and the WRT invariants. The algebra $\mathcal{A}$ is a finite-dimensional complex algebra satisfying the relations

$$\bar{K} = K^{-1}, \quad KX = sXK, \quad KY = \bar{s}YK,$$

$$[X, Y] = \frac{K^2 - K^{-2}}{s - \bar{s}}, \quad s = e^{\pi i/r}.$$

There are irreducible $\mathcal{A}$-modules V^i in each dimension $i > 0$. If we put $i = 2m + 1$, then V^i has a basis $\{e_m, \ldots, e_{-m}\}$. The action of $\mathcal{A}$ on the basis vectors is given by

$$Xe_j = [m + j + 1]_s e_{j+1}, \quad Ye_j = [m - j + 1]_s e_{j-1}, \quad Ke_j = s^j e_j.$$

The $\mathcal{A}$-modules V^i are self-dual for $0 < i < r$. The structure of their tensor products is similar to that in the classical case. The algebra $\mathcal{A}$ has the additional structure of a quasitriangular Hopf algebra with Drinfeld's universal R-matrix R satisfying the Yang–Baxter equation. One has an explicit formula for $R \in \mathcal{A} \otimes \mathcal{A}$ of the form

$$R = \sum c_{nab} X^a K^b \otimes Y^n K^b .$$

If V, W are $\mathcal{A}$-modules then R acts on $V \otimes W$. Composing with the permutation operator we get the operator $R' : V \otimes W \rightarrow W \otimes V$. These are the operators used in the definition of our link invariants. Let L be a framed link with n components L_i colored by $\mathbf{k} = \{k_1, \ldots, k_n\}$. Let $J_{L,\mathbf{k}}$ be the corresponding colored Jones polynomial. The colors are restricted to Lie in a family of irreducible modules V^i, one for each dimension $0 < i < r$. Let σ denote the signature of the linking matrix of L. Define τ_L by

$$\tau_L = \left(\sqrt{2/r} \sin(\pi/r) \right)^n e^{3(2-r)\sigma/(8r)} \sum [\mathbf{k}] J_{L,\mathbf{k}} ,$$

where the sum is over all admissible colors. Every 3-manifold can be obtained by surgery on a link in S^3. Two links give isomorphic manifolds if they are related by Kirby moves. It can be shown that the invariant τ_L is preserved under Kirby moves and hence defines an invariant of the 3-manifold M_L obtained by surgery on L. With suitable normalization it coincides with the WRT invariant. WRT invariants do not belong to the class of polynomial invariants or other known 3-manifold invariants. They arose from topological quantum field theory applied to calculate the partition functions in the Chern–Simons gauge theory.

A number of other mathematicians obtained invariants closely related to the Witten invariant. The equivalence of these invariants, defined by different

methods, was a folk theorem until a complete proof was given by Piunikhin in
[317]. Another approach to WRT invariants is via Hecke algebras and related
special categories. A detailed construction of modular categories from Hecke
algebras at roots of unity is given in [42]. For a special choice of the framing parameter one recovers the Reshetikhin–Turaev invariants of 3-manifolds
constructed from the representations of the quantum groups $U_q(sl(N))$ by
Reshetikhin, Turaev, and Wenzl [326, 384, 400]. These invariants were constructed by Yokota [415] using skein theory. As we discussed earlier the quantum invariants were obtained by Witten [404] using path integral quantization
of Chern–Simons theory. In *Quantum Invariants of Knots and 3-Manifolds*
[381], Turaev showed that the idea of modular categories is fundamental in
the construction of these invariants and that it plays an essential role in extending them to a topological quantum field theory. (See also the book [306]
by Ohtsuki for a comprehensive introduction to quantum invariants.) Since
the time of the early results discussed above, the WRT invariants for several
other manifolds and gauge groups have been obtained. We collect some of
these results below.

Theorem 10.9 *The WRT invariant for the lens space $L(p,q)$ in the canonical framing is given by*

$$W_k(L(p,q)) = -\frac{i}{\sqrt{2p(k+2)}} \exp\left(\frac{6\pi is}{k+2}\right)$$

$$\cdot \sum_{\delta \in \{-1,1\}} \sum_{n=1}^{p} \delta \exp\left(\frac{\delta}{2p(k+2)} + \frac{2\pi iqn^2(k+2)}{p} + \frac{2\pi in(q+\delta)}{p}\right),$$

where $s = s(q,p)$ is the Dedekind sum defined by

$$s(q,p) := \frac{1}{4p} \sum_{k=1}^{p-1} \cot\left(\frac{\pi k}{p}\right) \cot\left(\frac{\pi kq}{p}\right).$$

In all of these the invariant is well-defined only at roots of unity and perhaps near roots of unity if a perturbative expansion is possible. This situation
occurs in the study of classical modular functions and Ramanujan's mock
theta functions. Ramanujan introduced his mock theta functions in a letter
to Hardy in 1920 (the famous last letter) to describe some power series in
variable $q = e^{2\pi iz}$, $z \in \mathbf{C}$. He also wrote down (without proof, as was usual in
his work) a number of identities involving these series, which were completely
verified only in 1988 by Hickerson [191]. Recently, Lawrence and Zagier obtained several different formulas for the Witten invariant $\mathcal{W}_{SU(2),k}(M)$ of the
Poincaré homology sphere $M = \Sigma(2,3,5)$ in [246]. Using the work of Zwegers
[419], they show how the Witten invariant can be extended from integral k
to rational k and give its relation to the mock theta function. In particular,
they obtain the following fantastic formula, in the spirit of Ramanujan, for
the Witten invariant $\mathcal{W}_{SU(2),k}(M)$ of the Poincaré homology sphere

$$\mathcal{W}_{SU(2),k}(\Sigma(2,3,5)) = 1 + \sum_{n=1}^{\infty} x^{-n^2}(1+x)(1+x^2)\cdots(1+x^{n-1})$$

where $x = e^{\pi i/(k+2)}$. We note that the series on the right hand side of this formula terminates after $k+2$ terms.

We have not discussed the Kauffmann bracket polynomial or the theory of skein modules in the study of 3-manifold invariants. An invariant that combines these two ideas has been defined in the following general setting. Let R be a commutative ring and let A be a fixed invertible element of R. Then one can define a new invariant, $S_{2,\infty}(M;R,A)$, of an oriented 3-manifold M called the **Kauffmann bracket skein module**. The theory of skein modules is related to the theory of representations of quantum groups. This connection should prove useful in developing the theory of quantum group invariants which can be defined in terms of skein theory as well as by using the theory of representations of quantum groups.

10.8 Chern–Simons and String Theory

The general question "What is the relationship between gauge theory and string theory?" is not meaningful at this time. So, following the strong admonition by Galileo, we avoid "*disputar lungamente delle massime questioni senza conseguir verità nissuna*".[2] However, interesting special cases where such relationship can be established are emerging. For example, Witten [408] argued that Chern–Simons gauge theory on a 3-manifold M can be viewed as a string theory constructed by using a topological sigma model with target space T^*M. The perturbation theory of this string should coincide with Chern–Simons perturbation theory. The coefficient of k^{-r} in the perturbative expansion of $SU(n)$ theory in powers of $1/k$ comes from Feynman diagrams with r loops. Witten shows how each diagram can be replaced by a Riemann surface Σ of genus g with h holes (boundary components) with $g = (r - h + 1)/2$. Gauge theory would then give an invariant $\Gamma_{g,h}(M)$ for every topological type of Σ. Witten shows that this invariant would equal the corresponding string partition function $Z_{g,h}(M)$.

We now give an example of gauge theory to string theory correspondence relating the non-perturbative WRT invariants in Chern–Simons theory with gauge group $SU(n)$ and with topological string amplitudes which generalize the GW (Gromov–Witten) invariants of Calabi–Yau 3-folds [261]. The passage from real 3-dimensional Chern–Simons theory to the 10-dimensional string theory and further onto the 11-dimensional M-theory can be schematically represented by the following:

---

[2] lengthy discussions about the greatest questions that fail to lead to any truth whatever.

$$3 + 3 = 6 \text{ (real symplectic 6-manifold)}$$
$$= 6 \text{ (conifold in } \mathbf{C}^4 \text{)}$$
$$= 6 \text{ (Calabi–Yau manifold)}$$
$$= 10 - 4 \text{ (string compactification)}$$
$$= (11 - 1) - 4 \text{ (M-theory)}.$$

Let us consider the significance of the various lines of the above equation array. Recall that string amplitudes are computed on a 6-dimensional manifold which in the usual setting is a complex 3-dimensional Calabi–Yau manifold obtained by string compactification. This is the most extensively studied model of passing from the 10-dimensional space of supersymmetric string theory to the usual 4-dimensional space-time manifold. However, in our work we do allow these so called extra dimensions to form an open or a symplectic Calabi–Yau manifold. We call these the **generalized Calabi–Yau manifolds**. The first line suggests that we consider open topological strings on such a generalized Calabi–Yau manifold, namely, the cotangent bundle T^*S^3, with Dirichlet boundary conditions on the zero section S^3. We can compute the open topological string amplitudes from the $SU(n)$ Chern–Simons theory. Conifold transition [355] has the effect of closing up the holes in open strings to give closed strings on the Calabi–Yau manifold obtained by the usual string compactification from 10 dimensions. Thus, we recover a topological gravity result starting from gauge theory. In fact, as we discussed earlier, Witten had anticipated such a gauge theory/string theory correspondence by many years. Significance of the last line is based on the conjectured equivalence of M-theory compactified on S^1 to type IIA strings compactified on a Calabi–Yau 3-fold. We do not consider this aspect here. The crucial step that allows us to go from a real, non-compact, symplectic 6-manifold to a compact Calabi–Yau manifold is the conifold or geometric transition. Such a change of geometry and topology is expected to play an important role in other applications of string theory as well. A discussion of this example from physical point of view may be found in [6, 158].

Conifold Transition

To understand the relation of the WRT invariant of S^3 for $SU(n)$ Chern–Simons theory with open and closed topological string amplitudes on "Calabi–Yau" manifolds we need to discuss the concept of conifold transition. From the geometrical point of view this corresponds to symplectic surgery in six dimensions. It replaces a vanishing Lagrangian 3-sphere by a symplectic S^2. The starting point of the construction is the observation that T^*S^3 minus its zero section is symplectomorphic to the cone $z_1^2 + z_2^2 + z_3^2 + z_4^2 = 0$ minus the origin in $\mathbf{C}^4$, where each manifold is taken with its standard symplectic structure. The complex singularity at the origin can be smoothed out by the manifold M_τ defined by $z_1^2 + z_2^2 + z_3^2 + z_4^2 = \tau$, producing a Lagrangian S^3 vanishing cycle. There are also two so called small resolutions $M^\pm$ of the singularity with exceptional set $\mathbf{CP}^1$. They are defined by

$$M^\pm := \left\{ z \in \mathbf{C}^4 \;\middle|\; \frac{z_1 + iz_2}{z_3 \pm iz_4} = \frac{-z_3 \pm iz_4}{z_1 - iz_2} \right\}.$$

Note that $M_0 \setminus \{0\}$ is symplectomorphic to each of $M^\pm \setminus \mathbf{CP}^1$. Blowing up the exceptional set $\mathbf{CP}^1 \subset M^\pm$ gives a resolution of the singularity, which can be expressed as a fiber bundle F over $\mathbf{CP}^1$. Going from the fiber bundle T^*S^3 over S^3 to the fiber bundle F over $\mathbf{CP}^1$ is referred to in the physics literature as the **conifold transition**. We note that the holomorphic automorphism of $\mathbf{C}^4$ given by $z_4 \mapsto -z_4$ switches the two small resolutions $M^\pm$ and changes the orientation of S^3. Conifold transition can also be viewed as an application of mirror symmetry to Calabi–Yau manifolds with singularities. Such an interpretation requires the notion of symplectic Calabi–Yau manifolds and the corresponding enumerative geometry. The geometric structures arising from the resolution of singularities in the conifold transition can also be interpreted in terms of the symplectic quotient construction of Marsden and Weinstein.

10.8.1 WRT Invariants and String Amplitudes

To find the relation between the large n limit of $SU(n)$ Chern–Simons theory on S^3 to a special topological string amplitude on a Calabi–Yau manifold we begin by recalling the formula for the partition function (vacuum amplitude) of the theory $\mathcal{T}_{SU(n),k}(S^3)$, or simply $\mathcal{T}$. Up to a phase, it is given by

$$\mathcal{T} = \frac{1}{\sqrt{n(k+n)^{(n-1)}}} \prod_{j=1}^{n-1} \left[2\sin\left(\frac{j\pi}{k+n}\right) \right]^{n-j}. \qquad (10.68)$$

Let us denote by $F_{(g,h)}$ the amplitude of an open topological string theory on T^*S^3 of a Riemann surface of genus g with h holes. Then the generating function for the free energy can be expressed as

$$-\sum_{g=0}^{\infty}\sum_{h=1}^{\infty} \lambda^{2g-2+h} n^h F_{(g,h)}. \qquad (10.69)$$

This can be compared directly with the result from Chern–Simons theory if we expand the $\log \mathcal{T}$ as a double power series in λ and n. Instead of that we use the conifold transition to get the topological amplitude for a closed string on a Calabi–Yau manifold. We want to obtain the large n expansion of this amplitude in terms of parameters λ and τ, which are defined in terms of the Chern–Simons parameters by

$$\lambda = \frac{2\pi}{k+n}, \quad \tau = n\lambda = \frac{2\pi n}{k+n}. \tag{10.70}$$

The parameter λ is the string coupling constant and τ is the 't Hooft coupling $n\lambda$ of the Chern–Simons theory. The parameter τ entering in the string amplitude expansion has the geometric interpretation as the Kähler modulus of a blown up S^2 in the resolved $M^{\pm}$. If $F_g(\tau)$ denotes the amplitude for a closed string at genus g then we have

$$F_g(\tau) = \sum_{h=1}^{\infty} \tau^h F_{(g,h)}. \tag{10.71}$$

So summing over the holes amounts to filling them up to give the closed string amplitude.

The large n expansion of $\mathcal{T}$ in terms of parameters λ and τ is given by

$$\mathcal{T} = \exp\left[-\sum_{g=0}^{\infty} \lambda^{2g-2} F_g(\tau)\right], \tag{10.72}$$

where F_g defined in (10.71) can be interpreted on the string side as the contribution of closed genus g Riemann surfaces. For $g > 1$ the F_g can be expressed in terms of the Euler characteristic χ_g and the Chern class c_{g-1} of the Hodge bundle over the moduli space $\mathcal{M}_g$ of Riemann surfaces of genus g as follows:

$$F_g = \int_{\mathcal{M}_g} c_{g-1}^3 - \frac{\chi_g}{(2g-3)!} \sum_{n=1}^{\infty} n^{2g-3} e^{-n(\tau)}. \tag{10.73}$$

The integral appearing in the formula for F_g can be evaluated explicitly to give

$$\int_{\mathcal{M}_g} c_{g-1}^3 = \frac{(-1)^{(g-1)}}{(2\pi)^{(2g-2)}} 2\zeta(2g-2)\chi_g. \tag{10.74}$$

The Euler characteristic is given by the Harer–Zagier [185] formula

$$\chi_g = \frac{(-1)^{(g-1)}}{(2g)(2g-2)} B_{2g}, \tag{10.75}$$

where B_{2g} is the $(2g)$th Bernoulli number. We omit the special formulas for the genus 0 and genus 1 cases. The formulas for F_g for $g \geq 0$ coincide with those of the g-loop topological string amplitude on a suitable Calabi–Yau manifold. The change in geometry that leads to this calculation can be thought of as the result of coupling to gravity. Such a situation occurs in the quantization of Chern–Simons theory. Here the classical Lagrangian does not depend on the metric; however, coupling to the gravitational Chern–Simons term is necessary to make it a TQFT.

We have mentioned the following four approaches that lead to the WRT invariants:

1. Witten's TQFT calculation of the Chern–Simons partition function;
2. Quantum group (or Hopf algebraic) computations initiated by Reshetikhin and Turaev;
3. Kohno's special functions corresponding to representations of mapping class groups in the space of conformal blocks and a similar approach by Crane;
4. open or closed string amplitudes in suitable Calabi–Yau manifolds.

These methods can also be applied to obtain invariants of links, such as the Jones polynomial. Indeed, this was the objective of Witten's original work. WRT invariants were a byproduct of this work. Their relation to topological strings came later.

The WRT to string theory correspondence was extended by Gopakumar and Vafa (see hep-th/9809187, 9812127) by using string theoretic arguments to show that the expectation value of the quantum observables defined by the Wilson loops in the Chern–Simons theory has a similar interpretation in terms of a topological string amplitude. This led them to conjecture a correspondence exists between certain knot invariants (such as the Jones polynomial) and Gromov–Witten type invariants of generalized Calabi–Yau manifolds. Gromov–Witten invariants of a Calabi–Yau 3-fold X, are in general, rational numbers, since one has to get the weighted count by dividing by the order of automorphism groups. Using M-theory Gopakumar and Vafa argued that the generating series F_X of Gromov–Witten invariants in all degrees and all genera is determined by a set of integers $n(g, \beta)$. They give the following remarkable formula for F_X:

$$F_X(\lambda, q) = \sum_{g \geq 0} \sum_{k \geq 1} \sum \frac{1}{k} n(g, \beta)(2\sin(k\lambda/2))^{2g-2} q^{k\beta},$$

where λ is the string coupling constant and the first sum is taken over all nonzero elements β in $H_2(X)$. We note that for a fixed genus there are only finitely many nonzero integers $n(g, \beta)$. A mathematical formulation of the Gopakumar–Vafa conjecture (GV conjecture) is given in [312]. Special cases of the conjecture have been verified (see, for example, [313] and references therein). In [251] a new geometric approach relating the Gromov–Witten invariants to equivariant index theory and 4-dimensional gauge theory was used to compute the string partition functions of some local Calabi–Yau spaces and to verify the GV conjecture for them.

A knot should correspond to a Lagrangian D-brane on the string side and the knot invariant would then give a suitably defined count of compact holomorphic curves with boundary on the D-brane. To understand a proposed proof, recall first that a categorification of an invariant I is the construction of a suitable homology such that its Euler characteristic equals I. A well known example of this is Floer's categorification of the Casson invariant. We will discuss in Chapter 11 Khovanov's categorification of the Jones polynomial $V_\kappa(q)$ by constructing a bi-graded $sl(2)$-homology $H_{i,j}$ determined by the knot κ. Its quantum or graded Euler characteristic equals the Jones polynomial, i.e.,

$$V_\kappa(q) = \sum_{i,j} (-1)^j q^i \dim H_{i,j}.$$

Now let L_κ be the Lagrangian submanifold corresponding to the knot κ of a fixed Calabi–Yau space X. Let r be a fixed relative integral homology class of the pair (X, L_κ). Let $\mathcal{M}_{g,r}$ denote the moduli space of pairs (Σ_g, A), where Σ_g is a compact Riemann surface in the class r with boundary S^1 and A is a flat $U(1)$ connection on Σ_g. These data together with the cohomology groups $H^k(\mathcal{M}_{g,r})$ determine a tri-graded homology. It generalizes the Khovanov homology. Its Euler characteristic has a physical interpretation as a generating function for a class of invariants in string theory and these invariants can be used to obtain the Gromov–Witten invariants. Taubes has given a construction of the Lagrangians in the Gopakumar–Vafa conjecture. We note that counting holomorphic curves with boundary on a Lagrangian manifold was introduced by Floer in his work on the Arnold conjecture. The tri-graded homology is expected to unify knot homologies of the Khovanov type as well as knot Floer homology constructed by Ozswáth and Szabó [309], which provides a categorification of the Alexander polynomial. Knot Floer homology is defined by counting pseudo-holomorphic curves and has no known combinatorial description. An explicit construction of a tri-graded homology for certain torus knots was recently given by Dunfield, Gukov, and Rasmussen [math.GT/0505662].

Chapter 11
Knot and Link Invariants

11.1 Introduction

In this chapter we make some historical observations and comment on some early work in knot theory. Invariants of knots and links are introduced in Section 11.2. Witten's interpretation of the Jones polynomial via the Chern–Simons theory is discussed in Section 11.3. A new invariant of 3-manifolds is obtained as a byproduct of this work by an evaluation of a certain partition function of the theory. We already met this invariant, called the Witten–Reshetikhin–Turaev (or WRT) invariant in Chapter 10. In Section 11.4 we discuss the Vassiliev invariants of singular knots. Gauss's formula for the linking number is the starting point of some more recent work on self-linking invariants of knots by Bott, Taubes, and Cattaneo. We will discuss their work in Section 11.5. The self-linking invariants were obtained earlier by physicists using Chern–Simons perturbation theory. This work now forms a small part of the program initiated by Kontsevich [235] to relate topology of low-dimensional manifolds, homotopical algebras, and non-commutative geometry with topological field theories and Feynman diagrams in physics. See also the book [176] by Guadagnini. Khovanov's categorification of the Jones polynomial by Khovanov homology is the subject of Section 11.6. We would like to remark that in recent years many applications of knot theory have been made in chemistry and biology (for a brief of discussion of these and further references see, for example, [260]). Some of the material in this chapter is from my article [263]).

In the second half of the nineteenth century a systematic study of knots in $\mathbf{R}^3$ was made by Tait. He was motivated by Kelvin's theory of atoms modeled on knotted vortex tubes of ether. It was expected that physical and chemical properties of various atoms could be expressed in terms of properties of knots such as the knot invariants. Even though Kelvin's theory did not hold up, the theory of knots grew as a subfield of combinatorial topology. Tait classified the knots in terms of the crossing number of a regular projection.

K. Marathe, *Topics in Physical Mathematics*, DOI 10.1007/978-1-84882-939-8_11, 351
© Springer-Verlag London Limited 2010

A **regular projection** of a knot on a plane is an orthogonal projection of the knot such that at any crossing in the projection exactly two strands intersect transversely. Tait made a number of observations about some general properties of knots which have come to be known as the "Tait conjectures." In its simplest form the classification problem for knots can be stated as follows. Given a projection of a knot, is it possible to decide in finitely many steps if it is equivalent to an unknot? This question was answered affirmatively by Haken [182] in 1961. He proposed an algorithm that could decide if a given projection corresponds to an unknot. However, because of its complexity the algorithm has not been implemented on a computer even after a half century. We add that in 1974 Haken and Appel solved the famous four-color problem for planar maps by making essential use of a computer program to study the thousands of cases that needed to be checked. A very readable, non-technical account of their work can be found in [12].

11.2 Invariants of Knots and Links

Let M be a closed orientable 3-manifold. A smooth embedding of S^1 in M is called a **knot** in M. A **link** in M is a finite collection of disjoint knots. The number of disjoint knots in a link is called the number of **components** of the link. Thus, a knot can be considered a link with one component. Two links L, L' in M are said to be **equivalent** if there exists a smooth orientation-preserving automorphism $f : M \to M$ such that $f(L) = L'$. For links with two or more components we require f to preserve a fixed given ordering of the components. Such a function f is called an **ambient isotopy** and L and L' are called ambient isotopic. Ambient isotopy is an equivalence relation on the set of knots in M, and the equivalence class $[\kappa]$ of a knot κ under this relation is called the **isotopy class** of κ. If $h : S^1 \to M$ is a knot, then its image $\kappa = h(S^1)$ is also called a knot. A **knot type** $[\kappa]$ of a knot κ is defined to be the isotopy class of κ. As in all classification problems we try to find a set of invariants of knots that will distinguish different knot types. However, at this time no such set of invariants of knots is known. Another approach is to define a suitable "energy" functional on the space of knots and study the critical points of its gradient flow to identify the knot type of a given knot. Even though there are several candidates for energy functional there is no known example which produces an energy minimizer for every knot type.

Knots and links in $\mathbf{R}^3$ can also be obtained by using braids. A **braid** on n strands (or with n strings, or simply an n-braid) can be thought of as a set of n pairwise disjoint strings joining n distinct points in one plane with n distinct points in a parallel plane in $\mathbf{R}^3$. Making a suitable choice of planes and points we make the following definition.

Definition 11.1 *Fix a number $n \in \mathbf{N}$ and let i be an integer such that $1 \le i \le n$. Let A_i (resp., B_i) denote the point in $\mathbf{R}^3$ with x-coordinate i lying* *on the x-axis (resp., line $y = 0$, $z = 1$). Let $\sigma \in S_n$, the symmetric group on the first n natural numbers. A strand from A_i to $B_{\sigma(i)}$ is a smooth curve from A_i to $B_{\sigma(i)}$ such that the z coordinate of a point on the curve decreases monotonically from 1 to 0 as it traces out the curve from A_i to $B_{\sigma(i)}$. A set of n non-intersecting strands from A_i to $B_{\sigma(i)}$, $1 \leq i \leq n$, is called an n-braid in standard form. Two n-braids are said to be equivalent if they are ambient isotopic. The set of equivalence classes of n-braids is denoted by $\mathcal{B}_n$. It is customary to call an element of $\mathcal{B}_n$ also an n-braid. It is clear from the context whether one is referring to an equivalence class or its particular representative.*

The multiplication operation on $\mathcal{B}_n$ is induced by concatenation of braids. If A_i, B_i (resp., A_i', B_i') are the endpoints of braid b_1 (resp., b_2) then the braid $b_1 b_2$ is obtained by gluing the ends B_i of b_1 to the starting points A_i' of b_2. This can be put in the standard braid form going from $z = 1$ to $z = 0$ by reparametrization so that b_1 runs from $z = 1$ to $z = 1/2$ and b_2 runs from $z = 1/2$ to $z = 0$. A representative of the unit element is the braid consisting of n parallel strands from A_i to B_i, $1 \leq i \leq n$. Taking the mirror image of b_1 in the plane $z = 0$ gives a braid equivalent to b_1^{-1} (its parallel translation along the z-axis by 1 puts it in the standard form. A nontrivial braid has at least two strands joining A_i and B_j for $i \neq j$. For each i with $1 \leq i < n$ let σ_i be an n-braid with a strand from A_i to B_{i+1} crossing over the strand from A_{i+1} to B_i and with vertical strands from A_j to B_j for $j \neq i, i+1$. Then it can be shown that σ_i and σ_j are not equivalent whenever $i \neq j$ and that every n-braid can be represented as a product of braids σ_i and σ_i^{-1} for $1 \leq i < n$; i.e., the $n - 1$ braids σ_i generate the **braid group** $\mathcal{B}_n$. Analysis of the equivalence relation on braids leads to the relations

$$\sigma_i \sigma_{i+1} \sigma_i = \sigma_{i+1} \sigma_i \sigma_{i+1}, \qquad 1 \leq i \leq n - 2. \tag{11.1}$$

Equation (11.1) is called the **braid relation**, and .

$$\sigma_i \sigma_j = \sigma_j \sigma_i, \qquad 1 \leq i, j \leq n - 1 \text{ and } |i - j| > 1 \tag{11.2}$$

is called the **far commutativity** relation, for it expresses the fact that the generators σ_i, σ_j commute when the indices i, j are not immediate neighbors. This discussion leads to the following well-known theorem of M. Artin.

Theorem 11.1 *The set $\mathcal{B}_n$ with multiplication operation induced by concatenation of braids is a group generated by the elements σ_i, $1 \leq i \leq n-1$ subject to the braid relations (11.1) and the far commutativity relations (11.2).*

The braid group $\mathcal{B}_2$ is the infinite cyclic group generated by σ_1 and hence is isomorphic to **Z**. For $n > 2$ the group $\mathcal{B}_n$ contains a subgroup isomorphic to the free group on two generators and is therefore, non-Abelian. The group $\mathcal{B}_3$ is related to the modular group $PSL(2, \mathbf{Z})$ in the following way. It is easy to check that the elements $a = \sigma_1 \sigma_2 \sigma_1$, $b = \sigma_1 \sigma_2$ generate $\mathcal{B}_3$ and that the element $c = a^2 = b^3$ generates its center $Z(\mathcal{B}_3)$. Let $f : \mathcal{B}_3 \rightarrow SL(2, \mathbf{Z})$ be the mapping defined by

$$f(a) := \begin{pmatrix} 0 & 1 \\ -1 & 0 \end{pmatrix}, \qquad f(b) := \begin{pmatrix} 0 & 1 \\ -1 & 1 \end{pmatrix}.$$

One can show that f is a surjective homomorphism that maps c to $-I_2$, where I_2 is the identity element. Let $\hat{f} = \pi \circ f$, where π is the canonical projection of $SL(2, \mathbf{Z})$ to $PSL(2, \mathbf{Z})$. Then $\hat{f} : \mathcal{B}_3 \to PSL(2, \mathbf{Z})$ is also a surjective homomorphism with kernel the center $Z(\mathcal{B}_3)$. Thus, $\mathcal{B}_3/Z(\mathcal{B}_3)$ is isomorphic to the modular group $PSL(2, \mathbf{Z})$.

The map that sends the braid σ_i to the transposition $(i, i + 1)$ in the symmetric group S_n on n letters induces a canonical map $\pi : \mathcal{B}_n \to S_n$. The kernel, $\mathcal{P}_n$ of the surjective map π is called the **pure braid group** on n strands. Thus, we have an exact sequence of groups

$$\mathbf{1} \to P_n \to B_n \to S_n \to \mathbf{1},$$

where $\mathbf{1}$ denotes the group containing only the identity element. Braid groups can also be defined in terms of configuration spaces. Recall that the configuration space of n points in $\mathbf{R}^2$ is the space of unordered sequences of n pairwise distinct points. It is denoted by $C_n(\mathbf{R}^2)$. General discussion of configuration spaces is given later in this chapter in defining self-linking invariants. Each braid $b \in \mathcal{B}_n$ defines a homotopy class of loops in this space. In fact, we have an isomorphism of $\mathcal{B}_n$ with the fundamental group $\pi_1(C_n(\mathbf{R}^2))$ of the configuration space $C_n(\mathbf{R}^2)$. The group $\mathcal{B}_n$ is also isomorphic to the mapping class group of the unit disk in $\mathbf{R}^2 = \mathbf{C}$ with n punctures or marked points. Magnus pointed out that this interpretation of the braid group was known to Hurwitz. Braids are implicit in his work on monodromy done in 1891. However, they were defined explicitly by E. Artin in 1925. Hurwitz's interpretation was rediscovered by Fox and Neuwirth in 1962. Birman's book [40] is a classic reference for this and related material. For an updated account, see Kassel and Turaev [217].

Given an n-braid b with endpoints A_i, B_i we can obtain a link by joining each A_i and B_i by a smooth curve so that these n curves are non-intersecting and do not intersect the strands of b. The resulting link is denoted by $c(b)$ and is said to be obtained by the operation of the **closure of braid** b. The question whether every link in $\mathbf{R}^3$ is the closure of some braid is answered by the following classical theorem of Alexander.

Theorem 11.2 *The closure map from the set of braids to the set of links is surjective, i.e., any link (and, in particular, knot) is the closure of some braid. Moreover, if braids b and b' are equivalent, then the links $c(b)$ and $c(b')$ are equivalent.*

The minimum n such that a given link L is the closure of an n-braid is called the **braiding number** or **braid index** of L and is denoted by $b(L)$. To answer the question of when the closure of a braid is a knot, we consider the canonical homomorphism $\pi : \mathcal{B}_n \to S_n$ defined above. The following proposition characterizes when the closure of a braid is a knot.

Proposition 11.3 *The closure $c(b)$ of an n-braid b is a knot if and only if the permutation $\pi(b)$ has order n in the symmetric group S_n.*

In view of the above discussion it is not surprising that invariants of knots and links should be closely related to representations of the braid group. The earliest such representation was obtained in the 1930s by Burau. The Burau representation is a matrix representation over the ring $\mathbf{Z}[t, t^{-1}]$ of Laurent polynomials in t with integer coefficients. We now indicate the construction of the Burau representation of $\mathcal{B}_n$ by matrices of order n. The representation is trivial for $n = 1$. For $n = 2$ the generator σ_1 is mapped to the matrix $\hat{\sigma}_1 = U$ defined by

$$U := \begin{pmatrix} 1 - t & t \\ 1 & 0 \end{pmatrix},$$

For $n > 2$ the generators σ_i are mapped to the matrix $\hat{\sigma}_i$ with block diagonal entries I_{i-1}, U, and I_{n-i-1}, where I_k is the identity matrix of order k for $k \neq 0$ and I_0 is regarded as the empty block and is omitted. For example, the Burau representation of $\mathcal{B}_3$ is defined by

$$\hat{\sigma}_1 := \begin{pmatrix} 1 - t & t & 0 \\ 0 & 1 & 0 \\ 0 & 0 & 1 \end{pmatrix}, \qquad \hat{\sigma}_2 := \begin{pmatrix} 1 & 0 & 0 \\ 0 & 1 - t & t \\ 0 & 1 & 0 \end{pmatrix},$$

It is a classical result that the Burau representation is faithful for $n = 3$. After extensive work by several mathematicians, it is now known that the representation is not faithful for $n > 4$. The case $n = 4$ is open as of this writing. The Burau representation of $\mathcal{B}_n$ can be used to construct the Alexander polynomial invariant of the link obtained by the closure $c(b)$ of the braid $b \in \mathcal{B}_n$. After nearly 60 years Jones obtained his link invariant taking values in Laurent polynomials $\mathbf{Z}[t, t^{-1}]$ in t with integer coefficients. The new representations underlying his invariant are obtained by studying certain finite-dimensional von Neumann algebras. A special case of these representations was discovered earlier by Temperley and Lieb in their study of certain models in statistical mechanics. Many surprising interpretations of the Jones polynomial have been obtained using ideas coming from conformal field theory and quantum field theory. Thus we have another example of a result in physical mathematics. The Lawrence–Krammer–Bigelow representation of the braid group is a matrix representation of $\mathcal{B}_n$ over the ring of Laurent polynomials $\mathbf{Z}[t, t^{-1}, q, q^{-1}]$ in two variables t, q with integer coefficients. This representation is faithful for all $n \geq 1$ and shows that the group $\mathcal{B}_n$ is linear, i.e., it admits an injective homomorphism into the general linear group over $\mathbf{R}$.

Every link L in M has a **tubular neighborhood** $N(L)$, which is the union of smoothly embedded disjoint solid tori $D^2 \times S^1$, one for each component, so that the cores $\{0\} \times S^1$ of the tori form the given link L. The set $N(L)$ is called a **thickening** of L. Unless otherwise stated we shall take M to be $S^3 \cong \mathbf{R}^3 \cup \{\infty\}$ and write simply a link instead of a link in S^3. The diagrams of links are drawn as links in R^3. Every link L in S^3 or $\mathbf{R}^3$ can be thickened. A **link diagram** of L is a plane projection with crossings marked as over or under. Two links are equivalent if a link diagram of one can be changed into the given diagram of the other link by a finite sequence of moves called the **Reidemeister moves**. There are three types of Reidemeister moves as indicated in Figures 11.1, 11.2, and 11.3.

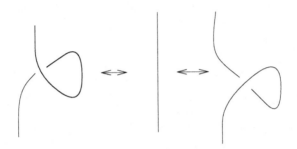

Fig. 11.1 Reidemeister moves of type I

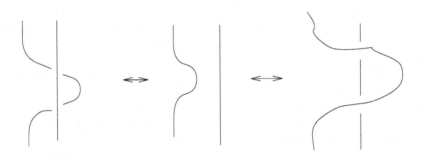

Fig. 11.2 Reidemeister moves of type II

The simplest combinatorial invariant of a knot κ is the **crossing number** $c(\kappa)$. It is defined as the minimum number of crossings in any projection of the knot κ. The classification of knots up to crossing number 17 is now known [195]. The crossing numbers for some special families of knots are known; however, the question of finding the crossing number of an arbitrary knot is still unanswered. Another combinatorial invariant of a knot κ that is easy to define is the **unknotting number** $u(\kappa)$, the minimum number of crossing changes in any projection of the knot κ that makes it into a projection of

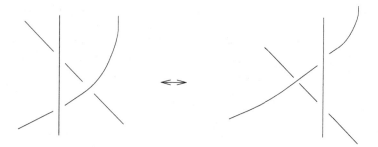

Fig. 11.3 Reidemeister moves of type III

the unknot. Upper and lower bounds for $u(\kappa)$ are known for any knot κ. An explicit formula for $u(\kappa)$ for a family of knots called torus knots, conjectured by Milnor nearly 40 years ago, has been proved recently by a number of different methods. The 3-manifold $S^3 \setminus \kappa$ is called the **knot complement** of κ. The fundamental group $\pi_1(S^3 \setminus \kappa)$ of the knot complement is an invariant of the knot κ. It is called the fundamental group of the knot (or simply the **knot group**) and is denoted by $\pi_1(\kappa)$. Equivalent knots have homeomorphic complements and conversely. However, this result does not extend to links.

A **Seifert surface** Σ_L (or simply Σ) for a link L is a connected compact orientable surface smoothly embedded in S^3 such that its boundary $\partial\Sigma = L$.

Theorem 11.4 *Every link in S^3 bounds a Seifert surface.*

The least genus of all Seifert surfaces of a given link L is called the **genus** of the link L. For example, the genus of the unknot is 0.

We define an integer invariant of a knot κ, called the **signature**, by using its Seifert surface Σ_κ. Let α_1, α_2 be two oriented simple loops on Σ_κ. Let α_2^+ be the loop obtained by moving α_2 away from Σ_κ along its positive normal in S^3. We can now associate an integer with the pair (α_1, α_2) to be the linking number $\mathbf{Lk}(\alpha_1, \alpha_2^+)$. This induces a bilinear form on $H_1(\Sigma; \mathbf{Z})$. Let Q be the matrix of this bilinear form with respect to some basis of $H_1(\Sigma; \mathbf{Z})$. Then the signature $\mathrm{sgn}(\kappa)$ of the knot κ is defined to be the signature of the symmetric matrix $Q + Q^t$. It can be shown that the signature of a knot is always an even integer. In fact, $\mathrm{sgn}(\kappa) = 2h(\kappa)$, where $h(\kappa)$ is an integer-valued invariant of the knot κ defined by studying a special class of representations of the knot group $\pi_1(S^3 \setminus \kappa)$ into the group $SU(2)$ along the lines of the construction of the Casson invariant [253].

In the 1920s Alexander gave an algorithm for computing a polynomial invariant $\Delta_\kappa(t)$ (a Laurent polynomial in t) of a knot κ, called the **Alexander polynomial**, by using its projection on a plane. He also gave its topological interpretation as an annihilator of a certain cohomology module associated to the knot κ. In the 1960s Conway defined his polynomial invariant and gave its relation to the Alexander polynomial. This polynomial is called the

Alexander–Conway polynomial, or simply the Conway polynomial. The Alexander–Conway polynomial of an oriented link L is denoted by $\nabla_L(z)$, or simply by $\nabla(z)$ when L is fixed. By changing a link diagram at one crossing we can obtain three diagrams corresponding to links L_+, L_-, and L_0, which are identical except for this crossing (see Figure 11.4).

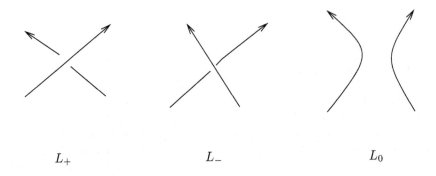

$$L_+ \qquad\qquad L_- \qquad\qquad L_0$$

Fig. 11.4 Altering a link at a crossing

We denote the corresponding polynomials of L_+, L_-, and L_0 by ∇_+, ∇_- and ∇_0 respectively. The Alexander–Conway polynomial is uniquely determined by the following simple set of axioms.

AC1. Let L and L' be two oriented links that are ambient isotopic. Then

$$\nabla_{L'}(z) = \nabla_L(z). \tag{11.3}$$

AC2. Let S^1 be the standard unknotted circle embedded in S^3. It is usually referred to as the **unknot** and is denoted by $\mathcal{O}$. Then

$$\nabla_{\mathcal{O}}(z) = 1. \tag{11.4}$$

AC3. The polynomial satisfies the **skein relation**

$$\nabla_+(z) - \nabla_-(z) = z\nabla_0(z). \tag{11.5}$$

We note that the original Alexander polynomial Δ_L is related to the Alexander–Conway polynomial of an oriented link L by the relation

$$\Delta_L(t) = \nabla_L(t^{1/2} - t^{-1/2}).$$

Despite these and other major advances in knot theory, the Tait conjectures remained unsettled for more than a century after their formulation. Then, in the 1980s, Jones discovered his polynomial invariant $V_L(t)$, called the **Jones polynomial**, while studying von Neumann algebras [210] and gave its interpretation in terms of statistical mechanics. A state model for the Jones polynomial was then given by Kauffman using his bracket polynomial. These new polynomial invariants led to the proofs of most of the Tait conjectures. As with the earlier invariants, Jones' definition of his polynomial invariant is algebraic and combinatorial in nature and was based on representations of the braid groups and related Hecke algebras.

The Jones polynomial $V_\kappa(t)$ of κ is a Laurent polynomial in t, which is uniquely determined by a simple set of properties similar to the axioms for the Alexander–Conway polynomial. More generally, the Jones polynomial can be defined for any oriented link L as a Laurent polynomial in $t^{1/2}$ so that reversing the orientation of all components of L leaves V_L unchanged. In particular, V_κ does not depend on the orientation of the knot κ. For a fixed link, we denote the Jones polynomial simply by V. Recall that there are three standard ways to change a link diagram at a crossing point. The Jones polynomials of the corresponding links are denoted by V_+, V_- and V_0 respectively. Then the Jones polynomial is characterized by the following properties:

JO1. Let L and L' be two oriented links that are ambient isotopic. Then

$$V_{L'}(t) = V_L(t). \qquad (11.6)$$

JO2. Let $\mathcal{O}$ denote the unknot. Then

$$V_{\mathcal{O}}(t) = 1. \qquad (11.7)$$

JO3. The polynomial satisfies the skein relation

$$t^{-1}V_+ - tV_- = (t^{1/2} - t^{-1/2})V_0. \qquad (11.8)$$

An important property of the Jones polynomial not shared by the Alexander–Conway polynomial is its ability to distinguish between a knot and its mirror image. More precisely, we have the following result. Let κ_m be the mirror image of the knot κ. Then

$$V_{\kappa_m}(t) = V_\kappa(t^{-1}). \qquad (11.9)$$

Since the Jones polynomial is not symmetric in t and t^{-1}, it follows that in general

$$V_{\kappa_m}(t) \neq V_\kappa(t). \qquad (11.10)$$

We note that a knot is called **amphichiral** (**achiral** in biochemistry) if it is equivalent to its mirror image. We shall use the simpler biochemistry terminology, so a knot that is not equivalent to its mirror image is called **chiral**. The condition expressed by (11.10) is sufficient but not necessary for chirality of a knot. The Jones polynomial did not resolve the following conjecture by Tait concerning chirality.

The chirality conjecture: *If the crossing number of a knot is odd, then it is chiral.*

A 15-crossing knot which provides a counterexample to the chirality conjecture is given in [195].

There was an interval of nearly 60 years between the discovery of the Alexander polynomial and the Jones polynomial. Since then a number of polynomial and other invariants of knots and links have been found. A particularly interesting one is the two-variable polynomial generalizing the Jones polynomial V which was defined in [211]. This polynomial is called the **HOM-FLY polynomial** (its name formed from the initials of authors of the article [140]) and is denoted by P. The HOMFLY polynomial $P(\alpha, z)$ satisfies the skein relation

$$\alpha^{-1}P_+ - \alpha P_- = zP_0. \tag{11.11}$$

Both the Jones polynomial V_L and the Alexander–Conway polynomial ∇_L are special cases of the HOMFLY polynomial. The precise relations are given by the following theorem.

Theorem 11.5 *Let L be an oriented link. Then the polynomials P_L, V_L, and ∇_L satisfy the relations:*

$$V_L(t) = P_L(t, t^{1/2} - t^{-1/2}) \quad and \quad \nabla_L(z) = P_L(1, z).$$

After defining his polynomial invariant, Jones also established the relation of some knot invariants with statistical mechanical models [208]. Since then this has become a very active area of research. We now recall the construction of a typical statistical mechanics model. Let X denote the configuration space of the model and let S denote the set (usually with some additional structure) of internal symmetries. The set S is also called the spin space. A state of the statistical system (X, S) is an element $s \in \mathcal{F}(X, S)$. The energy $\mathcal{E}_k$ of the system (X, S) is a functional

$$\mathcal{E}_k : \mathcal{F}(X, S) \to \mathbf{R}, \qquad k \in K$$

where the subscript $k \in K$ indicates the dependence of energy on the set K of auxiliary parameters, such as temperature, pressure, etc. For example, in the simplest lattice models the energy is often taken to depend only on the nearest neighboring states and on the ambient temperature, and the spin space is taken to be $S = \mathbf{Z}_2$, corresponding to the up and down directions. The weighted partition function of the system is defined by

$$Z_k := \sum \mathcal{E}_k(s)w(s),$$

where $w : \mathcal{F}(X, S) \to \mathbf{R}$ is a weight function and the sum is taken over all states $s \in \mathcal{F}(X, S)$. The partition functions corresponding to different weights are expected to reflect the properties of the system as a whole. Calculation of the partition functions remains one of the most difficult problems in statistical mechanics. In special models the calculation can be carried out by auxiliary relations satisfied by some subsets of the configuration space. The star-triangle relations or the corresponding Yang–Baxter equations are examples of such relations. One obtains a state-model for the Alexander or the Jones polynomial of a knot by associating to the knot a statistical system whose partition function gives the corresponding polynomial.

However, these statistical models did not provide a geometrical or topological interpretation of the polynomial invariants. Such an interpretation was provided by Witten [405] by applying ideas from quantum field theory (QFT) to the Chern–Simons Lagrangian. In fact, Witten's model allows us to consider the knot and link invariants in any compact 3-manifold M. Witten's ideas led to the creation of a new area—which we discussed in Chapter 7—called topological quantum field theory. TQFT, at least formally, allows us to express topological invariants of manifolds by considering a QFT with a suitable Lagrangian. An excellent account of several aspects of the geometry and physics of knots may be found in the books by Atiyah [14] and Kauffman [218].

We conclude this section by discussing a knot invariant that can be defined for a special class of knots. In 1978 Bill Thurston [373] created the field of hyperbolic 3-manifolds. A **hyperbolic manifold** is a manifold that admits a metric of constant negative curvature or equivalently a metric of constant curvature -1. The application of hyperbolic 3-manifolds to knot theory arises as follows. A knot κ is called hyperbolic if the knot complement $S^3 \setminus \kappa$ is a hyperbolic 3-manifold. It can be shown that the knot complement $S^3 \setminus \kappa$ of the hyperbolic knot κ has finite hyperbolic volume $v(\kappa)$. The number $v(\kappa)$ is an invariant of the knot κ and can be computed to any degree of accuracy; however the arithmetic nature of $v(\kappa)$ is not known. This result also extends to links. It is known that torus knots are not hyperbolic. The figure eight knot is the knot with the smallest crossing number that is hyperbolic. Thurston made a conjecture that effectively states that almost every knot is hyperbolic. Recently Hoste and Weeks made a table of knots with crossing number 16 or less by making essential use of hyperbolic geometry. Their table has more than 1.7 million knots, all but 32 of which are hyperbolic. Thistlethwaite has obtained the same table without using any hyperbolic invariants. A fascinating account of their work is given in [195]. We would like to add that there is a vast body of work on the topology and geometry of 3-manifolds which was initiated by Thurston. At present the relation of this work to the methods and results of the gauge theory, quantum groups, or statistical mechanics approaches to the study of 3-manifolds remains a mystery. There are some conjectures relating invariants obtained by different methods. For example, the **volume conjecture** (due to R. Kashaev, H. Murakami, and J. Murakami, see [306]) states that the invariant $v(\kappa)$ of the hyperbolic knot κ is equal to the limit of a certain function of the colored Jones polynomial of the knot κ. There is also an extension of the volume conjecture to links and to non-hyperbolic knots. This extension has been verified for special knots and links such as the figure eight knot and the Borromean rings.

11.3 TQFT Approach to Knot Invariants

In this section we discuss a surprising application of the Chern–Simons theory to the calculation of some invariants of knots and links via TQFT. We begin with some historical observations. The earliest example of TQFT is the "derivation" of the Ray–Singer analytic torsion [323, 324] in terms of Chern–Simons theory given in [341], although TQFT was not introduced until later. The analytic torsion is defined by using determinants of Laplacians on forms that are regularized by zeta function regularization. It is an anlogue of the classical Reidemeister torsion of 3-manifolds as generalized to arbitrary manifolds by Franz and de Rham. Reidemeister torsion was the first topological invariant that could distinguish between spaces that are homotopic but not homeomorphic. It can be used to classify lens spaces. Ray and Singer conjectured that these two invariants should be the same for compact Riemannian manifolds. This conjecture was proved independently by Cheeger and Müller in the late 1970s. See the book [382] by Turaev for further details.

Quantization of classical fields is an area of fundamental importance in modern mathematical physics. Although there is no satisfactory mathematical theory of quantization of classical dynamical systems or fields, physicists have developed several methods of quantization that can be applied to specific problems. Most important among these is Feynman's path integral method of quantization, which has been applied with great success in QED (quantum electrodynamics), the theory of quantization of electromagnetic fields. On the other hand the recently developed TQFT has been very useful in defining, interpreting, and calculating new invariants of manifolds. We note that at present TQFT cannot be considered a mathematical theory and our presentation is based on a development of the infinite-dimensional calculations by formal analogy with finite dimensional results. Nevertheless, TQFT has provided us with new results as well as a fresh perspective on invariants of low-dimensional manifolds. For example, at this time a geometric interpretation of polynomial invariants of knots and links in 3-manifolds such as the Jones polynomial can be given only in the context of TQFT.

Recall from Chapter 7 that a quantum field theory may be considered an assignment of the **quantum expectation** $\langle \Phi \langle_\mu$ to each gauge-invariant function (i.e., a quantum observable) $\Phi : \mathcal{A}(M) \to \mathbf{R}$. In the Feynman path integral approach to quantization the quantum expectation $\langle \Phi \rangle_\mu$ of an observable is expressed as a ratio of partition functions as follows:

$$\langle \Phi \rangle_\mu = \frac{Z_\mu(\Phi)}{Z_\mu(1)}. \tag{11.12}$$

There are several examples of gauge-invariant functions. For example, primary characteristic classes evaluated on suitable homology cycles give an important family of gauge-invariant functions. The instanton number k of $P(M, G)$ belongs to this family, as it corresponds to the second Chern class evaluated on the fundamental cycle of M representing the fundamental class $[M]$. The pointwise norm $|F_\omega|_x$ of the gauge field at $x \in M$, the absolute value $|k|$ of the instanton number k, and the Yang–Mills action are also gauge-invariant functions. Another important example of a quantum observable is given by the **Wilson loop functional**, which we will use later in this section. It is gauge-invariant and hence defines a quantum observable. Regarding a knot κ as a loop we get a quantum observable $\mathcal{W}_{\rho,\kappa}$ associated to the knot. For a link L with ordered components $\kappa_1, \kappa_1, \dots, \kappa_j$ and corresponding representations $\rho_1, \rho_1, \dots, \rho_j$ of G we define the Wilson functional $\mathcal{W}_L$ by

$$\mathcal{W}_L(\omega) := \mathcal{W}_{\rho_1, \kappa_1}(\omega) \mathcal{W}_{\rho_2, \kappa_2}(\omega) \dots \mathcal{W}_{\rho_j, \kappa_j}(\omega), \qquad \forall \omega \in \mathcal{A}_M.$$

When M is 3-dimensional, P is trivial (in a non-canonical way). We fix a trivialization to write $P(M, G) = M \times G$ and write $\mathcal{A}_M$ for $\mathcal{A}_{P(M,G)}$. Then the group of gauge transformations $\mathcal{G}_P$ can be identified with the group of smooth functions from M to G and we denote it simply by $\mathcal{G}_M$. The gauge theory used by Witten in his work is the Chern–Simons theory on a 3-manifold with gauge group $SU(n)$. The Chern–Simons Lagrangian L_{CS} is defined by

$$L_{CS} := \tfrac{k}{4\pi} \operatorname{tr}(A \wedge F - \tfrac{1}{3} A \wedge A \wedge A) = \tfrac{k}{4\pi} \operatorname{tr}(A \wedge dA + \tfrac{2}{3} A \wedge A \wedge A). \tag{11.13}$$

Recall that in this case the Chern–Simons action $\mathcal{A}_{CS}$ takes the form

$$\mathcal{A}_{CS} := \int_M L_{CS} = \frac{k}{4\pi} \int_M \operatorname{tr}(A \wedge dA + \tfrac{2}{3} A \wedge A \wedge A), \tag{11.14}$$

where $k \in \mathbf{R}$ is a coupling constant, A denotes the pull-back to M of the gauge potential (connection) ω by a section of P, and $F = F_\omega = d^\omega A$ is the gauge field (curvature of ω) on M corresponding to the gauge potential A. A local expression for (11.14) is given by

$$\mathcal{A}_{CS} = \frac{k}{4\pi} \int_M \epsilon^{\alpha\beta\gamma} \operatorname{tr}(A_\alpha \partial_\beta A_\gamma + \tfrac{2}{3} A_\alpha A_\beta A_\gamma), \tag{11.15}$$

where $A_\alpha = A_\alpha^a T_a$ are the components of the gauge potential with respect to the local coordinates $\{x_\alpha\}$, $\{T_a\}$ is a basis of the Lie algebra $su(n)$ in the fundamental representation, and $\epsilon^{\alpha\beta\gamma}$ is the totally skew-symmetric Levi-Civita symbol with $\epsilon^{123} = 1$. Let $g \in \mathcal{G}_M$ be a gauge transformation regarded as a function from M to $SU(n)$ and define the 1-form θ by

$$\theta := g^{-1} dg = g^{-1} \partial_\mu g \, dx^\mu.$$

Then the gauge transformation A^g of A by g has (local) components

$$A^g_\mu = g^{-1} A_\mu g + g^{-1} \partial_\mu g, \qquad 1 \le \mu \le 3. \tag{11.16}$$

In the physics literature the connected component of the identity $\mathcal{G}_{id} \subset \mathcal{G}_M$ is called the group of small gauge transformations. A gauge transformation not belonging to $\mathcal{G}_{id}$ is called a large gauge transformation. By a direct calculation, one can show that the Chern–Simons action is invariant under small gauge transformations, i.e.,

$$\mathcal{A}_{CS}(A^g) = \mathcal{A}_{CS}(A), \qquad \forall g \in \mathcal{G}_{id}.$$

Under a large gauge transformation g the action (11.15) transforms as follows:

$$\mathcal{A}_{CS}(A^g) = \mathcal{A}_{CS}(A) + 2\pi k \mathcal{A}_{WZ}, \tag{11.17}$$

where

$$\mathcal{A}_{WZ} := \frac{1}{24\pi^2} \int_M \epsilon^{\alpha\beta\gamma} \operatorname{tr}(\theta_\alpha \theta_\beta \theta_\gamma) \tag{11.18}$$

is the Wess–Zumino action functional. It can be shown that the Wess–Zumino functional is integer-valued and hence, if the Chern–Simons coupling constant k is taken to be an integer, then we have

$$e^{i\mathcal{A}_{CS}(A^g)} = e^{i\mathcal{A}_{CS}(A)}.$$

The integer k is called the **level** of the corresponding Chern–Simons theory. The action enters the Feynman path integral in this exponential form. It follows that the path integral quantization of the Chern–Simons model is gauge-invariant. This conclusion holds more generally for any compact simple group G if the coupling constant $c(G)$ is chosen appropriately. The action is manifestly covariant since the integral involved in its definition is independent of the metric on M and this implies that the Chern–Simons theory is a topological field theory. It is this aspect of the Chern–Simons theory that plays a fundamental role in our study of knot and link invariants. For $k \in \mathbf{N}$, the transformation law (11.17) implies that the Chern–Simons action descends to the quotient $\mathcal{B}_M = \mathcal{A}_M / \mathcal{G}_M$ as a function with values in $\mathbf{R}/\mathbf{Z}$. $\mathcal{B}_M$ is called the moduli space of gauge equivalence classes of connections. We denote this function by f_{CS}, i.e.,

$$f_{CS} : \mathcal{B}_M \to \mathbf{R}/\mathbf{Z}$$

is defined by

$$[\omega] \mapsto \mathcal{A}_{CS}(\omega), \qquad \forall [\omega] = \omega \mathcal{G}_M \in \mathcal{B}_M. \tag{11.19}$$

The field equations of the Chern–Simons theory are obtained by setting the first variation of the action to zero as

$$\delta \mathcal{A}_{CS} = 0.$$

The field equations are given by

$$* F_\omega = 0 \quad \text{or, equivalently,} \quad F_\omega = 0. \tag{11.20}$$

The calculations leading to the field equations (11.20) also show that the gradient vector field of the function f_{CS} is given by

$$\text{grad } f_{CS} = \frac{1}{2\pi} * F. \tag{11.21}$$

The gradient flow of f_{CS} plays a fundamental role in the definition of Floer homology. A discussion of Floer homology and its extensions is given in Chapter 10, following the article [275]. The solutions of the field equations (11.20) are called the **Chern–Simons connections**. They are precisely the flat connections.

We take the state space of the Chern–Simons theory to be the moduli space of gauge potentials $\mathcal{B}_M$. The partition function Z_k of the theory is defined by

$$Z_k(\Phi) := \int_{\mathcal{B}_M} e^{-i\mathcal{A}_{CS}(\omega)} \Phi(\omega) \mathcal{D}\mathcal{A},$$

where $\Phi : \mathcal{A}_P \to \mathbf{R}$ is a quantum observable (i.e., a gauge-invariant function) of the theory and $\mathcal{A}_{CS}$ is defined by (11.14). Gauge invariance implies that Φ defines a function on $\mathcal{B}_M$, and we denote this function by the same letter. The expectation value $\langle \Phi \rangle_k$ of the observable Φ is given by

$$\langle \Phi \rangle_k := \frac{Z_k(\Phi)}{Z_k(1)} = \frac{\int_{\mathcal{B}_M} e^{-i\mathcal{A}_{CS}(\omega)} \Phi(\omega) \mathcal{D}\mathcal{A}}{\int_{\mathcal{B}_M} e^{-i\mathcal{A}_{CS}(\omega)} \mathcal{D}\mathcal{A}}.$$

If $Z_k(1)$ exists, it provides a numerical invariant of M. For example, for $M = S^3$ and $G = SU(2)$, using the action (11.14) Witten obtains the following expression for this partition function as a function of the level k:

$$Z_k(1) = \sqrt{\frac{2}{k+2}} \sin\left(\frac{\pi}{k+2}\right). \tag{11.22}$$

This result is a special case of the WRT invariants that we discussed in Chapter 10. These new invariants of 3-manifolds arose as a byproduct of Witten's TQFT interpretation of the Jones polynomial as shown next. Taking for Φ the Wilson loop functional $\mathcal{W}_{\rho,\kappa}$, where ρ is the fundamental representation of G and κ is the knot under consideration, leads to the following interpretation of the Jones polynomial (up to a phase factor):

$$\langle \Phi \rangle_k = V_\kappa(q), \quad \text{where } q = e^{2\pi i/(k+2)}.$$

For a framed link L, we denote by $\langle L \rangle$ the expectation value of the corresponding Wilson loop functional for the Chern–Simons theory of level k and gauge group $SU(n)$ and with ρ_i the fundamental representation for all i. To verify the defining relations for the Jones polynomial of a link L in S^3, Witten starts by considering the Wilson loop functionals for the associated links L_+, L_-, L_0 and uses TQFT with Chern–Simons Lagrangian to obtain the relation

$$\alpha \langle L_+ \rangle + \beta \langle L_0 \rangle + \gamma \langle L_- \rangle = 0, \tag{11.23}$$

where the coefficients α, β, γ are given by the expressions

$$\alpha = -\exp\left(\frac{2\pi i}{n(n+k)}\right), \tag{11.24}$$

$$\beta = -\exp\left(\frac{\pi i(2-n-n^2)}{n(n+k)}\right) + \exp\left(\frac{\pi i(2+n-n^2)}{n(n+k)}\right), \tag{11.25}$$

$$\gamma = \exp\left(\frac{2\pi i(1-n^2)}{n(n+k)}\right). \tag{11.26}$$

We note that the calculation of the coefficients α, β, γ is closely related to the Verlinde fusion rules [393] and $2d$ conformal field theories. Substituting the values of α, β, γ into equation (11.23) and canceling a common factor $\exp\left(\frac{\pi i(2-n^2)}{n(n+k)}\right)$, we get

$$-t^{n/2}\langle L_+ \rangle + (t^{1/2} - t^{-1/2})\langle L_0 \rangle + t^{-n/2}\langle L_- \rangle = 0, \tag{11.27}$$

where we have put

$$t = \exp\left(\frac{2\pi i}{n+k}\right).$$

The Laurent polynomial determined by the skein relation (11.27) is called the Jones–Witten polynomial for gauge group $SU(n)$. For $SU(2)$ Chern–Simons theory equation (11.27), under the transformation $\sqrt{t} \to -1/\sqrt{t}$, goes over into equation (11.8), which is the skein relation characterizing the Jones polynomial. The significance of this transformation is related to the choice of a square root for t. It is more transparent when the quantum group definition of the Jones polynomial is used. We note that recently the Alexander–Conway polynomial has also been obtained by the TQFT methods in [146].

If $V^{(n)}$ denotes the Jones–Witten polynomial corresponding to the skein relation (11.27), then the family of polynomials $\{V^{(n)}\}$ can be shown to be equivalent to the two variable HOMFLY polynomial $P(\alpha, z)$.

We remark that the vacuum expectation values of Wilson loop observables in the Chern–Simons theory have been computed recently up to second order of the inverse of the coupling constant. These calculations have provided a quantum field theoretic definition of certain invariants of knots and links in 3-manifolds [87, 178]. A geometric formulation of the quantization of Chern–Simons theory is given in [25]. Another important approach to link invariants is via solutions of the Yang–Baxter equations and representations of the corresponding quantum groups (see, for example, [221, 223, 294, 317, 326, 380, 383]). For relations between link invariants, conformal field theories and 3-dimensional topology see, for example, [88, 230, 328].

11.4 Vassiliev Invariants of Singular Knots

We define a **singular knot** s as a knot with at most a finite number n of double point singularities. Each such singularity can be resolved in two ways by deforming one of the two crossing threads to an overcrossing or an undercrossing. We fix a double point p and use orientation to define a knot s_+ (resp., s_-) by overcrossing the first (resp., second) time the path passes the point p. This gives us two singular knots with $n-1$ singular points. Let v denote a knot invariant of a regular (i.e., non-singular) knot. We define a Vassiliev invariant v of the singular knot s with n singular points inductively by the formula

$$v(s) := v(s_+) - v(s_-) . \qquad (11.28)$$

This definition allows us to express v as a finite sum of the invariant v on a set of regular knots obtained by the complete resolution of the singular knot s into regular knots. The definition can be extended to oriented links by applying it to each component knot.

Definition 11.2 *We say that a Vassiliev invariant v is of order at most n if v is zero on any singular knot with more than n double points. The **order** of a Vassiliev invariant v denoted by $\mathrm{ord}(v)$ is the largest natural number m such that v is not zero on some singular knot with exactly m double points.*

The existence of nonzero Vassiliev invariants of a given finite order was established in [391]. In fact, knot polynomials such as the Jones polynomial lead to Vassiliev invariants as follows:

Theorem 11.6 *Let $V_K(q)$ be the Jones polynomial of the knot K. Put $q = e^x$ and expand V_K as a power series in x:*

$$V_K(e^x) = \sum_{n=0}^{\infty} v_n x^n.$$

Then the coefficient v_n of x^n in the series expansion is a Vassiliev invariant of order n.

The set V of Vassiliev invariants has a natural structure of an infinite-dimensional real vector space. It is filtered by a sequence of finite-dimensional vector spaces $V_n := \{v \in V \mid \mathrm{ord}(v) \le n\}$, $n \ge 0$. That is

$$V = \bigoplus V_n \quad \text{and} \quad V_0 \subset V_1 \subset V_2 \cdots .$$

We also have the following theorem due to Kontsevich.

Theorem 11.7 *The set V_n is non-empty and forms a real vector space. Furthermore the quotient space V_n/V_{n-1} is isomorphic to the space of linear functions on the n-chord diagrams modulo certain relations (one and four term relations).*

There are several open questions about Vassiliev invariants. For example: What can one say about $d_n = \dim V_n$ and the quantum dimension $\dim_q V = \sum d_n q^n$ of the real vector spaces V?

11.5 Self-linking Invariants of Knots

As we indicated in Chapter 7, one of the earliest investigations in combinatorial knot theory is contained in several unpublished notes written by Gauss between 1825 and 1844 and published posthumously. In obtaining a topological invariant by using a physical field theory, Gauss had anticipated topological field theory by almost 150 years. Even the term "topology" was not used then. Gauss's linking number formula can also be interpreted as the equality of topological and analytic degree of a suitable function. Starting with this a far-reaching generalization of the Gauss integral to higher self-linking integrals was obtained by Bott, Taubes, and Cattaneo. We now discuss their work.

Let us recall the relevant definitions. Let X, Y be two closed oriented n-manifolds. Let $q \in Y$ be a regular value of a smooth function $f : X \to Y$. Then $f^{-1}(q)$ has finitely many points $p_1, p_2, \ldots, p_j$. For each i, $1 \le i \le j$, define $\sigma_i = 1$ (resp., $\sigma_i = -1$) if the differential $Df : TX \to TY$ restricted to the tangent space at p_i is orientation-preserving (resp., reversing). Then the differential topological definition of the **mapping degree**, or simply **degree**, of f is given by

$$\deg(f) := \sum_{i=1}^{j} \sigma_i. \tag{11.29}$$

For the analytic definition we choose a volume form v on Y and define

$$\deg(f) := \frac{\int_X f^* v}{\int_Y v}. \tag{11.30}$$

In the analytic definition one often takes a normalized volume form so that $\int v = 1$. This gives a simpler formula for the degree. It follows from the well known de Rham's theorem that the topological and analytic definitions give the same result. To apply this result to deduce the Gauss formula, denote the two curves by C, C'. Then the map

$$\lambda : C \times C' \to S^2$$

defined by

$$\lambda(\mathbf{r}, \mathbf{r}') := \frac{(\mathbf{r} - \mathbf{r}')}{|\mathbf{r} - \mathbf{r}'|}, \qquad \forall (\mathbf{r}, \mathbf{r}') \in C \times C'$$

is well-defined by the disjointness of C and C'. If ω denotes the standard volume form on S^2, then we have

$$\omega = \frac{x\,dy \wedge dz + y\,dz \wedge dx + z\,dx \wedge dy}{(x^2 + y^2 + z^2)^{3/2}}.$$

The pull-back $\lambda^*(\omega)$ of ω to $C \times C'$ is precisely the integrand in the Gauss formula and $\int \omega = 4\pi$. It is easy to check that the topological degree of λ equals the linking number m. Let us define the **Gauss form** ϕ on $C \times C'$ by

$$\phi := \frac{1}{4\pi} \lambda^*(\omega).$$

Then the Gauss formula for the linking number can be rewritten as

$$\int \phi = m.$$

Now the map λ is easily seen to extend to the 6-dimensional space $C_2^0(\mathbf{R}^3)$ defined by

$$C_2^0(\mathbf{R}^3) := \mathbf{R}^3 \times \mathbf{R}^3 \setminus \{(x, x) \mid x \in \mathbf{R}^3\} = \{(x_1, x_2) \in \mathbf{R}^3 \times \mathbf{R}^3 \mid x_1 \neq x_2\}.$$

The space $C_2^0(\mathbf{R}^3)$ is called the **configuration space** of two distinct points in $\mathbf{R}^3$. Denoting by λ_{12} the extension of λ to the configuration space we can define the Gauss form ϕ_{12} on the space $C_2^0(\mathbf{R}^3)$ by

$$\phi_{12} := \frac{1}{4\pi} \lambda_{12}^*(\omega). \tag{11.31}$$

The definition of the space $C_2^0(\mathbf{R}^3)$ extends naturally to define $C_n^0(X)$, the configuration space of n distinct points in the manifold X as follows:

$$C_n^0(X) := \{(x_1, x_2, \ldots, x_n) \in X^n \mid x_i \neq x_j \text{ for } i \neq j, 1 \leq i, j \leq n\}.$$

In [149] it is shown how to obtain a functorial compactification $C_n(X)$ of the configuration space $C_n^0(X)$ in the algebraic geometry setting. A detailed account of the geometry and topology of configuration spaces of points on a manifold with special attention to $\mathbf{R}^n$ and S^n is given by Fadell and Husseini in [121]. In his lecture at the Geometry and Physics Workshop at MSRI in Berkeley (January 1994), Bott explained how the configuration spaces enter in the study of embedding problems and, in particular, in the calculation of embedding invariants. Let $f : X \hookrightarrow Y$ be an embedding. Then f induces embeddings of Cartesian products $f^n : X^n \hookrightarrow Y^n, n \in \mathbf{N}$. The maps f^n give embeddings of configuration spaces $C_n^0(X) \hookrightarrow C_n^0(Y), n \in \mathbf{N}$ by restriction. These maps in turn extend to the compactifications giving a family of maps

$$C_n^f : C_n(X) \to C_n(Y), \qquad n \in \mathbf{N}.$$

It is these maps C_n^f that play a fundamental role in the study of embedding invariants. As we have seen above, the Gauss formula for the linking number is an example of such a calculation. These ideas are used in [54] to obtain self-linking invariants of knots. Configuration spaces have also been used in the work of Bott and Cattaneo [52, 53] to study integral invariants of 3-manifolds. The first step is to observe that the λ_{12} defined in (11.31) can be defined on any two factors in the configuration space $C_n^0(\mathbf{R}^3)$ to obtain a family of maps λ_{ij} and these in turn can be used to define the Gauss forms $\phi_{ij}, i \neq j, 1 \leq i, j \leq n$,

$$\phi_{ij} := \frac{1}{4\pi} \lambda_{ij}^*(\omega), \quad \text{where } \lambda_{ij}(x_1, x_2, \ldots, x_n) := \frac{x_i - x_j}{|x_i - x_j|} \in S^2. \quad (11.32)$$

Let K_f denote the parametrized knot

$$f : S^1 \to \mathbf{R}^3 \text{ with } \left|\frac{df}{dt}\right| = 1, \qquad \forall t \in S^1.$$

Then we can use C_n^f to pull back forms ϕ_{ij} to $C_n^0(S^1)$ as well as to the spaces $C_{n,m}^0(\mathbf{R}^3)$ of $n + m$ distinct points in $\mathbf{R}^3$ of which only the first n are on S^1. These forms extend to the compactifications of the respective spaces and we continue to denote them by the same symbols. Integrals of forms obtained by products of the ϕ_{ij} over suitable spaces are called the self-linking integrals. In the physics literature self-linking integrals and invariants for the case $n = 4$ have appeared in the study of perturbative aspects of the Chern–Simons field theory in [30, 31, 177, 178]. A detailed study of the Chern–Simons perturbation theory from a geometric and topological point of view may be found in [25, 24]. The self-linking invariant for $n = 4$ can be obtained by using the Gauss forms ϕ_{ij} as follows: Let $\mathcal{K}$ denote the space of all parametrized knots. Then the Gauss forms pull back to the product $\mathcal{K} \times C_4(S^1)$, which fibers over $\mathcal{K}$ by the projection π_1 on the first factor. Let α denote the result of integrating the 4-form $\phi_{13} \wedge \phi_{24}$ along the fibers of π_1. While α is well-defined, it is not locally constant (i.e., $d\alpha \neq 0$) and hence does not define a knot invariant. The necessary correction term β is obtained by integrating the 6-form $\phi_{14} \wedge \phi_{24} \wedge \phi_{34}$ over the space $C_{3,1}(\mathbf{R}^3)$. In [54] it is shown that $\alpha/4 - \beta/3$ is locally constant on $\mathcal{K}$ and hence defines a knot invariant. It turns out that this invariant belongs to a family of knot invariants, called **finite-type invariants**, defined by Vassiliev [391], which we discussed in Section 11.4 (see also [13]). In [235] Gauss forms with different normalization are used in the formula for this invariant and it is stated that the invariant is an integer equal to the second coefficient of the Alexander–Conway polynomial of the knot. Kontsevich views the self-linking invariant formula as forming a small part of a very broad program to relate the invariants of low-dimensional manifolds, homotopical algebras, and non-commutative geometry with topological field theories and the calculus of Feynman diagrams. It seems that the full realization of this program will require the best efforts of mathematicians and physicists in the new millennium.

11.6 Categorification of the Jones Polynomial

As we discussed earlier, the discovery of the Jones polynomial invariant of links renewed interest and greatly increased research activity towards finding new invariants of links. Witten's work gave a new interpretation in terms of topological quantum field theory and in the process led to new invariants of 3-manifolds. Reshetikhin and Turaev gave a precise mathematical definition of these invariants (now called WRT invariants) in terms of representations of the quantum group $\mathfrak{sl}_q(2, \mathbf{C})$ (a Hopf algebra deformation of the Lie algebra $\mathfrak{sl}(2, \mathbf{C})$). Quantum groups were discovered independently by Drinfeld and Jimbo. By the early 1990s a number of invariants were constructed starting with the pair $(L, \mathfrak{g})$, where L is a link with components colored by representations of the complex simple Lie algebra $\mathfrak{g}$. Many (but not all) of these invariants are expressible as Laurent polynomials in a formal variable q. These polynomial invariants have representation-theoretic interpretation in terms of intertwiners between tensor products of irreducible representations of the quantum group $U_q(\mathfrak{g})$ (a Hopf algebra deformation of the universal enveloping algebra of $\mathfrak{g}$). These invariants are often referred to as quantum invariants of links and 3-manifolds. They form part of a new (rather loosely defined) area of mathematics called **quantum topology**.

In modern mathematics the language of category theory is often used to discuss properties of different mathematical structures in a unified way. In recent years category theory and categorical constructions have found applications in other branches of mathematics and also in theoretical physics. This has developed into an extensive area of research. In fact, as we siaw in Chapter 7, the axiomatic formulation of TQFT is given by the use of cobordism categories. We will use a special case of it in our discussion of Khovanov homology.

We begin by recalling that a categorification of an invariant I is the construction of a suitable (co)homology H^* such that its Euler characteristic $\chi(H^*)$ (the alternating sum of the ranks of (co)homology groups) equals I. Historically, the Euler characteristic was defined and understood well before the advent of algebraic topology. *Theorema egregium* of Gauss and the closely related Gauss–Bonnet theorem and its generalization by Chern give a geometric interpretation of the Euler characteristic $\chi(M)$ of a manifold M. They can be regarded as precursors of Chern–Weil theory as well as index theory. Categorification $\chi(H^*(M))$ of this Euler characteristic $\chi(M)$ by various (co)homology theories $H^*(M)$ came much later. A well-known recent example that we have discussed is the categorification of the Casson invariant by the Fukaya–Floer homology. Categorification of quantum invariants such as knot polynomials requires the use of quantum Euler characteristic and multi-graded knot homologies. Khovanov [219] has obtained a categorification of the Jones polynomial $V_L(q)$ by constructing a bi-graded $sl(2)$-homology $H_{i,j}$ determined by the link L. It is called the **Khovanov homology** of the link L and is denoted by $KH(L)$. The **Khovanov polynomial** $Kh_L(t, q)$ is defined by

$$Kh_L(t, q) = \sum_{i,j} t^j q^i \dim H_{i,j}. \qquad (11.33)$$

It can be thought of as a two-variable generalization of the Poincaré polynomial. The quantum or graded Euler characteristic of the Khovanov homology equals the non-normalized Jones polynomial. That is,

$$\hat{V}_L(q) = \chi_q(KH(L)) = \sum_{i,j} (-1)^j q^i \dim H_{i,j}.$$

Khovanov's construction follows Kauffman's state-sum model of the link L and his alternative definition of the Jones polynomial. Let $\hat{L}$ be a regular projection of L with $n = n_+ + n_-$ labeled crossings. At each crossing we can define two resolutions or states, the vertical or 1-state and horizontal or 0-state. Thus, there are 2^n total resolutions of $\hat{L}$ which can be put into a one-to-one correspondence with the vertices of an n-dimensional unit cube. For each vertex x let $|x|$ be the sum of its coordinates and let $c(x)$ be the number of disjoint circles in the resolution $\hat{L}_x$ of $\hat{L}$ determined by x. Kauffman's state-sum expression for the non-normalized Jones polynomial $\hat{V}(L)$ can be written as follows:

$$\hat{V}_L(q) = (-1)^{n_-} q^{(n_+ - 2n_-)} \sum (-q)^{|x|} (q + q^{-1})^{c(x)}, \qquad (11.34)$$

where the sum is taken over all the vertices x of the cube. Dividing this by the unknot value $(q + q^{-1})$ gives the usual normalized Jones polynomial $V(L)$. The Khovanov complex can be constructed using 2-dimensional TQFT and related Frobenius algebra as follows: Let V be a graded vector space over a fixed ground field K, generated by two basis vectors $v_\pm$ with respective degrees ± 1. The total resolution associates to each vertex x a 1-dimensional manifold M_x consisting of $c(x)$ disjoint circles. We can construct a $(1 + 1)$-dimensional TQFT (along the lines of Atiyah–Segal axioms discussed in Chapter 7) by associating to each edge of the cube a cobordism as follows: If xy is an edge of the cube we can get a pair of pants cobordism from M_x to M_y by noting that a circle at x can split into two at y or two circles at x can fuse into one at y. If a circle goes to a circle then the cylinder provides the cobordism. To the manifold M_x at each vertex x we associate the graded vector space

$$V_x(L) := V^{\otimes c(x)}\{|x|\}, \tag{11.35}$$

where $\{k\}$ is the degree shift by k. We define the Frobenius structure (see the book [227] by Kock for Frobenius algebras and their relation to TQFT) on V as follows. Multiplication $m : V \otimes V \to V$ is defined by

$$m(v_+ \otimes v_+) = v_+, \qquad m(v_+ \otimes v_-) = v_-,$$

$$m(v_- \otimes v_+) = v_-, \qquad m(v_- \otimes v_-) = 0.$$

Co-multiplication $\Delta : V \to V \otimes V$ is defined by

$$\Delta(v_+) = v_+ \otimes v_- + v_- \otimes v_+, \qquad \Delta(v_-) = v_- \otimes v_-.$$

Thus, v_+ is the unit. The co-unit $\delta \in V^*$ is defined by mapping v_+ to 0 and v_- to 1 in the base field. The rth chain group $C_r(L)$ in the Khovanov complex is the direct sum of all vector spaces $V_x(L)$, where $|x| = r$, and the differential is defined by the Frobenius structure. Thus,

$$C_r(L) := \oplus_{|x|=r} V_x(L). \tag{11.36}$$

We remark that the TQFT used here corresponds to the Frobenius algebra structure on V defined above. The rth homology group of the Khovanov complex is denoted by KH_r. Khovanov proved that the homology is independent of the various choices made in defining it. Thus, we have the next theorem.

Theorem 11.8 *The homology groups KH_r are link invariants. In particular, the Khovanov polynomial*

$$Kh_L(t,q) = \sum_j t^j \dim_q(KH_j)$$

is a link invariant that specializes to the non-normalized Jones polynomial. The Khovanov polynomial is strictly stronger than the Jones polynomial.

We note that the knots 9_{42} and 10_{125} are chiral. Their chirality is detected by the Khovanov polynomial but not by the Jones polynomial. Also, there are several pairs of knots with the same Jones polynomials but different Khovanov polynomials. For example $(5_1, 10_{132})$ is such a pair. Using equations (11.35) and (11.36) and the algebra structure on V, we can reduce the calculation of the Khovanov complex to an algorithm. A computer program implementing such an algorithm is discussed in [32]. A table of Khovanov polynomials for knots and links up to 11 crossings is also given there. We now illustrate Khovanov's categorification of the Jones polynomial of the right-handed trefoil knot 3_1.

Example 11.1 (Categorification of $V(3_1)$) *For the standard diagram of the trefoil 3_1, we have $n = n_+ = 3$ and $n_- = 0$. The quantum dimensions of the non-zero terms of the Khovanov complex with the shift factor included are given by*

$$C_0 = (q + q^{-1})^2, \ C_1 = 3q(q + q^{-1}), \ C_2 = 3q^2(q + q^{-1})^2, \ C_3 = q^3(q + q^{-1})^3,$$

and $C_i = 0$ for all other $i \in \mathbf{Z}$. The non-normalized Jones polynomial can be obtained by using the above chain complex (with differential induced by the Frobenius algebra defined earlier) or directly from (11.34) giving

$$\hat{V}_L(q) = (q + q^3 + q^5 - q^9). \tag{11.37}$$

The normalized or standard Jones polynomial is then given by

$$V_L(q) = \frac{q + q^3 + q^5 - q^9}{q + q^{-1}} = q^2 + q^6 - q^8.$$

By direct computation or using the program in [32] we obtain the following formula for the Khovanov polynomial of the trefoil

$$Kh(t, q) = q + q^3 + t^2 q^5 + t^3 q^9, \quad Kh(-1, q) = \chi_q = \hat{V}_L(q).$$

Based on computations using the program described in [32], Khovanov, Garoufalidis, and Bar-Natan (BKG) have formulated some conjectures on the structure of Khovanov polynomials over different base fields. We now state these conjectures.

The BKG Conjectures: For any prime knot κ there exists an even integer $s = s(\kappa)$ and a polynomial $Kh'_\kappa(t, q)$ with only non-negative coefficients such that

1. over the base field $K = \mathbf{Q}$,

$$Kh_\kappa(t, q) = q^{s-1}[1 + q^2 + (1 + tq^4)Kh'_\kappa(t, q)]$$

2. over the base field $K = \mathbf{Z}_2$,

$$Kh_\kappa(t, q) = q^{s-1}(1 + q^2)[1 + (1 + tq^2)Kh'_\kappa(t, q)]$$

3. moreover, if the knot κ is alternating, then $s(\kappa)$ is the signature of the knot and $Kh'_\kappa(t, q)$ contains only powers of tq^2.

The conjectured results are in agreement with all the known values of the Khovanov polynomials.

If $S \subset \mathbf{R}^4$ is an oriented surface cobordism between links L_1 and L_2 then it induces a homomorphism of Khovanov homologies of links L_1 and L_2. These homomorphisms define a functor from the category of link cobordisms to the category of bi-graded Abelian groups [205]. Khovanov homology extends to colored links (i.e., oriented links with components labeled by irreducible finite-dimensional representations of $sl(2)$) to give a categorification of the colored Jones polynomial. Khovanov and Rozansky have defined an $sl(n)$-homology for links colored by either the defining representation or its dual. This gives categorification of the specialization of the HOMFLY polynomial $P(\alpha, q)$ with $a = q^n$. The sequence of such specializations for $n \in \mathbf{N}$ would categorify the two variable HOMFLY polynomial $P(\alpha, q)$. For $n = 0$ the theory coincides with the Heegaard–Floer homology of Ozsváth and Szabo [309].

In the 1990s Reshetikhin, Turaev, and other mathematicians obtained several quantum invariants of triples $(\mathfrak{g}, L, M)$, where $\mathfrak{g}$ is a simple Lie algebra, $L \subset M$ is an oriented, framed link with components labeled by irreducible representations of $\mathfrak{g}$, and M is a 2-framed 3-manifold. In particular, there are polynomial invariants $\langle L \rangle$ that take values in $\mathbf{Z}[q^{-1}, q]$. Khovanov has conjectured that at least for some classes of Lie algebras (e.g., simply laced) there exists a bi-graded homology theory of labeled links such that the polynomial invariant $< L >$ is the quantum Euler characteristic of this homology. It should define a functor from the category of framed link cobordisms to the category of bigraded Abelian groups. In particular, the homology of the unknot labeled by an irreducible representation U of $\mathfrak{g}$ should be a Frobenius algebra of $\dim(U)$.

Epilogue

It is well known that the roots of "physical mathematics" go back to the very beginning of human attempts to understand nature. The abstraction of observations in the motion of heavenly bodies led to the early developments in mathematics. Indeed, mathematics was an integral part of natural philosophy. Rapid growth of the physical sciences aided by technological progress and increasing abstraction in mathematical research caused a separation of the sciences and mathematics in the twentieth century. Physicists' methods were often rejected by mathematicians as imprecise, and mathematicians' approaches to physical theories were not understood by physicists. We have already given many examples of this. However, theoretical physics did influence development of some areas of mathematics. Two fundamental physical theories, relativity and quantum theory, now over a century old, sustained interest in geometry and functional analysis and group theory. Yang–Mills theory, now over half a century old, was abandoned for many years before its relation to the theory of connections in a fiber bundle was found. It has paid rich dividends to the geometric topology of low-dimensional manifolds in the last quarter century. Secondary characteristic classes were given less than secondary attention when they were introduced. These classes turn out to be the starting point of Chern–Simons theory.

A major conference celebrating 20 years of Chern–Simons theory organized by the Max Planck and the Hausdorff institutes in Bonn was held in August 2009. It drew research workers in mathematics and theoretical physics from around the world. The work presented under the umbrella of this single topic shows that the 30 year old marriage between theoretical physics and mathematics is still going strong. Many areas such as statistical mechanics, conformal field theory, and string theory not included in this work have already led to new developments in mathematics.

The scope of physical mathematics continues to expand rapidly. Even for the topics that we have considered in this book a number of new results are appearing and new connections between old results are emerging. In fact, the recent lecture[1] by Curtis McMullen (Fields Medal, ICM 1998, Berlin) was entitled "From Platonic Solids to Quantum Topology." McMullen weaves a fascinating tale from ancient to modern mathematics pointing out unexpected links between various areas of mathematics and theoretical physics. He concludes with the statement of a special case of the volume conjecture, interpreting it as the equality between a gauge-theoretic invariant and a topological gravity invariant.

In view of all this activity we might liken this book with a modern tour through many countries in a few days. When you return home you can look at the pictures, think of what parts you enjoyed and then decide where you would like to spend more time. I hope that readers found several parts enjoyable and perhaps some that they may want to explore further. The vast and exciting landscape of physical mathematics is open for exploration.

We began this book with a poem dedicated to the memory of my mother and we conclude with a poem which touches upon some of the great discoveries that have shaped our understanding of nature.

In the beginning

God said:
Let there be gauge theories!

And there was light,
clear and bright
followed by particles
fundamental or otherwise.

They jostled along merrily,
sometimes weakly, sometimes strongly,
creating and annihilating things
small and big.

They hurtled along paths
as straight as could be.
And it was called
the miracle gravity.

Three fundamental forces
united in the standard model
now wait patiently to be
tied up with the fourth.

---

[1] The 2009 Reimar Lüst lecture in Bonn delivered on June 12.

Appendix A
Correlation of Terminology

As we remarked earlier, gauge theories and the theory of connections were developed independently by physicists and mathematicians, and as such have no standard notation. This is also true of other theories. To help the reader we present a table describing the correlation of terminology between physics and mathematics, prepared along the lines of Trautman [378] and [409].

Physics	Mathematics
Space-time	Lorentz 4-manifold M
Euclidean space-time	Riemannian 4-manifold M
Gauge group G	Structure group of a principal bundle $P(M, G)$ over M
Space of phase factors	Total space of the bundle
Gauge group bundle	$\mathrm{Ad}(P) = P \times_{\mathrm{Ad}} G$, where Ad is the adjoint action of G on itself
Gauge transformation	A section of the bundle $\mathrm{Ad}(P)$
Gauge algebra bundle	$\mathrm{ad}(P) = P \times_{\mathrm{ad}} \mathbf{g}$, where ad is the adjoint action of G on $\mathbf{g}$
Infinitesimal gauge transformation	A section of the bundle $\mathrm{ad}(P)$
Gauge potential ω on P	Connection 1-form $\omega \in \Lambda^1(P, \mathbf{g})$
Global gauge s	A section of the bundle $P(M, G)$
Gauge potential A_s on M	$A = s^*(\omega)$
Local gauge t	A section of the bundle $P(M, G)$, over an open set $U \subset M$.
Local gauge potential A_t	$A_t = t^*(\omega)$
Gauge field Ω on P	Curvature $d^\omega \omega = \Omega \in \Lambda^2(P, \mathbf{g})$
Gauge field F_ω on M	The 2-form $F_\omega \in \Lambda^2(M, \mathrm{ad}(P))$ associated to Ω
Group of projectable transformations	$\mathrm{Diff}_M(P) = \{f \in \mathrm{Diff}(P) \mid f \text{ covers } f_M \in \mathrm{Diff}(M)\}$

Correlation of Terminology (continued)

Physics	Mathematics
Group of generalized gauge transformations	$\text{Aut}(P) = \{f \in \text{Diff}(P) \mid f$ is G-equivariant$\}$
Group $\mathcal{G}$ of gauge transformations	$\text{Aut}_0(P) = \{f \in \text{Aut}(P) \mid f_M = id \in \text{Diff}(M)\}$
Group $\mathcal{G}_0 \subset \mathcal{G}$, of based gauge transformations	Subgroup of $\text{Aut}_0(P)$ of based bundle automorphisms
Group $\mathcal{G}_c$, of effective gauge transformations	The quotient group of $\mathcal{G}$ by its center $Z(\mathcal{G})$
Gauge algebra $\mathcal{LG}$	Lie algebra $\Gamma(\text{ad}\, P)$
Generalized Higgs field ϕ_r	A section of the associated bundle $E(M, F, r, P)$
Higgs field ϕ	ϕ_r, with r the fundamental representation
Bianchi identities for the gauge field F_ω	$d^\omega F_\omega = 0$
Yang–Mills equations for F_ω	$\delta^\omega F_\omega = 0$
Instanton (resp., anti-instanton) equations on a 4-manifold M	$*F_\omega = \pm F_\omega$, where $*$ is the Hodge operator
Instanton number of $P(M, G)$	The Chern class $c_1(P)$
BPST instanton	Canonical $SU(2)$-connection on the quaternionic Hopf fibration of S^7 over S^4
Dirac monopole	Canonical $U(1)$-connection on the Hopf fibration of S^3
Inertial frames on a space-time manifold (M, g)	The bundle $O(M, g)$ of orthonormal frames on M
Gravitational potential	Levi-Civita connection λ on $O(M, g)$
Gravitational field	The curvature R^λ of λ
Gravitational instanton equations	$[R^\lambda, *] = 0$, where $*$ is the Hodge operator
Gravitational instanton	Einstein space

Appendix B
Background Notes

This appendix contains material that might have hindered the exposition of various arguments in the text. It contains some historical observations which I found interesting and which throw additional light on the origin of concepts, nomenclature, and notation. It also includes biographical notes on some of the scientists whose work has influenced me over the years or are mentioned in the main body of the text.

1. **Bernoulli numbers**

   Bernoulli numbers B_n for $n \geq 0$ are defined by the formal power series

   $$\frac{x}{e^x - 1} = \sum_{n=0}^{\infty} \frac{B_n x^n}{n!}.$$

   Expanding the power series on the left and comparing its coefficients with those on the right, we get the Bernoulli numbers. Note that $B_n = 0$ for odd $n > 1$.

   $$B_0 = 1, \ B_1 = -\frac{1}{2}, \ B_2 = \frac{1}{6}, \ B_4 = -\frac{1}{30}, \ B_6 = \frac{1}{42}, \ B_8 = -\frac{1}{30},$$

   $$B_{10} = \frac{5}{66}, \ B_{12} = -\frac{691}{2730}, \ B_{14} = \frac{7}{6}, \ B_{16} = -\frac{3617}{510}, \ B_{18} = \frac{43{,}867}{798}.$$

   Bernoulli numbers satisfy the following recurrence relation

   $$B_0 = 1, \ B_n = -\frac{1}{n+1} \sum_{k=0}^{n-1} \binom{n+1}{k} B_k.$$

   This recurrence relation provides a simple but slow algorithm for computing B_n. Bernoulli numbers seem to attract both amateur and professional mathematician. Srinavasa Ramanujan's first paper was on Bernoulli numbers. He proved several interesting results but also made a conjecture that turned out to be false. For more information about the Bernoulli numbers, see the website (www.bernoulli.org).

An important property of the Bernoulli numbers is their appearence in the values of Euler's zeta function for even natural numbers $2n, n \in \mathbf{N}$. Euler defined and proved a product formula for the zeta function

$$\zeta(s) := \sum_{m=1}^{\infty} m^{-s} = \prod_{p}(1 - p^{-s}), \qquad s > 1,$$

where the product is over all prime numbers p. He evaluated the zeta function for all even numbers to obtain the following formula:

$$\zeta(2n) = -\frac{(2\pi i)^{2n}}{2(2n)!}B_{2n}.$$

He also computed approximate values of the zeta function at odd numbers. There is no known formula for these values similar to the formula for even numbers. Even the arithmetic nature of $\zeta(3)$ was unknown until Apery's proof of its irrationality. The nature of other zeta values is unknown.

Riemann generalized Euler's definition to complex values of the variable s. He proved that this zeta function can be extended to the entire complex plane as a meromorphic function and obtained a functional equation for it. The extended zeta function has zeros at negative even integers. These zeros are called the trivial zeros. Riemann conjectured that all the non-trivial zeros of the zeta function have real part $1/2$. This conjecture is known as the **Riemann hypothesis**. It is one of the open problems on the Clay Prize list. For a complete list of Clay mathematics prize problems, see the website (www.claymath.org).

2. **Bourbaki**

The first world war had a devastating effect on science and mathematics in France. Jean Dieudonné told me that he and his best friend Henri Cartan followed the course on differential geometry by Elie Cartan, but they did not understand much of what he was doing. They turned to their senior friend André Weil for help. Even so Dieudonné decided to opt for taking the examination in synthetic geometry. Graduates had large gaps in their knowledge. After graduation the friends were scatterred around France. Weil and H. Cartan were teaching calculus in Strasbourg using E. Goursat's well known text. Cartan turned to Weil for advice on the course so frequently that Weil decided it was time to write a new treatise for the course. Weil called some of his old friends (Jean Delsarte, Dieudonné, and Claude Chevalley) to join him and H. Cartan with the simple idea to write a new text to replace the text by Goursat.

This is the story of the birth of Bourbaki. The young men eager to finish this simple project had no idea what they were getting into. After a few meetings they realized that they had to start from scratch and present all the essentials of mathematics then known. They believed that they would have the first draft of this work in three years. They met for their first congress in 1935. Eventually, they chose to use the name "Nicholas Bourbaki" for their group, an invented name. It went on to become the most famous group of mathematicians in the history of mathematics. Bourbaki's founders were among the greatest mathematicians of the twentieth century. Their work changed the way mathemtics was done and presented. Their active participation in the Bourbaki group ended when a member turned 50 (this was the only rule of Bourbaki). The first chapter of Bourbaki's nine volumes (consisting of 40 books) came out four years later, rather than three. The last volume "Spectral Theory" was published in 1983. For a long time the membership and the work of the Bourbaki group remained a well guarded secret. Jean Dieudonné broke the silence in his article [100]. All the members were called upon to write various drafts of the chapters but the final version for printing was prepared by Dieudonné. (This explains the uniformity of the style through most of the forty books written by Bourbaki.) Over the years membership in the Bourbaki group changed to include some of the most influential mathematicians of the twentieth century such as A. Grothendieck, S. Lang, J. P. Serre, and L. Schwartz.

The rapid progress in the physical sciences and increasing abstraction in mathematics caused an almost complete separation of physics and mathematics. Even the mathematics used in theoretical physics did not have the rigor required in modern mathematics. Extreme generality in Bourbaki infurianted many mathematicians. For example, nobody had ever studied Euclidean geometry as a special case of the theory of Hermitian operators in Hilbert space even though Dieudonné says in [100] that this is well known. In his article [99] on the development of modern mathematics Dieudonné was stressing the fact that mathematics in the second half of the twentieth century had become a self sustaining field of knowledge, not depending on applications to other sciences. Thus, while the development of mathematics since antiquity to the first half of the twentieth century was strongly influenced by developments in the physical sciences, a new chapter in the history of mathematics had now begun.

As we pointed out in the preface, this article is often quoted to show that the abstraction stressed by Bourbaki was the major cause of the split between mathematics and the physical sciences. In fact, in the same article Dieudonné clearly stated that a dialogue with other sciences, such as theoretical physics, may be beneficial to all parties. He did not live to see such a dialogue or to observe that any dialogue has been far more beneficial to the mathematics in the last quarter century. I would like to add that in 2001 a physics seminar "Le Séminaire Poincaré" modeled after the well-known "Séminaire Bourbaki" was created by the Association des Collaborateurs de Nicolas Bourbaki.

I had several long discussions with Prof. Dieudonne. Most of the time, I asked the questions and he never tired of giving detailed answers. He was an embodiment of history. He spoke many languages fluently and had phenomenal memory. He enjoyed Indian food and was a frequent dinner guest at our home. This created a rather informal setting for our discussions. Early on I asked him about a statement in the preface to the first volume of Bourbaki. It said that no background in university mathematics was necessary to read these books. I asked him if there was any example of this. He immediately replied in the affirmative and gave the following account. There was a high school student in Belgium whom the mathematics teacher found quite a handful. So he gave him Bourbaki's book on set theory to read over the holidays. When the school reopened the student reported that he enjoyed the book very much but found out that he could prove one of the stated axioms from the others. The teacher was totally unprepared for this and decided to send a letter to Bourbaki with his student's proof. Dieudonné (the unofficial secretary of Bourbaki) knew that the group had not started from a minimal set of axioms. He checked and found the proof correct. The student's name is **Pierre Degline** (b. 1944, Fields medal, ICM 1978, Helsinki) who is now a professor of mathematics at IAS (the Institute for Advanced Study), Princeton. I came to know Prof. Degline during my visits to IAS, especially during the special year on QFT and string theory. During a social evening, I finally found the courage to ask him about the above story. He was very surprised and asked me how I knew it. I explained that I had heard it from Prof. Dieudonné during one of his visits to my home. To this day I have never heard of any other person who started to read Bourbaki before entering the university.

3. Fields Medals

It is well known that there is no Nobel Prize in mathematics. There are several stories about why this queen of sciences was omitted, most of them unsupported by evidence. Many mathematicians wanted an internationally recognized prize comparable in importance to the highly regarded Nobel Prize in the sciences and other areas. Canadian mathematician John Charles Fields actively pursued this idea as chairman of the committee set up for the purpose of organizing the ICM 1924 in Toronto. The idea of awarding a medal was well supported by several countries. The Fields' committee prepared an outline of principles for awarding the medals. The current rule that the awardee not be more than 40 years old at the time the award is granted stems from the principle that the award was to be in recognition of work already done, as well as to encourage further achievement on the part of the recipient. Another statement indicated that the medals should be as purely international and impersonal as possible and not be attached in any way to the name of any country, institution, or person.

Fields died before the opening of the ICM 1932 in Zürich opened, but the proposal was accepted by the congress. The medal became known as the Fields Medal (against Fields's wishes) since it was awarded for the first time at ICM 1936 in Oslo to Lars Ahlfors and Jesse Douglas. On one side of the medal is engraved a laurel branch and a diagram of a sphere contained in a cylinder from an engraving thought to have been on Archimedes' tomb. The Latin inscription on it may be translated as "Mathematicians, having congregated from the whole world, awarded (this medal) because of outstanding writings." On the obverse is the head of Archimedes surrounded by the Latin inscription from the first century Roman poet Manilius's Astronomica. It may be translated as "to pass beyond your understanding and make yourself master of the universe." The complete passage from which this phrase is taken is strikingly similar to verses in many parts of the Vedas, the ancient Indian scriptures. Manilius writes:

> The object of your quest is god; you are seeking to scale the skies and though born beneath the rule of fate, to gain knowledge of that fate; you are seeking to pass beyond your understanding and make yourself master of the universe. The toil involved matches the reward to be won, nor are such high attainments secured without a price.

The Fields medalists have certainly fulfilled the expectations for continued achievements. However, the monetary value (currently about $15,000 Canadian) makes it a poor cousin to the Nobel prize, but is consistent with that old adage: "One does not go into mathematics to become rich." Here, I am also reminded of the remark made by the famous American comic Will Rogers when he was awarded the Congressional Medal of Honor. He said: "With this medal and 25 cents I can now buy a cup of coffee." In fact, several other prizes for which mathematicians are eligible far exceed the monetary value of the Fields Medal. In particular, the Abel prize instituted in 2003 and given annually is closer in spirit and monetary value to the Nobel prize. Jean-Pierre Serre (the youngest person to receive the Fields Medal at ICM 1954, in Amsterdam, at age 27) received the first Abel Prize in 2003. The Abel Prize for 2009 has been awarded to the Fields medalist M. L. Gromov. More information on the Abel Prize may be found at the official website (www.abelprisen.no/en/).

I would like to add that of the 28 Fields medalists since the ICM 1978 to the ICM 2006, 14 have made significant contribution to advancing the interaction of mathematics and theoretical physics. Ed Witten is the only physicist to be awarded the Fields Medal (ICM 1990, Keyoto). More information about the Fields medal can be found at the IMU website (www.mathunion.org).

4. **Harmonic Oscillator**

The quantization of the classical harmonic oscillator was one of the earliest results in quantum mechanics. It admits an exact closed-form solution in terms of the well-known Hermite polynomials. The oscillator's discrete energy spectrum also showed that the lowest energy level need not have zero energy and it can be used to explain the spectrum of diatomic molecules and to provide a tool for approximation of the spectra of more complex molecules. It was used in the early study of the black body radiation in the theory of heat and it can be applied to understand the motion of atoms in lattice models of crystals. We remark that the classical harmonic oscillator was the first dynamical system that was quantized through the canonical quantization principle.

We now discuss the canonical quantization of a single classical harmonic oscillator. The Hamiltonian H of the particle of mass m vibrating with frequency ω is given by

$$H(p, q) = \frac{p^2}{2m} + \frac{1}{2}m\omega^2 q^2,$$

where p, q are the conjugate variables momentum and position. In canonical quatization they are replaced by operators $\hat{p}, \hat{q}$ on the space of wave functions defined by

$$\hat{p}(\psi) := -i\hbar\frac{d\psi}{dq}, \quad \hat{q}(\psi) := q(\psi), \quad \psi \in L^2(q).$$

These operators satisfy the cononical commutation relation

$$[\hat{q}, \ \hat{p}] = i\hbar.$$

This prescription gives us the Schrödinger equation $\hat{H}(\psi) = E(\psi)$ for the wave function ψ (eigenfunction of $\hat{H}$) with energy level E (eigenvalue of $\hat{H}$). It is a second order ordinary differential equation

$$\frac{-\hbar^2}{2m}\ddot{\psi} + \frac{1}{2}m\omega^2 q^2\psi = E\psi.$$

The set of values of E for which the Schrödinger equation admits a solution is called the spectrum of the quantum harmonic oscillator. It can be shown that the spectrum is discrete and is given by

$$E_n = \hbar\omega(n + \tfrac{1}{2}), \qquad n \geq 0.$$

The corresponding wave function ψ_n is given by the explicit formula

$$\psi_n(x) = \left(\frac{\alpha}{\pi}\right)^{1/4}\frac{H_n(x)e^{-x^2/2}}{\sqrt{2^n n!}}, \quad \text{where } \alpha = m\omega/\hbar, \ x = q\sqrt{\alpha}\ .$$

In the above formula $H_n(x)$ is the nth Hermite polynomial, defined by

$$H_n(x) = (-1)^n e^{x^2}(\frac{d}{dx})^n e^{-x^2}.$$

Dirac defined **creation** (or raising) and **annihilation** (or lowering) operators, collectively called the ladder operators, which allow one to find the spectrum without solving the Schrödinger equation. (Feynman used this result to test his path integral quantization method.) These operators can be defined for more general quantum mechanical systems and also generalize to quantum field theory. We now indicate their use for the harmonic oscillator. The creation and annihilation operators $a^\dagger$ and a are related to the position and momentum operators by the relations

$$a^\dagger + a = \sqrt{\alpha/2}\,\hat{q}, \qquad a^\dagger - a = \frac{\sqrt{2/\alpha}\,\hat{p}}{i\hbar}.$$

From these relations it is easy to show that

$$a^\dagger(\psi_n) = \sqrt{n+1}\,\psi_{n+1}, \qquad a(\psi_n) = \sqrt{n}\,\psi_{n-1}.$$

It is these relations that led to the names "creation" (or raising) and "annihilation" (or lowering) operators. It follows that

$$[a, a^\dagger] = I, \qquad (a^\dagger a)(\psi_n) = n\psi_n \ .$$

In view of the second relation the operator $(a^\dagger a)$ is called the **number operator**. Using the definitions of $a^\dagger$ and a we can write the quantum Hamiltonian $\hat{H}$ as

$$\hat{H} = \hbar\omega(a^\dagger a + \tfrac{1}{2}).$$

The energy spectrum formula follows immediately from the above expression. It is these energy levels that appear in Witten's supersymmetric Hamiltonian and its relation to the classical Morse theory. The harmonic oscillator serves as a test case for other methods of quantization as well. The geometric quantization method applied to coupled harmonic oscillators leads to the same results as shown in [271,272]. This method requires the use of manifolds with singularities.

5. **Parallel Postulate** The parallel postulate was the fifth postulate or axiom in Euclid's geometry. For those readers who have not seen it recently, here is the statement:

> Let L be a straight line in a plane P. Through every point $x \in P$ not lying on the line L, there passes one and only one straight line L_x that is parallel to L.

The parallel postulate is much more complicated and non-intuitive than the other postulates of Euclidean geometry. Euclid attempted to prove this postulate on the basis of other postulates. He did not use it in his early books. Eventually, however, it was necessary for him to use it, but he wrote his 13 books without finding a proof. Finding its proof became a major challenge in geometry. Over the next 2000 years, many leading mathematicians attempted to prove it and announced their "proofs." There are more false proofs of this statement than any other statement in the history of mathematics. It turns out that most of these attempts contain some hidden assumption which itself is equivalent to the parallel postulate. Perhaps the most well known of these equivalent statements is: "The sum of the angles of a triangle is π." If we take the surface Σ of constant negative curvature as a model of a plane then its straight lines are the geodesics. Gauss's theorem applied to a geodesic triangle implies that the sum of the angles of a triangle is less than π. All other postulates of Euclid hold in this model. In this geometry the parallel postulate is replaced by the statement:

> Let L be a straight line in a plane P. Through every point $x \in P$ not lying on the line L, there pass at least two straight lines that are parallel to L.

The geometry based on the above parallel postulate is called **non-Euclidean geometry**. It was developed independently by the Russian mathematician Lobachevski and the Hungarian mathematician J. Bolyai. Gauss discovered this geometry earlier but he never published his work, although he had indicated it in his letter to F. Bolyai, J. Bolyai's father. Euclidean and non-Euclidean geometries became special cases of Riemannian geometry whose foundations were unveiled in Riemann's famous lecture[1] " Über de Hypothesen welche der Geometrie zu grunde liegen" in Göttingen delivered in the presence of his examiner Gauss. Riemann's work paved the way for Einstein's theory of gravity generalizing his special theory of relativity.

6. **Platonic Solids**

A regular convex polyhedron in $\mathbf{R}^3$ all of whose faces are congruent regular polygons is called a Platonic solid. The icosahedron, which we met in Chapter 2, is one of the five Platonic solids. These five regular polyhedra have been known since antiquity. Their names derive from the number of faces they have: tetrahedron (4), hexahedron (6) (more commonly called a cube), octahedron (8), dodecahedron (12), and icosahedron (20). Taking the centers of faces of a polyhedron as vertices generates its **dual** polyhedron. A tetrahedron is **self-dual** since taking the centers of its faces as vertices generates again a tetrahedron. Note that the cube and the octahedron are dual as are the dodecahedron and the icosahedron. Each of the Platonic solids is homeomorphic to S^2 and so has Euler characteristic $\chi = 2$. Euler gave a combinatorial definition of χ by $\chi = v - e + f$, where v is the number of vertices, e the number of edges, and f the number of faces of the given polyhedron. In table B.1 we also list b, the number of edges bounding a face, and n, the **valence**, or the number of edges meeting a vertex. The Platonic solids exhaust all the possibilities of pairs (b, n) that can occur in a regular polyhedron in $\mathbf{R}^3$.

Symmetry groups of the Platonic solids occur in many applications. As we have seen in Chapter 2, the moduli space of icosahedra inscribed in a sphere turns out to be a homology 3-sphere that is not homeomorphic to S^3. This example led Poincaré to reformulate his famous conjecture (now

---

[1] On the hypotheses which lie at the foundations of geometry

Table B.1   Platonic solids or regular polyhedra in $\mathbf{R}^3$

Name	v	e	f	b	n
Tetrahedron	4	6	4	3	3
Cube	8	12	6	4	3
Octahedron	6	12	8	3	4
Dodecahedron	20	30	12	5	3
Icosahedron	12	30	20	3	5

a theorem). The interest in these solids named after Plato (around 360 BCE) has continued to this day. In fact, the title of the 2009 Reimar Lüst lecture in Bonn delivered on June 12 by Curtis McMullen (Fields Medal, ICM 1998, Berlin) was "From Platonic Solids to Quantum Topology."

7. **Ramanujan**

India's greatest contribution to modern mathematics came in the last century in the work of Srinivasa Ramanujan. Its story is one of the most fascinating chapters in the history of mathematics. It began about a hundred years ago with a long letter from Ramanujan, a poor unknown clerk in Madras to Prof. G. H. Hardy, a well known English mathematician. It was a desperate cry for help. Ramanujan wrote that he had no university education but was striking a new path for himself in mathematics research. The letter contained over 100 results in mathematics that Ramanujan had obtained on his own without any formal university education in mathematics. His earlier attempts to communicate his work to other mathematicians had been unsuccessful. Hardy decided to show the letter to his close friend J. E. Littlewood who was also an eminent mathematician. They immediately realized that the writer had no exposure to standard mathematics. The letter contained some well known results and some that were known to be false. Some of the results seemed familiar. But it took Hardy far more time to prove them than he anticipated. The letter also contained several results the likes of which they had never seen before. Ramanujan had given no indication of how he obtained these results. After a few hours of work, they could not prove any of these new results and they were the best in the field. Hardy wrote later: A single look at them is enough to show that they could only be written down by a mathematician of the highest class. We concluded that these strange looking results must be true since no one would have had the imagination to invent them. Hardy replied immediately: I was exceedingly interested by your letter and by the theorems you state. However, I must see the proofs of some of your assertions. Ramanujan wrote back: I have found in you a friend who views my labors sympathetically. I am a half starving man who needs nourishment to preserve my brains.

Hardy was shocked to find out that Ramanujan had failed the first Arts examination of Indian universities. He could not get him a satisfactory position in India to continue his research in mathematics. He therefore used all his influence to get him a position at Cambridge University which involved no teaching duties. In March 1914 Ramanujan sailed to England to take up his new position. In 1916 he was awarded BSc by research (the degree was later called Phd). Winter months were very difficult for his health. But he continued to produce first rate mathematics working with Hardy almost every day that he was not ill. In 1917 he fell seriously ill and his doctors felt that he did not have much time. Hardy worked tirelessly to make sure that Ramanujan received the highest honor in Britain, namely election as a fellow of the Royal Society of London in 1918. Hardy wrote: He will return to India with a scientific standing and reputation such as no Indian has enjoyed before, and I am confident that India will regard him as the treasure he is. Ramanujan returned to India in March 1919 in poor health. His illness was never properly diagnosed and his health continued to deteriorate. On April 26, 1920 at age 32 Ramanujan passed away.

Ramanujan rarely gave detailed and rigorous proofs of his formulas and this led him sometimes to wrong results. In his notebooks Ramanujan listed a large number of results. Some of these notebooks were lost and were found quite accidentally. They have now been published by the American Mathematical Society. Ramanujan discovered nearly 4000 results by himself. Most of the new results are now proved but there are still several which are unproved. In the last year of his life spent in India, Ramanujan's health continued to deteriorate but his mathematical abilities remained undiminished. This is clearly shown by Ramanujan's famous last letter to Hardy written shortly before his death. In this letter he announced his discovery of the mock theta functions and discussed some of their remarkable properties. We are just beginning to understand these functions 90 years after their discovery. Recently, my friend Prof. Don Zagier, director of the Max Planck Institute for Mathematics in Bonn gave a research seminar on the mock theta functions and their surprising relation to string theory in Physics. Ramanujan would not have been surprised by this. He always felt that all his results were part of universal knowledge. If they appear where we don't expect to see them, then this is because of our limited vision. Let us hope and pray that we improve our record of recognizing great visionaries and helping them fulfill their destiny.

Let me conclude this brief biography of Ramanujan by recounting the most well known story about Ramanujan's friendship with numbers. Hardy went to visit Ramanujan when he was hostitalized. Hardy said that he came in a taxi bearing a rather dull and uninteresting number and he took it as an omen that he would find Ramanujan in poor health. On hearing the number, Ramanujan's eyes lit up. He said: No Mr. Hardy, in fact it is one of the most interesting numbers that I know. It is the smallest number that can be written as the sum of two cubes in two different ways. This number 1729 is now known as the Ramanujan number. For a fascinating account of Ramanujan's life and work I refer the reader to Kanigel's book "The man who knew infinity." It is a definitive biography of Ramanujan.

Appendix C
Categories and Chain Complexes

C.1 Categories

In modern mathematics the language of category theory is often used to discuss properties of different mathematical structures in a unified way. In recent years category theory and categorical constructions have been used in algebraic and geometric theories and have also found surprising applications to invariants of low-dimensional manifolds. The axiomatic development of TQFT is also based on a special category. Furthermore, applications to string theory and symplectic field theory have developed into an extensive area of research. We give here a mere outline of the basic ideas of category theory.

A **category** $\mathcal{C}$ consists of

1. a class $Ob(\mathcal{C})$ whose elements are called the **objects** of $\mathcal{C}$;
2. a correspondence that associates to eacth ordered pair (A, B) of objects of $\mathcal{C}$ a set $Mor_{\mathcal{C}}(A, B)$ (or simply $Mor(A, B)$) such that

$$A \neq C \text{ or } B \neq D \Rightarrow Mor(A, B) \cap Mor(C, D) = \emptyset.$$

   The elements of $Mor_{\mathcal{C}}(A, B)$ are called $\mathcal{C}$-**morphisms** (or simply morphisms or **arrows**) of A to B;
3. a correspondence that associates to each ordered triple (A, B, C) of objects of $\mathcal{C}$ a map $\circ$, called the **composition law**,

$$\circ : Mor(B, C) \times Mor(A, B) \to Mor(A, C),$$

   such that $(f, g) \mapsto f \circ g$, satisfying the following two properties:

   a. Existence of identities: $\forall A \in Ob(\mathcal{C})$, there exists an **identity morphism** $1_A \in Mor(A, A)$ such that, $\forall B \in Ob(\mathcal{C}), \forall f \in Mor(A, B), \forall g \in Mor(B, A)$,
$$f \circ 1_A = f \quad \text{and} \quad 1_A \circ g = g;$$

b. Associativity: $(h \circ g) \circ f = h \circ (g \circ f)$, $\forall (f, g, h) \in Mor(A, B) \times Mor(B, C) \times Mor(C, D)$.

As is customary we have used the same notation for morphisms as for maps.

A morphism $f \in Mor_{\mathcal{C}}(A, B)$ is said to be a $\mathcal{C}$-**isomorphism**, or simply an isomorphism, of A with B if there exists a morphism $g \in Mor_{\mathcal{C}}(B, A)$ such that

$$f \circ g = 1_B \quad \text{and} \quad g \circ f = 1_A.$$

Two objects A, B are said to be isomorphic if there exists an isomorphism of A with B. One frequently refers to only $Ob(\mathcal{C})$, when speaking of a category whose morphisms and composition law are understood. For example, we speak of the category Gr of groups with the understanding that the morphisms are the homomorphisms of groups with the usual composition law. A similar convention holds for other categories such as the category Top of topological spaces whose morphisms are continuous functions and Top^0 of pointed topological spaces whose morphisms are base point-preserving continuous functions.

A category $\mathcal{A}$ is said to be a **subcategory** (resp., **full subcategory**) of the category $\mathcal{B}$ if $Ob(\mathcal{A}) \subset Ob(\mathcal{B})$ and for each $X, Y \in Ob(\mathcal{A})$ we have $Mor_{\mathcal{A}}(X, Y) \subset Mor_{\mathcal{B}}(X, Y)$ (resp., $Mor_{\mathcal{A}}(X, Y) = Mor_{\mathcal{B}}(X, Y)$). As an example we note that the category Ab of Abelian groups is a full subcategory of Gr, while the category $DIFF$ of differential manifolds is a subcategory of the category TOP of topological manifolds which is not a full subcategory.

Let $\mathcal{C}$ be a category. An object $A \in Ob(\mathcal{C})$ is called a **universal initial** (resp. **universal final**) or simply **initial** (resp., **final**) **object** of $\mathcal{C}$ if, $\forall B \in Ob(\mathcal{C})$, $Mor(A, B)$ (resp. $Mor(B, A)$) contains only one element. Universal initial (resp. final) objects of a category, if they exist, are isomorphic. Universal objects play an important role in many constructions in modern mathematics, even though in applications it is customary to consider some concrete realization of such objects. We now give some examples of universal objects.

Let K be a field and let V, W denote fixed K-vector spaces and let $Z, Z_1, \ldots$ denote K-vector spaces. Consider the category $\mathcal{B}$ such that

1. the set of objects is $Ob(\mathcal{B}) = \{(Z, g) \mid g \in L(V, W; Z)\}$, where $L(V, W; Z)$ is the K-vector space of all bilinear maps of $V \times W$ to Z;
2. for $g_1 \in L(V, W; Z_1)$, $g_2 \in L(V, W; Z_2)$ a morphism f from (Z_1, g_1) to (Z_2, g_2) is a linear map $f : Z_1 \rightarrow Z_2$ such that $f \circ g_1 = g_2$, i.e., the following diagram commutes:

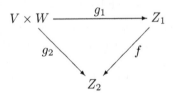

It can be shown that in this category a universal initial object exists. It is called the **tensor product** of V and W and is denoted by $(V \otimes W, t)$ or simply by $V \otimes W$. It is customary to write $v \otimes w$ for $t(v, w)$ and to call it the tensor product of v and w. If V and W are finite-dimensional, then a realization of $V \otimes W$ is given by $L(V^*, W^*; K)$, where $v \otimes w$, $v \in V$, $w \in W$ is identified with the map defined by

$$(\alpha, \beta) \mapsto \alpha(v)\beta(w), \quad \forall \alpha, \beta \in V^* \times W^*.$$

If in the above example we take $W = V$ and require $g \in L(V, V; Z)$ to be skew-symmetric (resp., symmetric) then in the resulting category the universal initial object is called the **exterior product** (resp., **symmetric product**) of V with itself and is denoted by $\Lambda^2(V)$ (resp., $S^2(V)$). The map $V \times V \to \Lambda^2(V)$ is denoted by $\wedge$ and $\wedge(\alpha, \beta)$ is denoted by $\alpha \wedge \beta$ and is called the **exterior product** of α with β. The map $V \times V \to S^2(V)$ is denoted by $\vee$, and $\vee(\alpha, \beta)$ is denoted by $\alpha \vee \beta$ and is called the **symmetric product** of α with β. Both the above constructions can be extended to any finite number of vector spaces. In particular, the tensor product

$$\underbrace{V \otimes V \otimes \cdots \otimes V}_{r \text{ times}} \otimes \underbrace{V^* \otimes V^* \otimes \cdots \otimes V^*}_{s \text{ times}},$$

where V^* is the dual of V, is called the **tensor space** of type (r, s) over V and is denoted by $T^r_s(V)$. The exterior product

$$\underbrace{V \wedge V \wedge \cdots \wedge V}_{k \text{ times}}$$

is called the **space of exterior k-forms** on V and is denoted by $\Lambda^k(V)$. The symmetric product

$$\underbrace{V \vee V \vee \cdots \vee V}_{k \text{ times}}$$

is called the **space of symmetric k-tensors** on V and is denoted by $S^k(V)$.

Another important example of an initial object is the direct limit of modules, which we used to define singular cohomology with compact support. Let I be a **directed set**, i.e., a set with a partial order $\leq$ such that, $\forall i_1, i_2 \in I$, there exists $i \in I$ such that $i_1 \leq i, i_2 \leq i$. If I is a directed set, let us denote by I_0^2 the subset of $I^2 := I \times I$ defined by

$$I_0^2 := \{(i, j) \in I^2 \mid i \leq j\}.$$

A **direct system** in a category $\mathcal{C}$ relative to the directed set I is a couple

$$D = (\{A_i \mid i \in I\}, \{f_i^j \mid (i, j) \in I_0^2\})$$

such that $\{A_i \mid i \in I\}$ is a family of objects of $\mathcal{C}$ and $f_i^j : A_i \to A_j$ is a family of $\mathcal{C}$-morphisms such that

1. $f_i^i = id_{A_i}$, $\forall i \in I$;
2. $f_j^k \circ f_i^j = f_i^k$, $\forall i, j, k \in I$, $i \le j \le k$.

Given the direct system D, let $\mathcal{C}(D)$ be the category such that:

Ob: The objects of $\mathcal{C}(D)$ are the pairs $(A, \{g_i \mid i \in I\})$ where $A \in Ob(\mathcal{C})$ and $g_i : A_i \to A$, $i \in I$, is a $\mathcal{C}$-morphism such that $g_i = g_j \circ f_i^j$, $\forall (i,j) \in I_0^2$.

Mor: If $\mathcal{A} = (A, \{g_i \mid i \in I\})$, $\mathcal{B} = (B, \{h_i \mid i \in I\})$ are objects of $\mathcal{C}(D)$, a morphism of $\mathcal{A}$ to $\mathcal{B}$ is a $\mathcal{C}$-morphism $f : A \to B$ such that $f \circ g_i = h_i$, $\forall i \in I$.

Given the direct system D as above, the **direct limit** of D is a universal initial object of the category $\mathcal{C}(D)$, i.e., an object $(A, \{g_i \mid i \in I\})$ of $\mathcal{C}(D)$ such that, $\forall (B, \{h_i \mid i \in I\}) \in Ob(\mathcal{C}(D))$, there exists a unique $\mathcal{C}$-morphism $f : A \to B$ satisfying the relations $f \circ g_i = h_i$, $\forall i \in I$. By definition, two direct limits of a direct system D, if they exist, are isomorphic objects of the category $\mathcal{C}(D)$. Hence, we can speak of "the" direct limit of D. If $(A, \{g_i \mid i \in I\})$ is a direct limit of D, we also say that A is the direct limit of the A_i and write

$$A = \lim_{\longrightarrow} A_i.$$

Direct limits exist in the categories of sets, topological spaces, groups, and R-modules. As an example we give the direct limit for a direct system D in the category of R-modules. Let $A = \oplus_{i \in I} A_i$ be the direct sum of the modules A_i and $u_i : A_i \to A$ the natural injection. Let us denote by B the submodule of A generated by all elements

$$u_j f_i^j (x_i) - u_i(x_i), \qquad (i,j) \in I_0^2, \ x_i \in A_i.$$

Let $\pi : A \to A/B$ be the quotient map and, $\forall i \in I$, g_i be the map defined by

$$g_i := \pi \circ u_i : A_i \to A/B.$$

Then $(A/B, \{g_i \mid i \in I\})$ is a direct limit of D.

Remark. If I has a largest element m, i.e., $i \le m$, $\forall i \in I$, then

$$(A_m, \{g_i = f_i^m \mid i \in I\})$$

is a direct limit of D.

Let $\mathcal{A}, \mathcal{B}$ be two categories. A **covariant** (resp., **contravariant**) functor F from $\mathcal{A}$ to $\mathcal{B}$ is a correspondence which associates with each $X \in Ob(\mathcal{A})$ an object $F(X) \in Ob(\mathcal{B})$ and to each $f \in Mor_{\mathcal{A}}(X, Y)$ a morphism $F(f) \in Mor_{\mathcal{B}}(F(X), F(Y))$ (resp., $F(f) \in Mor_{\mathcal{B}}(F(Y), F(X))$) such that

1. $F(1_X) = 1_{F(X)}$,

2. $F(g \circ f) = F(g) \circ F(f)$,   (resp., $F(g \circ f) = F(f) \circ F(g)$), $\forall (f, g) \in Mor_{\mathcal{A}}(X, Y) \times Mor_{\mathcal{A}}(Y, Z)$).

An example of a covariant functor is provided by the nth homotopy functor $\pi_n : Top^0 \to Gr$, which associates with each pointed topological space its nth homotopy group as defined in Chapter 2.

Let $F : \mathcal{A} \to \mathcal{B}$, $G : \mathcal{A} \to \mathcal{B}$ be two covariant functors. A **natural transformation** $\tau : F \to G$ is a correspondence τ which associates to each object $X \in \mathcal{A}$ a morphism $\tau(X) \in Mor_{\mathcal{B}}(F(X), G(X))$ such that

$$G(f) \circ \tau(X) = \tau(Y) \circ F(f), \quad \forall f \in Mor_{\mathcal{A}}(X, Y).$$

The defining equation of the natural transformation τ may be indicated by the commutativity of the following diagram:

$$
\begin{array}{ccc}
F(X) & \xrightarrow{\tau(X)} & G(X) \\
{\scriptstyle F(f)} \downarrow & & \downarrow {\scriptstyle G(f)} \\
F(Y) & \xrightarrow[\tau(Y)]{} & G(Y)
\end{array}
$$

A similar definition may be given for a natural transformation between contravariant functors.

C.2 Chain Complexes

Any homology (cohomology) theory of topological spaces is based on the construction of a structure called a chain (cochain) complex. In this section we discuss the general theory of chain complexes over a principal ideal domain. Let A be a commutative ring with unit element. We recall that A is called an **integral domain** if it has no zero divisors, i.e.,

$$\forall a, b \in A, \; ab = 0 \Rightarrow a = 0 \text{ or } b = 0.$$

A subring I of A is called an **ideal** if $AI = I = IA$. An ideal I is called a **principal ideal** if $I = aA$, for some $a \in A$. A **principal ideal domain** is an integral domain A such that every ideal in A is principal. Clearly, every field is a principal ideal domain. The ring of integers $\mathbf{Z}$ is a principal ideal domain that is not a field. In what follows we shall be concerned with modules over a fixed principal ideal domain $\mathbf{P}$. Recall that a **module** M over $\mathbf{P}$ or a $\mathbf{P}$-module is a generalization of the notion of a vector space with base field replaced by $\mathbf{P}$. For example, any ideal in $\mathbf{P}$ is a $\mathbf{P}$-module. The class of $\mathbf{P}$-modules and $\mathbf{P}$-linear maps forms a category. It is called the category of $\mathbf{P}$-modules and is denoted by $\mathcal{M}_{\mathbf{P}}$. Note that every Abelian group can be regarded as a **Z**-module. Thus, the category $\mathcal{M}_{\mathbf{Z}}$ is the same as the category of Abelian groups. As we will see in the next paragraph, the category of **P**-modules provides a general setting for defining a (co)chain complex and (co)homology modules generalizing the classical theories. The notion of an **Abelian category**, which is a generalization of the category of **P**-modules, arose from Grothendieck's attempt to unify different cohomology theories such as sheaf and group cohomologies by identifying the basic properties needed in their definitions. His attempt culminated in K-theory, which has found applications in algebraic and analytic geometry and more recently in theoretical physics.

A **chain complex** over **P** is a pair (C_*, δ), where $C_* = \{C_q \mid q \in \mathbf{Z}\}$ is a family of **P**-modules and $\delta = \{\delta_q : C_q \to C_{q-1} \mid q \in \mathbf{Z}\}$ is a family of **P**-linear maps such that

$$\delta_{q-1} \circ \delta_q = 0, \qquad \forall q \in \mathbf{Z}. \tag{C.1}$$

An element of C_q is called a q-**chain**. The **P**-linear map δ_q is called the qth **boundary operator**. One usually omits the subscript for the δ_q's and also writes $\delta^2 := \delta \circ \delta = 0$ to indicate that equation (C.1) is true. The chain complex (C_*, δ) is also represented by the following diagram:

$$\cdots \xleftarrow{\partial_{q-1}} C_{q-1} \xleftarrow{\partial_q} C_q \xleftarrow{\partial_{q+1}} C_{q+1} \xleftarrow{\partial_{q+2}} \cdots$$

Let (C_*, δ) be a chain complex over **P**. The **P**-module

$$Z_q(C_*, \delta) := \operatorname{Ker} \delta_q$$

is called the **P**-module of q-**cycles** and the **P**-module

$$B_q(C_*, \delta) := \operatorname{Im} \delta_{q+1}$$

is called the **P**-module of q-**boundaries**. The **P**-module

$$H_q(C_*, \delta) = Z_q(C_*, \delta)/B_q(C_*, \delta)$$

is called the qth **homology P-module** of the chain complex. We will simply write Z_q, B_q, and H_q instead of $Z_q(C_*, \delta)$, $B_q(C_*, \delta)$, and $H_q(C_*, \delta)$, respectively, when the complex (C_*, δ) is understood. The q-cycles $z, z' \in Z_q$ are said to be **homologous** if $z - z'$ is a q-boundary, i.e., $z - z' \in B_q$. The family $H_* := \{H_q \mid q \in \mathbf{Z}\}$ is called the **Z-graded homology module** or simply the **homology** of the chain complex (C_*, δ). A chain complex (C_*, δ), is said to be **exact** at C_q, if

$$\operatorname{Ker} \delta_q = \operatorname{Im} \delta_{q+1}.$$

The chain complex is said to be **exact** if it is exact at C_q, $\forall q \in \mathbf{Z}$. Thus the homology of a chain complex is a measure of the lack of exactness of the chain complex.

Given the chain complexes (C^1_*, δ^1), (C^2_*, δ^2) over $\mathbf{P}$, a **chain morphism** of (C^1_*, δ^1) into (C^2_*, δ^2) is a family $f_* = \{f_q : C^1_q \to C^2_q \mid q \in \mathbf{Z}\}$ of $\mathbf{P}$-linear maps such that, $\forall q \in \mathbf{Z}$, the following diagram commutes:

$$
\begin{array}{ccc}
C^1_q & \xrightarrow{\ \delta^1_q\ } & C^1_{q-1} \\
\downarrow{\scriptstyle f_q} & & \downarrow{\scriptstyle f_{q-1}} \\
C^2_q & \xrightarrow[\ \delta^2_q\ ]{} & C^2_{q-1}
\end{array}
$$

That is,

$$
\delta^2_q \circ f_q = f_{q-1} \circ \delta^1_q, \qquad \forall q \in \mathbf{Z}. \tag{C.2}
$$

It is customary to indicate the above chain morphism f_* by the diagram

$$
C^1_* \xrightarrow{\ f_*\ } C^2_*.
$$

Using equation (C.2) we can show that a chain morphism $f_* = \{f_q : C^1_q \to C^2_q \mid q \in \mathbf{Z}\}$ induces a family of $\mathbf{P}$-linear maps

$$
H_*(f_*) = \{H_q(f_*) : H_q(C^1_*, \delta^1) \to H_q(C^2_*, \delta^2) \mid q \in \mathbf{Z}\}
$$

among the homology modules. We observe that chain complexes and chain morphisms form a category and H_* is a covariant functor from this category to the category of graded $\mathbf{P}$-modules and $\mathbf{P}$-linear maps.

Let (C_*, δ) be a chain complex and let $D_* = \{D_q \subset C_q \mid q \in \mathbf{Z}\}$ be a family of submodules such that $\delta_q(D_q) \subset D_{q-1}$, $\forall q \in \mathbf{Z}$. Then $(D_*, \delta_{|D_*})$ is called a chain **subcomplex** of (C_*, δ). Let

$$
C_*/D_* = \{C_q/D_q \mid q \in \mathbf{Z}\}, \quad \bar{\delta} = \{\bar{\delta}_q \mid q \in \mathbf{Z}\},
$$

where $\bar{\delta}_q : C_q/D_q \to C_{q-1}/D_{q-1}$ is the map induced by passage to the quotient, i.e., $\bar{\delta}_q(\alpha + D_q) = \delta_q(\alpha) + D_{q-1}$. Then $(C_*/D_*, \bar{\delta})$ is called the **quotient chain complex** of (C_*, δ) by D_*.

Recall that a **short exact sequence** of $\mathbf{P}$-modules is a sequence of $\mathbf{P}$-modules and $\mathbf{P}$-linear maps of the type

$$
0 \longrightarrow C \xrightarrow{\ f\ } D \xrightarrow{\ g\ } E \longrightarrow 0
$$

such that f is injective, g is surjective and $\operatorname{Im} f = \operatorname{Ker} g$. A **short exact sequence of chain complexes** is a sequence of chain complexes and chain morphisms

$$
0 \longrightarrow C^1_* \xrightarrow{\ f_*\ } C^2_* \xrightarrow{\ g_*\ } C^3_* \longrightarrow 0
$$

such that, $\forall q \in \mathbf{Z}$, the following sequence of modules is exact:

$$0 \longrightarrow C_q^1 \xrightarrow{f_q} C_q^2 \xrightarrow{g_q} C_q^3 \longrightarrow 0.$$

We observe that if D_* is a subcomplex of C_*, then we have the following short exact sequence of complexes:

$$0 \longrightarrow D_* \xrightarrow{i_*} C_* \xrightarrow{\pi_*} C_*/D_* \longrightarrow 0$$

where i_* is the inclusion morphism and π_* is the canonical projection morphism. Let

$$0 \longrightarrow C_*^1 \xrightarrow{f_*} C_*^2 \xrightarrow{g_*} C_*^3 \longrightarrow 0$$

be a short exact sequence of chain complexes and let us denote by H_q^i the homology modules $H_q(C_*^i, \delta^i)$, $i = 1, 2, 3$. It can be shown that there exists a family

$$h_* = \{h_q : H_q^3 \to H_{q-1}^1 \mid q \in \mathbf{Z}\}$$

of linear maps such that the following homology sequence is exact:

$$\cdots \longrightarrow H_q^1 \xrightarrow{f_q} H_q^2 \xrightarrow{g_q} H_q^3 \xrightarrow{h_q} H_{q-1}^1 \longrightarrow \cdots$$

The family h_* is called the family of **connecting morphisms** associated to the short exact sequence of complexes. The corresponding homology sequence indicated in the above diagram is called the **long exact homology sequence**.

If we consider the construction of chain complexes, but with arrows reversed, we have cochain complexes, which we now define. A **cochain complex** over $\mathbf{P}$ is a pair (C^*, d), where $C^* = \{C^q \mid q \in \mathbf{Z}\}$ is a family of $\mathbf{P}$-modules and $d = \{d^q : C^q \to C^{q+1} \mid q \in \mathbf{Z}\}$ is a family of $\mathbf{P}$-linear maps such that

$$d^{q+1} \circ d^q = 0, \qquad \forall q \in \mathbf{Z}. \tag{C.3}$$

The $\mathbf{P}$-linear map d^q is called the qth **coboundary operator**. As with chain complexes, one usually omits the superscript for the d^q and also writes $d^2 := d \circ d = 0$ to indicate that equation (C.3) is true. The cochain complex (C^*, d) is also represented by the diagram

$$\cdots \xrightarrow{d^{q-2}} C^{q-1} \xrightarrow{d^{q-1}} C^q \xrightarrow{d^q} C^{q+1} \xrightarrow{d^{q+1}} \cdots$$

Let (C^*, d) be a cochain complex over $\mathbf{P}$. The $\mathbf{P}$-module

$$Z^q(C^*, d) := \operatorname{Ker} d^q$$

is called the $\mathbf{P}$-module of q-**cocycles** and the $\mathbf{P}$-module

$$B^q(C^*, d) := \operatorname{Im} d^{q-1}$$

is called the $\mathbf{P}$-module of q-**coboundaries**. The $\mathbf{P}$-module

$$H^q(C^*, d) = Z^q(C^*, d)/B^q(C^*, d)$$

is called the qth **cohomology P-module** of the cochain complex. Cochain morphisms are defined by an obvious modification of the definition of chain morphisms. We observe that cochain complexes and cochain morphisms form a category and H^* is a contravariant functor from this category to the category of graded **P**-modules and **P**-linear maps.

Let (C_*, δ) be a chain complex. Let C_q' be the algebraic dual of C_q and ${}^t\delta$ the transpose of δ. Then (C^*, d) defined by

$$C^* = \{C_q' \mid q \in \mathbf{Z}\}, \qquad d = \{{}^t\delta_q \mid q \in \mathbf{Z}\},$$

is a cochain complex. This cochain complex is called the **dual cochain complex** of the the chain complex (C_*, δ).

As we indicated earlier, when $\mathbf{P} = \mathbf{Z}$, the **P**-modules are Abelian groups. In this case one speaks of homology and cohomology groups instead of **Z**-modules.

In many mathematical and physical applications the notions of chain and cochain complexes that we have introduced are not sufficient and one must consider **multiple complexes**. We shall only consider the **double complexes**, also called **bicomplexes**, which may combine the chain and cochain complex in one double complex. We consider this situation in detail. A **mixed double complex** is a triple (C_*^*, d, δ) where $C_*^* = \{C_q^p \mid p, q \in \mathbf{Z}\}$ is a family of **P**-modules and $\delta = \{\delta_q^p : C_q^p \to C_{q-1}^p \mid p, q \in \mathbf{Z}\}$, $d = \{d_q^p : C_q^p \to C_q^{p+1} \mid p, q \in \mathbf{Z}\}$ are families of **P**-linear maps such that

$$\delta_{q-1}^p \circ \delta_q^p = 0, \quad d_q^{p+1} \circ d_q^p = 0, \quad \delta_q^{p+1} \circ d_q^p = d_{q-1}^p \circ \delta_q^p, \qquad \forall p, q \in \mathbf{Z}.$$

The double chain complex and double cochain complex can be defined similarly (see Bott and Tu [55] for further details). It is customary to associate to this double complex a cochain complex (T^*, D) called the **total cochain complex**, where

$$T^k = \bigoplus \{C_q^p \mid p - q = k\},$$

and

$$D^k = d^k + (-1)^k \delta_k,$$

where

$$d^k = \bigoplus \{d_q^p \mid p - q = k\}, \ \delta_k = \bigoplus \{\delta_q^p \mid p - q = k\}.$$

Since $d^k(\oplus \omega_q^p) = \oplus d_q^p \omega_q^p \in T^{k+1}$ and $\delta_k(\oplus \omega_q^p) = \oplus \delta_q^p \omega_q^p \in T^{k+1}$, we have that D^k maps T^k to T^{k+1}. It is easy to verify that $D^2 = 0$, i.e., $D^{k+1} \circ D^k = 0$, $\forall k$. D is called the **total differential**. A slight modification of the above construction is used in [236] to give an interpretation of the BRST operator in quantum field theory as a total differential D of a suitable double complex.

Appendix D
Operator Theory

D.1 Introduction

It is well known that pure gauge theories cannot describe interactions that have massive carrier particles. A resolution of this problem requires the introduction of associated fields. We discussed these fields and their couplings in Chapter 6. As with pure gauge fields, the coupled fields are solutions of partial differential equations on a suitable manifold. The theory of differential operators necessary for studying these field equations is a vast and very active area in mathematics and physics. In this appendix we give a brief introduction to the parts of operator theory that are relevant to the applications to field theories. Special operators and corresponding index theorems are also discussed here. We are interested in the situation where the operators are linear differential or pseudo-differential operators on modules of smooth sections of vector bundles over a suitable base manifold. Standard references for this material are Gilkey [154], Palais [310], and Wells [398]. In what follows we consider complex vector bundles over a compact base manifold. This simplifies many considerations, although the corresponding theory over non-compact base manifold and for real vector bundles can be developed along similar lines. Sobolev spaces play a fundamental role in the study of differential operators and in particular, in the study of operators arising in gauge theories. The analysis of differential operators is greatly simplified by the use of these spaces, essentially because they are Hilbert spaces in which differential operators, which are not continuous in the usual L^2 spaces, are continuous. For an introduction to the theory of Sobolev spaces, see for example, Adams [5]. We now give a brief account of some important aspects of these spaces, which are needed in this appendix.

D.2 Sobolev Spaces

Recall that a **multi-index** α in dimension n is an ordered n-tuple of non-negative integers, i.e., $\alpha = (\alpha_1, \alpha_2, \ldots, \alpha_n)$. The number $|\alpha| = \alpha_1 + \alpha_2 + \cdots + \alpha_n$ is called the length of α. We will use the notation

$$\partial_x^\alpha = \partial^{|\alpha|}/\partial x_1^{\alpha_1} \cdots \partial x_n^{\alpha_n} \quad \text{and} \quad D_x^\alpha := (-i)^{|\alpha|} \partial_x^\alpha$$

to indicate that derivatives are taken with respect to the variable $x = (x_1, \ldots, x_n) \in \mathbf{R}^n$, and we will write ∂^α and D^α when this variable is understood. Let ν be a measure on $\mathbf{R}^n$ and let $L^2(\nu)$ be the Hilbert space of complex-valued functions on $\mathbf{R}^n$, which are square integrable with respect to ν, with the inner product defined by

$$(f, g) \mapsto (f|g) := \int_{\mathbf{R}^n} \bar{f} g \, d\nu.$$

In the above definition of inner product on $L^2(\nu)$ we have followed the convention commonly used in the physics literature. We alert the reader that most mathematical works use the convention of linearity in the first variable and semi-linearity in the second. However, both induce the same norm on $L^2(\nu)$ and the use of one or the other convention is simply a matter of choice. Let μ_n or simply μ denote the Lebesgue measure on $\mathbf{R}^n$. Let us denote by $\hat{f}$ or Ff the **Fourier transform** of the Lebesgue integrable function $f : \mathbf{R}^n \to \mathbf{C}$. By definition,

$$Ff(x) \equiv \hat{f}(x) := (2\pi)^{-n/2} \int_{\mathbf{R}^n} f(y) e^{-ix \cdot y} d\mu(y),$$

where $x \cdot y = x_1 y_1 + \cdots + x_n y_n$. Let $\mathcal{S}_n$ denote the **Schwartz space** of (rapidly decreasing) smooth complex-valued functions on $\mathbf{R}^n$ such that, $\forall p \in \mathbf{N}$ and for every multi-index α, there exists a real positive constant $C_{\alpha,p}$ such that

$$(1 + |x|^2)^p |D^\alpha f(x)| \leq C_{\alpha,p}, \quad \forall x \in \mathbf{R}^n.$$

The space $\mathcal{S}_n$ is a dense subspace of $L^2(\mu)$. The **Parseval relation** $(f|g) = (Ff|Fg)$ implies that $\mathcal{S}_n$ is mapped isometrically onto itself by the Fourier transform F. Using this property, it can be shown that F has a unique extension to a unitary operator on $L^2(\mu)$. If $\alpha = (\alpha_1, \ldots, \alpha_n)$ is a multi-index, let us denote by M^α the operator of multiplication on $\mathcal{S}_n$ defined by

$$(M^\alpha f)(x) = x^\alpha f(x), \quad \forall x \in \mathbf{R}^n,$$

where $x^\alpha := x_1^{\alpha_1} x_2^{\alpha_2} \cdots x_n^{\alpha_n}$. Then one can show that, on $\mathcal{S}_n$,

$$D^\alpha \circ F = (-1)^{|\alpha|} F \circ M^\alpha, \tag{D.1}$$

$$F \circ D^\alpha = M^\alpha \circ F. \tag{D.2}$$

We note that relation (D.2) is used in passing from coordinate representation to the momentum representation in quantum mechanics.

Let ν_s, $s \in \mathbf{R}$, denote the measure on $\mathbf{R}^n$ defined by

$$d\nu_s(x) = (1 + |x|^2)^s d\mu(x).$$

For $f \in \mathcal{S}_n$ the **Sobolev s-norm** is defined by

$$|f|_s := \int_{\mathbf{R}^n} |Ff| d\nu_s.$$

The **Sobolev space** $H_s(\mathbf{R}^n)$ is the completion of $\mathcal{S}_n$ in this norm. It is customary to refer to $H_s(\mathbf{R}^n)$ as the **Sobolev s-completion**, or simply the **Sobolev completion**, of $\mathcal{S}_n$. We observe that if $s < t$ then $L^2(\nu_t) \subset L^2(\nu_s)$ (in particular, $L^2(\nu_0) = L^2(\mu)$) and, for $s \geq 0$,

$$H_s(\mathbf{R}^n) := \{ f \in L^2(\mu) \mid Ff \in L^2(\nu_s) \}.$$

Furthermore, if $f, g \in H_s(\mathbf{R}^n)$, $s \geq 0$, then their inner product is defined by

$$(f|g)_s = \int_{\mathbf{R}^n} \overline{Ff} \cdot Fg \, d\nu_s.$$

Thus, for $s \geq 0$, $H_s(\mathbf{R}^n) = F^{-1}(L^2(\nu_s))$ and F maps $H_s(\mathbf{R}^n)$ isometrically onto the Hilbert space $L^2(\nu_s)$. Due to the relation (D.2) we can consider $H_s(\mathbf{R}^n)$, $s \geq 0$, as the space of square summable complex-valued functions f on $\mathbf{R}^n$ whose distributional derivative $D^\alpha f$ is a function in $L^2(\mu)$, for all multi-indices α such that $|\alpha| \leq s$. In the following theorem we list three useful results concerning Sobolev spaces.

Theorem D.1 Let $H_s(\mathbf{R}^n), s \in \mathbf{R}$ denote the Sobolev s-completion of the Schwartz space $\mathcal{S}_n$. Then we have the following:

1. (Sobolev lemma) Let $B^k \subset C^k(\mathbf{R}^n)$ denote the Banach space of complex-valued functions f of class C^k with norm $|\cdot|_{B^k}$ defined by

$$|f|_{B^k}^2 := \sup_{\mathbf{R}^n} \sum_{|\alpha| \leq k} |D^\alpha f|^2.$$

   If $s > n/2 + k$ and $f \in H_s(\mathbf{R}^n)$, then f may be considered a function of class C^k, and the natural injection of $H_s(\mathbf{R}^n)$ into the Banach space B^k is continuous.

2. (Rellich lemma) Let $s < t$ and let $\{f_i\}$ be a sequence of functions in $\mathcal{S}_n$ with support in a compact subset of $\mathbf{R}^n$ and such that, for some constant K, $|f_i|_t \leq K$, $\forall i$. Then a subsequence of $\{f_i\}$ converges in $H_s(\mathbf{R}^n)$. Thus, the natural injection of $H_t(\mathbf{R}^n)$ into $H_s(\mathbf{R}^n)$ is compact when restricted to functions with support in a fixed compact subset of $\mathbf{R}^n$.

3. For all $s \in \mathbf{R}$, the restriction to $\mathcal{S}_n \times \mathcal{S}_n$ of the $L^2(\mu)$-inner product extends to a bilinear map of $H_{-s}(\mathbf{R}^n) \times H_s(\mathbf{R}^n) \to \mathbf{C}$, which gives a canonical identification of $H_{-s}(\mathbf{R}^n)$ with the dual of $H_s(\mathbf{R}^n)$.

The above construction of Sobolev spaces can be generalized to sections of vector bundles over a manifold as follows. Let E be a Hermitian vector bundle over an m-dimensional compact manifold M with n-dimensional fibers. Let $\{(U_\alpha, \phi_\alpha)\}$ be an atlas on M with a finite family of charts such that $\phi_\alpha(U_\alpha) = V_\alpha \subset \mathbf{R}^m$ and let $\{\rho_\alpha\}$ be a smooth partition of unity subordinate to this atlas. Let us suppose that the atlas is chosen to give a local representation $\{(U_\alpha, \psi_\alpha)\}$ of the bundle E so that $\psi_\alpha : U_\alpha \times \mathbf{C}^n \to \pi^{-1}(U_\alpha)$. If $\xi \in \Gamma(E)$, then $\psi_\alpha^{-1} \circ \rho_\alpha \xi$ induces, through the chart ϕ_α, a section of $V_\alpha \times \mathbf{C}^n$ over V_α, which can be considered a map of V_α into $\mathbf{C}^n$, denoted by f_α. This map has compact support and can be extended to a map, also denoted by f_α, of $\mathbf{R}^m$ into $\mathbf{C}^n$ with the same support. Thus, f_α is an n-tuple $f_{\alpha,i}$, $i = 1, \ldots, n$, of elements of $\mathcal{S}_m$. Therefore, we may define

$$\|\xi\|_s := \sum_{\alpha,i} |f_{\alpha,i}|_s.$$

It is easy to see that $\|\ \|_s$ is a norm on $\Gamma(E)$ and that $\Gamma(E)$ with this norm is a pre-Hilbert space. Let $H_s(E)$ be the Hilbert space obtained by completion of this normed vector space. One can show that norms related to different partitions of unity and to other choices made in the definition are equivalent. Thus, $H_s(E)$ is topologically a Hilbert space. The results of Theorem D.1 have obvious generalizations to this global case.

D.3 Pseudo-Differential Operators

The results on $U \subset \mathbf{R}^n$ are generally referred to as local, while those on a manifold (obtained by patching toghether these local results) are referred to as global. We use a vector notation which easily generalizes to the global case. Let $U \times \mathbf{C}^h$ be the trivial complex vector bundle over an open subset $U \subset \mathbf{R}^m$ and let Γ_m^h be the space of smooth sections of this bundle. A **differential operator** of order k from $U \times \mathbf{C}^h$ to $U \times \mathbf{C}^j$ is a linear map $P : \Gamma_m^h \to \Gamma_m^j$ which can be written as

$$P = \sum_{|\alpha| \leq k} a_\alpha D^\alpha, \tag{D.3}$$

where a_α is a $(j \times h)$-matrix-valued function on U. Let $U = \mathbf{R}^m$ and let

$$\mathcal{S}_m^h \equiv \underbrace{\mathcal{S}_m \oplus \cdots \oplus \mathcal{S}_m}_{h \text{ times}}$$

denote the space of sections of $\mathbf{R}^m \times \mathbf{C}^h$. An element of $\mathcal{S}_m^h$ is an h-tuple of functions in $\mathcal{S}_m$. Let $\mathcal{I}$ denote the operator on functions $f : \mathbf{R}^m \to \mathbf{C}$ defined by $(\mathcal{I}f)(x) := f(-x)$. Then $M^\alpha \circ \mathcal{I} = (-1)^{|\alpha|}\mathcal{I} \circ M^\alpha$. This relation and equations (D.1) and (D.2) imply that

$$D^\alpha = \mathcal{I} \circ F \circ M^\alpha \circ F.$$

Using this equation we obtain the following expression for the operator P on $f \in \mathcal{S}_m^h$:

$$(Pf)(x) = (2\pi)^{-m/2} \int_{\mathbf{R}^m} e^{ix \cdot y} p(x,y)(Ff)(y) d\mu(y), \qquad \text{(D.4)}$$

where $p(x,y)$ is the matrix-valued function defined by

$$p(x,y) = \sum_{|\alpha| \leq k} a_\alpha(x) y^\alpha.$$

The function p is called the **total symbol** of P, while the function $\sigma_k(P)$ defined by

$$\sigma_k(P)(x,y) = \sum_{|\alpha|=k} a_\alpha(x) y^\alpha$$

is called the k-**symbol** or the **principal symbol** (or simply the **symbol**) of P. Let E (resp., F) be a Hermitian vector bundle of rank h (resp., j) over an m-dimensional differential manifold M. Let $P : \Gamma(E) \to \Gamma(F)$ be a linear operator from sections of E to sections of F. The operator P is called a **(linear) differential operator of order** k from E to F if locally, in a coordinate neighborhood U, P has the expression (D.3). Alternatively we can express this definition by saying that P factors through the k-jet extension $J^k(E)$ of E, i.e., there exists a vector bundle morphism $f : J^k(E) \to F$ such that $P = f_* \circ j^k$, where $f_* : \Gamma(J^k(E)) \to \Gamma(F)$ is the map induced by f, i.e., the following diagram commutes

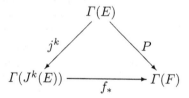

In fact, this formulation can easily be extended to define a non-linear differential operator of order k between sections of arbitrary fiber bundles. We denote by $D_k(E,F)$ the **space of linear differential operators of order** k from E to F. We now give a direct definition of the symbol of an operator $P \in D_k(E,F)$, which generalizes the definition of k-symbol given above in the local case. Let

$$T_0^* M = T^* M \setminus \{\text{the image of the zero section}\}$$

and let $p_0 : T_0^*M \rightarrow M$ be the restriction of the canonical projection $p : T^*M \rightarrow M$ to T_0^*M. Let p_0^*E (resp., p_0^*F) be the pull-back of the bundle E (resp., F) to the space T_0^*M with projection π_E^* (resp., π_F^*). Thus, we have the diagram

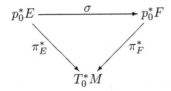

where σ is a morphism of vector bundles. The set of k-symbols is the set $\mathrm{Sm}_k(E, F)$ defined by

$$\mathrm{Sm}_k(E, F) := \{\sigma \in \Gamma(\mathrm{Hom}(p_0^*E, p_0^*F)) \mid \sigma(c\alpha_x, e) = c^k \sigma(\alpha_x, e),$$
$$c > 0, \ \alpha_x \in T_0^*M\}.$$

The k-**symbol** (or simply the symbol) of P denoted by $\sigma_k(P) \in \mathrm{Sm}_k(E, F)$ is defined by

$$\sigma_k(P)(\alpha_x, e) := P(t)(x),$$

where $(\alpha_x, e) \in p_0^*E$ and $t \in \Gamma(E)$ is defined as follows. We can choose $s \in \Gamma(E)$ and $f \in \mathcal{F}(M)$ such that $s(x) = e$ and $df(x) = \alpha_x$; then

$$t = \frac{i^k}{k!}(f - f(x))^k s.$$

It can be shown that $P(t)(x)$ depends only on α_x and e and is independent of the various local choices made. We say that an operator $P \in D_k(E, F)$ is **elliptic** if $\sigma_k(P)$ is an isomorphism. In particular, if P is elliptic then $\mathrm{rank}\, E = \mathrm{rank}\, F$.

The relation (D.4) suggests how to enlarge the class of differential operators to a class (the class of pseudo-differential operators) large enough to contain also the parametrix (a quasi-inverse modulo an operator of low order) of every elliptic operator. Let $k \in \mathbf{R}$; a **total symbol** of order k in $\mathbf{R}^m$ is a $(j \times h)$-matrix-valued smooth function p, with components $p_{r,s} : \mathbf{R}^m \times \mathbf{R}^m \rightarrow \mathbf{C}$ with compact support in the first m-dimensional variable, such that, for all multi-indices α, β, there exists a constant $C_{\alpha,\beta}$ such that

$$|D_x^\alpha D_y^\beta p(x, y)| \leq C_{\alpha,\beta}(1 + |y|)^{k - |\beta|}.$$

We denote by $\mathrm{Sm}_k(j, h)$ the vector space of these total symbols of order k represented by $j \times h$ matrices. Let $p \in \mathrm{Sm}_k(j, h)$; we define the operator P on $\mathcal{S}_m^h$ with the relation (D.4). One can show that P is a linear continuous operator from $\mathcal{S}_m^h$ into $\mathcal{S}_m^j$ and that P can be extended to a continuous operator $P_s : H_s^h(\mathbf{R}^m) \rightarrow H_{s-k}^j(\mathbf{R}^m)$, $\forall s \in \mathbf{R}$, where $H_s^h(\mathbf{R}^m)$ denotes vectorial Sobolev space with h components in $H_s(\mathbf{R}^m)$. We will often write P

instead of P_s. The operator P is called the **pseudo-differential operator** of order k associated with $p \in \mathrm{Sm}_k(j,h)$ and p is called the total symbol of P. We observe that if $l < k$ then $\mathrm{Sm}_l(j,h)$ is a subspace of $\mathrm{Sm}_k(j,h)$. Let us denote as $\mathrm{Sm}_{<k}(j,h)$ the space $\cup_{l<k} \mathrm{Sm}_l(j,h)$. If $k \in \mathbf{Z}$ then $\mathrm{Sm}_{<k}(j,h) = \mathrm{Sm}_{k-1}(j,h)$. The k-symbol of P, denoted $\sigma_k(P)$, is the class $\sigma_k(P)$ of p in the quotient $\mathrm{Sm}_k(j,h)/\mathrm{Sm}_{<k}(j,h)$. We denote by $\mathrm{Sm}_{-\infty}(j,h)$ the space

$$\mathrm{Sm}_{-\infty}(j,h) := \bigcap_k \mathrm{Sm}_k(j,h).$$

We observe that $p \in \mathrm{Sm}_{-\infty}(j,h)$ is a matrix-valued function with components in the Schwartz space $\mathcal{S}_{2m}$, with compact support in the first (m-dimensional) variable. We say that P, with total symbol p, is an **infinitely smoothing** pseudo-differential operator if $p \in \mathrm{Sm}_{-\infty}(j,h)$. We denote by $PD_k(j,h)$ (resp., $PD_{-\infty}(j,h)$) the space of pseudo-differential operators of order k (resp., infinitely smoothing pseudo-differential operators) associated with total symbols in $\mathrm{Sm}_k(j,h)$ (resp., $\mathrm{Sm}_{-\infty}(j,h)$). We observe that, by the Sobolev lemma, if $P \in PD_{-\infty}(j,h)$ then the j components of Pf are smooth, $\forall f \in H_s^h(\mathbf{R}^m)$, $\forall s \in \mathbf{R}$. We say that two operators $P, Q \in PD_k(j,h)$ are **equivalent** if $P - Q \in PD_{-\infty}(j,h)$. One can show that, if $P \in PD_k(h,i)$, $Q \in PD_l(i,j)$ then $PQ \in PD_{k+l}(h,j)$ and $P^* \in PD_k(i,h)$.

The above theory can be extended to vector bundles as follows. Let E (resp., F) be a Hermitian vector bundle of rank h (resp., j) over an m-dimensional compact differential manifold M. Let $P : \Gamma(E) \to \Gamma(F)$ be a linear operator from sections of E to sections of F. The operator P is called a **(linear) pseudo-differential operator of order k** from E to F if locally, in any trivialization

$$\psi_E : U \times \mathbf{C}^h \to \pi_E^{-1}(U), \quad \psi_F : U \times \mathbf{C}^j \to \pi_F^{-1}(U)$$

and any chart $\phi : U \to \mathbf{R}^m$, the induced operator from sections of $\phi(U) \times \mathbf{C}^h \to \phi(U)$ to sections of $\phi(U) \times \mathbf{C}^j \to \phi(U)$ extends to an operator $\tilde{P}$ that is in $PD_k(j,h)$ modulo infinitely smoothing operators. We denote by $PD_k(E,F)$ the **space of pseudo-differential operators** from E to F. A total symbol of order k from E to F is an element $p \in \Gamma(\mathrm{Hom}(\pi^*E, \pi^*F))$ that, in any trivialization ψ_E, ψ_F, ϕ as above, induces $\tilde{p} \in \mathrm{Sm}_k(j,h)$. We denote by $\mathrm{Sm}_k(E,F)$ the **space of symbols of order k** from E to F. In analogy with the definition of $\mathrm{Sm}_{<k}(j,h)$ we may define $\mathrm{Sm}_{<k}(E,F)$. One can show that for any operator $P \in PD_k(E,F)$ there exists a well-defined class $\sigma_k(P) \in \mathrm{Sm}_k(E,F)/\mathrm{Sm}_{<k}(E,F)$, which is called the k-**symbol** (or the principal symbol or simply the symbol) of P. For differential operators from E to F this definition coincides with the previous one. We say that $P \in PD_k(E,F)$ is elliptic if in the class of $\sigma_k(P)$ there is an element which, for each $(x,y) \in T_0^*M$, is an isomorphism of E_x onto F_x. Thus, if P is elliptic, $\mathrm{rank}\, E = \mathrm{rank}\, F$. We denote by $El_k(E,F)$ the space of elliptic operators from E to F. A **parametrix** for $P \in El_k(E,F)$ is a pseudo-differential operator

$Q \in PD_k(F, E)$ such that

$$QP = id - S_1 \quad \text{and} \quad PQ = id - S_2,$$

where S_1 and S_2 are infinitely smoothing operators on E and F, respectively, i.e.,

$$S_1 \in PD_{-\infty}(E, E) \text{ and } S_2 \in PD_{-\infty}(F, F).$$

We observe that a parametrix of P is an inverse of P modulo infinitely smoothing operators. One can show that every elliptic operator over a compact manifold admits a parametrix which is unique up to equivalence. An important consequence of the existence of the parametrix for an operator $P \in El_k(E, E)$ with $k > 0$ is the fact that every eigenfunction u of P is smooth.

In order to state some fundamental properties of elliptic operators we now recall some basic facts about the theory of continuous linear operators between Hilbert spaces. Let H, K be Hilbert spaces and $T : H \to K$ a linear continuous map. The **adjoint** T^* of T is the (unique) linear continuous operator $T^* : K \to H$ such that

$$(x|Ty) = (T^*x|y), \quad \forall x \in K, \ \forall y \in H.$$

If H, K are only pre-Hilbert spaces, then an operator T^* satisfying the above relation does not necessarily exist, and if it does exist, it need not be unique. The operator T is said to be **compact** if, for every bounded sequence $\{u_n\}$ in H, the sequence $\{Tu_n\}$ contains a convergent subsequence. The adjoint of a compact operator is compact and the composition of a compact operator with a continuous operator is compact. Furthermore, if the range $\text{Im}\,T$ of T is closed, then $\dim \text{Im}\,T < \infty$. The operator T is said to be of **trace class** if it is compact and the sequence of eigenvalues of $(T^*T)^{1/2}$, counted with their multiplicity, is summable. The operator T from a Hilbert space H to a Hilbert space K is said to be a **Fredholm operator** if $\ker T$ and $\ker T^*$ are finite-dimensional. It follows that, if K is infinite-dimensional, a continuous operator T cannot be compact and Fredholm. A continuous operator $P : H \to K$ is Fredholm if and only if there exists a continuous operator $Q : K \to H$ and compact operators $S_1 : H \to H$, $S_2 : K \to K$ such that

$$QP = id - S_1 \quad \text{and} \quad PQ = id - S_2. \tag{D.5}$$

Thus, Fredholm operators are the continuous operators that are invertible modulo compact operators. In particular, from equation (D.5) it follows that the operator Q is Fredholm. If P is a Fredholm operator the **index** of P, denoted by $\text{Ind}(P)$, is defined by

$$\text{Ind}(P) = \dim \ker P - \dim \ker P^*. \tag{D.6}$$

If $P : H \to K$ and $Q : L \to H$ are Fredholm operators then $PQ : L \to K$ is a Fredholm operator and

$$\mathrm{Ind}(PQ) = \mathrm{Ind}(P) + \mathrm{Ind}(Q).$$

If S is a compact operator then $id - S$ is a Fredholm operator and a simple calculation shows that $\mathrm{Ind}(id - S) = 0$. Thus, for the Fredholm operator Q in (D.5) we have

$$\mathrm{Ind}(PQ) = 0$$

and hence

$$\mathrm{Ind}(Q) = -\mathrm{Ind}(P).$$

We observe that, by definition, if P is Fredholm then P^* is also Fredholm and

$$\mathrm{Ind}(P^*) = -\mathrm{Ind}(P).$$

If $P \in PD_k(E, F)$, then P has a well-defined extension to a continuous operator $P_s : H_s(E) \to H_{s-k}(F)$, $\forall s$. It is customary to denote P_s simply by P when the extension is clear from the context. The following theorem contains some important properties of pseudo-differential operators.

Theorem D.2 *Let E (resp., F) be a Hermitian vector bundle over a compact Riemannian manifold M and let $P \in PD_k(E, F)$. Then we have the following:*

1. *There exists a unique $P^* \in PD_k(F, E)$ such that*

$$\int_M \langle v, Pu \rangle_F = \int_M \langle P^* v, u \rangle_E, \qquad u \in \Gamma(E), \; v \in \Gamma(F),$$

   where $\langle \, , \, \rangle_E$ (resp. $\langle \, , \, \rangle_F$) denotes the inner product in the fibers of E (resp., F). P^* is called the **formal adjoint** *of P. Moreover, $\sigma_k(P^*) = (\sigma_k(P))^*$.*

2. *The operator P is elliptic if and only if its formal adjoint P^* is elliptic. In this case, for all $s \in \mathbf{R}$, P_s is a Fredholm operator, $\ker P$ and $\ker P^*$ are finite-dimensional and if we define $\mathrm{Ind}(P) := \dim \ker P - \dim \ker P^*$, we have*

$$\mathrm{Ind}(P_s) = \mathrm{Ind}(P).$$

3. *We say that two k-symbols σ_0, σ_1 are* **regularly homotopic** *if there exists a homotopy σ_t, $0 \le t \le 1$, such that $\sigma_t(\alpha_x)$ is an isomorphism for all $\alpha_x \in T_0^* M$. Then, if $P \in El_k(E, F)$, its index depends only on the regular homotopy class of its symbol.*

4. *Let $F = E$, let $E_\lambda := \ker(P - \lambda \, id)$ denote the eigenspace associated to the eigenvalue $\lambda \in \mathbf{C}$ and let $\mathcal{V}$ be the set of eigenvalues of P. If $P \in El_k(E, F)$ is self-adjoint and $k > 0$, then each eigenspace of P is finite-dimensional and each eigenvector of P is a smooth section of E. Furthermore, $\mathcal{V}$ is not bounded and the Hilbert space $L^2(E)$ is a direct sum of the eigenspaces of*

P, i.e.,

$$L^2(E) = \bigoplus_{\lambda \in \mathcal{V}} E_\lambda.$$

The above theory for a single operator can be generalized to apply to differential complexes that we now define. Let $E_0, \ldots, E_n$ be a set of Riemannian vector bundles over a compact connected Riemannian manifold (M, g) and let

$$L_q : \Gamma(E_q) \to \Gamma(E_{q+1}), \qquad q = 0, \ldots, n-1$$

be a set of pseudo-differential operators of some fixed order k, such that

$$L_{q+1} \circ L_q = 0, \qquad 0 \le q \le n-2. \tag{D.7}$$

The finite cochain complex (see Appendix C)

$$0 \longrightarrow \Gamma(E_0) \xrightarrow{L_0} \Gamma(E_1) \xrightarrow{L_1} \cdots \xrightarrow{L_{n-1}} \Gamma(E_n) \longrightarrow 0$$

is called a **differential complex of order** k (or simply a differential complex) and is denoted by $(\Gamma(E), L)$ or simply by (E, L). We observe that the generalized de Rham sequence (see Chapter 4) is an example of a sequence which fails to be a complex, the obstruction being given by the curvature. If (E, L) is a differential complex, then a q-cocycle (resp., q-coboundary) of (E, L) is an element of $Z^q(E, L) = \ker L_q$ (resp., $B^q(E, L) = \operatorname{Im} L_{q-1}$). A qth cohomology class is an element of the q-th cohomology space $H^q(E, L) = Z^q(E, L)/B^q(E, L)$ of the complex (E, L). The **symbol sequence** associated to the differential complex (E, L) of order k is the following sequence of vector bundles and homomorphisms of vector bundles

$$0 \longrightarrow p_0^*(E_0) \xrightarrow{\sigma_k(L_0)} p_0^*(E_1) \xrightarrow{\sigma_k(L_1)} \cdots \xrightarrow{\sigma_k(L_{n-1})} p_0^*(E_n) \longrightarrow 0.$$

The differential complex is said to be **elliptic** if its associated symbol sequence is exact, i.e.,

$$\operatorname{Im} \sigma_k(L_{q-1}) = \ker \sigma_k(L_q), \qquad q = 0, 1, \ldots, n.$$

A single elliptic operator may be considered an elliptic complex. Given the elliptic complex (E, L) we define the jth **Laplacian**, or Laplace operator Δ_j, of the complex (E, L) by

$$\Delta_j := L_j^* L_j + L_{j-1} L_{j-1}^*, \qquad 0 \le j \le n. \tag{D.8}$$

The operator Δ_j is an elliptic operator for all j. It is self-adjoint (i.e., $\Delta_j = \Delta_j^*$) with respect to the inner product defined by

$$\langle\langle \sigma_1, \sigma_2 \rangle\rangle := \int_M \langle \sigma_1(x), \sigma_2(x) \rangle_j \, dv_g, \qquad \sigma_1, \sigma_2 \in \Gamma(E_j),$$

where $\langle\ ,\ \rangle_j$ is the metric on E_j. Some properties of Laplace operators that are relevant to physical applications are given in [315, 323]. The solutions of $\Delta_j(s) = 0$ are called **harmonic** sections as in the particular case of the de Rham complex. One can show that the Hodge decomposition theorem is valid also for general elliptic complexes (see, for example, Gilkey [154] or Wells [398]). Thus, the space of all harmonic sections is a finite-dimensional subspace of $\Gamma(E_j)$ isomorphic to the jth cohomology space $H^j(E, L)$. The **index** of the elliptic complex (E, L) is defined by

$$\mathrm{Ind}_a(E, L) = \sum_{j=0}^{n} (-1)^j \dim H^j(E, L). \qquad (D.9)$$

In the case of an elliptic complex consisting of a single elliptic operator P this definition reduces to the definition of the index of P given earlier. Definition (D.9) of the index is formulated using the spaces of solutions of differential equations and is therefore called the **analytic index**. The subscript "a" in $\mathrm{Ind}_a(E, L)$ refers to this analytic aspect of the definition. One can also define the **topological index** $\mathrm{Ind}_t(E, L)$, which can be expressed in terms of topological invariants (characteristic classes) associated to the complex (E, L). It turns out that the analytic index and the topological index of the elliptic complex (E, L) coincide, i.e.,

$$\mathrm{Ind}_a(E, L) = \mathrm{Ind}_t(E, L). \qquad (D.10)$$

This is the content of the classical Atiyah–Singer index theorem and its extensions. In the following example we apply this discussion to the de Rham complex.

Example D.1 *Recall that the de Rham complex of an m-dimensional manifold M is the differential complex*

$$0 \longrightarrow \Lambda^0(M) \overset{d}{\longrightarrow} \Lambda^1(M) \overset{d}{\longrightarrow} \cdots \overset{d}{\longrightarrow} \Lambda^n(M) \longrightarrow 0.$$

The operator d is a differential operator of order 1 and its symbol is given by

$$\sigma_1(d)(\alpha_x, \beta_x) = i\alpha_x \wedge \beta_x, \quad \beta_x \in \Lambda^j(M)_x, \ \forall j.$$

It is easy to verify that the corresponding symbol sequence is exact. Thus, the de Rham complex $(\Lambda(M), d)$ is an elliptic complex. The topological index $\mathrm{Ind}_t(\Lambda, d)$ of the de Rham complex is defined by

$$\mathrm{Ind}_t(\Lambda, d) = \sum_{k=0}^{m} (-1)^k b_k(M) = \chi(M),$$

where $b_k(M)$ is the kth Betti number of M and $\chi(M)$ is the Euler characteristic of M. From equation (D.9), we have

$$\mathrm{Ind}_a(\Lambda, d) = \sum_{j=0}^{n} (-1)^j \dim H^j(\Lambda, d).$$

By Hodge theory, $H^j(\Lambda, d)$ can be identified with the space of harmonic j-forms and hence with the corresponding de Rham cohomology spaces. Now recall from Chapter 2 that, by the classical de Rham theorem, the de Rham cohomology is in fact topological, i.e., does not depend on the differential structure and is isomorphic to the singular cohomology with real coefficients, i.e., $H^q_{\mathrm{deR}}(M) \cong H^q(M; \mathbf{R}), \forall q$. Thus, we have a topological characterization of the analytic index $\mathrm{Ind}_a(\Lambda, d)$. This may be regarded as a very special case of the index theorem.

One of the most extensively studied differential operators in mathematical physics is the Dirac operator. We devote the next section to a study of this operator by using the bundle of spinors on a Riemannian manifold.

D.4 The Dirac Operator

Spinors were first introduced by physicists to study representations of the universal covering group $Spin(3) = SU(2)$ of the rotation group $SO(3)$. The non-relativistic theory of spin was developed by Pauli. Pauli's spin matrices are the generators of the spin group $SU(2)$, which was interpreted as representing an intrinsic angular momentum of the electron. The relativistic wave equation for the electron was discovered by Dirac [102]. The wave functions of the electron occurring in Dirac's equation belong to the 4-dimensional representation of $Spin(3, 1) = SL(2, \mathbf{C})$, which is the universal covering group, of the connected component of the identity of the Lorentz group. The formal treatment of spinors in these and other related works was replaced by a geometrical definition by E. Cartan. In the introduction to his famous book on the theory of spinors [71], Cartan writes:

> ... because of this geometrical origin, the matrices used by physicists in Quantum Mechanics appear of their own accord, and we can grasp the profound origin of the property, possessed by Clifford algebras, of representing rotations in space having any number of dimensions. Finally this geometrical origin makes it very easy to introduce spinors into Riemannian geometry, and particularly to apply the idea of parallel transport to these geometrical entities.

It is this generalized Dirac operator that plays a fundamental role in index theory and in particular in the study of moduli spaces of gauge fields and in the Seiberg–Witten theory.

Our discussion of the Dirac operator is based on Clifford bundles over Riemannian manifolds. A typical fiber of a **Clifford bundle** is a **Clifford**

algebra. Let $n = r + s$ and let g_r be the standard pseudo-metric on $\mathbf{R}^n$ of signature (r, s). Let $\mathrm{Pin}(r, s)$ be the group of the elements of $C(r, s)$ that are products of a finite number of elements $\gamma(v) \in \gamma(\mathbf{R}^n) \subset C(r, s)$, such that $g_r(v, v) = \pm 1$. The group $\mathrm{Pin}(r, s)$ is called the **Pin group** of $(\mathbf{R}^n, g_r)$. The subgroup of $\mathrm{Pin}(r, s)$ consisting of the elements which are products of an even number of elements in $\gamma(\mathbf{R}^n)$ is called the **spinor group** of $(\mathbf{R}^n, g_r)$ and is denoted by $Spin(r, s)$. Let $\alpha : \mathbf{R}^n \to C(r, s)$ be the map defined by

$$\alpha(v) = -v, \qquad \forall v \in \mathbf{R}^n.$$

This map extends to a unique algebra automorphism of $C(r, s)$ such that $\alpha^2 = id$. Let $u \in \mathrm{Pin}(r, s)$. By identifying $\mathbf{R}^n$ with $\gamma(\mathbf{R}^n)$ we define the map $\widetilde{\mathrm{Ad}}_u$

$$\widetilde{\mathrm{Ad}}_u : \mathbf{R}^n \to \mathbf{R}^n \quad \text{by } \widetilde{\mathrm{Ad}}_u(v) := \alpha(u)vu^{-1} \in \gamma(\mathbf{R}^n) \cong \mathbf{R}^n,$$

One can show that $\widetilde{\mathrm{Ad}}_u$ is an invertible linear map of $\mathbf{R}^n$ into itself. In fact, it turns out that

$$\widetilde{\mathrm{Ad}}_u \in O(r, s), \qquad \forall u \in \mathrm{Pin}(r, s).$$

The map

$$\widetilde{\mathrm{Ad}} : u \mapsto \widetilde{\mathrm{Ad}}_u \text{ of } \mathrm{Pin}(r, s) \to O(r, s)$$

is called the **twisted adjoint representation** of $\mathrm{Pin}(r, s)$. It is a surjective homomorphism whose restriction to $Spin(r, s)$ is, for $n > 2$, a two-sheeted universal covering of $SO(r, s)$. In what follows we shall restrict ourselves to the case of signature $(n, 0)$ and denote $Spin(n, 0)$ by $Spin(n)$. The general case can be treated along similar lines.

Let E be an oriented Riemannian vector bundle of rank $n > 2$ and let $SO(E)$ be the bundle of oriented orthonormal frames in E. A **spin structure** $Spin(E)$ on E is a universal extension of the bundle $SO(E)$ to the group $Spin(n)$. The bundle $Spin(E)$ is a principal $Spin(n)$-bundle over M and is called the **spin frame bundle of E**.

Theorem D.3 *Let E be an oriented Riemannian vector bundle over a manifold M. E admits a spin structure if and only if its second Stiefel–Whitney class $w_2(E)$ is zero. Furthermore, if $w_2(E) = 0$, then the spin structures on E are in one-to-one correspondence with the elements of $H^1(M; \mathbf{Z}_2)$.*

If M is an oriented Riemannian manifold we define a spin structure on M to be a spin structure on TM. A **spin manifold** is an oriented Riemannian manifold M with $w_2(M) := w_2(TM) = 0$, together with a spin structure on M. We will denote by $Spin(M)$ the corresponding spin structure on TM. By abuse of language $Spin(M)$ is called the **spin frame bundle** of M.

Let E be an oriented Riemannian vector bundle of rank n; the **Clifford bundle** of E, denoted by $\mathrm{Cl}(E)$, is the associated bundle $(C(n) \equiv C(n, 0))$

$$\mathrm{Cl}(E) := SO(E) \times_r C(n),$$

where $r : SO(n) \to \mathrm{Aut}(C(n))$ is the canonical representation of $SO(n)$ into the automorphism group of $C(n)$. It is easy to see that $\mathrm{Cl}(E)_x = C(E_x)$, for all x in the base space of E. Thus, $\mathrm{Cl}(E)$ is a bundle of Clifford algebras. If the bundle E admits a spin structure with $Spin(n)$-bundle $Spin(E)$, a **real** (resp., **complex**) **spinor bundle** of E is an associated vector bundle of the form

$$Spin(E) \times_\rho V,$$

where ρ is a representation of $Spin(n)$ in the homomorphisms of the real (resp., complex) vector space V. Different spinor bundles of E may be obtained with different choices of $Spin(E)$ and ρ. We will denote by $S(E, \rho)$ (or simply $S(E)$) the spinor bundle of E defined by ρ. We observe that the vector space V of the representation ρ is a left module over $C(n)$. Thus, $S(E)_x$ is a left module over $C(E_x) \cong C(n)$ and $S(E)$ is a bundle of left modules. We observe that the Clifford bundle $\mathrm{Cl}(E)$ is a particular case of spinor bundle because

$$\mathrm{Cl}(E) = Spin(E) \times_{\mathrm{Ad}} C(n),$$

where Ad is the representation of $Spin(n)$ into the automorphisms of $C(n)$ such that $\mathrm{Ad}_u v = uvu^{-1}$, $u \in Spin(n) \subset C(n)$. It follows that $\mathrm{Ad}_{-1} = id$. Furthermore, the canonical vector space isomorphism of the exterior algebra $\Lambda(\mathbf{R}^n)$ with $C(n)$ (not an algebra isomorphism) gives rise to a canonical isomorphism

$$\Lambda(E) \cong \mathrm{Cl}(E). \tag{D.11}$$

Let M be an oriented Riemannian m-manifold with $w_2(M) = 0$ and let $S(M) := S(TM)$ be a real spinor bundle of TM. Let us suppose that $S(M)$ is Riemannian and has a Riemannian connection with covariant derivative ∇. The **Dirac operator** $\mathcal{D}$ relative to ∇ is the first order differential operator $\mathcal{D}$ on sections of spinor bundle $\Gamma(S(M))$ defined by

$$\mathcal{D}\sigma := \sum_{j=1}^m e_j \cdot \nabla_{e_j}\sigma, \qquad \sigma \in \Gamma(S(M)), \tag{D.12}$$

where, $\{e_1, \dots, e_m\}$ is an orthonormal basis of $T_x M \ \forall x \in M$, and $\cdot$ denotes Clifford multiplication. In Minkowski space the Dirac operator has the expression

$$\mathcal{D} = \sum \gamma_k \partial_k. \tag{D.13}$$

From this it follows that $\mathcal{D}^2 = g^{jk}\partial_j\partial_k$, which is the Laplacian in Minkowski space. Thus, the Dirac operator can be regarded as the square root of the Laplacian.

The Dirac operator is an elliptic operator with symbol

$$\sigma_1(\mathcal{D})(u_x, v_x) = iu_x \cdot v_x, \qquad u_x, v_x \in C(T_x M).$$

We observe that a Riemannian connection always exists on $SO(M)$. In fact, by the fundamental theorem of Riemannian geometry, $SO(M)$ (M is a Riemannian manifold) admits a canonical torsion-free Riemannian connection which can be lifted to $Spin(M)$ and thus to any spinor bundle on M. Thus, we have a canonical Riemannian connection also on $Cl(M) := Cl(TM)$. Under the canonical isomorphism (D.11) with $E = TM$, the Dirac operator $\mathcal{D}$ relative to the canonical Riemannian connection on $Cl(M)$ satisfies the relation

$$\mathcal{D} \cong d + \delta.$$

A **Dirac bundle** over a Riemannian manifold M is a Riemannian spinor bundle $S(M)$ of TM with a Riemannian connection ω on $S(M)$ such that:

1. for all $x \in M$ and for all unit vectors $e \in T_x M$ we have

$$\langle e \cdot u_x, e \cdot v_x \rangle = \langle u_x, v_x \rangle, \qquad \forall u_x, v_x \in S(M)_x;$$

2. if ∇ is the covariant derivative of the canonical Riemannian connection on M, then

$$\nabla^\omega(\phi \cdot \sigma) = \nabla \phi \cdot \sigma + \phi \cdot \nabla^\omega \sigma.$$

The space $\Gamma_0(S(M))$ of sections of $S(M)$ with compact support has an inner product defined by

$$(\sigma_1, \sigma_2) := \int_M \langle \sigma_1, \sigma_2 \rangle. \tag{D.14}$$

With respect to this inner product, the Dirac operator of a Dirac bundle is formally self-adjoint, i.e.

$$(\mathcal{D}\sigma_1, \sigma_2) = (\sigma_1, \mathcal{D}\sigma_2), \qquad \sigma_1, \sigma_2 \in \Gamma_0(S(M)).$$

The Dirac operator of a Dirac bundle $S(M)$ over a complete Riemannian manifold M can be extended to the space $L^2(S(M))$ of square integrable sections of $S(M)$ and one can show that this extension is essentially self-adjoint, i.e., the closure of $\mathcal{D}$ is self-adjoint. We have given several examples of index theorems in Chapter 5.

References

1. Abraham, R., Marsden, J.: Foundations of Mechanics. W. A. Benjamin, New York (1980)
2. Abraham, R., Marsden, J., Ratiu, T.: Manifolds, Tensor Analysis and Applications. Addison-Wesley, New York (1983)
3. Adams, C.: The Knot Book. Amer. Math. Soc., Providence (2004)
4. Adams, J.F.: Infinite Loop Spaces. Princeton University Press, Princeton (1978)
5. Adams, R.A.: Sobolev Spaces. Academic Press, New York (1975)
6. Aganagic, M., Mariño, M., Vafa, C.: All loop topological string amplitudes from Chern–Simons theory. Comm. Math. Phys. **247**, 467–512 (2004)
7. Aharonov, Y., Bohm, D.: Significance of electromagnetic potentials in the quantum theory. Phys. Rev. **115**, 485–491 (1959)
8. Aharonov, Y., Bohm, D.: Further considerations on electromagnetic potentials in the quantum theory. Phys. Rev. **123**, 1511–1524 (1961)
9. Albeverio, S., Jost, J., Paycha, S., Scarlatti, S.: A Mathematical Introduction to String Theory. Cambridge University Press, Cambridge (1997)
10. Amaldi, U., et al.: Comparison of grand unified theories with electroweak and strong coupling constants measured at LEP. Phys. Lett. **260B**, 447–455 (1991)
11. Amsler, C., et al.: Review of particle physics. Phys. Lett. **667B**, 1–13,405 (2008)
12. Appel, K., Haken, W.: The solution of the four-color-map problem. Sci. Amer. **Sept.**, 108–121 (1977)
13. Arnold, V.I.: The Vassiliev theory of discriminants and knots. In: First European Cong. Math. vol. I Prog. in Math., # 119, pp. 3–29. Birkhauser, Berlin (1994)
14. Atiyah, M.: The Geometry and Physics of Knots. Cambridge University Press, Cambridge (1990)
15. Atiyah, M.F.: The Geometry of Yang–Mills Fields. Fermi Lectures, Scuola Normale Superiore. Acad. Naz. Lincei, Pisa (1979)
16. Atiyah, M.F.: Topological quantum field theories. Publ. Math. Inst. Hautes Etudes Sci. **68**, 175–186 (1989)
17. Atiyah, M.F., Bott, R.: The Yang–Mills equations over Riemann surfaces. Phil. Trans. R. Soc. Lond. **A 308**, 523–615 (1982)
18. Atiyah, M.F., Hitchin, N.J.: Geometry and Dynamics of Magnetic Monopoles. Princeton University Press, Princeton (1988)
19. Atiyah, M.F., Hitchin, N.J., Drinfeld, V.G., Manin, Y.I.: Construction of instantons. Phys. Lett. **65A**, 185–187 (1978)
20. Atiyah, M.F., Hitchin, N.J., Singer, I.M.: Self-duality in four dimensional Riemannian geometry. Proc. Roy. Soc. Lond. **A362**, 425–461 (1978)
21. Atiyah, M.F., Jeffrey, L.: Topological Lagrangians and cohomology. J. Geo. Phys. **7**, 119–136 (1990)

22. Atiyah, M.F., Jones, J.D.S.: Topological aspects of Yang–Mills theory. Comm. Math. Phys. **61**, 97–118 (1978)

23. Atiyah, M.F., Singer, I.M.: Dirac operators coupled to vector potentials. Proc. Nat. Acad. Sci.(U.S.A.) **81**, 2597–2600 (1984)

24. Axelrod, S., Singer, I.: Chern–Simons perturbation theory. J. Diff. Geom. **39**, 787–902 (1994)

25. Axelrod, S., et al.: Geometric quantization of Chern–Simons gauge theory: I. J. Diff. Geom. **33**, 787–902 (1991)

26. van Baal, P.: Some results for $SU(N)$ gauge-fields on the hypertorus. Comm. Math. Phys. **85**, 529–547 (1982)

27. Babelon, O., Viallet, C.M.: The Geometrical Interpretation of the Faddeev–Popov Determinant. Phys. Lett. **85B**, 246–248 (1979)

28. Bakalov, B., A. Kirillov jr.: Lectures on Tensor Categories and Modular Functors. University Lect. Series, volume 21. Amer. Math. Soc., Providence (2001)

29. Banyaga, A., Hurtubise, D.: Lectures on Morse Homology. Texts in Math. Sci., volume 29. Kluwer Acad. Publ., Dordrecht (2004)

30. Bar-Natan, D.: Perturbative aspects of the Chern–Simons topological quantum field theory. Ph.D. thesis, Princeton University (1991)

31. Bar-Natan, D.: Vassiliev's knot invariant. Topology **34**, 423–472 (1995)

32. Bar-Natan, D.: On Khovanov's categorification of the jones polynomial. Alg. & Geo. Top. **2**, 337–370 (2002)

33. Baulieu, L., Singer, I.M.: Topological Yang–Mills symmetry. Nucl. Phys. B (Proc. Suppl.) **15B**, 12–19 (1988)

34. Beem, J.K., Ehrlich, P.E.: Global Lorentzian Geometry. Marcel-Dekker, New York (1981)

35. Beilinson, A., Drinfeld, V.: Chiral Algebras. Coll. Pub., vol. 51. Amer. Math. Soc., Providence (2004)

36. Belavin, A., Polyakov, A., Schwartz, A., Tyupkin, Y.: Pseudoparticle solutions of the Yang–Mills equations. Phys. Lett. **59B**, 85–87 (1975)

37. Belavin, A., Polyakov, A., Zamolodchikov, A.: Infinite conformal symmetries in two dimensional quantum field theories. Nucl. Phys. **B 241**, 333–380 (1984)

38. Berezin, F.A.: Introduction to Superanalysis. Reidel, Dordrecht (1987)

39. Besse, A.: Einstein Manifolds. Springer-Verlag, Berlin (1986)

40. Birman, J.S.: Braids, links and the mapping class groups. Ann. Math. Studies, # 82. Princeton University Press, Princeton (1994)

41. Birmingham, D., et al.: Topological field theory. Phy. Rep. **209**, 129–340 (1991)

42. Blanchet, C.: Hecke algebras, modular categories and 3-manifolds quantum invariants. Topology **39**, 193–223 (2000)

43. Bleecker, D.: Gauge Theory and Variational Principles. Addison-Wesley, Reading (1981)

44. Boden, H.U., Herald, C.M.: A connected sum formula for the $SU(3)$ Casson invariant. J. Diff. Geom. **53**, 443–464 (1998)

45. Boden, H.U., Herald, C.M.: The SU(3) Casson invariant for integral homology 3-spheres. J. Diff. Geom. **50**, 147–206 (1998)

46. Bogomol'nyi, E.B.: The stability of classical solutions. Sov. J. Nucl. Phys. **24**, 449–454 (1976)

47. Booss, B., Bleecker, D.D.: Topology and Analysis. Springer-Verlag, Berlin (1985)

48. Borcherds, R.E.: Monstrous moonshine and monstrous Lie superalgebras. Inventiones Math. **109**, 405–444 (1992)

49. Bott, R.: The stable homotopy of the classical groups. Ann. Math. **70**, 313–337 (1959)

50. Bott, R.: Lectures on characteristic classes and foliations. In: Lect. Notes in Math., # 279, pp. 1–94. Springer-Verlag, Berlin (1972)

51. Bott, R.: Equivariant Morse theory and the Yang–Mills equations on Riemann surfaces. In: The Chern Symposium 1979, pp. 11–22. Springer-Verlag, Berlin (1980)

52. Bott, R., Cattaneo, A.S.: Integral invariants of 3-manifolds. J. Diff. Geom. **48**, 357–361 (1998)
53. Bott, R., Cattaneo, A.S.: Integral invariants of 3-manifolds. II. J. Diff. Geom. **53**, 1–13 (1999)
54. Bott, R., Taubes, C.: On the self-linking of knots. J. Math. Phys. **35**, 5247–5287 (1994)
55. Bott, R., Tu, L.: Differential Forms in Algebraic Topology. Springer-Verlag, Berlin (1982)
56. Bourbaki, N.: Groupes et algèbres de Lie, Chapitres 2 et 3, Chapitres 7 et 8. Springer-Verlag, Berlin (2006)
57. Bourbaki, N.: Groupes et algèbres de Lie, Chapitre 1, Chapitres 4 à 6, Chapitre 9. Springer-Verlag, Berlin (2007)
58. Bourguignon, J.P.: Harmonic curvature for gravitational and Yang–Mills fields. In: Lect. Notes in Math.#949, pp. 35–47 (1982)
59. Bourguignon, J.P.: Yang–Mills theory: the differential geometric side. In: Lect. Notes in Math.#1263, pp. 13–54 (1987)
60. Bourguignon, J.P., Lawson, Jr., H.B.: Yang–Mills theory: Its physical origin and differential geometric aspects. In: S.T. Yau (ed.) Seminar on Differential Geometry, Annals of Mathematics Studies # 102, pp. 395–422 (1982)
61. Boyer, C.P., Mann, B.M.: Homology Operations on Instantons. J. Diff. Geom. **28**, 423–465 (1988)
62. Boyer, C.P., Mann, B.M.: Monopoles, non-linear σ models, and two-fold loop spaces. Comm. Math. Phys. **115**, 571–594 (1988)
63. Boyer, C.P., et al.: The topology of instanton moduli spaces. I: The Atiyah–Jones conjecture. Ann. of Math. **137**, 561–609 (1993)
64. Boyer, S., Nicas, A.: Varieties of group representations and Casson's invariant for rational homology 3-spheres. Trans. Amer. Math. Soc. **322**, 507–522 (1990)
65. Braam, P.J., Donaldson, S.K.: Floer's work on instanton homology, knots and surgery. In: H. Hofer, C.H. Taubes, E. Zehnder (eds.) Memorial Volume to Andreas Floer. Birkhauser, Boston (1994)
66. Braam, P.J., Donaldson, S.K.: Fukaya–Floer homology and gluing formulae for polynomial invariants. In: H. Hofer, C.H. Taubes, E. Zehnder (eds.) Memorial Volume to Andreas Floer. Birkhauser, Boston (1994)
67. Buchdahl, N.P.: Instantons on $\mathbf{cp}_2$. J. Diff. Geom. **24**, 19–52 (1986)
68. Burde, G., Zieschang, H.: Knots. de Gruyter, Berlin (1986)
69. Canarutto, D.: Marathe's generalized gravitational fields and singularities. Il Nuovo Cimento **75B**, 134–144 (1983)
70. Cao, H.D., Zhu, X.P.: A complete proof of the Poincaré and geometrization conjectures - application of the Hamilton–Perelman theory of the Ricci flow. Asian J. Math. **10, No. 2**, 165–492 (2006)
71. Cartan, E.: The Theory of Spinors. Hermann, Paris (1966)
72. Chang, K.C.: Infinite Dimensional Morse Theory and Multiple Solution Problems. Birkhäuser, Boston (1993)
73. Cheng, T., Li, L.: Gauge theory of elementary particle physics. Oxford University Press, Oxford (1984)
74. Chern, S.S.: On the curvatura integra of a riemannian manifold. Ann. Math. **46**, 674–684 (1945)
75. Chern, S.S., Simons, J.: Some cohomology classes in principal fiber bundles and their applications to Riemannian geometry. Proc. Nat. Acad. Sci. (U.S.A.) **68**, 791–794 (1971)
76. Chern, S.S., Simons, J.: Characteristic forms and geometric invariants. Ann. Math. **99**, 48–69 (1974)
77. Chevalley, C.: Theory of Lie Groups. Princeton University Press, Princeton (1946)
78. Choquet-Bruhat, Y., DeWitt-Morette, C.: Analysis, Manifolds and Physics. Part II. North-Holland, Amsterdam (1989)

79. Choquet-Bruhat, Y., DeWitt-Morette, C., Dillard-Bleick, M.: Analysis, Manifolds and Physics. North-Holland, Amsterdam (1982)

80. Coleman, S.: Aspects of Symmetry. Cambridge University Press, Cambridge (1986)

81. Collin, O., Steer, B.: Instanton Floer homology for knots via 3-orbifolds. J. Diff. Geom. **51**, 149–202 (1999)

82. Conway, J.H., Norton, S.P.: Monstrous moonshine. Bull. London Math. Soc. **11**(3), 308–339 (1979)

83. Conway, J.H., Smith, D.A.: On Quaternions and Octonions: Their Geometry, Arithmetic, and Symmetry. A. K. Peters, Ltd., Natick, MA (2003)

84. Conway, J.H., et al.: Atlas of Finite Groups. Oxford University Press, Eynsham (1985)

85. Coquereax, R., Jadczyk, A.: Riemannian Geometry, Fiber Bundles, Kaluza–Klein Theories and All That. Lect. Notes in Phys., vol. 16. World Scientific, Singapore (1988)

86. Cotta-Ramusino, P., Reina, C.: The action of the group of bundle automorphisms on the spaces of connections and the geometry of gauge theories. J. Geo. Phys. **1**, 121–155 (1984)

87. Cotta-Ramusino, P., et al.: Quantum field theory and link invariants. Nuc. Phy. **330B**, 557–574 (1990)

88. Crane, L.: 2-d physics and 3-d topology. Comm. Math. Phys. **135**, 615–640 (1991)

89. Crane, L., Frenkel, I.: Four dimensional topological quantum field theory, Hopf categories, and the canonical bases. J. Math. Phys. **35**, 5136–5154 (1994)

90. Croom, F.H.: Basic Concepts of Algebraic Topology. Springer-Verlag, Berlin (1978)

91. Cummins, C.J., Gannon, T.: Modular equations and the genus zero property of moonshine functions. Invent. Math. **129**, 413–443 (1997)

92. Cuntz, J., et al.: Topological and Bivariant K-Theory. Birkhäuser, Boston (2007)

93. Curtis, W.D., Miller, F.R.: Differential Manifolds and Theoretical Physics. Academic Press, New York (1978)

94. Dadhich, N.K., Marathe, K.B., Martucci, G.: Electromagnetic resolution of curvature and gravitational instantons. Il Nuovo Cim. **114 B**, 793–806 (1999)

95. Deligne, P., et al. (eds.): Quantum Fields and Strings: A Course for Mathematicians, vol. 1. Amer. Math. Soc., Providence (1999)

96. Deligne, P., et al. (eds.): Quantum Fields and Strings: A Course for Mathematicians, vol. 2. Amer. Math. Soc., Providence (1999)

97. Dell'Antonio, G., Zwanziger, D.: Every gauge orbit passes inside the Gribov horizon. Comm. Math. Phys. **138**, 291–299 (1991)

98. tom Dieck, T.: Algebraic Topology. Eur. Math. Soc., Zürich (2008)

99. Dieudonné, J.: Recent developments in mathematics. Amer. Math. Monthly **71**, 239–248 (1964)

100. Dieudonné, J.: The work of Nicolas Bourbaki. Amer. Math. Monthly **77**, 134–145 (1970)

101. Dijkgraaf, R., Witten, E.: Topological gauge theories and group cohomology. Comm. Math. Phys. **129**, 393–429 (1990)

102. Dirac, P.A.M.: The Principles of Quantum Mechanics. Clarendon Press, Oxford (1947)

103. Doi, H., Matsumoto, Y., Matumoto, T.: An explicit formula of the metric on the moduli space of BPST-instantons over s^4. In: A Fete of Topology, pp. 543–556. Academic Press, New York (1988)

104. Donaldson, S.K.: An application of gauge theory to four-dimensional topology. J. Diff. Geom. **18**, 279–315 (1983)

105. Donaldson, S.K.: A new proof of a theorem of Narasimhan and Seshadri. J. Diff. Geom. **18**, 269–277 (1983)

106. Donaldson, S.K.: Self-dual connections and the topology of 4-manifolds. Bull. Amer. Math. Soc. **8**, 81–83 (1983)

107. Donaldson, S.K.: Connections, cohomology and the intersection forms of 4-manifolds. J. Diff. Geom. **24**, 275–341 (1986)
108. Donaldson, S.K.: The geometry of four-manifolds. In: A. Gleason (ed.) Proc. Int. Cong. Math., Berkeley 1986, pp. 43–54. Amer. Math. Soc., Providence (1987)
109. Donaldson, S.K.: Polynomial invariants for smooth four-manifolds. Topology **29**, 257–315 (1990)
110. Donaldson, S.K.: Gluing techniques in the cohomology of moduli spaces. In: L. Goldberg, A. Phillips (eds.) Topological Methods in Modern Mathematics, pp. 137–170. Publish or Perish Inc., Houston (1993)
111. Donaldson, S.K.: Floer homology groups in Yang–Mills theory. Cambridge University Press, Cambridge (2002)
112. Donaldson, S.K., Kronheimer, P.: The Geometry of Four-Manifolds. Oxford University Press, Oxford (2001)
113. Donaldson, S.K., Sullivan, D.P.: Quasiconformal 4-manifolds. Acta Math. **163**, 181–252 (1989)
114. Douady, A., Verdier, J.L.: Les équations de Yang–Mills, astérisque 71-72. Société Math. de France, Paris (1980)
115. Dyson, F.: Missed opportunities. Bull. Amer. Math. Soc. **78**, 635–652 (1972)
116. Eguchi, T., Gilkey, P.B., Hanson, A.J.: Gravitation, gauge theories and differential geometry. Physics Reports **66**, 213–393 (1980)
117. Ehresmann, C.: Les connexions infinitésimales dans un espace fibré différentiable. In: Colloque de Topologie, Bruxelles (1950), pp. 29–55. Masson, Paris (1951)
118. Eilenberg, S., Steenrod, N.: Foundations of Algebraic Topology. Princeton University Press, Princeton (1952)
119. Evans, D.E., Kawahigashi, Y.: Quantum Symmetry on Operator Algebras. Oxford University Press, Oxford (1998)
120. Faddeev, L., Slavnov, A.A.: Gauge Fields, an Introduction to Quantum Theory. Benjamin, Reading (1980)
121. Fadell, E.R., Husseini, S.Y.: Geometry and Topology of Configuration Spaces. Springer-Verlag, Berlin (2001)
122. Feehan, P.M.N., Leness, T.G.: $PU(2)$ monopoloes, IV: And transversality. Comm. Anal. Geom. **5**(3), 685–791 (1997)
123. Feehan, P.M.N., Leness, T.G.: $PU(2)$ monopoloes, I: Regularity, Uhlenbeck compactness, and transversality. J. Diff. Geom. **49**, 265–410 (1998)
124. Felsager, B.: Geometry, Particles and Fields. Odense University Press, Odense (1981)
125. Feynman, R.P., Hibbs, A.R.: Quantum Mechanics and Path Integrals. McGraw-Hill, New York (1965)
126. Fineschi, F., Giannetti, R., Marathe, K.: Generalized products and associated structures on Euclidean spaces. Int. J. Math. Edu. in Sci. and Tech. **27**, 493–505 (1996)
127. Fintushel, R., Stern, R.: Instanton homology of Seifert fibered homology three spheres. Proc. London Math. Soc. **61**, 109–137 (1990)
128. Fintushel, R., Stern, R.: Integer graded instanton homology groups for homology three spheres. Topology **31**, 589–604 (1992)
129. Fischer, A.E.: The internal symmetry group of a connection on a principal fiber bundle with applications to gauge field theories. Comm. Math. Phys. **113**, 231–262 (1987)
130. Floer, A.: An instanton-invariant for 3-manifolds. Comm. Math. Phys. **118**, 215–240 (1988)
131. Floer, A.: Witten's complex and infinite dimensional Morse theory. J. Diff. Geom. **30**, 207–221 (1989)
132. Frankel, T.: The Geometry of Physics. Cambridge University Press, Cambridge (2004)
133. Freed, D., Morrison, D., Singer, I. (eds.): Quantum Field Theory, Supersymmetry, and Enumerative Geometry. IAS/Park City math.; vol. 11. Amer. Math. Soc., Providence (2006)

134. Freed, D., Uhlenbeck, K. (eds.): Geometry and quantum field theory. IAS/Park City math.; vol. 1. Amer. Math. Soc., Providence (1995)
135. Freed, D.S., Uhlenbeck, K.K.: Instantons and Four-manifolds. Springer-Verlag, Berlin (1991)
136. Freedman, M.H.: The topology of four dimensional manifolds. J. Diff. Geom. **17**, 357–453 (1982)
137. Freedman, M.H.: There is no room to spare in four-dimensional space. Notices Amer. Math. Soc. **31**, 3–6 (1984)
138. Frenkel, I., Huang, Y., Lepowsky, J.: On Axiomatic Approaches to Vertex Operator Algebras and Modules. Mem. AMS, vol. 104. Amer. Math. Soc., Providence (1993)
139. Frenkel, I., Lepowsky, J., Meurman, A.: Vertex Operator Algebras and the Monster. Pure and App. Math., #134. Academic Press, New York (1988)
140. Freyd, R., et al.: A new polynomial invariant of knots and links. Bull. Amer. Math. Soc. (N.S.) **12**, 239–246 (1985)
141. Friedman, R., Morgan, J.W.: Algebraic surfaces and 4-manifolds: some conjectures and speculations. Bull. Amer. Math. Soc. (N.S.) **18**, 1–19 (1988)
142. Friedman, R., Morgan, J.W.: Smooth Four-manifolds and Complex Surfaces. Springer-Verlag, Berlin (1994)
143. Friedman, R., Morgan, J.W. (eds.): Gauge theory and the topology of four-manifolds. IAS/Park City math.; vol. 4. Amer. Math. Soc., Providence (1998)
144. Friedrich, T.: Dirac operators in Riemannian Geometry. Grad. studies in math.; vol. 25. Amer. Math. Soc., Providence (2000)
145. Fröhlich, J.: Two-dimensional conformal field theory and three-dimensional topology. Int. J. Mod. Phys. **4-20**, 5321–5399 (1989)
146. Frohman, C., Nicas, A.: The alexander polynomial via topological quantum field theory. In: Differential Geometry, Global Analysis and Topology. Can. Math. Soc. Conf. Proc. vol. 12, pp. 27–40. Am. Math. Soc., Providence (1992)
147. Fukaya, K.: Floer homology for oriented 3-manifolds. Advanced Studies in Pure Math. **20**, 1–92 (1992)
148. Fulp, R.O., Norris, L.K.: Splitting of the connection in gauge theories with broken symmetry. J. Math. Phys. **24**, 1871–1887 (1983)
149. Fulton, W., MacPherson, R.: Compactification of configuration spaces. Ann. Math. **139**, 183–225 (1994)
150. Furuta, M.: Perturbation of moduli spaces of self-dual connections. J. Fac. Sci. Univ. Tokyo Sect. IA Math. **34**, 275–297 (1987)
151. Gannon, T.: Monstrous moonshine: The first twenty-five years. Bull. London Math. Soc. **38**, 1–33 (2006)
152. Gannon, T.: Moonshine beyond the Monster. Cambridge University Press, Cambridge (2006)
153. Garcia, P.: Gauge algebras, curvature and symplectic structure. J. Diff. Geom. **12**, 209–227 (1977)
154. Gilkey, P.: Invariance theory, the heat equation, and the Atiyah–Singer index theorem. Publish or Perish Press, Boston (1984)
155. Glashow, S.L.: Towards a unified theory: threads in a tapestry. Rev. Mod. Phys. **92**, 539–543 (1980)
156. Gocho, T.: The topological invariant of 3-manifolds based on the $U(1)$ gauge theory. J. of Fac. Sci. University Tokyo **39**, 165–184 (1992)
157. Gompf, R.: An infinite set of exotic R^4's. J. Diff. Geom. **21**, 283–300 (1985)
158. Gopakumar, R., Vafa, C.: On the gauge theory/geometry correspondence. Ad. Theor. Math. Phys. **3**, 1415–1443 (1999)
159. Gorenstein, D.: Finite Simple Groups. Plenum Press, New York (1982)
160. Greenberg, M.J., Harper, J.R.: Algebraic Topology: A First Course. Benjamin, Reading (1981)

161. Greub, W.: Complex line bundles and the magnetic field of a monopole. In: K. Bleuler, A. Reetz (eds.) Differential Geometrical Methods in Mathematical Physics, Proc. Bonn 1975, Lect. Notes in Math. # 570, pp. 350–354. Springer-Verlag, Berlin (1977)

162. Greub, W., Halperin, S., Vanstone, R.: Connections, Curvature, and Cohomology, vol. I. Academic Press, New York (1972)

163. Greub, W., Halperin, S., Vanstone, R.: Connections, Curvature, and Cohomology, vol. II. Academic Press, New York (1973)

164. Greub, W., Halperin, S., Vanstone, R.: Connections, Curvature, and Cohomology, vol. III. Academic Press, New York (1976)

165. Greub, W., Petry, H.R.: On the lifting of structure groups. In: K. Bleuler, H.R. Petry, A. Reetz (eds.) Differential Geometrical Methods in Mathematical Physics II, Proc. Bonn 1977, Lect. Notes in Math. # 676, pp. 217–246. Springer-Verlag, Berlin (1978)

166. Gribov, V.N.: Instability of non-Abelian gauge theories and impossibility of choice of Coulomb gauge. SLAC Translation **176** (1977)

167. Gribov, V.N.: Quantization of non-Abelian gauge theories. Nucl. Phys. **B139**, 1–19 (1978)

168. Griess, R.: The friendly giant. Invent. Math. **69**, 1–102 (1982)

169. Groisser, D.: $SU(2)$ Yang–Mills–Higgs theory on $\mathbf{R}^3$. Ph.D. thesis, Harvard University (1983)

170. Groisser, D.: Integrality of the monopole number in $SU(2)$ Yang–Mills–Higgs theory on $\mathbf{R}^3$. Comm. Math. Phys. **93**, 367–378 (1984)

171. Groisser, D.: The geometry of the moduli space of $\mathbf{CP}^2$ instantons. Inventiones Mathematicae **99**, 393–409 (1990)

172. Groisser, D., Parker, T.H.: The Riemannian geometry of the Yang–Mills moduli space. Comm. Math. Phys. **112**, 663–689 (1987)

173. Groisser, D., Parker, T.H.: The geometry of the Yang–Mills moduli space for definite manifolds. J. Diff. Geom. **29**, 499–544 (1989)

174. Groisser, D., Parker, T.H.: Semiclassical Yang–Mills theory I: Instantons. Comm. Math. Phys. **135**, 101–140 (1990)

175. Grossman, B., Kephart, T.W., Stasheff, J.D.: Solutions to Yang–Mills field equations in eight dimensions and the last Hopf map ((erratum) Comm. Math. Phys. 100 (1985), 311). Comm. Math. Phys. **97**, 431–437 (1984)

176. Guadagnini, E.: The Link Invariants of the Chern–Simons Field Theory. de Gruyter, Berlin (1993)

177. Guadagnini, E., et al.: Perturbative aspects of Chern–Simons field theory. Phys. Lett. **B 227**, 111–117 (1989)

178. Guadagnini, E., et al.: Wilson lines in Chern–Simons theory and link invariants. Nuc. Phy. **330B**, 575–607 (1990)

179. Guillemin, V., Sternberg, S.: Symplectic Techniques in Physics. Cambridge University Press, Cambridge (1984)

180. Guillemin, V.W., Sternberg, S.: Supersymmetry and Equivariant de Rham Theory. Springer-Verlag, Berlin (1999)

181. Habermann, L.: On the geometry of the space of $Sp(1)$-instantons with Pontrjagin index 1 on the 4-sphere. Ann. Global Anal. Geo. **6**, 3–29 (1988)

182. Haken, W.: Theorie der Normalflachen. Acta Math. **105**, 245–375 (1961)

183. Hamilton, R.S.: Three manifolds with positive Ricci curvature. J. Diff. Geo. **17**, 255–306 (1982)

184. Hamilton, R.S.: Four manifolds with positive curvature operator. J. Diff. Geo. **24**, 153–179 (1986)

185. Harer, J., Zagier, D.: The Euler characteristic of the moduli space of curves. Inven. Math. **85**, 457–485 (1986)

186. Harvey, R., H. B. Lawson, Jr.: Calibrated geometries. Acta Math. **148**, 47–157 (1982)

187. Heil, A., et al.: Anomalies from the point of view of G-theory. J. Geo. Phys. **6**, 237–270 (1989)

188. Helgason, S.: Geometric Analysis on Symmetric Spaces, 2nd ed. Amer. Math. Soc., Providence (2008)

189. Herald, C.M.: Flat connections, the Alexander invariant and Casson's invariant. Comm. Anal. Geom. **5**, 93–120 (1997)

190. Hermann, R.: Yang–Mills, Kaluza–Klein and the Einstein Program. Mat. Sci. Press, Brookline (1978)

191. Hickerson, D.: A proof of the mock theta conjectures. Invent. Math. **94**, 639–660 (1988)

192. Higgs, P.: Spontaneous symmetry breakdown without massless bosons. Phys. Rev. **145**, 1156–1163 (1966)

193. Hitchin, N.J.: The geometry and topology of moduli spaces. In: Lect. Notes in Math. # 1451, pp. 1–48. Springer-Verlag, Berlin (1989)

194. Horváthy, P.A., Rawnsley, J.H.: Monopole charges for arbitrary compact gauge groups and Higgs fields in any representation. Comm. Math. Phys. **99**, 517–540 (1985)

195. Hoste, J., Thistlethwaite, M., Weeks, J.: The first 1,701,936 knots. Math. Intelligencer **20**, # **4**, 33–48 (1998)

196. H.Taubes, C.: A simplicial model for Donaldson-Floer theory. In: H. Hofer, C.H. Taubes, E. Zehnder (eds.) Memorial Volume to Andreas Floer. Birkhauser, Boston (1995)

197. Huang, Y.: Two-Dimensional Conformal Geometry and Vertex Operator Algebras. Progress in Math., #148. Birkhäuser, Basel (1997)

198. Husemoller, D.: Fiber Bundles, 2nd edition. Springer-Verlag, Berlin (1975)

199. Ikeda, M., Miyachi, Y.: On an extended framework for the description of elementary particles. Prog. Theor. Phys. **16**, 537–547 (1956)

200. Ikeda, M., Miyachi, Y.: On the static and spherically symmetric solutions of the Yang–Mills field. Prog. Theor. Phys. **27**, 537–547 (1962)

201. Itoh, M.: The moduli space of Yang–Mills connections over a Kähler surface is a complex manifold. Osaka J. Math. **22**, 845–862 (1985)

202. Itoh, M.: Quaternionic structure on the moduli space of Yang–Mills connections. Math. Ann. **276**, 581–593 (1987)

203. Itoh, M.: Geometry of anti-self-dual connections and Kuranishi map. J. Math. Soc. Japan **40**, 9–33 (1988)

204. Itoh, M.: Yang–Mills connections and the index bundles. Tsukuba J. Math. **13**, 423–441 (1989)

205. Jacobsson, M.: An invariant of link cobordisms from Khovanov homology. Alg. & Geo. Top. **4**, 1211–1251 (2004)

206. Jadczyk, A.: Symmetry of Einstein-Yang–Mills systems and dimensional reduction. J. Geo. Phys. **1**, 97–126 (1984)

207. Jaffe, A., Taubes, C.: Vortices and Monopoles: Structure of Static Gauge Theories. Birkhauser, Boston (1980)

208. Jones, V.: On knot invariants related to some statistical mechanical models. Pac. J. Math. **137**, 311–334 (1989)

209. Jones, V., Sunder, V.S.: Introduction to subfactors. LMS Lect. Notes, vol. 234. Cambridge University Press, Cambridge (1997)

210. Jones, V.F.R.: A polynomial invariant for knots via von Neumann algebras. Bull. Amer. Math. Soc. **12**, 103–111 (1985)

211. Jones, V.F.R.: Hecke algebra representations of braid groups and link polynomials. Ann. Math. **126**, 335–388 (1987)

212. Jost, J.: Bosonic Strings: A Mathematical Treatment. AMS/IP Studies in Advanced Mathematics, vol. 21. Amer. Math. Soc., Providence (2001)

213. Jost, J.: Compact Riemannian Surfaces. Universitext, 2nd edition. Springer-Verlag, Berlin (2002)

214. Jost, J.: Riemannian Geometry and Geometric Analysis. Universitext, 5th edition. Springer-Verlag, Berlin (2008)

215. Kamber, F.W., Tondeur, P.: Foliated Bundles and Characteristic Classes. Lect. Notes in Math., #493. Springer-Verlag, Berlin (1975)
216. Karoubi, M.: K-Theory. Springer-Verlag, Berlin (1978)
217. Kassel, C., Turaev, V.: Braid Groups. Grad. Texts in Math., #247. Springer-Verlag, New York (2008)
218. Kauffman, L.H.: Knots and Physics. Series on Knots and Everything - vol. 1, 3rd Edition. World Scientific, Singapore (2001)
219. Khovanov, M.: A categorification of the Jones polynomial. Duke Math. J. **101**, 359–426 (2000)
220. Killingback, T.P.: The Gribov ambiguity in gauge theories on the four-torus. Phys. Lett. **138B**, 87–90 (1984)
221. Kirby, R., Melvin, P.: Evaluation of the 3-manifold invariants of Witten and Reshetikhin–Turaev. In: S.K. Donaldson, C.B. Thomas (eds.) Geometry of Low-dimensional Manifolds, vol. II, Lect. Notes # 151, pp. 101–114. London Math. Soc., London (1990)
222. Kirby, R., Melvin, P.: The 3-manifold invariants of Witten and Reshetikhin–Turaev for $sl(2, \mathbf{C})$. Inven. Math. **105**, 473–545 (1991)
223. Kirillov, A.N., Reshetikhin, N.Y.: Representations of the algebra $U_q(SL(2, \mathbf{C}))$, q-orthogonal polynomials and invariants of links. In: V.G. Kac (ed.) Infinite dimensional Lie algebras and groups, pp. 285–339. World Sci., Singapore (1988)
224. Knizhnik, V.G., Zamolodchikov, A.B.: Current algebra and Wess–Zumino models in two dimensions. Nucl. Phys. B **247**, 83–103 (1984)
225. Kobayashi, S., Nomizu, K.: Foundations of Differential Geometry, vol. 1. Wiley-Interscience, New York (1963)
226. Kobayashi, S., Nomizu, K.: Foundations of Differential Geometry, vol. 2. Wiley-Interscience, New York (1969)
227. Kock, J.: Frobenius Algebras and 2D Topological Quantum Field Theories. LMS Student Texts, vol. 59. Cambridge University Press, Cambridge (2004)
228. Kock, J., Vainsencher, I.: An invitation to quantum cohomolgy: Kontsevich's formula for rational plane curves. Prog. in Math., vol. 249. Birkhäuser, Boston (2007)
229. Kodiyalam, V., Sunder, V.S.: Topological quantum field theories from subfactors. Res. Notes in Math., vol. 423. Chapman & Hall/CRC, London (2001)
230. Kohno, T.: Topological invariants for three manifolds using representations of the mapping class groups I. Topology **31**, 203–230 (1992)
231. Kohno, T.: Topological invariants for three manifolds using representations of the mapping class groups II: Estimating tunnel number of knots. In: Mathematical Aspects of Conformal and Topological Field Theories and Quantum Groups, *Contemporary Math.*, vol. 175, pp. 193–217. Amer. Math. Soc., Providence (1994)
232. Kohno, T.: Conformal Field Theory and Topology. Trans. math. monographs, vol. 210. Amer. Math. Soc., Providence (2002)
233. Kondracki, W., Rogulski, J.: On the stratification of the orbit space for the action of automorphisms on connections. Dissertationes Math. (Warszawa) **CCL**, 1–62 (1986)
234. Kondraski, W., Sadowski, P.: Geometric structure on the orbit space of gauge connections. J. Geo. Phys. **3**, 421–434 (1986)
235. Kontsevich, M.: Feynman diagrams and low-dimensional topology. In: First European Cong. Math. vol. II Prog. in Math., # 120, pp. 97–121. Birkhäuser, Berlin (1994)
236. Kostant, B., Sternberg, S.: Symplectic reduction, BRS cohomology, and infinite-dimensional Clifford algebras. Ann. Phys. **176**, 49–113 (1987)
237. Kotschick, D., Morgan, J.W.: $SO(3)$-invariants for 4-manifolds with $b_2^+ = 1$. II. J. Diff. Geom. **39**, 433–456 (1994)
238. Kronheimer, P.B., Mrówka, T.S.: Gauge theory for embedded surfaces I. Topology **32**, 773–826 (1993)
239. Kronheimer, P.B., Mrówka, T.S.: Recurrence relations and asymptotics for four-manifold invariants. Bull. Amer. Math. Soc. **30**, 215–221 (1994)

240. Kronheimer, P.B., Mrówka, T.S.: Gauge theory for embedded surfaces II. Topology **34**, 37–97 (1995)

241. Kronheimer, P.B., Mrówka, T.S.: Witten's conjecture and property P. Geometry and Topology **8**, 295–310 (2004)

242. Lanczos, C.: A remarkable property of the Riemann-Christoffel tensor in four dimensions. Ann. Math. **39**, 842–850 (1938)

243. Landi, G.: The natural spinor connection on S_8 is a gauge field. Lett. Math. Phys. **11**, 171–175 (1986)

244. Lang, S.: Algebra. Springer, New York (2002)

245. Lang, S.: Introduction to differentiable Manifolds. Springer, New York (2002)

246. Lawrence, R., Zagier, D.: Modular forms and quantum invariants of 3-manifolds. Asian J. Math. **3**, 93–108 (1999)

247. Lawson, Jr., H.B.: The Theory of Gauge Fields in Four Dimensions. Regional Conference Series in Mathematics # 58. Amer. Math. Soc., Providence (1985)

248. Lawson, Jr., H.B., Michelsohn, M.L.: Spin Geometry. Princeton University Press, Princeton (1989)

249. LeBrun, C.: Weyl curvature, Einstein metrics, and Seiberg–Witten theory. Math. Res. Lett. **5**, 423–438 (1998)

250. Lepowsky, J., Li, H.: Introduction to Vertex Operator Algebras and Their Representations. Progress in Math., #227. Birkhäuser, Basel (2004)

251. Li, J., Liu, K., Zhou, J.: Topological string partition functions as equivariant indices. Asian J. Math. **10, No. 1**, 81 – 114 (2006)

252. Li, W.: Casson-Lin's invariant for a knot and Floer homology. J. Knot Theo. Ramifications **6**, 851–877 (1997)

253. Lin, X.S.: A knot invariant via representation spaces. J. Diff. Geom. **35**, 337–357 (1992)

254. Lusztig, G.: Introduction to Quantum Groups. Birkhäuser, Basel (1993)

255. Maciocia, A.: Metrics on the moduli spaces of instantons over Euclidean 4-space. Comm. Math. Phys. **135**, 467–482 (1991)

256. Manin, Y.I.: New exact solutions and cohomological analysis of ordinary and supersymmetric Yang–Mills equations (in Russian). Trudy Mat. Inst. Steklov **165**, 98–114 (1984)

257. Manin, Y.I.: Gauge Field Theory and Complex Geometry. SpringerVerlag, Berlin (1988)

258. Manin, Y.I.: Frobenius manifolds, quantum cohomology, and moduli spaces. Coll. Pub., vol. 47. Amer. Math. Soc., Providence (1999)

259. Manturov, V.: Knot Theory. Chapman & Hall/CRC, London (2004)

260. Marathe, K.: A chapter in physical mathematics: Theory of knots in the sciences. In: B. Engquist and W. Schmidt (ed.) Mathematics Unlimited - 2001 and Beyond, pp. 873–888. Springer-Verlag, Berlin (2001)

261. Marathe, K.: Topological quantum field theory as topological gravity. In: B. Fauser and others (ed.) Mathematical and Physical Aspects of Quantum Gravity, pp. 189–205. Birkhauser, Berlin (2006)

262. Marathe, K.: The review of symmetry and the monster by Marc Ronan (Oxford). Math. Intelligencer **31**, 76–78 (2009)

263. Marathe, K.: Geometric topology and field theory on 3-manifolds. In: M. Banagl and D. Vogel (ed.) The Mathematics of Knots, Contributions in the Mathematical and Computational Sciences, Vol. 1, pp. 151–207. Springer-Verlag, Berlin (2010)

264. Marathe, K.B.: Structure of relativistic spaces. Ph.D. thesis, University of Rochester (1971)

265. Marathe, K.B.: A condition for paracompactness of a manifold. J. Diff. Geom. **7**, 571–572 (1972)

266. Marathe, K.B.: Generalized field equations of gravitation. Rend. Mat. (Roma) **6**, 439–446 (1972)

267. Marathe, K.B.: Spaces admitting gravitational fields. J. Math. Phys. **14**, 228–233 (1973)
268. Marathe, K.B.: The mean curvature of gravitational fields. Physica **114A**, 143–145 (1982)
269. Marathe, K.B.: Generalized gravitational instantons. In: Proc. Coll. on Diff. Geom., Debrecen (Hungary) 1984, pp. 763–775. Colloquia Math Soc. J. Bolyai, Hungary (1987)
270. Marathe, K.B.: Gravitational instantons with source. In: Particles, Fields, and Gravitation, Lodz, Poland 1998, *AIP Conf. Proc.*, vol. 453, pp. 488–497. Amer. Inst. Phys., Woodbury, NY (1998)
271. Marathe, K.B., Martucci, G.: Geometric quantization of the nonisotropic harmonic oscillator. Il Nuovo Cim. **79B, N. 1**, 1–12 (1984)
272. Marathe, K.B., Martucci, G.: Quantization on V-manifolds. Il Nuovo Cim. **86B, N. 1**, 103–109 (1985)
273. Marathe, K.B., Martucci, G.: The geometry of gauge fields. J. Geo. Phys. **6**, 1–106 (1989)
274. Marathe, K.B., Martucci, G.: The Mathematical Foundations of Gauge Theories. Studies in Mathematical Physics, vol. 5. North-Holland, Amsterdam (1992)
275. Marathe, K.B., Martucci, G., Francaviglia, M.: Gauge theory, geometry and topology. Seminario di Matematica dell'Università di Bari **262**, 1–90 (1995)
276. Marathe, K.B., et al.: A geometric setting for field theories. In: G. M. Rassias (ed.) Topology, Analysis and Applications. World Sci., Singapore (1992)
277. Marcolli, M.: Seiberg–Witten Gauge Theory. Texts and Readings in Math., vol. 17. Hindustan Book Agency, New Delhi (1999)
278. Marsden, J.: Applications of Global Analysis in Mathematical Physics. Publish or Perish, Inc., Boston (1974)
279. Massey, W.S.: A basic course in algebraic topology. Springer, New York (1991)
280. Mathai, V., Quillen, D.: Superconnections, thom classes and equivariant differential forms. Topology **25**, 85–110 (1986)
281. Matumoto, T.: Three Riemannian metrics on the moduli space of BPST-instantons over S^4. Hiroshima Math. J. **19**, 221–224 (1989)
282. McDuff, D., Salamon, D.: J-holomorphic Curves and Quantum Cohomology. University Lect. Series, # 6. Amer. Math. Soc., Providence (1994)
283. McDuff, D., Salamon, D.: Introduction to Symplectic Topology. Oxford University Press, Oxford (1995)
284. Milnor, J.: Morse Theory. Ann. of Math. Studies, No. 51. Princeton University Press, Princeton (1973)
285. Milnor, J., Husemoller, D.: Symmetric Bilinear Forms. Springer-Verlag, Berlin (1973)
286. Milnor, J.W., Stasheff, J.D.: Characteristic Classes. Ann. of Math. Studies, No. 76. Princeton University Press, Princeton (1974)
287. Modugno, M.: Sur quelques propriétés de la double 2-forme gravitationnelle w. Ann. Inst. Henri Poincaré **XVIII**, 251–262 (1973)
288. Moore, G., Seiberg, N.: Classical and quantum conformal field theory. Comm. Math. Phys. **123**, 177–254 (1989)
289. Morgan, J., Tian, G.: Ricci Flow and the Poincar'e Conjecture. Clay Mathematics Monographs vol. 3. Amer. Math. Soc., Providence (2007)
290. Morgan, J.W.: Comparison of Donaldson polynomial invariants with their algebro-geometric analogues. Topology **32**, 449–488 (1993)
291. Morgan, J.W., O'Grady, K.: Differential Topology of Complex Surfaces. Lect. Notes in Math. # 1545. Springer-Verlag, Berlin (1993)
292. Morse, M., Cairns, S.: Critical Point Theory in Global Analysis and Differential Topology. Academic Press, New York (1969)
293. Mrówka, T., Ozsváth, P., Yu, B.: Seiberg–Witten monopoles on Seifert fibered spaces. Comm. Anal. Geom. **5**(3), 685–791 (1997)

294. Murakami, H.: Quantum $SU(2)$-invariants dominate Casson's $SU(2)$-invariant. Math. Proc. Camb. Phil. Soc. **115**, 253–281 (1993)

295. Nakahara, M.: Geometry, Topology and Physics. Grad. Student series in Phy., 2nd ed. Inst. Phy., Philadelphia (2003)

296. Narasimhan, M.S., Ramanan, S.: Existence of universal connections I. Amer. J. Math. **83**, 563–752 (1961)

297. Narasimhan, M.S., Ramanan, S.: Existence of universal connections II. Amer. J. Math. **85**, 223–231 (1963)

298. Nash, C.: Differential Topology and Quantum Field Theory. Acad. Press, London (1991)

299. Nicolaescu, L.I.: Notes on Seiberg–Witten Theory. Grad. Studies in Math., volume 28. Amer. Math. Soc., Providence (2000)

300. Norton, S.P.: The uniqueness of the Fischer–Griess monster. In: Finite Groups— Coming of Age, Contemp. Math., 45, pp. 271–285. Amer. Math. Soc., Providence (1985)

301. Novikov, S.P., Taimanov, I.A.: Modern Geometric Structures and Fields. Grad. Studies in Math., volume 71. Amer. Math. Soc., Providence (2006)

302. Nowakowski, J., Trautman, A.: Natural connections on Stiefel bundles are sourceless gauge fields. J. Math. Phys. **19**, 1100–1103 (1978)

303. noz, V.M.: Fukaya–Floer homology of $\Sigma \times S^1$ and applications. J. Diff. Geom. Soc. **53**, 279–326 (1999)

304. Ocneanu, A.: Quantized groups, string algebras and Galois theory for algebras. In: Operator Algebras and Applications, LMS Lect. Notes, vol. 136, pp. 119–172. Cambridge University Press, Cambridge (1988)

305. Ogg, A.P.: Modular functions. In: Santa Cruz Conference on Finite Groups, Proc. Sympos. Pure Math., 37, pp. 521–532. Amer. Math. Soc., Providence (1980)

306. Ohtsuki, T.: Quantum Invariants. Series on Knots and Everything - vol. 29. World Scientific, Singapore (2002)

307. Omohundro, S.M.: Geometric Perturbation Theory in Physics. World Scientific, Singapore (1986)

308. O'Neill, B.: Semi-Riemannian Geometry. Academic Press, New York (1983)

309. Ozsváth, P., Szabó, Z.: On knot Floer homology and the four-ball genus. Geom. Topol. **7**, 225–254 (2003)

310. Palais, R.S.: Foundations of global non-linear analysis. Benjamin, New York (1968)

311. Palais, R.S.: Seminar on the Atiyah–Singer Index Theorem. Ann. of Math. Studies, No. 57. Princeton University Press, Princeton (1974)

312. Pandharipande, R.: Hodge integrals and degenerate contributions. Comm. Math. Phys. **208**, 489–506 (1999)

313. Pandharipande, R.: Three questions in Gromov–Witten theory. In: Proc. ICM 2002, Beijing, vol. II, pp. 503–512 (2002)

314. Parker, T.: Gauge theories on four dimensional Riemannian manifolds. Comm. Math. Phys. **85**, 563–602 (1982)

315. Patodi, V.K.: Curvature and the eigenvalues of the Laplace operator. J. Diff. Geom. **5**, 233–249 (1971)

316. Petrov, A.Z.: Einstein Spaces. Pergamon Press, New York (1969)

317. Piunikhin, S.: Reshetikhin–Turaev and Kontsevich-Kohno-Crane 3-manifold invariants coincide. J. Knot Theory **2**, 65–95 (1993)

318. Pizer, A.: A note on a conjecture of Hecke. Pacific J. Math. **79**, 541–548 (1978)

319. Porteous, I.R.: Topological Geometry, 2nd edition. Cambridge University Press, Cambridge (1981)

320. Prasolov, V.V., Sossinsky, A.B.: Knots, Links, Braids and 3-Manifolds. Translations of Math. Mono., volume 154. Amer. Math. Soc., Providence (1997)

321. Pressley, A., Segal, G.: Loop Groups. Clarendon Press, Oxford (1986)

322. Quigg, C.: Gauge Theories of the Strong, Weak, and Electromagnetic Interactions. Benjamin, Reading (1983)

323. Ray, D.B., Singer, I.M.: R-torsion and the Laplacian on Riemannian manifolds. Adv. in Math. **7**, 145–210 (1971)

324. Ray, D.B., Singer, I.M.: Analytic torsion for complex manifolds. Ann. Math. **98**, 154–177 (1973)

325. Rennie, R.: Geometry and topology of chiral anomalies in gauge theories. Adv. in Phy. **39**, 617–779 (1990)

326. Reshetikhin, N., Turaev, V.G.: Invariants of 3-manifolds via link polynomials and quantum groups. Invent. Math. **103**, 547–597 (1991)

327. Ronan, M.: Symmetry and the Monster. Oxford University Press, Oxford (2006)

328. Rozansky, L., Saleur, H.: Quantum field theory for the multi-variable Alexander–Conway polynomial. Nucl. Phy. **B 376**, 461–509 (1992)

329. Rozansky, L., Witten, E.: Hyper-Kähler geometry and invariants of three-manifolds. Selecta Math. (N.S.) **3**(3), 401–458 (1997)

330. Rubbia, C., Jacob, M.: The Z^0. Amer. Scientist **78**, 502–519 (1990)

331. Sachs, R.K., Wu, H.: General Relativity for Mathematicians. SpringerVerlag, Berlin (1977)

332. Sadun, L., Segert, J.: Non-self-dual Yang–Mills connections with nonzero Chern number. Bull. Amer. Math. Soc. **24**, 163–170 (1991)

333. Salam, A.: Gauge unification of fundamental forces. Rev. Mod. Phys. **92**, 525–536 (1980)

334. Salamon, D.: Morse theory, the Conley index and Floer homology. Bull. London Math. Soc. **22**, 113–140 (1990)

335. Saveliev, N.: Floer homology of Brieskorn homology spheres. J. Diff. Geom. Soc. **53**, 15–87 (1999)

336. Saveliev, N.: Lectures on the Topology of 3-Manifolds. de Gruyter, Berlin (1999)

337. Sawin, S.: Links, quantum groups and TQFTs. Bull. Amer. Math. Soc. **33**, 413–445 (1996)

338. Schmidt, B.G.: A new definition of singular points in general relativity. Gen. Relat. and Gravitation **1**, 269–280 (1971)

339. Schoen, R., Yau, S.T.: Positivity of the total mass of a general space-time. Phys. Rev. Lett. **43**, 1457–1459 (1979)

340. Schulman, C.S.: Techniques and Applications of Path Integrals. Wiley, New York (1981)

341. Schwarz, A.S.: The partition function of degenerate quadratic functional and Ray–Singer invariants. Lett. Math. Phys. **2**, 247–252 (1978)

342. Schwarz, M.: Morse Homology. Prog. in Math., vol. 111. Birkhauser, Boston (1993)

343. Scorpan, A.: The Wild World of 4-Manifolds. Amer. Math. Soc., Providence (2005)

344. Sedlacek, S.: A direct method for minimizing the Yang–Mills functional over 4-manifolds. Comm. Math. Phys. **86**, 515–527 (1982)

345. Seeley, R.: Complex powers of an elliptic operator. Proc. Symp. Pure Math. **10**, 515–527 (1973)

346. Segal, G.: The definition of conformal field theory. In: K. Bleuler, M. Werner (eds.) Differential Geometric Methods in Theoretical Physics, pp. 165–171. Academic Press, Boston (1988)

347. Segal, G.: Two-dimensional conformal field theories and modular functors. In: Proc. IXth Int. Cong. on Mathematical Physics, pp. 22–37. Adam Hilger, Bristol (1989)

348. Shanahan, P.: The Atiyah–Singer Index Theorem. Springer-Verlag, Berlin (1978)

349. Shifman, M.A.: Anomalies in gauge theories. Phy. Rep. **209**, 341–378 (1991)

350. Sibner, L.M., Sibner, R.J., Uhlenbeck, K.: Solutions to Yang–Mills Equations which are not self-dual. Proc. Natl. Acad. Sci. USA **86**, 8610–8613 (1989)

351. Singer, I.M.: Some remarks on the Gribov ambiguity. Comm. Math. Phys. **64**, 7–12 (1978)

352. Singer, I.M.: Some problems in the quantization of gauge theories and string theories. In: R. Wells, Jr. (ed.) The Mathematical Heritage of Hermann Weyl, Proc. Symp. Pure Math. vol. 48, pp. 199–216. Am. Math. Soc., Providence (1988)

353. Singer, I.M., Thorpe, J.: The curvature of 4-dimensional Einstein spaces. In: Global Analysis, Papers in Honor of K. Kodaira, pp. 355–365. Princeton University Press, Princeton (1969)

354. Smale, S.: An infinite dimensional version of Sard's theorem. Ann. Math. **87**, 213–221 (1973)

355. Smith, I., Thomas, R.P., Yau, S.T.: Symplectic conifold transitions. J. Diff. Geom. **62**, 209–242 (2002)

356. Snaith, V.P.: Stable homotopy around the Arf-Kervaire invariant. Prog. in Math. vol. 273. Birkhäuser, Basel (2009)

357. Spanier, E.H.: Algebraic Topology. McGraw-Hill, New York (1966)

358. Spivak, M.: A Comprehensive Introduction to Differential Geometry, 5 volumes. Publish or Perish, Inc., Boston (1979)

359. Steenrod, N.E.: The Topology of Fibre Bundles. Princeton University Press, Princeton (1951)

360. Stong, R.E.: Notes on Cobordism Theory. Princeton University Press, Princeton (1968)

361. Stredder, P.: Natural differential operators on Riemannian manifolds and representations of the orthogonal and special orthogonal groups. J. Diff. Geom. **10**, 657–660 (1975)

362. Taubes, C.: Seiberg–Witten and Gromov invariants for symplectic 4-manifolds. International Press, Sommerville (2000)

363. Taubes, C.H.: Stability in Yang–Mills theory. Comm. Math. Phys. **91**, 235–263 (1983)

364. Taubes, C.H.: Min-max theory for the Yang–Mills–Higgs equations. Comm. Math. Phys. **97**, 473–540 (1985)

365. Taubes, C.H.: The stable topology of self-dual moduli spaces. J. Diff. Geom. **29**, 163–230 (1989)

366. Taubes, C.H.: Casson's invariant and gauge theory. J. Diff. Geom. **31**, 547–599 (1990)

367. Taubes, C.H.: The geometry of the Seiberg–Witten invariants. Surveys in Diff. Geom. **2**, 221–238 (1996)

368. Taubes, C.H.: $GR = SW$: Counting curves and connections. J. Diff. Geom. Soc. **52**, 453–609 (1999)

369. Thaddeus, M.: Conformal field theory and the cohomology of the moduli space of stable bundles. J. Diff. Geom. **35**, 131–150 (1992)

370. Thompson, J.G.: Finite groups and modular functions. Bull. London Math. Soc. **11**(3), 347–351 (1979)

371. Thompson, J.G.: Some numerology between the Fischer–Griess monster and the elliptic modular function. Bull. London Math. Soc. **11**(3), 352–353 (1979)

372. Thompson, J.G.: Uniqueness of the Fischer–Griess monster. Bull. London Math. Soc. **11**(3), 340–346 (1979)

373. Thurston, W.: Three-Dimensional Geometry and Topology. Princeton University, Princeton (1997)

374. Thurston, W.P.: A norm for the homology of 3-manifolds. Mem. Amer. Math. Soc. **59 no. 339**, 99–130 (1986)

375. Tian, Y.: The based $SU(n)$-instanton moduli spaces. Ann. Math. **298**, 117–139 (1994)

376. Tillmann, U. (ed.): Topology, Geometry and Quantum Field Theory. LMS lecture note, vol. 308. Cambridge University Press, Cambridge (2004)

377. Trautman, A.: Solutions of the Maxwell and Yang–Mills equations associated with Hopf fiberings. Intern. J. of Theor. Phys. **16**, 561–565 (1977)

378. Trautman, A.: Differential Geometry for Physicists. Bibliopolis, Naples (1984)

379. Turaev, V.: A combinatorial formulation for the Seiberg–Witten invariants of 3-manifolds. Math. Res. Lett. **5**, 583–598 (1998)

380. Turaev, V.G.: The Yang–Baxter equation and invariants of link. Invent. Math. **92**, 527–553 (1988)

381. Turaev, V.G.: Quantum Invariants of Knots and 3-Manifolds. Studies in Math. #18. de Gruyter, Amsterdam (1994)

382. Turaev, V.G.: Torsions of 3-Dimensional Manifolds, *Progress in Mathematics*, vol. 208. Birkhäuser, Basel (2002)

383. Turaev, V.G., Viro, O.Y.: State sum invariants of 3-manifolds and quantum $6j$-symbols. Topology **31**, 865–895 (1992)

384. Turaev, V.G., Wenzl, H.: Quantum invariants of 3-manifolds associated with classical simple Lie algebras. Int. J. Math. **4**, 323–358 (1993)

385. Uhlenbeck, K.: Connections with L^p bounds on curvature. Comm. Math. Phys. **83**, 31–42 (1982)

386. Uhlenbeck, K.: Removable singularities in Yang–Mills fields. Comm. Math. Phys. **83**, 11–30 (1982)

387. Uhlenbeck, K.: Variational problems for gauge fields. In: S.T. Yau (ed.) Seminar on Differential Geometry, Annals of Mathematics Studies # 102, pp. 455–464 (1982)

388. Unger, F.R.: Invariant $Sp(1)$-instantons on $S^2 \times S^2$. Geometriae Dedicata **23**, 365–368 (1987)

389. Urakawa, H.: Equivariant theory of Yang–Mills connections over Riemannian manifolds of cohomogeneity one. Indiana Univ. Math. J. **37**, 753–788 (1988)

390. Vaisman, I.: Symplectic Geometry and Secondary Characteristic Classes. Progress in Math., No. 72. Birkhäuser, Boston (1987)

391. Vasiliev, V.: Cohomology of knot spaces. In: V.I. Arnold (ed.) Theory of Singularities and its Applications, Adv. Sov. Math. # 1, pp. 23–70. Amer. Math. Soc. (1990)

392. Vassiliev, V.A.: Complements of Descriminants of Smooth Maps: Topology and Applications. Translations of Math. Mono., volume 98. Amer. Math. Soc., Providence (1994)

393. Verlinde, E.: Fusion rules and modular transformations in $2d$ conformal field theory. Nucl. Phys. **B300**, 360–376 (1988)

394. Walker, K.: An Extension of Casson's Invariant. Ann. Math. Studies. Princeton University Press, Princeton (1991)

395. Wang, H.Y.: The existence of nonminimal solutions to the Yang–Mills equation with group $SU(2)$ on $S^2 \times S^2$ and $S^1 \times S^3$. J. Diff. Geom. **34**, 701–767 (1991)

396. Warner, F.W.: Foundations of Differential Manifolds and Lie Groups. Scott, Foresman, Glenview (1971)

397. Weinberg, S.: Conceptual foundations of the unified theory of weak and electromagnetic interactions. Rev. Mod. Phys. **92**, 515–524 (1980)

398. Wells, Jr., R.O.: Differential Analysis on Complex Manifolds. Springer-Verlag, Berlin (1980)

399. Wells, Jr., R.O.: Complex Geometry in Mathematical Physics. Université de Montréal, Montréal (1982)

400. Wenzl, H.: Braids and invariants of 3-manifolds. Invent. Math. **114**, 235–275 (1993)

401. Weyl, H.: The Classical Groups. Princeton University Press, Princeton (1946)

402. Wigner, E.P.: The unreasonable effectivness of mathematics in the natural sciences. Comm. Pure and App. Math. **XIII**, 1–14 (1960)

403. Witten, E.: Supersymmetry and Morse theory. J. Diff. Geom. **17**, 661–692 (1982)

404. Witten, E.: Topological quantum field theory. Comm. Math. Phys. **117**, 353–386 (1988)

405. Witten, E.: Quantum field theory and the Jones polynomial. Comm. Math. Phys. **121**, 359–399 (1989)

406. Witten, E.: Quantization of Chern–Simons gauge theory with complex gauge group. Comm. Math. Phys. **137**, 29–66 (1991)

407. Witten, E.: Monopoles and four-manifolds. Math. Res. Lett. **1**, 769–796 (1994)

408. Witten, E.: Chern–Simons gauge theory as string theory. Prog. Math. **133**, 637–678 (1995)

409. Wu, T.T., Yang, C.N.: Concept of nonintegrable phase factors and global formulation of gauge fields. Phys. Rev. **12D**, 3845–3857 (1975)

410. Yang, C.N.: Magnetic monopoles, fiber bundles and gauge fields. Ann. N. Y. Acad. Sci. **294**, 86–97 (1977)

411. Yang, C.N.: Fiber bundles and the physics of the magnetic monopole. In: The Chern Symposium 1979, pp. 247–253. Springer-Verlag, Berlin (1980)

412. Yang, C.N., Mills, R.L.: Conservation of isotopic spin and isotopic gauge invariance. Phys. Rev. **96**, 191–195 (1954)

413. Yang, C.N., Mills, R.L.: Isotopic spin conservation and a generalized gauge invariance. Phys. Rev. **95**, 631 (1954)

414. Yetter, D.N.: Functorial Knot Theory. Series on Knots and Everything - vol. 26. World Scientific, Singapore (2001)

415. Yokota, Y.: Skeins and quantum $SU(N)$ invariants of 3-manifolds. Math. Ann. **307**, 109–138 (1997)

416. Yoshida, T.: Floer homology and splittings of manifolds. Ann. Math. **134**, 277–323 (1991)

417. Zeidler, E.: Quantum Field Theory I: Basics in Mathematics and Physics. Springer-Verlag, Berlin (2006)

418. Zeidler, E.: Quantum Field Theory II: Quantum Electrodynamics. Springer-Verlag, Berlin (2008)

419. Zwegers, S.P.: Mock θ-functions and real analytic modular forms. In: q-Series with Applications to Combinatorics, Number Theory, and Physics, Contemp. Math. # 291, pp. 269–277. Amer. Math. Soc., Providence (2001)

Index